BASIC

Programs for Land Surveying

BASIC
Programs for Land Surveying

P. H. MILNE
Department of Civil Engineering,
University of Strathclyde, UK

LONDON NEW YORK
E. &. F. N. SPON

First published 1984 by
E. & F. N. Spon Ltd
11 New Fetter Lane, London EC4P 4EE
Published in the USA by
E. & F. N. Spon
733 Third Avenue, New York NY 10017

Typeset by Keyset Composition, Colchester
Printed in Great Britain at the
University Press, Cambridge

ISBN 0 419 13000 4 (cased)
ISBN 0 419 13010 1 (paperback)

British Library Cataloguing in Publication Data

Milne, P. H.
BASIC programs for land surveying.
1. Surveying—Data processing 2. Basic
(Computer program language)
I. Title
526.9′028′542 TA562

ISBN 0-419-13000-4
ISBN 0-419-13010-1 Pbk

Library of Congress Cataloging in Publication Data

Milne, P. H.
BASIC programs for land surveying.
Bibliography: p.
Includes index.
1. Surveying—Computer programs. 2. Basic (Computer program language). 3. Apple II (Computer)—Programming.
4. Apple IIe (Computer)—Programming.
I. Title. II. Title: B.A.S.I.C. programs for land surveying.
TA549.M5374 1984 526.9′028′5425 84-1296
ISBN 0-419-13000-4
ISBN 0-419-13010-1 (pbk.)

Contents

Preface

The advent of the low-cost microcomputer in the late 1970s, with its accessibility to students and small surveying firms, has now created a need for BASIC programs in land surveying. Whilst finishing the BASIC programs for a previous book on underwater surveying, *Underwater Acoustic Positioning Systems*, I entered into discussions with Mr Phillip Read, Editor of E. & F. N. Spon Ltd, for a similar book on land surveying. Accordingly, I would like to thank both Mr Phillip Read (Editor) and Mary Ann Ommanney (Editorial Manager) for their encouragement and patience in producing this book.

All the programs are written in Applesoft BASIC for the Apple II+/e microcomputer, rather than Algol or FORTRAN, due to the increased teaching of BASIC in schools, colleges and universities. To ease transfer to other microcomputers, only routines and commands generally available in Microsoft BASIC have been used. The author has been grateful for the opportunity in the Department of Civil Engineering at Strathclyde University to handle numerous computers, including the Hewlett-Packard HP-85 and HP-87, Apple II+/e and Commodore PET. Due to the versatility of the Apple II+, this was the microcomputer chosen for the Department's microcomputer laboratory, and therefore also for the programs in this book. A table is included for the conversion, where required, of commands for other popular microcomputers, and it is hoped these programs will be useful to land surveyors involved in building and construction, civil engineering, highway engineering, mining engineering or quantity surveying.

It has been my good fortune to receive very generous assistance from fellow research workers and surveying equipment manufacturers who have either given demonstrations or supplied equipment for evaluation. I would like to thank all those who kindly provided information or illustrations to complement the text (all the photographs provided are duly acknowledged in the captions).

I am extremely grateful to Professor D. I. H. Barr of the Department of Civil Engineering at the University of Strathclyde for his continued encouragement in publishing the results of my research work. I would like to express my thanks to all the undergraduates and post-graduate students who have assisted directly or indirectly in developing or debugging these programs. Gratitude is also due to

John Gilligan and the technicians in the Department for their help and assistance, both in the laboratory and on site at students' survey exercises, etc.

I also wish to thank Mrs Sheena Nelson of the Mechanical, Civil and Chemical Engineering Drawing Office at the University of Strathclyde for assistance in the preparation of the illustrations for the book, especially Mrs Karen Struthers for the figures and Mrs Isabel Mungall for the photographs.

Finally I would like to thank my family for their support and understanding. It was my father, the late Henry Milne, FRICS, initially with Messrs Morham and Brotchie, Chartered Quantity Surveyors in Edinburgh, and latterly, District Valuer for the Inland Revenue Glasgow 1 Office, who introduced me to surveying. My sincere thanks go to my wife, Helen, for her helpful suggestions and perseverance in typing the complete manuscript.

Department of Civil Engineering,
University of Strathclyde,
John Anderson Building,
Glasgow G4 0NG,
Scotland.

P. H. Milne
September 1983

PROGRAM DISCS AVAILABLE

This book contains many lengthy program listings. To save typing them into your computer the programs are available on 5¼″ discs for the Apple II+/e 48K system. The programs are contained on four discs, one for each of Chapters 3 to 6, under the same program names as in the book. Other microcomputer software/hardware formats are also available, including CP/M. For details of prices of discs, etc., write for information to

Strathand Ltd,
2.02 Kelvin Campus,
West of Scotland Science Park,
Glasgow G20 0SP,
Scotland.

Glossary of abbreviations and technical terms used with microcomputers

Access time The length of time between a request for information and its availability.
Address A number designating a specific storage location in the device.
Algorithm A set of procedures for solving a problem. Derived from the name of a 9th-century Arab mathematician called Al Khwaruznu.
Alphanumeric A descriptive term for a set of characters consisting of alphabetic, numeric and special characters.
Applesoft A floating-point version of BASIC used on the Apple microcomputer.
Application program A set of computer instructions written for a specific user application.
Array An arrangement of elements (numbers, characters, etc.) in rows and columns.
ASCII Acronym for American Standard Code for Information Interchange. The ASCII code is used extensively in data communications.
Assembly language see **Language.**
Backup It is normal practice to keep a backup copy of all discs to cover in the event of damage or failure.
BASIC Beginner's All-purpose Symbolic Instruction Code. A high-level computer language, with most commands in recognizable English.
Batch mode A method of working whereby all data for the application is input to the computer in a continuous stream. After all the data has been input, the program is run, and the data processed.
Baud The rate of data transmission, often designed so that one baud equals one binary bit of data.
Binary A number system with a base of 2 using only the digits 0 and 1. Computer devices work either in the presence or absence of a signal, either 0 or

1. Binary digits are combined to form number systems and codes, such as the ASCII code, octal and hexidecimal.

Bit A contraction of 'binary digit' where a bit is represented by 0 or 1 depending upon the 'on' or 'off' state of the bit.

Bubble This type of memory uses a form of magnetic technology for mass storage of data which is faster and more reliable than disc storage.

Buffer storage Any device that temporarily holds data during data transfer, usually between internal and external forms of storage, to free the CPU for other tasks.

Bug A mistake in a program causing it to give erroneous results.

Byte A sequence of eight bits which may be used to represent one character of information.

Cathode Ray Tube (CRT) Essentially a visual display unit (VDU) for display of text or graphics.

Central Processing Unit (CPU) This component, right at the heart of the microprocessor, performs the tasks of data manipulation and the organization of peripheral components.

Chaining A clever programming technique, which splits a task into several separate programs that run in sequence. Useful for running large programs which would exceed memory capacity available.

Character A letter, digit, symbol or punctuation mark used to represent one printable character of information. A character code normally consists of seven binary bits.

Colour card An interface card which when plugged into an I/O port in a microcomputer (e.g. Apple II+) enables colour to be output on to a colour monitor or standard colour TV.

Command A code or series of codes which when processed by the CPU causes one of the peripherals to perform some action.

Compiler A utility which converts a high-level language program, which needs to be interpreted every time it is run, into a machine code program, which runs faster, needing less or no interpretation.

Computer Aided Design (CAD) Special software to generate graphics shapes with special terminals to display or print them.

Configure Design and set up a system containing elements of hardware and/or software.

Control Program/Microcomputers (CP/M) A disc operating system for microcomputers based on the 8080, 8085 and the Z-80 type microprocessor chip. It is a software package developed and distributed by Digital Research, Inc., of Pacific Grove, California.

Cursor An electronically generated symbol (usually a line or a square) appearing on the screen of a video display terminal to indicate where the next character will appear.

Daisy-chained Peripheral devices that are attached to a bus (e.g., HP-IB) one after the other are said to be daisy-chained. Devices can also be star connected.

Data Information stored in numerical or text format, used as transients in programs, for calculations or information storage.

Database A pool of shared data, supported by utilities for editing, sorting, entering new data and so on.
Data processing The manipulation of data by the execution of a sequence of instructions. Synonymous with information processing.
Debug To detect, locate and eliminate mistakes in a program or a malfunction in a microcomputer's electronic circuits.
Disc A magnetic storage device, either flexible (floppy) or hard (Winchester), which can store data or programs in digital format.
Disc drive A unit which contains a reading and writing head for loading data on to a disc, or reading data from a disc. The drive also contains the motor for rotating the discs. Hard discs of much greater storage capacity are usually housed in sealed units, whereas flexible discs are easily swapped.
DOS (Disc Operating System) A series of routines which need to be loaded into the CPU to enable it to initialize, save to and read from the disc, plus numerous other associated refinements.
Dump Transfer amounts of data straight to a peripheral, like a printer or disc.
Edit To modify data by inserting, changing or eliminating characters.
Editor An editor is a program that allows a user to edit data or instructions.
Erase To wipe out information stored in the microcomputer memory or in other storage media, e.g., disc.
Execute To carry out an operation in a program, or 'run' a program.
Firmware Computer instructions stored in read only memory (ROM).
FORTRAN (*For*mula *Tran*slation language) A high-level computer language used frequently for scientific calculations.
GP-IB General Purpose-Interface Bus, same as IEEE-488.
Hard copy A printed paper copy of a program or its graphic results produced by a printer or plotter connected to the microcomputer.
Hardware Generic term for all manufactured computer equipment, i.e., the physical parts as contrasted with the programs (software).
Hexidecimal A number system based upon 16, that is common to some microcomputers. Hexidecimal digits that are more than decimal 9 are expressed as characters *A* to *F* (*A* representing 10). Four binary bits are required to express one hexidecimal digit.
High-level language see **Language.**
HP Hewlett-Packard.
HP-IB Hewlett-Packard-Interface Bus, same as IEEE-488.
IEEE-488 Standard digital interface for parallel transmission of data adopted by the Institute of Electrical and Electronic Engineers in 1975, although originally developed as HP-IB.
Integer BASIC A form of BASIC which stores its numbers in integer format (no decimals).
Interface A device for linking one component with another, such as a printer and a microcomputer, to permit transfer of data.
Interpreter A program that translates each high-level language program statement into executable machine instructions each time the high-level statement is encountered during the running of the user's program.

I/O (Input/output) Interface cards are placed normally at the back of the microcomputer, in the I/O ports and connected by cable to peripherals.

K (Kilo) A convenient notation for describing volume. However, where computer memory is concerned, K represents 1024. Accordingly, 64 K represents 65536 bytes.

Language A set of rules and symbols for passing instructions into a computer.

Assembly language A language which allows the use of mnemonic symbols instead of patterns '0' and '1', thereby easing the task of the programmer.

High-level language A language which allows computer instructions to be written in a restricted form of the English language. High-level languages have to be converted to machine language, either by compiling or interpreting, before they can be used.

Machine language A set of rules or symbols which may be executed immediately by the computer. The symbols will normally be restricted to '0' and '1' and the rules will define the allowed patterns of the symbols.

Load To transfer data from an external data storage device, e.g., a disc, into a microcomputer.

Loop A sequence of instructions repeated until the loop is terminated.

Machine language see **Language.**

Macro A series of instructions which can be linked together to be operated by one or two key strokes, or instructions.

Mainframe A very large computer, capable of handling many jobs at any one time and many terminals.

Memory Any device used to store data or instructions for the computer. Memory devices are compared in terms of storage capacity, access time and cost.

Menu A list of options presented to the operator during execution of a program.

Microcomputer A small computer employing a microprocessor as the central processing unit (CPU).

Microprocessor Electronic circuitry on a single integrated chip that can perform data manipulation.

Mother-board The large printed circuit board (PCB) of a microcomputer which holds all the silicon chips, the central processing unit (CPU) and the input/output (I/O) ports.

Off-line Data collected online during a real-time program is often rerun off-line for post-survey analysis.

Online Data collected during an interactive real-time program is often stored for subsequent processing, printing and plotting, i.e. no longer in a real-time mode.

Operating system A program which organizes the internal operation of the computer – see **CP/M.**

Parallel communications The standard character and ASCII code transmission method where bits are sent on eight lines at a time in parallel.

Pascal A high-level structured language which needs compiling to run.

Peripheral An external accessory connected to the microcomputer and used for input/output or storage of data, e.g., visual display units (VDU), keyboards, line printers, plotters, floppy and hard discs.

Pixel The smallest addressable picture element or point on a VDU, generally given as number of horizontal and vertical points, e.g., 280 × 192 for Apple II High Resolution Graphics.

Program A list of computer instructions connected in a logical format directing the computer to perform specific operations.

Random Access Memory (RAM) A medium for data or program storage, the contents of which may be changed during the operation of the computer. Sometimes called read/write memory. Contents of RAM are lost when the microcomputer is switched off, i.e., volatile memory.

Read Only Memory (ROM) A component of the microcomputer which contains certain standard and custom-designed instructions required for its internal use, where they are only available for reading data.

Real-time A real-time program is one in which, during its execution, it is subjected to real-time constraints, i.e., input of data, e.g., distance or time, generated by some external activity, with which the program interacts by means of online devices.

RS-232C Serial communications interface for printers, etc.

Sequential access Accessing memory in a linear as opposed to a random fashion. Cassette files are restricted to very slow sequential access.

Serial communications The standard character and ASCII code transmission method where bits are sent, one at a time, in sequence.

Software Generic term for computer programs and digitized information which is used to issue instructions to the computer hardware and peripherals for specific applications.

Star connected A bus (e.g., HP-IB) configuration in which one peripheral device is like the hub of a wheel with the other devices on the rim, connected to the first device along the spokes. This configuration is not recommended since more than three HP-IB connectors would place extreme stress on the host device at the centre of the wheel.

Terminal An input/output device, normally consisting of a visual display unit and connected to a central mainframe computer for use in time-sharing mode.

Time-sharing A system in which CPU time and system resources, e.g., printer and plotter, are shared between a number of tasks.

Utilities Programs which have been developed to make life easier for those writing software. These include editors, compilers, character generators and so on. Some can be incorporated into programs to improve their running.

Visual Display Unit (VDU) A cathode ray tube (CRT) display screen for text and/or graphics, often called a 'Monitor'.

Volatile memory Memory whose content is lost when electrical power is removed, e.g., RAM, whereas bubble memory is non-volatile.

Z-80 Card A very popular microprocessor, similar to the 8080 and 8085 systems, which are required for running a CP/M operating system.

1

Introduction to computers for surveying

To the land surveyor, the computer is an extremely useful tool for data acquisition and the laborious task of data processing. It will never replace the practical surveyor on site, but it can bring substantial benefits by both speeding up the data recording and processing and eliminating arithmetical errors.

Land surveying can be described as the art of making measurements of the relative positions of natural and man-made features on the earth's surface. These measurements may then be used to calculate areas, delineate boundaries, make plans or drawings to some suitable scale, or for setting out construction works in association with architects, builders, civil engineers, quantity surveyors, structural engineers or town planners. Land surveying is also associated with the preparation of maps for geographical and geological purposes.

In the late 1970s the development of small microcomputers with their ease of programming has now given students and small firms access to computer power only previously available to firms with mainframe machines. As these larger machines were often operated in a batch-mode, it was necessary to prepare a large number of punched cards for subsequent processing, with little personal involvement. With the development of interactive BASIC microcomputer systems, the era of the 'personal computer' has arrived.

With the upsurge in the teaching of BASIC in both schools and universities, BASIC was chosen as the language for the programs in this book. Despite claims that BASIC is an unstructured language compared with Pascal, it has considerable advantages in being readily adaptable or portable to various microcomputers. An introduction to BASIC programming is included in Chapter 2 for those unfamiliar with some of the terminology and techniques.

Initial microcomputers developed in the late 1970s were of the 8-bit variety with a maximum memory capability of 64 K (where K is taken to be 1000 bytes, actually 1024 bytes). Advances in microprocessor technology in the early 1980s have seen the introduction of 16 and 32 bit machines with memories of 256 K and upwards. Such vast memory capabilities can now be used to provide facilities comparable with those previously associated with mainframes and minicomputers, programmed in Algol and FORTRAN.

With such a vast choice of 8, 16 and 32 bit machines, the selection of a microcomputer for use in land surveying is a complex business. Sometimes it is recommended that the software be obtained first and then a machine found to run it. The reason for this is that the language dialects of the various microcomputers are often not compatible, which means programs cannot be transferred easily from one machine to another. To simplify this task, various operating systems such as CP/M and UNIX have been developed to surmount this obstacle.

In addition to the microcomputer's software, careful attention should be paid to future interfacing requirements. The one essential is a printer for obtaining printouts of the data. Other useful features are the ability to connect colour graphic screens, graphic tablets and plotters for producing drawings, etc. The three most common interfaces used to link such peripherals are Centronics parallel, IEEE-488, and RS-232C. The first is a standard printer interface and avoids the handshaking problems with RS-232C interfaces. For general plotting facilities, the IEEE-488, sometimes referred to as HP-IB, since it was developed originally by Hewlett-Packard before being adopted by the computer industry, allows up to fourteen peripherals to be linked together.

To obtain sufficient memory capacity for data processing in surveying, the majority of microcomputers are mains supplied and are suitable for installation in a site office. A few examples in the 8-bit category are the well-known Apple II+, Apple IIe, Commodore Pet, Hewlett-Packard 80 series, North Star Advantage and Tandy TRS-80. In the 16-bit category, two popular business machines are the ACT Sirius 1 and the IBM Personal Computer. Moving up the scale, greater memory capacities and capabilities are available on the Hewlett-Packard HP-9845, the Olivetti P6066 and the Wang 2200.

All the BASIC programs listed in this book were developed on the Apple II+ using the Applesoft language (Fig. 1.1). Where possible, only routines available in Microsoft BASIC have been used to ensure portability to other machines and most of the included programs have also been run on the Commodore Pet, the Epson HX-20 and Hewlett-Packard Series-80 machines (both the HP-85 and HP-87, the former using cassette software, and the latter disc software transferred using an HP-IB interface). With the vast number of Apple II+/e models worldwide, several commercial surveying software packages are now available, including: general co-ordinate geometry (COGO) programs, traverse surveys and contour plotting. In addition to Applesoft, CP/M programs can also be run on the Apple II+/e with the addition of a Z-80 card, and these Applesoft programs are directly transferrable. Unfortunately, where the programs re-use the screen display for printing using the 'PRINT' statement, new routines have to be written for CP/M since the correct command is then 'LPRINT'.

Several commercial survey firms have standardized their land surveying software to run on Hewlett-Packard Series-80 machines. The earliest machines were the HP-83 and HP-85 which were introduced with a basic 16 K expandable

Fig. 1.1. Basic Apple II+ with twin disc drives, 12 inch monitor and Epson MX-80 printer.

to 32 K. This was followed by the HP-87 (Fig. 1.2) the HP-86 and then the HP-85B, each with a possible memory expansion capability to 544 K. With each of the Series-80 microcomputers, it is possible to connect numerous peripherals. The standard interface is of course HP-IB but additional cards are available to use peripherals on either Centronics parallel or RS-232C. Both the HP-83/85

Fig. 1.2. Hewlett-Packard HP-87 with twin disc drives (HP-82901M), 80-column screen, 80-column printer (HP-82905B) and A3 plotter (HP-9872A).

Fig. 1.3. Business-oriented Hewlett-Packard HP-86 with single disc drive (HP-9130A), 12-inch monitor and 80-column printer (HP-82905B). (Photograph by courtesy of D. J. Herriott Ltd.)

have small (5 inch) 32-column screens, which is increased to 80 columns on the HP-87. The more business-oriented HP-86 (Fig. 1.3) has a much larger 12 inch screen, and is the one now recommended by several surveying software firms, due to the ease with which the host microcomputer can be interfaced to a vast array of printers, plotters, graphics tablets, etc.

As mentioned previously, the above portable microcomputers are suitable for site or head offices with a mains electricity supply. Before the advent of modern battery-operated computers, to facilitate computations in the field, some hand-held calculators could be fitted with surveying software modules, for example the Texas Instruments' TI Programmable 58/59 and the Hewlett-Packard HP-41CV (Fig. 1.4). The surveying software for both of these instruments, in the latter case using RPN (Reverse Polish Notation), thus allowed comprehensive computations in the field. For record purposes, there is also a battery-operated printer for use with the HP-41CV. However, until 1982, it was still necessary to transcribe data by hand to the larger microcomputers for processing. With the introduction of an HP-IL (Hewlett-Packard-Interface Loop) it is now possible to collect data in the field with the HP-41CV and then transfer the data directly via an HP-IL interface to an HP Series-80 machine (Fig. 1.5). The HP-IL interface can also be used to link small printers, cassette drives and extra display monitors (Fig. 1.6).

Fig. 1.4. The Texas Instruments' TI 58/59 and Hewlett-Packard HP-41CV hand-held calculators with surveying modules.

Several surveying equipment manufacturers of electronic theodolites and EDMs now provide data interfaces for recording observations (Fig. 1.7), as dealt with more specifically in Chapter 3 (Section 3.7).

With recent developments in miniaturization in the domain of microprocessors and electronics, there are now several small hand-held microcomputers running BASIC which can be used very effectively in the field for data analysis. One of these is the compact HP-75C with a single-line LCD window, a built-in

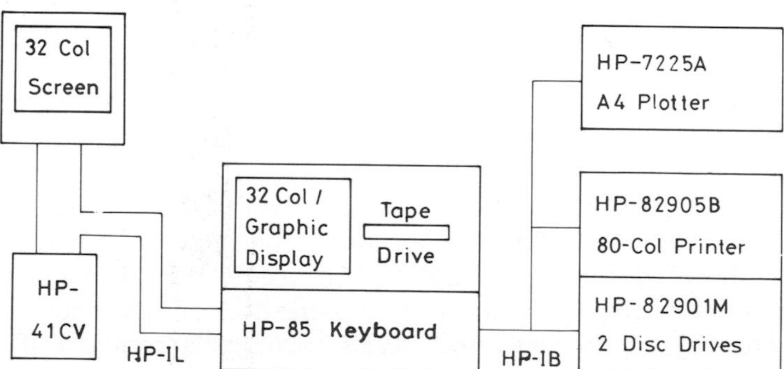

Fig. 1.5. Block diagram for Hewlett-Packard HP-41CV connected by HP-IL interface to HP-85 and additional monitor.

Fig. 1.6. Hewlett-Packard's HP-41CV and HP-85 connected by HP-IL interface for display, printout and plotting (HP-7225A).

HP-IL interface and 48 K operating system (Fig. 1.8). This is a very versatile machine with surveying software in ROM and the ability to link directly with a Series-80 machine, like the HP-41CV mentioned previously. With a resident 16 K CMOS user memory, expandable to 24 K, the programs are available instantaneously.

Another portable microcomputer is the Epson HX-20 with a 4-line 20-column display (Fig. 1.9). This machine has a standard 16 K RAM and 32 K ROM with the possibility of adding another 32 K ROM or 16 K ROM/16 K RAM. It also has a built-in printer and uses either a micro-cassette deck or cassette tape recorder. This microcomputer uses Microsoft BASIC, and several commercial surveying programs are already available for numerous field surveying tasks. Like the HP-75C, the Epson HX-20 has continuous memory which can be divided into five areas for storing programs. With its ability to link directly to an 80- or

Fig. 1.7. Geodat data recorder used in conjunction with Aga Geodimeter 122 for automatic recording of field observations. (Photograph by courtesy of Aga Geotronics.)

132-column printer, it is thus possible to obtain field data and results very quickly. Also available is a large screen interface giving either a 32-column by 16-line display or a colour graphics facility (Fig. 1.9).

To land surveyors on site, the availability of microcomputing power in the field is going to be of inestimable benefit since it will now be possible to compute traverse data, station co-ordinates by intersection or resection, calculate setting out tables for road curves in the field, etc., etc. The choice of such a portable computer with so many new machines coming on to the market, for example the Tandy TRS-80 Model 100 with a large 40-column by 8-line display (Fig. 1.10), will depend on the software availability and its interfacing capabilities. Compact machines like the Epson HX-20 (Fig. 1.9) with integral micro-cassette and printer will therefore be more suitable than those linked by cables, etc., especially on an exposed site.

The ability to record field observations electronically for subsequent data processing and analysis in head office has now led to the development of integrated surveying and plotting systems. In previous years, major advances in the use of computer graphics for computer-aided design and drafting (CAD) has now led to the availability of such systems for surveying.

Two such systems specifically for surveying are available from Survey and General Instruments and Wild. The first system from Survey and General Instruments called the SGI-HASP Digital IIS survey system (Fig. 1.11) com-

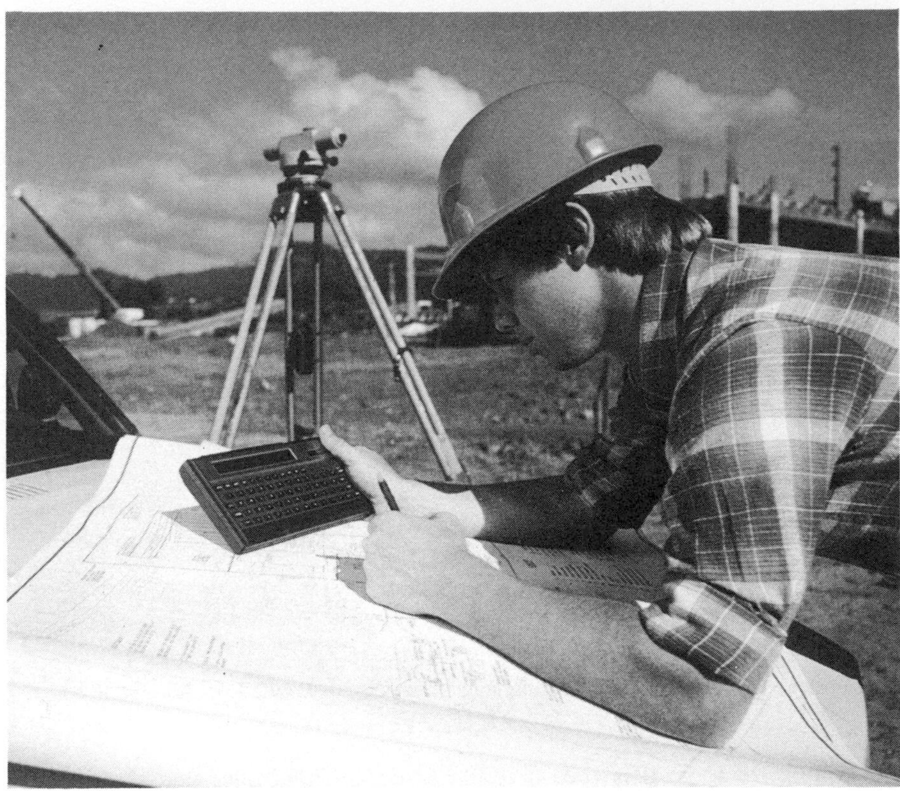

Fig. 1.8. Hewlett-Packard 75C, a battery-operated hand-held microcomputer for use on site. (Photograph by courtesy of Hewlett-Packard.)

prises a Hewlett-Packard HP-9816 microcomputer and associated peripherals – printer, plotter and digitizer. Data can be input either via disc, digitizer or keyboard for data analysis and the production of survey drawings, etc.

The second interactive graphics and surveying system called Geomap has recently been developed by Wild Heerbrugg. This integrated system will handle the whole process from data acquisition of field measurements through data processing and data output in either digital or graphic form for the production of maps, and is based on a Tektronix 4054 with two disc drives, monitor and plotting table. Such systems have often been developed in-house in the past by large companies like Wimpeys who have developed specialized software like CALSID and SIDS for the production of site plans, etc.

The cost of developing and installing such integrated graphic systems cannot often be considered cost effective by small firms or educational establishments unless available to third parties in a batch-processing mode. The recent introduction of 80-column boards for microcomputers to allow their use for word-processing, CP/M and greater VisiCalc display has presented designers with the

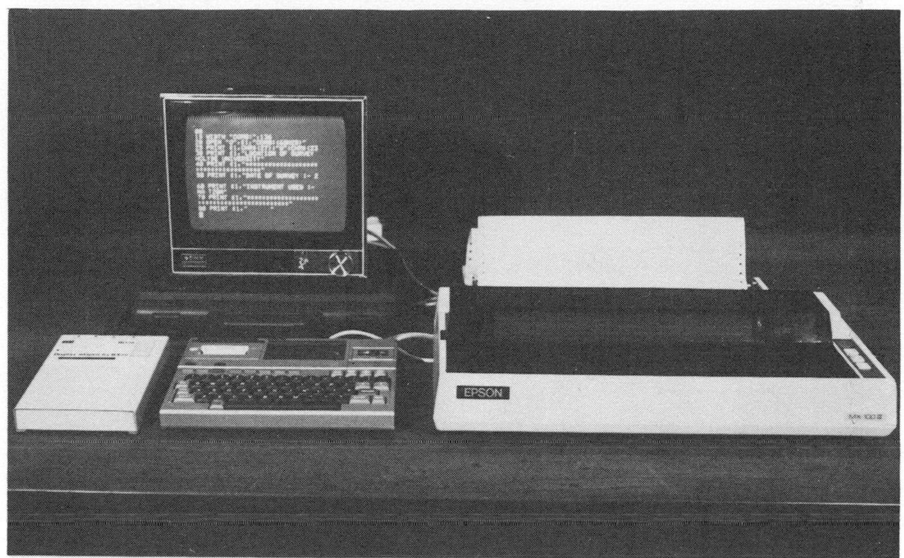

Fig. 1.9. Epson HX-20 battery-operated portable microcomputer shown with Epson MX-100 printer and HO-20 interface for additional text/graphic display.

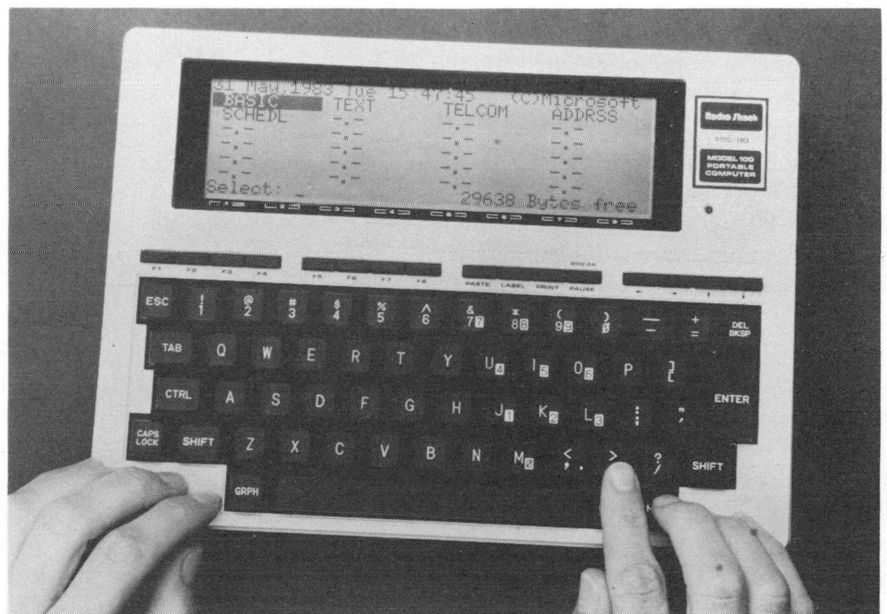

Fig. 1.10. Tandy TRS-80 Model 100 battery-operated microcomputer with 40-column, 8-line display. (Photograph by courtesy of Tandy Corporation.)

Fig. 1.11. The SGI-HASP Digical IIS survey system in use based on an HP-9816 microcomputer. (Photograph by courtesy of Survey & General Instrument Co. Ltd.)

opportunity to develop systems capable of simultaneous display screens with text and graphics.

The Apple II+ discussed earlier (Fig. 1.1) is one such microcomputer which can be readily adapted to display text or graphics on different screens. In the Apple II+ the 40-column or graphics screen has the output address 'PR#0' whilst the 80-column card is commonly 'PR#3'. It is thus possible to display all text descriptions, menus for the selection of programs, etc. on the text screen with the simultaneous presentation of graphics and drawings on the graphics screen.

One of the advantages of the Apple II/IIe microcomputer is the provision of eight slots on the mother-board for the connection of peripherals (Fig. 1.12). In the Apple II+ used at Strathclyde University for surveying, a language card is located in slot 0 to increase its memory to 64 K. The printer interface resides in slot 1 and hence in the program lists any hard copy printouts are directed to 'PR#1' the printer slot. Slot 2 has been used for a graphics tablet to allow cross-sectional areas and other features to be entered directly (Fig. 1.13). An 80-column card, if used for word-processing, must be in slot 3. The only two other specified slots are number 6 for the disc drives and number 7 for the colour graphics (280 × 192 pixel) card. However when plotting lines, this resolution tends to give ragged diagonal lines, commonly called 'jaggies' or stair-stepping.

Fig. 1.12. Interactive computer graphic workstation based on Apple II+ with dual screens, 80-column monochrome and colour graphics, together with printer, plotter and graphics tablet as developed at Strathclyde University.

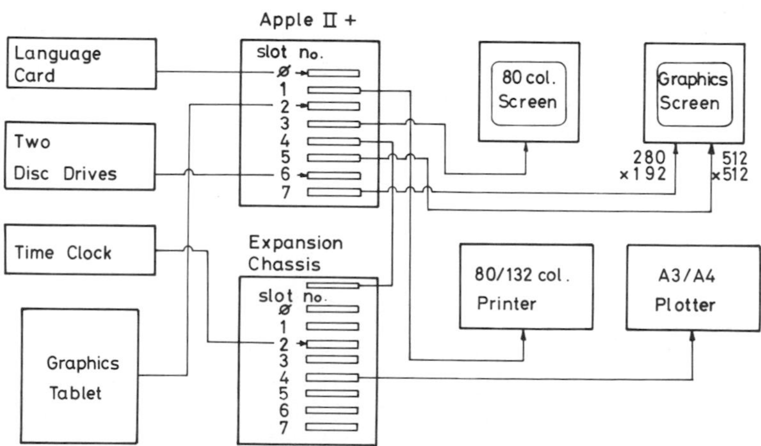

Fig. 1.13. Block diagram for Apple II+ interactive computer graphic workstation showing location of interface cards with addition of expansion chassis.

To overcome this, several manufacturers now produce graphic processor cards giving, in our case, a 512 × 512 pixel screen. As graphics is a machine-dependent topic, it is regrettably not dealt with in this book.

Since plotting is a major part of surveying, it is useful to have an IEEE-488 interface. Unfortunately, this card tends to overheat in the main Apple and hence the use of an Expansion Chassis (Fig. 1.12) which gives another eight slots, where the IEEE-488 interface is located in slot 4, and hence the 'PR#4' command. For real-time applications, a time-clock is also an advantage to annotate printouts, etc., but it is not used in any of the following BASIC listings in an attempt to keep them as simple as possible.

After an introduction to BASIC programming in Chapter 2, the remaining four chapters are split into related topics: theodolite/traverse surveying; areas/volumes/mass-haul diagrams; curve ranging and adjustment of observations. The over-all strategy has been to produce straightforward BASIC programs that could be transferred to other microcomputers. Hence the programming routines have been split into major blocks of 1000 lines, with smaller routines starting at multiples of 100 or 500 lines, depending on length. A large number of remarks (REM) have also been included to describe the function of each routine, but these could be omitted if memory space were limited. Throughout the book, regularly used routines have been given the same line numbers to allow an easy build-up of a library of subroutines which can be incorporated in any program.

All the listed programs have been written with a keyboard entry. However, in practice in the field, station co-ordinates, etc., could well be supplied from a data file and information available from either cassette or disc. The reason for the keyboard entry was to keep the listings simple and capable of adaptation by different programmers on varied microcomputers.

Every reasonable effort has been made during their development to ensure the accuracy of the computer programs in this book. The author acknowledges, however, that the possibility of either typographical or logical errors exists, and for this reason, those making use of these programs are advised to verify independently the accuracy and correctness of the methods and procedures presented in the following pages.

2

BASIC programming

2.1 INTRODUCTION TO BASIC

Although several different computer languages were mentioned in Chapter 1, all the programs in this book are written in BASIC. This is often the first language taught by computer science departments to university and college students and has gained widespread popularity due to the numerous microcomputers supporting BASIC.

Unfortunately, despite attempts over the years to establish a universal ANSI-standard version of BASIC for portability on various types of microcomputers, each manufacturer often offers improvements (enhancements) in the BASIC dialects available. This means that programs written exclusively for one machine may not be compatible with another. In order to assist in software portability the programs presented in this book have been written in a simple form of BASIC which can readily be adapted to specific machines.

To facilitate program portability, various computer operating systems have been developed with a standard structure, for example CP/M (trade mark of Digital Research, Inc.), which stands for Control Program/Microprocessors. The most common implementation of CP/M in microcomputers is using Microsoft's ANSI-standard BASIC-80. Some microcomputers are Z-80 based, whereas others require a Z-80 soft-card to be installed in the machine before CP/M can be run. However, since not all microcomputers have a Z-80 soft-card, it was decided to give printer listings using Applesoft BASIC developed by Apple Computers Ltd, since Applesoft BASIC is very similar in many cases to the BASIC dialects available on other popular microcomputers, e.g., Hewlett-Packard, North Star, PET and TRS-80. Where compatibility or changes are required for a specific microcomputer these are listed in Table 2.1. In the past BASIC was solely available on mains-operated machines, but the advent of small battery-operated portables, e.g., Epson HX-20 and Hewlett-Packard HP-75C, now allows computing to be carried out in the field, albeit with limited memory capability.

Table 2.1 Commands, functions and statements

Apple II+/e *Applesoft*	*Apple II+/e* *MBASIC*	*Epson HX-20* *Epson BASIC*	*HP Series-80* *BASIC*
ABS(*n*)	ABS(*n*)	ABS(*n*)	ABS(*n*)
ATN(*x*)	ATN(*x*)	ATN(*x*)	ATN(*x*)
CHR$(*n*)	CHR$(*n*)	CHR$(*n*)	CHR$(*n*)
CLEAR	CLEAR	CLEAR	RUN
COS(*x*)	COS(*x*)	COS(*x*)	COS(*x*)
DEF FN<name>(*p*)	DEF FN<name>(*p*)	DEF FN<name>(*p*)	DEF FN<name>(*p*)
DIM	DIM	DIM	DIM
END	END	END	END
EXP(*n*)	EXP(*n*)	EXP(*n*)	EXP(*n*)
FOR . . TO . . [STEP]	FOR . . TO . . [STEP]	FOR . . TO . . [STEP]	FOR . . TO . . [STEP]
GOSUB	GOSUB	GOSUB	GOSUB
GOTO	GOTO	GOTO	GOTO
HOME	HOME	CLS	CLEAR
IF . . GOTO	IF . . GOTO	IF . . GOTO . . [ELSE]	
IF . . THEN	IF . . THEN . . [ELSE]	IF . . THEN . . [ELSE]	IF . . THEN . . [ELSE]
INPUT	INPUT	INPUT	INPUT
INT(*n*)	INT(*n*)	INT(*n*)	INT(*n*)
INVERSE	INVERSE		
LET	LET	LET	LET
LOG(*n*)	LOG(*n*)	LOG(*n*)	LOG(*n*)
NEXT	NEXT	NEXT	NEXT
NORMAL	NORMAL		
NOTRACE	NOTRACE	TROFF	NORMAL
ON . . GOSUB	ON . . GOSUB	ON . . GOSUB	ON . . GOSUB
ON . . GOTO	ON . . GOTO	ON . . GOTO	ON . . GOTO
ONERR GOTO	ON ERROR GOTO	ON ERROR GOTO	ON ERROR GOTO
	OPTION BASE (*m*)	OPTION BASE (*m*)	OPTION BASE (*m*)
PEEK(*n*)	PEEK(*n*)	PEEK(*n*)	
PEEK(222)	ERR	ERR	ERRN
	ERL	ERL	ERRL
POKE(addr,*n*)	POKE(addr,*n*)	POKE(addr,*n*)	
PR#(*n*)			
"PR#0":PRINT	PRINT	PRINT	DISP
"PR#1":PRINT	LPRINT	LPRINT	PRINT
PRINT CHR$(7)	BEEP(*n*,*n*)	SOUND	BEEP
READ	READ	READ	READ
REM	REM	REM	REM
RESUME	RESUME	RESUME	CONT
RETURN	RETURN	RETURN	RETURN
SIN(*x*)	SIN(*x*)	SIN(*x*)	SIN(*x*)
SQR(*n*)	SQR(*n*)	SQR(*n*)	SQR(*n*)
STOP	STOP	STOP	STOP
TAB(*n*)	TAB(*n*)	TAB(*n*)	TAB(*n*)
TAN(*x*)	TAN(*x*)	TAN(*x*)	TAN(*x*)
TRACE	TRACE	TRON	TRACE

Footnotes: [. . .] indicates optional entry
addr is address
n is a numeric expression
p is a parameter
x is angle in radians

Table 2.1—*continued*

North Star BASIC	*PET/CBM MBASIC*	*TRS-80 BASIC*	*Comments*
ABS(*n*)	ABS(*n*)	ABS(*n*)	Absolute value
ATN(*x*)	ATN(*x*)	ATN(*x*)	Arctangent
CHR$(*n*)	CHR$(*n*)	CHR$(*n*)	Characters
RUN	CLR	CLEAR	Clear variables
COS(*x*)	COS(*x*)	COS(*x*)	Cosine
DEF FN<name>(*p*)	DEF FN<name>(*p*)	DEF FN<name>(*p*)	Define function
DIM	DIM	DIM	Dimension array
END	END	END	End of program
EXP(*n*)	EXP(*n*)	EXP(*n*)	Exponential
FOR . . TO . . [STEP]	FOR . . TO . . [STEP]	FOR . . TO . . [STEP]	Loop
GOSUB	GOSUB	GOSUB	Subroutine
GOTO	GOTO	GOTO	Branch to line
	PRINT "clr"	CLS	Clear screen
	IF . . GOTO		Conditional
IF . . THEN . . [ELSE]	IF . . THEN	IF . . THEN	Conditional
INPUT	INPUT	INPUT	Data entry
INT(*n*)	INT(*n*)	INT(*n*)	Integer
	PRINT "R"		Screen inverse
LET	LET	LET	Assignment
LOG(*n*)	LOG(*n*)	LOG(*n*)	Logarithm (e)
NEXT	NEXT	NEXT	Loop closure
	PRINT "-"		Screen normal
		TROFF	Tracing off
ON . . GOSUB	ON . . GOSUB	ON . . GOSUB	Conditional
ON . . GOTO	ON . . GOTO	ON . . GOTO	Conditional
ERRSET		ON ERROR GOTO	Error branch
			Option $m = 0$ or 1
EXAM(*n*)	PEEK(*n*)	PEEK(*n*)	Read memory
ERRSET			Error number
ERRSET			Error line
FILL(addr,*n*)	POKE(addr,*n*)	POKE(addr,*n*)	Stores data
PRINT#(*n*)	PRINT#(*n*)		I/O address
PRINT	PRINT	PRINT	Print (screen)
PRINT#(*n*)	PRINT	PRINT	Print (printer)
PRINT CHR$(7)			Bell (sound)
READ	READ	READ	Read data
REM	REM	REM	Remark
		RESUME	Resume at error
RETURN	RETURN	RETURN	Return to line
SIN(*x*)	SIN(*x*)	SIN(*x*)	Sine
SQRT(*n*)	SQR(*n*)	SQR(*n*)	Square root
STOP	STOP	STOP	Stop program
TAB(*n*)	TAB(*n*)	TAB(*n*)	Column tabbing
	TAN(*x*)	TAN(*x*)	Tangent
		TRON	Trace routine

Throughout this book it is assumed that the reader has had an introduction to BASIC through either formal or informal teaching, and this book should not be looked on as a BASIC instruction manual. Therefore the main aim is to apply the knowledge of computer programming in BASIC to assist in solving the many laborious and time-consuming data reduction tasks required of the surveyor. In spite of the above assertion there follows a general discussion of the BASIC language commands and their implementation and translation for different machines.

2.2 BASIC STATEMENTS

2.2.1 Program structure

A computer program is made up of a sequence of statements to define the precise procedure to be followed. In BASIC these statements are preceded by a line number to indicate the sequence of operations. The increment value of the lines is normally 10 but can be reduced to 1 if required. The advantage of programming in steps of 10, allows nine spare lines should an extra instruction or subsequent error trap be required. It also allows remark (REM) statements to be included to assist a reader to follow the reasoning behind the programming statements. In this book the programs have been written in a structured approach with all the programming routines split into blocks of 1000 lines with smaller subroutines starting at multiples of 100 or 500 lines depending on length.

2.2.2 Variables

As stated in Section 2.1, the surveyor is often called upon to reduce vast amounts of data. The data has to be entered into the microcomputer at the keyboard in the form of variables for storing. There are two kinds of variables, either 'numeric variables' or 'string variables'. Numeric variables can only store a number, e.g. 25, 37.625, etc., and these are designated by a single letter, e.g. A, B, C, etc., or a letter followed by a number, e.g. $A0$, $A1$, $A2$.

The values of these variables may be entered from the keyboard or may be assigned at the start of the program in the form:

line number [LET] variable = mathematical expression

where the word LET is often optional and thus if the number 37.625 is to be stored in variable $A1$, either of the following is acceptable:

```
100 LET A1 = 37.625
```

or

```
100 A1     = 37.625
```

String variables on the other hand are treated by the microcomputer as a string of characters, either numbers or letters or both, and are designated by a single letter followed by a '\$', *B*\$, etc. Some versions of BASIC allow word strings, e.g. 'name\$', for the storing of string variables. However, in the simpler BASIC programs in this book, only single letter and number designations are used wherever possible.

2.2.3 Array variables

It is often very convenient in a program, instead of storing information in ten separate numeric variables, e.g. $A1$, $A2$, $A3$, etc., to use an array variable with ten elements. An array with a single dimension, e.g. $A(10)$, is equivalent to a table with just one row of numbers, where the index identifies a number within the single row. An array with two dimensions, e.g. $A(3, 2)$, yields a table of numbers with rows and columns, and is often used in matrix algebra, where one index identifies the row, and the other index identifies the column.

In most microcomputer systems, the lower bound of a numeric array is 0, thus the array $A(10)$ would contain spaces for eleven variables $A(0)$, $A(1)$, $A(2)$ . . . $A(10)$, and the two-dimensional array $A(3, 2)$ would contain spaces for a matrix of 4 rows (0, 1, 2, 3) and 3 columns (0, 1, 2). Some microcomputers, like the Hewlett-Packard HP-80 series and those using Microsoft BASIC, allow an 'OPTION BASE' statement at the beginning through which one can use either the lower bound of 0, i.e. 'OPTION BASE 0', or, if 1, i.e. 'OPTION BASE 1'. The programs in this book use the 'OPTION BASE 0' format meaning that the third element in an array $A(n)$, where n is a numeric integer, will be stored in $A(2)$.

At the outset in a program, it is desirable to dimension the size of the arrays to be used in the program; with some microcomputers this is optional, but it is a good habit in laying down a structured format. The following is an example of the dimensioning statement for a single-dimension array P of 10 variables, a two-dimensional matrix Q with 4 rows and 3 columns, and a string variable R with 25 characters (it should be noted here that string variables start at character 1)

```
10 DIM P(9), Q(3, 2), R$(25)
```

2.2.4 Mathematical expressions

In each of the programs provided in this book, the algorithms required for solution are discussed and the algebraic expressions presented. It should be remembered that where in algebra there is implied multiplication between operators, this is not the case in BASIC and care must be taken in studying the exact hierarchy in which mathematical operators work. The list of arithmetic operators, in order of precedence, is given at the top of p. 18.

To avoid ambiguity, or to change the order in which the operations are performed, it is essential to use parentheses. Operations within parentheses are

Operator	Operation	Sample expression
^ or ↑	Exponentiation	X ^ Y
*	Multiplication	X * Y
/	Division	X / Y
+	Addition	X + Y
−	Subtraction	X − Y

performed first, but the hierarchy of the operators within the parentheses remains the same. For example, the function:

$$\frac{a+b}{c+d} \text{ becomes } (A+B)/(C+D),$$

where the parentheses are essential. If there are two denominators on the lower line with implied multiplication, these should also be included in parentheses to give the correct result. For example, the function:

$$\frac{x+y}{kz} \text{ becomes } (X+Y)/(K * Z)$$

2.2.5 Relational operators

The correct execution of a program often depends on the relation between two numeric variables, and depending upon whether the result is true or false, a decision is taken regarding the program flow. The following operators are commonly used:

Operator	Relation tested	Expression
=	Equality	X = Y
< >	Inequality	X < > Y
<	Less than	X < Y
>	Greater than	X > Y
< =	Less than or equal to	X < = Y
> =	Greater than or equal to	X > = Y

These relational operators are used in conjunction with a conditional statement of the type:

IF expression 1 < > expression 2 THEN line number.

This then allows the program to branch, or if the entered data is false, to return to the beginning to elicit a correct response.

2.2.6 Input and output

One of the considerable advantages of the modern microcomputer is its ability to be interactive, thus allowing 'conversational' programs where the user enters data in response to prompts from the computer. However, there is nothing more annoying to a user than having loaded a program to be presented with a '?' on the screen, and not knowing what response is expected. All programs should of course be documented to give the precise order and form of data entry, but this is often overlooked and the user has to search through the manuals to find the correct response.

Data entry input is often combined with an output such as the display on the screen of a statement requesting a data entry, or the display of results, e.g.

line number PRINT "text enclosed in quotes"
or line number PRINT list of variables

where the data entry routine takes the form:

line number INPUT variable (either string or numeric).

If an HP-80 series computer is used, the screen display statement is DISP rather than PRINT.

The programs in this book have been written from the viewpoint of screen documentation, that is, all information is provided on the screen as to the type of data, numeric or string, and to the format, for example when an angle is to be entered, a choice of radians or degrees is given:

```
200  PRINT "ARE ANGLES IN DEGREES <1>, RADIANS <2> OR DO YOU WISH TO
QUIT <3>."
210  INPUT U
220  IF U < 1 OR U > 3 THEN 200
230  ON U GOTO 240,290,9500
240  R = 57.2957795
```

In the above example the exact nature of the response '1', '2', or '3' is indicated in line 200 with the entry *U* on line 210. The two following lines therefore select degrees if '1' at line 240, or radians if '2' at line 290. Line 220 is there to provide an error trap just in case something else other than 1–3 has been entered, thus returning the user to the entry point to the program rather than causing an error to develop later in the program. If degrees is chosen, the format can be either decimal or degrees, minutes and seconds (DD.MMSS), as follows:

```
250  PRINT "ARE ANGLES IN DECIMALS <1> OR DD.MMSS <2>";
260  INPUT D9
270  IF D9 < 1 OR D9 > 2 THEN 250
```

The user therefore indicates his choice '1' or '2' which is entered on line 260, with line 270 providing an error trap.

In each of the above cases there was only one entry in each line, *U* or *D*9. However it is possible to include more than one entry at a time, e.g. INPUT *A*,*B*,*C*, where the data is separated by commas ','. Since the comma separates data items it cannot be used in a string entry unless inverted commas are used at either end of the entry to ensure the string is not split, otherwise a data error will occur.

An example of multiple entries occurs if the angle format selected is DD.MMSS, where it is essential to convert this format to decimal degrees before it can be used in calculations. To keep the programs simple, the following entry routine has been used in the early programs:

```
2050  PRINT "ENTER ANGLE AS DEGS, MINS, SECS"
2060  PRINT "ANGLE 1 = ";
2070  INPUT A4, A5, A6
2080  A1 = A4 + A5/60 + A6/3600
```

where the degrees are entered in $A4$, minutes in $A5$ and seconds in $A6$ in line 2070 and then the angle $A1$ converted to decimals in line 2080.

An alternative to the above would be to enter the data directly in DD.MMSS. However, the conversion to decimals is more complicated, namely:

```
2050  PRINT "ENTER ANGLE AS DD.MMSS"
2060  PRINT "ANGLE 1 = ";
2070  INPUT A1
2071  D0 = INT (A1)
2072  M1 = (A1−D0) * 100
2073  M0 = INT(M1 + .1)
2074  S0 = INT((M1 − M0) * 100 + .5)
2080  A1 = D0 + M0/60 + S0/3600
```

where for comparison with the three data inputs ($A4$, $A5$, $A6$) there is now only one ($A1$). However there are four extra lines 2071–2074 where to begin with, the integer part $D0$ is identified, then the minutes $M0$ and finally the seconds $S0$ determined before $A1$ can be converted to decimals in line 2080. The programs in this book use the simpler triple data entry format in the early programs. It has been assumed that the programs are being used by those familiar with angle formats; however, should the user wish to provide an error trap for angle entry, a considerable number of checks have to be made in the above angle entry routine, namely:

```
2075  IF A1 < 0 OR A1 > 360 THEN 2050
2076  IF M0 < 0 OR M0 > 60 THEN 2050
2077  IF S0 < 0 OR S0 > 60 THEN 2050
```

i.e. a check is carried out to ensure that no angles less than 0 or greater than 360 have been entered, together with checks to see if either of the entries for minutes and seconds is greater than 60. If any of these conditions are true, the program reverts to the input entry line of 2050. As this would have to be done each time an angie was entered it could add considerably to the number of program lines. A simpler method could be to use a subroutine as described in the next section.

2.2.7 Loops and subroutines

There are many ways in which the route through a computer program may be selected. The IF . . . THEN routine was discussed in Section 2.2.5. Another method is to jump out of the program sequence to another part of the program by specifying:

line number GOTO line number.

This unconditional branch can often cause an untidy spaghetti-like thread through a program making it difficult to follow. The programs in this book follow as structured a form as BASIC will allow, and GOTO is only used to jump over parts of the program where there has been a previous choice made and segments have to be avoided to return to the mainstream of the flow.

Where there is a repetition of data to be entered into a program or where a series of instructions are to be performed a given number of times, it is often useful to construct a loop using the following format:

line number FOR variable = expression 1 TO expression 2 [STEP expression 3],

where the default value for the STEP in brackets is 1, so this need only be included if the step is greater than 1. The end of the loop is signified by:

line number NEXT variable.

Where a certain sequence of events is repeated several times in a program, like the DD.MMSS angle input in Section 2.2.6, it is often convenient to use a GOSUB routine which can be accessed from any part of the program. The routine is completed with a RETURN statement. Earlier a simple routine was given for a three-part angle data input for DD.MMSS. Unfortunately this is not possible when returning from a decimal angle to a DD.MMSS format. Here it is best to use a GOSUB routine, namely:

```
2070  D = A1
2080  GOSUB 8050
2090  A1 = D
  ⋮
8050  D1 = INT (D)
8060  D2 = (D – D1) * 3600
8070  M = INT (D2 / 60)
8080  S = D2 – 60 * M + 0.5
8090  IF S < 60 THEN 8130
8100  M = M + 1 : S = 0
8110  IF M < 60 THEN 8130
8120  D1 = D1 + 1 : M = 0
8130  D = D1 * 10000 + M * 100 + S
8140  D = INT (D) / 10000
8150  RETURN
```

Before calling the GOSUB routine which requires the angle D, it is necessary to reassign the required angle $A1$ to D in line 2070 before going to the subroutine at line 8050. Here the degrees ($D1$) are abstracted as the integer of D, then the remaining fractional part ($D - D1$) is converted to seconds (line 8060) and the integer part of the minutes found in line 8070. The remainder will give the number of seconds. Nevertheless, care must be taken to see when the final integer is determined that values are rounded up and hence the addition of 0.5 seconds in line 8080. If S is less than 60 then all is well, but if S is greater than 60 due to the addition of the 0.5 seconds, the minutes M must be increased by 1 and S set to zero. A similar consideration is given to the minutes in line 8110 before finally collecting the degrees, minutes and seconds together in line 8130, whose integral is taken in line 8140 and divided by 10000 to give the DD.MMSS format before returning to the program, to reassign the subroutine angle D to the required angle $A1$.

Both the above GOTO and GOSUB statements can also be incorporated in multiple branching statements, which is very convenient when a menu is presented for program selection. If the menu asks for responses, say 1 to 5, and the required routines are located at line numbers 1000, 2000, 3000, 4000 and 5000 we can combine these five IF . . . THEN statements in the form:

```
400  INPUT J
410  IF J <1 OR J > 5 THEN 300
420  ON J GOTO 1000,2000,3000,4000,5000
```

where line 410 is an error trap. Assuming a key in the range 1 to 5 was pressed, then the program will branch to the appropriate routine. This is the normal approach where the program is later directed back to the original menu. An alternative format would be:

```
420  ON J GOSUB 1000,2000,3000,4000,5000
```

which would return to the next line number, so careful selection of the correct statement is required.

2.2.8 Other available statements

Most versions of BASIC support general scientific functions, for example:

SQR	Square root	
SIN	Sine	(angle in radians)
COS	Cosine	(angle in radians)
TAN	Tangent	(angle in radians)
ATN	Arctangent	(gives angle in radians)
LOG	Natural logarithm	(base e)
INT	Integer value	
ABS	Absolute value	

Some microcomputers like the Spectrum and Hewlett-Packard 80 series also offer:

ASN Arcsine
ACS Arccosine,

often with the choice of radians (RAD) or degrees (DEG). As the majority of microcomputers do not have these enhanced benefits, the programs in this book have been written to accept all angles either in degrees or radians, with an automatic conversion to radians before any arithmetic operations are performed. Since the arcsine and arccosine functions are extremely useful in the trigonometrical solution of triangles, these can be defined in terms of the arctangent (ATN). Since these may be required several times in a program it is often useful to use a DEF FN statement to define and name a function written by the user, where the format is:

line number DEF FN <name> (parameter) = function definition.

From a knowledge of the formulae for an

$$\text{angle } x, \qquad \tan(x) = \frac{\sin(x)}{\cos(x)}$$

$$\text{and} \qquad \sin^2(x) + \cos^2(x) = 1$$

The following BASIC statements may thus be used to find an angle in radians:

```
55  REM FNC9 = ARCCOSINE
60  DEF FNC9(C9) = (1.5707964–ATN(C9 / SQR(1 – C9 * C9))) * R
65  REM FNS9 = ARCSINE
70  DEF FNS9(S9) = ATN(S9 / SQR(1 – S9 * S9)) * R
```

Thus if $A2$ = arccosine ($X2$) we can write $A2$ = FNC9($X2$), where $X2$ is substituted for $C9$ on line 60. Similarly if $A3$ = arcsine ($X3$) we can write $A3$ = FNS9($X3$), where $X3$ is substituted for $S9$ in line 70.

Mention was made previously of the ability of the programs to manipulate angles in either degrees or radians. Thus at the outset, when asked if the angles are in degrees or radians, Section 2.2.6, the program was directed to lines 240 for degrees, e.g.

```
240  R = 57.2957795
```

or to line 290 for radians, e.g.

```
290  R = 1.
```

Thus if R is included in all functions as shown earlier on lines 60 and 70, if in radians, the operator $R = 1$ has no effect, whereas if in degrees the effect is to multiply the result by 57.2957795 to convert back to degrees from radians.

Other convenient statements are those to clear the screen before displaying headings or graphics and the ability to invert the image. In Applesoft and Microsoft BASIC programs, the command to clear the screen is 'HOME', whereas with the Hewlett-Packard Series-80 computers it is CLEAR, which in Applesoft or Microsoft BASIC would clear all variables! As there is often a need to regularly clear the screen, the statement has only been used once in each program at line 9000, so other machines' clear screen commands can easily be accommodated (see Table 2.1), and each time it is required it is accessed through a GOSUB routine, e.g. for Applesoft, etc.

```
100  GOSUB 9000
:
:
9000  HOME
9010  RETURN
```

To obtain specific effects on the screen it is often possible on some microcomputers like the Apple to invert the display using an INVERSE command with a NORMAL statement required to return the screen to normal output. This technique has not been applied in the general programs, but it is useful in the error trace routine described in the next section.

2.2.9 Error tracing

During the entry of a computer program at the keyboard, the interactive nature of BASIC has the advantage that if grammatical errors are incorporated, the machine will normally 'beep' or sound a 'bell' and display a syntax error. If this happens, then the program statements require modification by editing, and these techniques are not discussed as they vary from machine to machine. However, the user should be aware of the problems of entering false data, for example, in a triangle if three sides *S*1, *S*2, *S*3 are entered, then a triangle can only be formed if the sum of two sides is greater than the third. The following is a triangle input routine:

```
1020  PRINT "<1> SIDE/SIDE/SIDE SOLUTION"
1030  PRINT "****************************"
1040  PRINT "SIDE 1 = ";
1050  INPUT S1
1060  IF S1 < = 0 THEN 1040
1070  PRINT "SIDE 2 = ";
1080  INPUT S2
1090  IF S2< = 0 THEN 1070
1100  PRINT "SIDE 3 = ";
1110  INPUT S3
1120  IF S3 < = 0 THEN 1100
1130  IF S1 + S2 < = S3 THEN 1170
1140  IF S2 + S3 < = S1 THEN 1170
1150  IF S1 + S3 < = S2 THEN 1170
1160  GOTO 1210
1170  PRINT CHR$(7) :REM CTRL-B (BELL)
1180  PRINT "SORRY YOU DO NOT HAVE A TRIANGLE"
1190  PRINT "PLEASE CHECK AND RE-ENTER SIDES"
1200  GOTO 1030
```

After each keyboard entry for *S*1, *S*2, *S*3 the entry is checked to be greater than zero. After all sides are entered the lengths are checked, and if too short the program displays an error message, lines 1170 to 1190 and then reverts back to the start of the data entry. The 'beep' or 'bell' command will depend on the version of BASIC used (Table 2.1). For example, the Hewlett-Packard Series-80 use 'BEEP', whereas on the Apple II using Applesoft, it is necessary to use the control-B (CTRL-B) character which is stored in the ASCII code as CHR$(7) as shown on line 1170. The user should therefore check the sound capabilities of the particular machine in use.

The above data entry errors should be trapped before the main program computations commence. However, a programming slip in typing in a program often occurs, and it is very useful to incorporate an error tracing routine to determine the nature of the error, and the line number on which it appears. This is often invoked by using an ON ERROR routine near the start of the program, e.g.

```
90 ON ERROR GOTO 8500
```

The nature of the routine at 8500 will depend on the facilities available on the machine used (Table 2.1). For example, Microsoft BASIC conveniently has two useful features as given below:

```
8500 E = ERR: REM FIND ERROR TYPE
8510 L = ERL: REM FIND ERROR LINE NO.
8520 INVERSE
8530 PRINT "ERROR NO. ";E;" FOUND ON LINE NO. ";L
8540 NORMAL
8550 STOP
```

This routine will stop execution of the program and display the type of error and line number.

The Hewlett-Packard Series-80 computers also have a similar feature:

```
  90 ON ERROR GOTO 8500
8500 E = ERRN @ REM FIND ERROR TYPE
8510 L = ERRL @ REM FIND ERROR LINE NO.
8520 DISP
8530 DISP "ERROR NO. ";E;" FOUND ON LINE NO. ";L
8540 STOP
```

If an error occurs when using Applesoft on an Apple II, the screen will display the number at which the program stopped. The error number can also be found by using PEEK (222), but not the line number. A useful routine which can be incorporated in an Applesoft program is to find the error number E and then use the INVERSE mode with the TRACE command to find the line number where the error occurred using a RESUME statement as shown below:

```
  90 ONERR GOTO 8500
   :
8500 PRINT CHR$ (7): REM BELL
8510 E = PEEK (222): REM FIND ERROR NO.
8515 REM IF C = 0 PRINT ERROR NO.
8520 IF C = 1 THEN 8580
8530 INVERSE
8540 PRINT "ERROR NO. ";E;" FOUND"
8550 C = 1: REM ONLY PRINT E ONCE
8560 TRACE:REM TURN TRACE ROUTINE ON
8570 RESUME: REM GO BACK TO ERROR
8580 PRINT "ERROR ON SECOND LINE NO."
8590 NORMAL: REM RETURN TO NORMAL SCREEN
8600 NOTRACE: REM TURN TRACE ROUTINE OFF
8610 STOP
```

In long programs which are built up block by block, it is often useful to write the error trace routine at the outset, so that the program can be checked as it is typed in.

The final aspect of error tracing is the checking that, not only is the program grammatically correct, but that the program should provide the correct answers.

All programs in this book are provided with sample data giving the correct answers for checking the correct running of the programs. Mention has been made previously of error trapping to check for erroneous data entries and most programs include these. However for the purpose of simplicity, where the error trapping is extensive like the DD.MMSS error trap discussed in Section 2.2.6, with an additional three lines from 2075 to 2077, these have been omitted for reasons of clarity.

2.3 PRESENTATION OF OUTPUT

2.3.1 Screen format

In Section 2.2.6 the interactive nature of BASIC was discussed, where data was entered from the keyboard with the presentation of the results, etc., on the screen. The programs in this book have been written on the assumption that the user has available a small alphanumeric screen capable of several rows and columns, rather than just a single line entry. Many of the programs presented in this book have been run on one of the following:

(a) Hewlett-Packard HP-8532 column × 16 line display,
(b) Apple II+ 40 column × 24 line display,
(c) Apple II+ 80 column × 24 line display,
(d) Apple II+ (with CP/M) 80 column × 24 line display,
(e) Hewlett-Packard HP-87 80 column × 16 line display.

The computer program listings are shown for the Apple II+/e with its 40 column × 24 line display, as it was felt that this format could easily be adapted for other machines. Some of the programs have even been adapted to run on the 20 column by 4 line screen of the Epson HX-20 as a specific example. However, no screen graphics have been included in the programs since these are very machine dependent, especially as some microcomputers have the zero (0,0) in the top left-hand corner, and others use the bottom left-hand corner.

When using the Apple II+ with a cassette system, the 40 column display screen (the default screen for printing) is accessed by the command PRINT "PR#0" and a printer in slot 1 would be accessed by the command PRINT "PR#1". If a disc-based Apple II+/e system is used, that is a Disc Operating System (DOS), then it is necessary to use a control-D (CTRL-D) character, as listed below:

```
10  REM OUTPUT TEXT TO A PRINTER
20  REM CREATE CTRL-D CHARACTER
30  D$ = CHR$(4)
40  REM SELECT I/O CARD IN SLOT 1
50  PRINT D$; "PR#1"
60  PRINT "text"
 :
180 REM DESELECT PRINTER AND RETURN TO SCREEN
190 PRINT D$; "PR#0"
200 END
```

All the programs in this book have used a disc-based system using the above sequence of commands. To simplify conversion to cassette or other disc operating systems, the printer statements have been accessed using GOSUB routines to reduce the number of changes required.

2.3.2 Data printout

Just as the screen format is machine dependent, so are the various print commands to various manufacturers' printers, depending upon whether the printer is linked by a Centronics parallel interface, an IEEE-488 interface or an RS-232C interface.

To facilitate print formatting, many versions of BASIC in addition to the standard PRINT command also offer a PRINT USING command (Microsoft BASIC) allowing the user to print strings or numbers in a specified format. Such commands have not been used in the programs in this book to ease portability between machines. In most cases only commas ',' allowing a tab spacing every 16 columns or the semi-colon ';' causing the items to be closed up have been used. Occasionally the TAB command has been used where it is desired to close-up the columns.

In most cases the print functions to a printer echo the print format on the screen, that is if a 40 column Apple II screen is used, the printer, even if allowing 80 or 132 columns, will stop after 40 columns and continue on the next line. This is annoying, not only for data printouts, but also for program listings. Most printers therefore normally have a command which allows printing out beyond the 40 column mark. In the case of the Epson MX-80, a very common printer used in conjunction with the Apple II+/e, the command is:

line number POKE 1656 + S, 80,

where *S* is the slot number of the printer card, commonly 1. If the user is fortunate enough to have for example a 132 column Qume printer, on an RS-232C interface then the command is:

line number POKE 1784 + S, 132.

However, despite using the above commands to give an 80 column printout, if individual tabs are required in Applesoft with a Qume printer it is necessary to use the following command:

line number POKE 36, r

where *r* is an integer equal to the tab position required. An example of this is shown below:

line number variable 1; : POKE 36,8 : PRINT variable 2; : POKE 36,21 : PRINT variable 3; : POKE 36,36 : PRINT variable 4; : POKE 36,51 : PRINT variable 5 where the variables 1 to 5 are printed out at columns 1, 8, 21, 36 and 51 respectively.

If CP/M BASIC is used with an 80 column screen, then of course there is not the same problem in using the TAB statement. However in CP/M the PRINT command displays the text or variables on the screen, and the LPRINT command is used to give a hard copy printout. The LPRINT command assumes a 132 character printer, and the LPRINT USING command can be used if required. It is essential with CP/M that the printer interface is in slot 1 of the Apple II.

Each of the programs listed has used Applesoft DOS for data printouts, but these can be removed or adapted as required as they are included as GOSUB routines, and not as part of the main program.

2.3.3 Other peripherals

The increased random access memory (RAM) of microcomputers has spawned a whole host of peripherals for the computer graphics market, in the form of graphic tablets, joysticks, high-resolution colour graphic screens, improved pixel resolution screens (512 × 512) which is comparable with mainframe monitors, and finally graphics plotters (Figs 1.11–13). As the majority of these peripherals require commands which are machine-dependent, the same problem arises in BASIC programming, as discussed previously for graphic displays.

However, the small A4 plotter has become of increasing acceptance recently as prices have dropped. This is of course beneficial, especially to the surveyor who wishes to plot tacheometric surveys, traverse plots or road curves – and thus drawing routines have been included using the standard IEEE-488 interface. The various control codes, etc. for scaling and plotting are discussed in the relevant programs.

2.4 SUMMARY OF APPLESOFT BASIC STATEMENTS

Assignment	LET	Computes and assigns value
	DIM	Allocates space for array variables
Comment	REM	Allows remarks about program control
Input	INPUT	Reads data from keyboard
	READ	Reads data from DATA statements
	DATA	Storage area for data
	RESTORE	Restores pointer to top of data stack
Output	PRINT	Prints output list
Program Control	CLEAR	Clears all variables
	IF . . . GOTO	Conditional branching (line no.)
	IF . . . THEN	Conditional branching (statement)
	FOR.TO.STEP	Opens loop

	NEXT	Closes loop
	GOTO	Unconditional branching
	GOSUB	Transfers flow to subroutine
	RETURN	Return from subroutine
	RESUME	Return from error
	ON . . . GOSUB	Multiple subroutine branching
	ON . . . GOTO	Multiple branching
	ONERR GOTO	Error trapping control
	PEEK(n)	Reads number(n) in memory address
	POKE(n)	Stores number(n) in memory address
	TRACE	Trace line routine
	NO TRACE	Turns trace off
	STOP	Stops program
	END	End of program
Screen	HOME	Clears screen
Control	INVERSE	Inverts characters on screen
	NORMAL	Returns screen to normal from INVERSE
	TAB	Tabbing control for columns
Functions	ABS	Absolute value
	ATN	Arctangent (gives angle in radians)
	CHR$	Character function
	COS	Cosine (angle in radians)
	DEF FN	Defined function
	EXP	Exponential
	INT	Integer value
	LOG	Natural logarithm (base e)
	SIN	Sine (angle in radians)
	SQR	Square root
	TAN	Tangent (angle in radians)

The above list gives the BASIC statements used in the programs listed in this book. To assist readers who wish to convert these programs from the Apple to another machine, Table 2.1 gives the conversions required for other well-known microcomputers.

3

Theodolite and traverse surveying

3.1 INTRODUCTION TO ANGLE CONVENTIONS

Each of the programs in this book gives the operator an option at the beginning of the program to select either the decimal angle format or degrees, minutes, seconds format (DD.MMSS), as discussed earlier in Section 2.2.6. If the DD.MMSS format is chosen, the angle is converted to decimals prior to use in the computer program.

Where field angles are used in a program, for example in tacheometry, intersections or traverse work, it is essential to select at the outset either whole circle bearings (WCB), sometimes referred to as azimuth bearings, or reduced bearings where the angles are relative to the north–south meridian, in one of the four quadrants as shown in Fig. 3.1.

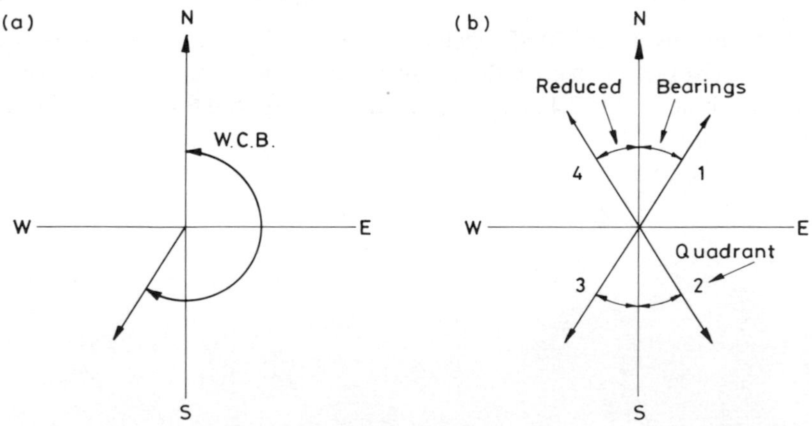

Fig. 3.1. (a) Whole circle bearings (WCBs) are quoted as clockwise angles from due north; (b) Reduced bearings can lie in one of four quadrants (1–4) where the angles are relative to the north-south meridian.

If the whole circle bearing option is chosen, Fig. 3.1(a), the angles are measured from due north in a clockwise direction, and thus by checking the angle to see if it is greater than 90°, 180°, 270°, the exact quadrant can be determined for further computations.

If the reduced bearing option is chosen, Fig. 3.1(b), then the angle between the north–south meridian and the bearing is entered, flanked by the quadrant directions, either '*N*', '*S*' and '*E*', '*W*' from which the exact quadrant can be calculated and the relevant whole circle bearing deduced.

3.2 TRIANGLE SOLUTIONS

In surveying there is often a need to determine the distance between two points which are not intervisible, but which can both be observed from a third point. A series of five routines are therefore included in the '<TRIANGLE>' solutions

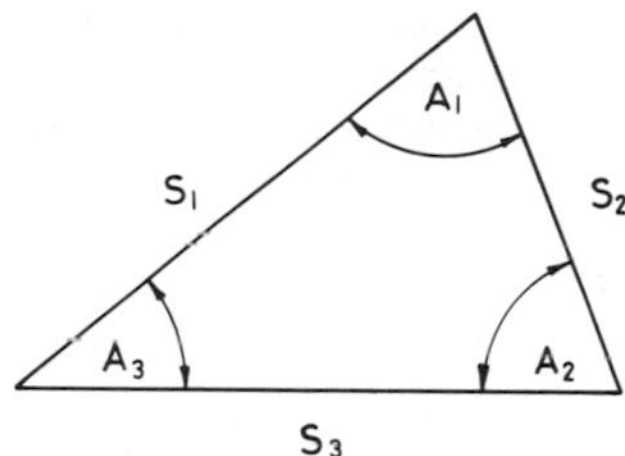

Fig. 3.2. Nomenclature for angles and sides of triangles in triangle solutions.

program where from a knowledge of all the sides, or a mixture of sides and angles, it is possible to solve the other sides and angles. The area of the triangle is also calculated. As the program is non-dimensional, the area is in square units of those used, either feet or metres. The program is run from a menu in lines 300–400 giving the following selection:

<1> SIDE/SIDE/SIDE	($S1, S2, S3$)
<2> ANGLE/SIDE/ANGLE	($A1, S1, A3$)
<3> SIDE/ANGLE/ANGLE	($S1, A1, A2$)
<4> SIDE/ANGLE/SIDE	($S1, A1, S2$)
<5> SIDE/SIDE/ANGLE	($S1, S2, A2$),

where in the computer program the angles $A1$, $A2$, $A3$ and sides $S1$, $S2$, $S3$ represent the angles A_1, A_2, A_3 and S_1, S_2, S_3 shown in Fig. 3.2. It should be noted that the angles and sides are numbered clockwise with A_1 adjacent to S_1 rather than in the triangle notation where A_1 is opposite side a_1. As only uppercase letters can be used in Applesoft, the notation shown in Fig. 3.2 has been adopted throughout the book for triangle solutions.

3.2.1 Triangle solutions – subroutine index

	Line numbers	*Function*
(a)	10–90	Initialization and control
(b)	100–190	Screen header display
(c)	200–290	Selection of degrees/radians, decimals/DD.MMSS
(d)	300–420	Program menu selection
(e)	500–630	Display of triangle notation on screen
(f)	700–750	Printout of triangle notation on printer
(g)	1000–1530	<1> Side/side/side ($S1$, $S2$, $S3$) solution
(h)	2000–2280	<2> Angle/side/angle ($A1$, $S1$, $A3$) solution
(i)	3000–3240	<3> Side/angle/angle ($S1$, $A1$, $A2$) solution
(j)	4000–4190	<4> Side/angle/side ($S1$, $A1$, $S2$) solution
(k)	5000–5320	<5> Side/side/angle ($S1$, $S2$, $A2$) solution
(l)	6000–6340	Conversion of figures to 3/4 decimals for display
(m)	6350–6460	Display of results
(n)	6500–6600	Routine to convert decimal angles to DD.MMSS
(o)	7000–7130	Printout of triangle solutions
(p)	8000–8010	Routine to clear screen
(q)	8500–8610	Error trace routine
(r)	9000–9500	Termination of program

(a) Initialization and control

Line numbers 10–90

The program is initialized in lines 10–90. The arccosine and arcsine functions are defined in lines 55–70 and the essential variables used as counters, etc. set to zero in line 80. The error tracing routine is set in line 90, causing the program to branch to the subroutine at line 8500, should an error occur.

(b) Screen header display

Line numbers 100–190

This is the screen display used for civil engineering students at Strathclyde University. The operator can alter this as required to suit another organization. Line 100 clears the screen, and the command is entered at line 8000 in a subroutine to simplify portability to other BASIC language dialects.

(c) Selection of degrees/radians, decimals/DD.MMSS

Line numbers 200–290

Line 200 gives the operator a choice of angles in degrees or radians with an input error trap on line 220. When degrees <1> is selected, the program sets R = 57.2957795 at line 240 and all subsequent angles entered are divided by R to convert the angle to radians. If degrees are selected, there is a choice of decimal angles or a degrees, minutes, seconds format. Thus if $D9 = 1$, angles are in decimals, and if $D9 = 2$, then angles are in degrees, minutes and seconds. If radians <2> is selected, the program branches to line 290 and sets $R = 1$.

(d) Program menu selection

Line numbers 300–420

The screen is cleared in line 300 and the triangle notation displayed (GOSUB 500) with a choice of five programs presented with a sixth option to end the program, 'QUIT'. An input error trap is provided on line 410 and the program branches to the various triangle solution subroutines in line 420.

(e) Display of triangle notation on screen

Line numbers 500–630

As mentioned earlier, no screen graphics are included in these programs since this is machine-dependent and creates portability problems. It has however been possible, using normal text lines and the '*' symbol, to display a triangle with the side and angle notation as defined in Fig. 3.2, to remind the operator each time a menu selection is required. For record purposes, if a printer is attached, the routine at line 610 branches to the subroutine at line 700 on the first occasion only. The counter '*C*' is used to determine whether a printout is required or not. At the outset, line 80, '*C*' is set to 0. After the first display '*C*' is set to 1, line 620, or 2 (line 710), depending upon whether a printout has been obtained.

(f) Printout of triangle notation on printer

Line numbers 700–750

The routine here is based on Applesoft DOS and may have to be altered to suit a different language and operating system. The advantage of the same 'PRINT' command to both the screen and the printer can be used to advantage here by re-using the previous routine for screen display which is now directed to the printer by the command 'PR#1'.

(g) <1> Side/side/side (*S*1, *S*2, *S*3) solution

Line numbers 1000–1530

In case <1>, if all sides are known (S_1, S_2, S_3) then the angles can be solved by first determining the half perimeter, P, and by substitution in the following formulae:

$$P = (S_1 + S_2 + S_3)/2 \quad (3.1)$$

$$A_3 = 2\cos^{-1}[P(P - S_2)/S_1S_3]^{1/2} \quad (3.2)$$

$$A_2 = 2\cos^{-1}[P(P - S_1)/S_2S_3]^{1/2} \quad (3.3)$$

$$A_1 = \cos^{-1}[-\cos(A_3 + A_2)] \quad (3.4)$$

However the scientific notation for the inverse cosine ($\cos^{-1}$) or arccosine is not programmable in Applesoft, and it is necessary to use the $\tan^{-1}$ or arctangent to solve for the arccosine. In the <TRIANGLE> program the arccosine is defined in a 'DEF FN' statement in line 60 and the angle A_3, given in Equation 3.2, is determined in lines 1230–1240 where the square root of the square bracket is first evaluated and stored in $T3$ and then A_3 calculated, i.e.

```
1230 T3 = SQR ( P* (P – S2) / (S1 * S3))
1240 A3 = 2 * FN C9 (T3)
```

At the outset of this subroutine, the screen is cleared and the triangle notation displayed (line 1000). Then each of the sides S_1, S_2, S_3 are entered with individual input error checks to trap a negative or zero entry. If three positive values have been entered, the program then checks in lines 1130–1150 that the three lengths form a triangle, and if not, the bell is rung (line 1170) to alert the operator's attention to an entry error with a request to check and re-enter the sides.

The half perimeter P, Equation 3.1, is calculated in line 1220 and then the angles A_3 and A_2 given by Equations 3.2 and 3.3 determined in lines 1230–1260. The angle A_1, Equation 3.4, is solved using the subroutine at lines 1500–1520. A counter 'K' is now used to check if it is necessary to use the subroutine at line 1500. At line 1210, K was set to zero before entering the side/side/side solution. Later program selections also use the routines in lines 1220–1260 where K is set to 1, thus skipping the subroutine at line 1500 before going to the screen display at line 6000 with a printout if required before returning to the menu.

(h) <2> Angle/side/angle (*A*1, *S*1, *A*3) solution

Line numbers 2000–2280

In case <2> two angles (A_1, A_3) and the included side S_1 are known (Fig. 3.2).

The third angle can now be found by either subtracting A_1 and A_3 from 180° or by calculation:

$$A_2 = \cos^{-1}[-\cos(A_3 + A_1)] \tag{3.5}$$

Having solved A_2, the other two sides S_2, S_3 can be found using the formulae:

$$S_2 = S_1 \sin A_3 / \sin A_2 \tag{3.6}$$

$$S_3 = S_1 \cos A_3 + S_2 \cos A_2 \tag{3.7}$$

The angles and sides are entered in lines 2040–2210, with the angle input in decimals if either $R = 1$ (radians) or $D9 = 1$ (decimal degrees), or in degrees, minutes and seconds if $D9 = 2$ (DD.MMSS). The angle A_2 is calculated in lines 2220–2230 with the sides S_1, S_2 in lines 2240–2250. The program then branches to the screen display at line 6000 with a printout if required, before returning to the menu.

(i) <3> Side/angle/angle (*S*1, *A*1, *A*2) solution

Line numbers 3000–3240

In case <3> with one side (S_1) and two angles (A_1, A_2), Fig. 3.2, it reverts after solving the third angle (A_3) to a similar solution to case <2> above, where

$$A_3 = \cos^{-1}[-\cos(A_1 + A_2)] \tag{3.8}$$

Thus after entering S_1, A_1, A_2 in lines 3040–3210, the angle A_3 is calculated in lines 3220–3230 before being directed back at line 3240 into the solution for case <2> at line 2240 to solve for S_1 and S_2.

(j) <4> Side/angle/side (*S*1, *A*1, *S*2) solution

Line numbers 4000–4190

This is the case of two sides (S_1, S_2) and the included angle, A_1 as shown in Fig. 3.2, where the third side S_3 is solved from the cosine rule:

$$S_3 = (S_1^{\,2} + S_2^{\,2} - 2S_1S_2\cos A_1)^{1/2} \tag{3.9}$$

After entering the two sides and the angle in lines 4040–70, S_3 is calculated in line 4180. The program then reverts back to case <1> where the three sides were known, but since A_1 is known, K is set to 1 at line 4190 before the program branches from line 4200 to line 1220, where the angles A_3 and A_2 are calculated.

(k) <5> Side/side/angle (*S*1, *S*2, *A*2) solution

Line numbers 5000–5320

In this case of two sides (S_1, S_2) and an adjacent angle (A_2) as shown in Fig. 3.2 there can be two solutions if S_2 is greater than S_1 as shown in Fig. 3.3. The angles and sides are entered in lines 5040–5170 and then the first solution, case <5> solved. Angle A_3 is first found using the sine rule:

$$A_3 = \sin^{-1}[S_2\sin A_2/S_1] \tag{3.10}$$

in lines 5180–5190. From the angles A_2 and A_3 and Equation 3.4, the subroutine at line 1500 can now be used to calculate A_1. Also the length of the side S_3 is calculated using Equation 3.7 at line 5210.

For the first solution, Z is set to zero and the decimal value of A_3 stored in D3 and A_2 stored in D4 at line 5220 for use in lines 5250–5280 if there is a second solution. If S_2 is less than or equal to S_1 (line 5230) the program branches to line 2260, and thence to the display and printout, returning to the menu. However if S_2 is greater than S_1, the program continues to line 5240 for display and printout and returns to line 5250 for the second solution, with $Z = 1$, which is set on line

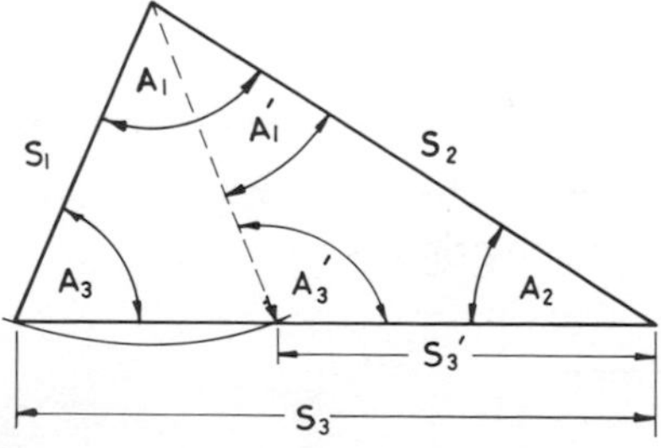

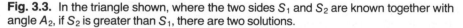

Fig. 3.3. In the triangle shown, where the two sides S_1 and S_2 are known together with angle A_2, if S_2 is greater than S_1, there are two solutions.

5290. The second solution for angle A_3 (A_3' in Fig. 3.3) is calculated in lines 5250–5260 and A_1 calculated as before from Equation 3.4, and then S_3 from Equation 3.7 at lines 5270–5280. The second solution for case <5> is designated '<6>' in both the display and printouts.

(l) Conversion of figures to 3/4 decimals for display

Line numbers 6000–6340

For presentation since 'PRINT USING' statements have not been used to specify the number of figures after the decimal point, it is necessary to use the integer (INT) command to tidy up the figures. Prior to this conversion, if the area of the triangle is required, this is calculated at line 6030 using the formula:

$$A = (S_1 S_3 \sin A_3)/2 \tag{3.11}$$

For distance, 3 decimal places (to the nearest mm) are usually sufficient and the following expression is used:

S1 = INT(S1 * 1000 + 0.5) / 1000

However a check has to be made on the angles to see if $D9$ equals 1 or 2 for either decimal or DD.MMSS format. If the former, then the figures can be converted as above for sides. If the DD.MMSS format is required, the subroutine at line 6500 is called for conversion.

(m) Display of results

Line numbers 6350–6460

The results of the three angles and sides together with the area are now presented on a cleared screen, with an option to obtain a printout if desired. If not, the program returns to the subroutine and back to the menu.

(n) Routine to convert decimal angles to DD.MMSS

Line numbers 6500–6600

This routine was described previously in Section 2.2.7.

(o) Printout of triangle solutions

Line numbers 7000–7130

This routine is based on Applesoft DOS and may require adaptation for different microcomputers and printers.

(p) Routine to clear screen

Line numbers 8000–8010

To simplify portability, the Applesoft 'HOME' command to clear the screen has been placed in a subroutine to allow an easy change to another microcomputer.

(q) Error trace routine

Line numbers 8500–8610

This Applesoft error trace routine was described previously in Section 2.2.9 and helps to identify the type of error and line number on which it occurs.

(r) Termination of program

Line numbers 9000–9500

When the operator desires to terminate entry to the program by selecting 'QUIT' from the menu, this routine will clear the screen and display an end-of-program message to advise the operator that computations are complete.

3.2.2 Triangle solutions – numeric variables

A	= Area of triangle
$A1–A3$	= Three angle subscripts of triangle
$A4–A9$	= Segments of degrees, minutes and seconds
C	= Counter for triangle notation printout
$C9$	= Defined function for arccosine
D	= Angle storage, decimal to DD.MMSS
$D1$	= Integer value of D degrees
$D2$	= Total number of seconds in fraction of D degrees
$D3$	= Storage for decimal angle $A3$
$D4$	= Storage for decimal angle $A2$
$D9$	= Menu selection for decimals <1>, DD.MMSS <2>
E	= Error number
J	= Menu selection for triangle solution routine
K	= Counter for use of subroutine at line 1500
M	= Number of minutes in angle
N	= Number of triangle solutions
P	= 1/2 perimeter of triangle
Q	= Counter for error trace routine
R	= 1 for radians, = 57.2957795 for degrees
S	= Number of seconds in angle
$S1–S3$	= Three side subscripts of triangle
$S9$	= Defined function for arcsine
$T1–T3$	= Temporary angle storage in defined function routines for arccosine and arcsine
U	= Menu selection for degrees <1>, or radians <2>
Z	= Counter for first and second solutions

3.2.3 Triangle solutions – string variables

A$	= Question response (Y/N)
D$	= CHR$(4)
M$	= Triangle solution type "<1>" to "<5>"

3.2.4 Triangle solutions – BASIC program

```
10  REM  <TRIANGLE> PROGRAM FOR APPLE II+.
20  REM  ASSUMES QUME PRINTER
30  REM
40  REM
50  REM  ALL ANGLES IN RADIANS
55  REM  FN C9 = ARCCOSINE
60  DEF  FN C9(C9) = (1.5707964 -  ATN (C9 /  SQR (1 - C9 * C9))) * R
65  REM  FN S9 = ARCSINE
70  DEF  FN S9(S9) =  ATN (S9 /  SQR (1 - S9 * S9)) * R
80 C = 0:D9 = 0:N = 0:Q = 0:Z = 0
90  ONERR  GOTO 8500
100  GOSUB 8000
110  PRINT "****************************************"
120  PRINT "*                                      *"
130  PRINT "*      UNIVERSITY  OF  STRATHCLYDE     *"
140  PRINT "*  DEPARTMENT  OF  CIVIL  ENGINEERING  *"
150  PRINT "*                                      *"
160  PRINT "*           SURVEYING  SECTION         *"
170  PRINT "*          TRIANGLE   SOLUTIONS        *"
180  PRINT "****************************************"
190  PRINT
200  PRINT "ARE ANGLES IN DEGREES <1>, RADIANS <2>  OR DO YOU WISH TO QUI
     T <3>."
210  INPUT U: IF U < 1 OR U > 3 THEN 100
220  ON U GOTO 230,290,9000
230 R = 57.2957795
240  PRINT "ARE ANGLES IN DECIMALS<1> OR DD.MMSS<2>"
250  INPUT D9: IF D9 < 1 OR D9 > 2 THEN 240
260  IF D9 = 1 THEN T$ = "(DECIMAL DEGS)"
270  IF D9 = 2 THEN T$ = "(DD.MMSS DEGS)"
280  GOTO 300
290 R = 1:T$ = "(RADIANS)"
300  GOSUB 8000
310  GOSUB 500
320  PRINT "SELECT PROGRAM FROM MENU :-"
330  PRINT
340  PRINT "     <1> SIDE/SIDE/SIDE    (S1,S2,S3)"
350  PRINT "     <2> ANGLE/SIDE/ANGLE  (A1,S1,A3)"
360  PRINT "     <3> SIDE/ANGLE/ANGLE  (S1,A1,A2)"
370  PRINT "     <4> SIDE/ANGLE/SIDE   (S1,A1,S2)"
380  PRINT "     <5> SIDE/SIDE/ANGLE   (S1,S2,A2)"
390  PRINT "     <6> QUIT";
400  INPUT J
410  IF J < 1 OR J > 6 THEN 300
420  ON J GOTO 1000,2000,3000,4000,5000,9000
495  REM  ******************************************
500  PRINT
510  PRINT "                A1        "
520  PRINT "                 *        "
530  PRINT "                * *       "
540  PRINT "            S1 *   * S2   "
550  PRINT "              *     *     "
560  PRINT "             *  S3   *    "
570  PRINT "          A3 *********** A2"
580  PRINT
590  IF C = 1 THEN 630
600  IF C = 2 THEN  RETURN
```

```
610  GOSUB 700
620 C = 1
630  RETURN
695  REM  *******************************************
700 D$ =  CHR$ (4): REM  CTRL-D
710 C = 2
720  PRINT D$;"PR#1"
730  GOSUB 500
740  PRINT D$;"PR#0"
750  RETURN
995  REM  *******************************************
1000  GOSUB 8000: GOSUB 500
1010 M$ = "<1>"
1020  PRINT "<1> SIDE/SIDE/SIDE (S1,S2,S3) SOLUTION"
1030  PRINT "****************************************"
1040  PRINT "SIDE 1 = ";
1050  INPUT S1
1060  IF S1 <  = 0 THEN 1040
1070  PRINT "SIDE 2 = ";
1080  INPUT S2
1090  IF S2 <  = 0 THEN 1070
1100  PRINT "SIDE 3 = ";
1110  INPUT S3
1120  IF S3 <  = 0 THEN 1100
1130  IF S1 + S2 <  = S3 THEN 1170
1140  IF S2 + S3 <  = S1 THEN 1170
1150  IF S1 + S3 <  = S2 THEN 1170
1160  GOTO 1210
1170  PRINT  CHR$ (7): REM  CTRL-B (BELL)
1180  PRINT "SORRY YOU DO NOT HAVE A TRIANGLE"
1190  PRINT "PLEASE CHECK AND RE-ENTER SIDES"
1200  GOTO 1030
1210 K = 0
1220 P = (S1 + S2 + S3) / 2
1230 T3 =  SQR (P * (P - S2) / (S1 * S3))
1240 A3 = 2 *  FN C9(T3)
1250 T2 =  SQR (P * (P - S1) / (S2 * S3))
1260 A2 = 2 *  FN C9(T2)
1270  IF K = 1 THEN 1290
1280  GOSUB 1500
1290  GOSUB 6000
1295  REM  RETURN TO MENU
1300  GOTO 100
1495  REM  *******************************************
1500 T1 =  -  COS ((A3 + A2) / R)
1510 A1 =  FN C9(T1)
1520 K = 0
1530  RETURN
1995  REM  *******************************************
2000  GOSUB 8000: GOSUB 500
2010 M$ = "<2>"
2020  PRINT "<2> ANGLE/SIDE/ANGLE (A1,S1,A3) SOLUTION"
2030  PRINT "****************************************"
2040  IF R = 1 OR D9 = 1 THEN 2100
2050  PRINT "ENTER ANGLE AS DEGS,MINS,SECS"
2060  PRINT "ANGLE 1 = ";
2070  INPUT A4,A5,A6
2080 A1 = A4 + A5 / 60 + A6 / 3600
2090  GOTO 2120
```

```
2100  PRINT "ANGLE 1 = ";
2110  INPUT A1
2120  PRINT "SIDE 1 = ";
2130  INPUT S1
2140  IF S1 <  = 0 THEN 2120
2150  IF R = 1 OR D9 = 1 THEN 2200
2160  PRINT "ANGLE 3 = ";
2170  INPUT A7,A8,A9
2180 A3 = A7 + A8 / 60 + A9 / 3600
2190  GOTO 2220
2200  PRINT "ANGLE 3 = ";
2210  INPUT A3
2220 T2 =  -  COS ((A3 + A1) / R)
2230 A2 =  FN C9(T2)
2240 S2 = S1 *  SIN (A3 / R) /  SIN (A2 / R)
2250 S3 = S1 *  COS (A3 / R) + S2 *  COS (A2 / R)
2260  GOSUB 6000
2270  PRINT
2275  REM  RETURN TO MENU
2280  GOTO 100
2995  REM  ********************************************
3000  GOSUB 8000: GOSUB 500
3010 M$ = "<3>"
3020  PRINT "<3> SIDE/ANGLE/ANGLE (S1,A1,A2) SOLUTION"
3030  PRINT "*****************************************"
3040  PRINT "SIDE 1 = ";
3050  INPUT S1
3060  IF S1 <  = 0 THEN 3040
3070  IF R = 1 OR D9 = 1 THEN 3130
3080  PRINT "ENTER ANGLE AS DEGS,MINS,SECS"
3090  PRINT "ANGLE 1 = ";
3100  INPUT A4,A5,A6
3110 A1 = A4 + A5 / 60 + A6 / 3600
3120  GOTO 3160
3130  PRINT "ANGLE 1 = ";
3140  INPUT A1
3150  GOTO 3200
3160  PRINT "ANGLE 2 = ";
3170  INPUT A7,A8,A9
3180 A2 = A7 + A8 / 60 + A9 / 3600
3190  GOTO 3220
3200  PRINT "ANGLE 2 = ";
3210  INPUT A2
3220 T3 =  -  COS ((A2 + A1) / R)
3230 A3 =  FN C9(T3)
3240  GOTO 2240
3995  REM  ********************************************
4000  GOSUB 8000: GOSUB 500
4010 M$ = "<4>"
4020  PRINT "<4> SIDE/ANGLE/SIDE (S1,A1,S2) SOLUTION"
4030  PRINT "*****************************************"
4040  PRINT "SIDE 1 = ";
4050  INPUT S1
4060  IF S1 <  = 0 THEN 4040
4070  IF R = 1 OR D9 = 1 THEN 4130
4080  PRINT "ENTER ANGLE AS DEGS,MINS,SECS"
4090  PRINT "ANGLE 1 = ";
4100  INPUT A4,A5,A6
4110 A1 = A4 + A5 / 60 + A6 / 3600
```

```
4120  GOTO 4150
4130  PRINT "ANGLE 1 = ";
4140  INPUT A1
4150  PRINT "SIDE 2 = ";
4160  INPUT S2
4170  IF S2 <  = Ø THEN 415Ø
418Ø S3 =  SQR (S1 * S1 + S2 * S2 - 2 * S1 * S2 *  COS (A1 / R))
419Ø K = 1
42ØØ  GOTO 122Ø
4995  REM  *********************************************
5ØØØ  GOSUB 8ØØØ: GOSUB 5ØØ
5Ø1Ø M$ = "<5>"
5Ø2Ø  PRINT "<5> SIDE/SIDE/ANGLE (S1,S2,A2) SOLUTION"
5Ø3Ø  PRINT "***************************************"
5Ø4Ø  PRINT "SIDE 1 = ";
5Ø5Ø  INPUT S1
5Ø6Ø  IF S1 <  = Ø THEN 5Ø4Ø
5Ø7Ø  PRINT "SIDE 2 = ";
5Ø8Ø  INPUT S2
5Ø9Ø  IF S2 <  = Ø THEN 5Ø7Ø
51ØØ  IF R = 1 OR D9 = 1 THEN 516Ø
511Ø  PRINT "ENTER ANGLE AS DEGS,MINS,SECS"
512Ø  PRINT "ANGLE 2 = ";
513Ø  INPUT A7,A8,A9
514Ø A2 = A7 + A8 / 6Ø + A9 / 36ØØ
515Ø  GOTO 518Ø
516Ø  PRINT "ANGLE 2 = ";
517Ø  INPUT A2
518Ø T3 = S2 / S1 *  SIN (A2 / R)
519Ø A3 =  FN S9(T3)
52ØØ  GOSUB 15ØØ
521Ø S3 = S1 *  COS (A3 / R) + S2 *  COS (A2 / R)
522Ø Z = Ø:D3 = A3:D4 = A2
523Ø  IF S2 <  = S1 THEN 226Ø
524Ø  GOSUB 6ØØØ
5245  REM  SECOND SOLUTION, Z=1
525Ø T3 =  - COS (D3 / R)
526Ø A3 =  FN C9(T3):A2 = D4
527Ø  GOSUB 15ØØ
528Ø S3 = S1 *  COS (A3 / R) + S2 *  COS (A2 / R)
529Ø Z = 1:M$ = "<6>"
53ØØ  GOSUB 6ØØØ
531Ø  PRINT
5315  REM  RETURN TO MENU
532Ø  GOTO 1ØØ
5995  REM  *********************************************
6ØØØ  GOSUB 8ØØØ
6Ø1Ø  PRINT M$;" TRIANGLE SOLUTION : ";T$
6Ø2Ø  PRINT "========================================"
6Ø25  REM  CONVERT TO 3 OR 4 DECIMAL PLACES
6Ø3Ø A = S1 * S3 *  SIN (A3 / R) / 2
6Ø4Ø A =  INT (A * 1ØØØ + .5) / 1ØØØ
6Ø5Ø S1 =  INT (S1 * 1ØØØ + .5) / 1ØØØ
6Ø6Ø  IF R = 1 OR D9 = 1 THEN 611Ø
6Ø7Ø D = A1
6Ø8Ø  GOSUB 65ØØ
6Ø9Ø A1 = D
61ØØ  GOTO 613Ø
611Ø A1 = A1 * 1ØØØØ + .5
```

```
6120 A1 =  INT (A1) / 10000
6130 S2 = S2 * 1000 + .5
6140 S2 =  INT (S2) / 1000
6150 :
6160  IF R = 1 THEN 6220
6170  IF D9 = 1 THEN 6220
6180 D = A2
6190  GOSUB 6500
6200 A2 = D
6210  GOTO 6240
6220 A2 = A2 * 10000 + .5
6230 A2 =  INT (A2) / 10000
6240 S3 = S3 * 1000 + .5
6250 S3 =  INT (S3) / 1000
6260  IF R = 1 THEN 6320
6270  IF D9 = 1 THEN 6320
6280 D = A3
6290  GOSUB 6500
6300 A3 = D
6310  GOTO 6340
6320 A3 = A3 * 10000 + .5
6330 A3 =  INT (A3) / 10000
6340  PRINT
6350  PRINT "SIDE  1 = ";S1
6360  PRINT "ANGLE 1 = ";A1
6370  PRINT "SIDE  2 = ";S2
6380  PRINT "ANGLE 2 = ";A2
6390  PRINT "SIDE  3 = ";S3
6400  PRINT "ANGLE 3 = ";A3
6410  PRINT "AREA    = ";A: PRINT
6420  PRINT : PRINT "DO YOU WISH PRINTOUT OF RESULTS (Y/N)";
6430  INPUT A$
6440  IF A$ = "Y" THEN 7000
6450  IF A$ = "N" THEN  RETURN
6460  IF A$ < > "N" THEN 6420
6495  REM  *********************************************
6500 D1 =  INT (D)
6510 D2 = (D - D1) * 3600
6520 M =  INT (D2 / 60)
6530 S = D2 - 60 * M + .5
6540  IF S < 60 THEN 6580
6550 M = M + 1:S = 0
6560  IF M < 60 THEN 6580
6570 D1 = D1 + 1:M = 0
6580 D = D1 * 10000 + M * 100 + S
6590 D =  INT (D) / 10000
6600  RETURN
6995  REM  *********************************************
7000 D$ =  CHR$ (4): REM  CTRL-D
7010  PRINT D$;"PR#1"
7020 N = N + 1
7030  PRINT M$;" TRIANGLE SOLUTION NO. ";N;" : ";T$
7040  PRINT "================================================"
7050  PRINT
7060  PRINT "SIDE  1 = ";S1;: POKE 36,25: PRINT "ANGLE 1 = ";A1
7070  PRINT "SIDE  2 = ";S2;: POKE 36,25: PRINT "ANGLE 2 = ";A2
7080  PRINT "SIDE  3 = ";S3;: POKE 36,25: PRINT "ANGLE 3 = ";A3
7090  PRINT "AREA    = ";A
7100  PRINT "************************************************"
```

```
711Ø  PRINT
712Ø  PRINT D$;"PR#Ø"
713Ø  RETURN
7995  REM  *******************************************
8ØØØ  HOME : REM  CLEAR SCREEN
801Ø  RETURN
8495  REM  *******************************************
85ØØ  PRINT  CHR$ (7): REM  CTRL-B (BELL)
851Ø E =  PEEK (222): REM  GET ERROR NO.
852Ø  IF Q = 1 THEN 858Ø
853Ø  INVERSE
854Ø  PRINT "ERROR NO. ";E;" FOUND"
855Ø Q = 1
856Ø  TRACE
857Ø  RESUME
858Ø  PRINT "ERROR ON SECOND LINE NO."
859Ø  NORMAL
8595  REM  INPUT ERROR DETECTED
86ØØ  NOTRACE
861Ø  STOP
8995  REM  *******************************************
9ØØØ  GOSUB 8ØØØ
9Ø1Ø  PRINT
9Ø2Ø  PRINT "END TRIANGLE PROGRAM"
9Ø3Ø  PRINT
9Ø4Ø  PRINT "****************************************"
9Ø5Ø  REM  PROGRAM PREPARED BY
9Ø6Ø  REM  DR. P.H. MILNE
9Ø7Ø  REM  DEPT. OF CIVIL ENGINEERING
9Ø8Ø  REM  UNIVERSITY OF STRATHCLYDE
9Ø9Ø  REM  GLASGOW G4 ONG
91ØØ  REM  SCOTLAND
911Ø  REM  *******************************************
95ØØ  END
```

3.2.5 Triangle solutions – computer printout

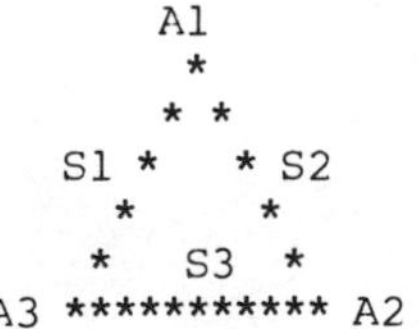


```
<1> TRIANGLE SOLUTION NO. 1 : (RADIANS)
===============================================

SIDE  1 = 3              ANGLE 1 = 1.57Ø8
SIDE  2 = 4              ANGLE 2 = .6435
SIDE  3 = 5              ANGLE 3 = .9273
AREA    = 6
***********************************************
```

```
<2> TRIANGLE SOLUTION NO. 2 : (RADIANS)
==================================================

SIDE  1 = 3                ANGLE 1 = 1.5708
SIDE  2 = 4                ANGLE 2 = .6435
SIDE  3 = 5                ANGLE 3 = .9273
AREA    = 6
**************************************************

<3> TRIANGLE SOLUTION NO. 3 : (RADIANS)
==================================================

SIDE  1 = 3                ANGLE 1 = 1.5708
SIDE  2 = 4                ANGLE 2 = .6435
SIDE  3 = 5                ANGLE 3 = .9273
AREA    = 6
**************************************************

<4> TRIANGLE SOLUTION NO. 4 : (RADIANS)
==================================================

SIDE  1 = 3                ANGLE 1 = 1.5708
SIDE  2 = 4                ANGLE 2 = .6435
SIDE  3 = 5                ANGLE 3 = .9273
AREA    = 6
**************************************************

<5> TRIANGLE SOLUTION NO. 5 : (RADIANS)
==================================================

SIDE  1 = 3                ANGLE 1 = 1.5708
SIDE  2 = 4                ANGLE 2 = .6435
SIDE  3 = 5                ANGLE 3 = .9273
AREA    = 6
**************************************************

<6> TRIANGLE SOLUTION NO. 6 : (RADIANS)
==================================================

SIDE  1 = 3                ANGLE 1 = .2838
SIDE  2 = 4                ANGLE 2 = .6435
SIDE  3 = 1.4              ANGLE 3 = 2.2143
AREA    = 1.68
**************************************************
               A1
                *
               * *
           S1 *   * S2
             *     *
            *  S3   *
         A3 *********** A2

<1> TRIANGLE SOLUTION NO. 1 : (DECIMAL DEGS)
==================================================

SIDE  1 = 5                ANGLE 1 = 90
SIDE  2 = 8.66             ANGLE 2 = 30
SIDE  3 = 10               ANGLE 3 = 60
AREA    = 21.651
**************************************************
```

```
<2> TRIANGLE SOLUTION NO. 2 : (DECIMAL DEGS)
==================================================

SIDE  1 = 5               ANGLE 1 = 9Ø
SIDE  2 = 8.66            ANGLE 2 = 3Ø
SIDE  3 - 1Ø              ANGLE 3 - 6Ø
AREA    = 21.651
**************************************************

<3> TRIANGLE SOLUTION NO. 3 : (DECIMAL DEGS)
==================================================

SIDE  1 = 5               ANGLE 1 = 9Ø
SIDE  2 = 8.66            ANGLE 2 = 3Ø
SIDE  3 = 1Ø              ANGLE 3 = 6Ø
AREA    = 21.651
**************************************************

<4> TRIANGLE SOLUTION NO. 4 : (DECIMAL DEGS)
==================================================

SIDE  1 = 5               ANGLE 1 = 9Ø
SIDE  2 = 8.66            ANGLE 2 = 3Ø
SIDE  3 = 1Ø              ANGLE 3 = 6Ø
AREA    = 21.651
**************************************************

<5> TRIANGLE SOLUTION NO. 5 : (DECIMAL DEGS)
==================================================

SIDE  1 = 5               ANGLE 1 = 9Ø.ØØØ1
SIDE  2 = 8.66            ANGLE 2 = 3Ø
SIDE  3 = 1Ø              ANGLE 3 = 6Ø
AREA    = 21.651
**************************************************

<6> TRIANGLE SOLUTION NO. 6 : (DECIMAL DEGS)
==================================================

SIDE  1 = 5               ANGLE 1 = 3Ø
SIDE  2 = 8.66            ANGLE 2 = 3Ø
SIDE  3 = 5               ANGLE 3 = 12Ø.ØØØ1
AREA    = 1Ø.825
**************************************************

                A1
                 *
                * *
            S1 *   * S2
              *     *
             *  S3   *
          A3 *********** A2

<1> TRIANGLE SOLUTION NO. 1 : (DD.MMSS DEGS)
==================================================

SIDE  1 = 4.5             ANGLE 1 = 1Ø7.513
SIDE  2 = 6.25            ANGLE 2 = 29.1829
SIDE  3 = 8.75            ANGLE 3 = 42.5
AREA    = 13.385
**************************************************
```

```
<2> TRIANGLE SOLUTION NO. 2 : (DD.MMSS DEGS)
==================================================

SIDE  1 = 4.5              ANGLE 1 = 1∅7.513
SIDE  2 = 6.25             ANGLE 2 = 29.183
SIDE  3 = 8.75             ANGLE 3 = 42.5
AREA    = 13.385
**************************************************

<3> TRIANGLE SOLUTION NO. 3 : (DD.MMSS DEGS)
==================================================

SIDE  1 = 4.5              ANGLE 1 = 1∅7.513
SIDE  2 = 6.25             ANGLE 2 = 29.1829
SIDE  3 = 8.75             ANGLE 3 = 42.5∅∅1
AREA    = 13.385
**************************************************

<4> TRIANGLE SOLUTION NO. 4 : (DD.MMSS DEGS)
==================================================

SIDE  1 = 4.5              ANGLE 1 = 1∅7.513
SIDE  2 = 6.25             ANGLE 2 = 29.1829
SIDE  3 = 8.75             ANGLE 3 = 42.5∅∅1
AREA    = 13.385
**************************************************

<5> TRIANGLE SOLUTION NO. 5 : (DD.MMSS DEGS)
==================================================

SIDE  1 = 4.5              ANGLE 1 = 1∅7.5131
SIDE  2 = 6.25             ANGLE 2 = 29.1829
SIDE  3 = 8.75             ANGLE 3 = 42.5
AREA    = 13.385
**************************************************

<6> TRIANGLE SOLUTION NO. 6 : (DD.MMSS DEGS)
==================================================

SIDE  1 = 4.5              ANGLE 1 = 13.3131
SIDE  2 = 6.25             ANGLE 2 = 29.1829
SIDE  3 = 2.15             ANGLE 3 = 137.1
AREA    = 3.289
**************************************************
```

3.3 TACHEOMETRY REDUCTION

After levelling in land surveying, students generally progress to field survey exercises incorporating optical distance measurement. The use of a level or transit theodolite with a vertical staff eliminates cumbersome tape measurements, and is commonly referred to as tacheometry. The distance is found by taking readings to two small lines marked on the diaphragm of the majority of modern instruments. These lines are called *stadia lines* or *stadia hairs* and are marked as shown in Fig. 3.4. The distance between the stadia hairs is fixed and is known as the *stadia interval*. In modern instruments the stadia interval is chosen so that with a horizontal sight on a vertical staff, the staff intercepts given by the

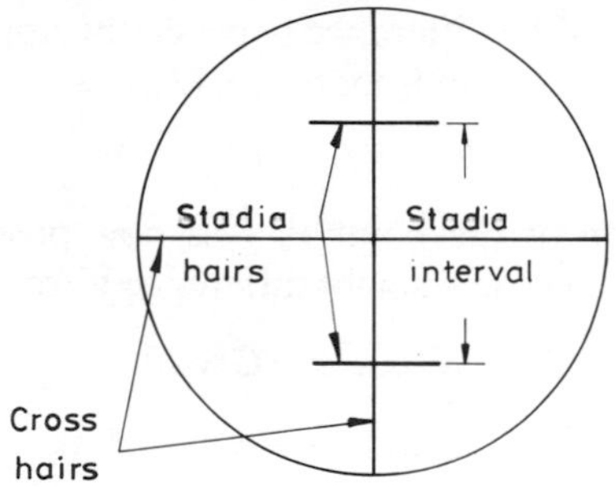

Fig. 3.4. Typical diaphragm on surveyor's telescope showing cross-hairs and stadia interval.

stadia hairs multiplied by a constant (K) gives the horizontal distance. Some older instruments use an externally focusing telescope, and to calculate the distance between two points there is also an additive constant (C) to be taken into consideration, so that the stadia equation for distance becomes:

$$D = Ks + C \tag{3.12}$$

Most modern theodolites and levels are now designed with an *anallactic lens* in place of the externally focusing telescope, so that C is reduced to zero. The stadia interval is also chosen so that K is 100 to simplify the calculations.

The advantage of the transit theodolite over the level is its ability to read vertical as well as horizontal angles to cover hilly terrain for contouring, etc. In Fig. 3.5, where a theodolite at survey station S is taking an inclined sight on to a vertical staff at survey point P, the distance D, given by Equation 3.12, will be the slope distance where s is the staff intercept perpendicular to the line of sight. However with vertical staff tacheometry, the staff intercept I is vertical, as shown in Fig. 3.5, where the relationship between I and s is given by:

$$s = I\cos\theta \tag{3.13}$$

where θ is angle of elevation or depression, and

$$I = U - L \tag{3.14}$$

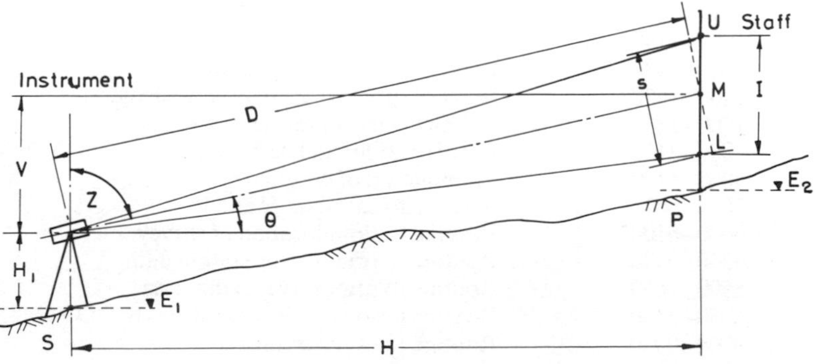

Fig. 3.5. Nomenclature for vertical staff tacheometry.

i.e., the staff intercept I, is the difference between the upper (U) and lower (L) staff readings. Substituting for s in Equation 3.12 gives:

$$D = KI\cos\theta + C \tag{3.15}$$

To find the true position of the point P, it is now possible to calculate the horizontal distance H and vertical height difference V from the angle θ:

$$H = KI\cos^2\theta + C\cos\theta \tag{3.16}$$

$$V = 1/2KI\sin 2\theta + C\,sin\,\theta \tag{3.17}$$

To determine the elevation or level of the point P (E_2) the elevation of the survey station S must be known (E_1), together with the height of the instrument (H_1):

$$E_2 = E_1 + H_1 \pm V - M \tag{3.18}$$

where V is positive for an angle of elevation, and negative for an angle of depression, and M is the mid-staff reading (Fig. 3.5).

Some of the older theodolites gave angles of elevation or depression indicated by a + or −, from the horizontal, and hence the angle θ shown in Fig. 3.5. Many of the modern theodolites, however, measure angles from the zenith, with a reading of 90° for the horizontal in face left and 270° for the horizontal in face right.

In the computer program for tacheometric stadia reduction <STADIA>, the program assumes a zenith angle. Angles can either be entered in face left (0°–180°), or face right (180°–360°) as the program checks the face prior to commencing the calculations. This program calculates both the elevation of the survey point and its distance from the survey station for subsequent plotting. If a 3-D position fix is requested, the program <TACHY.3-D> in Section 3.4 should be used.

3.3.1 Tacheometry reduction – subroutine index

	Line numbers	*Function*
(a)	10–300	Initialization and control
(b)	1000–1190	Entry of survey location, etc.
(c)	1200–1330	Entry of survey station information
(d)	1400–1650	Entry of survey point data and calculations
(e)	1700–1850	Screen display of results
(f)	2000–2150	Entry of additional data
(g)	4000–4100	Termination of program
(h)	5000–5020	Routine to calculate H and V
(i)	6000–6100	Routine to print location of survey, etc.
(j)	6250–6340	Routine to print survey station data
(k)	6500–6660	Routine to print survey point data
(l)	8050–8150	Routine to convert decimal angles to DD.MMSS
(m)	9000–9010	Routine to clear screen

(a) Initialization and control

Line numbers 10–300

The program is initialized and space reserved for string variables for keyboard entry and regular screen display. Since the Apple II+ uses radians for angle calculations, R is set to 57.2957795 in line 300 and all entries for angles in degrees divided by R to convert to radians.

(b) Entry of survey location

Line numbers 1000–1190

Line 1000 clears the screen and presents a screen header with requests for site information for record purposes, location, operator, date, etc. After entry, line 1160 gives a printout if required, in the subroutine at line 6000.

(c) Entry of survey station information

Line numbers 1200–1330

Line 1200 clears the screen and requests entry of the station number S, elevation Z, and the height of the theodolite above the station H_0. After entry, line 1300 gives a printout if required in the subroutine at line 6250.

(d) Entry of survey point data and calculations

Line numbers 1400–1650

Again the screen is cleared at line 1400 and data requested for the survey point, number, zenith angle, upper, mid and lower stadia hair readings on the staff, just in the order of field booking. Line 1495 checks to see if the zenith angle has been read in face left or face right and, if necessary, adjusted prior to the calculations. The angle A in the program represents θ in Fig. 3.5. The tacheometry calculations for H and V (Equations 3.16 and 3.17) are carried out in the subroutine at line 5000 and the level of the survey point (Z_0) calculated from Equation 3.18 at lines 1530 or 1620 depending upon whether the zenith angle gave an angle of depression or elevation respectively. Finally, the distances and levels are rounded to the third decimal place (lines 1630–1640) and the zenith angle converted to DD.MMSS for presentation.

(e) Screen display of results

Line numbers 1700–1850

The screen is again cleared at line 1700 and the horizontal distance and level of the survey point displayed, with an option to obtain a printout of the results in the subroutine at line 6500. The program continues at line 2000.

(f) Entry of additional data

Line numbers 2000–2150

An option is now given for more data from the existing survey station at line 1400, or for data from another survey station at line 1200. If a printout has been obtained from the previous station ($P1 = 1$) then a finishing line is printed (line 2130) prior to returning to line 1200. If, however, the data entry is complete, the program branches to line 4000.

(g) Termination of program

Line numbers 4000–4100

If a previous printout has been obtained, a finishing line is printed under the data (line 4030) and then an end-of-program message is displayed to advise the operator that computations are complete.

(h) Routine to calculate *H* and *V*

Line numbers 5000–5020

The tacheometry calculations for H and V from Equations 3.16 and 3.17 are provided in this subroutine where A is the angle θ in Fig. 3.5. Note A is divided by R to convert the angle from degrees to radians before calculation.

(i) Routine to print location of survey, etc.

Line numbers 6000–6100

This routine is provided for record purposes, and is based on Applesoft DOS and may require alteration for different microcomputers and printers.

(j) Routine to print survey station data

Line numbers 6250–6340

The same comments apply as above for (i).

(k) Routine to print survey point data

Line numbers 6500–6660

The printout routine assumes an 80-column printer to give a one-line display for each point.

(l) Routine to convert decimal angles to DD.MMSS

Line numbers 8050–8150

This routine was described previously in Section 2.2.7.

(m) Routine to clear screen

Line numbers 9000–9010

To simplify portability, the Applesoft 'HOME' command to clear the screen has been placed in a subroutine to allow an easy change to another microcomputer.

3.3.2 Tacheometry reduction – numeric variables

A	= Angle of elevation or depression
C	= Additive constant
D, *D*1, *D*2	= Number of degrees in angle
H	= Horizontal distance to survey point
*H*0	= Height of instrument above station
I	= Stadia intercept on staff = *U*–*L*
K	= Multiplying constant
L	= Lower stadia hair reading on staff
M	= Mid stadia hair reading on staff
*M*1	= Number of minutes in zenith angle
P	= Survey point number
*P*1	= Printer code
R	= 57.2957795, conversion to radians
S	= Survey station number
*S*1	= Number of seconds in zenith angle
U	= Upper stadia hair reading on staff
V	= Vertical height difference
*V*0	= Zenith angle
*V*1–*V*3	= Degrees, minutes, seconds of zenith angle
Z	= Elevation of survey station
*Z*0	= Elevation of survey point

3.3.3 Tacheometry reduction – string variables

A$	Storage for location of survey
B$	Storage for operator's name
C$	Storage for date of survey
D$	= CHR$ (4), i.e. CTRL-D
E$	= "SORRY, DATA ERROR . . . PLEASE RE-ENTER"
K$	Storage for instrument used
P$	= "ENTER SURVEY POINT NUMBER – NB NUMERALS ONLY"
Q$	= Question response (Y/N)
S$	= "IF IN DECIMAL DEGS, FILL MINS AND SECS WITH 0"
U$	= "ENTER UPPER(U), MID(M), LOWER(L) STADIA HAIR READINGS, SEPARATED BY COMMAS"
V$	= "ENTER ZENITH ANGLE IN DEGS, MINS, SECS SEPARATED BY COMMAS"

3.3.4 Tacheometry reduction – BASIC program

```
10  REM  <STADIA> PROGRAM FOR APPLE II+.
20  REM  SURVEYING SECTION - USES QUME PRINTER
30  REM  ****************************************
40  REM  PROGRAM PREPARED BY
50  REM  DR. P.H. MILNE
60  REM  DEPARTMENT OF CIVIL ENGINEERING
70  REM  UNIVERSITY OF STRATHCLYDE
80  REM  GLASGOW G4 ONG
90  REM  ****************************************
100  GOSUB 9000
110  DIM A$(50),B$(25),C$(25),D$(25)
120  DIM E$(40),P$(50),S$(50),U$(80),V$(60)
130 E$ = "SORRY, DATA ERROR ---- PLEASE RE-ENTER"
140 P$ = "ENTER SURVEY POINT NUMBER -             N.B.  NUMERALS ONLY"
150 U$ = "ENTER UPPER(U), MID(M), LOWER(L) STADIA HAIR READINGS, SEPARATE
    D BY COMMAS"
160 V$ = "ENTER ZENITH ANGLE IN DEGS, MINS, SECS  SEPARATED BY COMMAS,"
170 S$ = "IF IN DECIMAL DEGS, FILL MINS AND SECS  WITH 0"
180 D$ =  CHR$ (4): REM  CTRL-D
300  LET R = 57.2957795:P1 = 0
1000  GOSUB 9000
1010  PRINT "UNIVERSITY OF STRATHCLYDE"
1020  PRINT "STADIA REDUCTION PROGRAM."
1030  PRINT "****************************************"
1040  PRINT
1050  PRINT "ENTER LOCATION OF WORK, ETC. WHEN       REQUESTED, AND PRESS
     <RETURN>"
1060  PRINT "ENTER LOCATION OF SURVEY :-"
1070  INPUT A$
1080  PRINT "ENTER OPERATOR'S NAME :-"
1090  INPUT B$
1100  PRINT "ENTER DATE OF SURVEY AS DD/MM/YY :-"
1110  INPUT C$
1120  PRINT "ENTER INSTRUMENT USED :-"
1130  INPUT K$
1140  PRINT "ENTER INSTRUMENT CONSTANTS, K,C         SEPARATED BY A COMMA
    ,"
1150  INPUT K,C
1160  PRINT "DO YOU WISH A PRINTOUT OF LOCATION ETC., (Y/N);
1170  INPUT Q$
1180  IF Q$ = "Y" THEN 6000
1190  IF Q$ <  > "N" THEN 1160
1200  GOSUB 9000
1210  PRINT "ENTER SURVEY STATION INFORMATION WHEN   REQUESTED, AND PRESS
     <RETURN>"
1220  PRINT
1230  PRINT "ENTER SURVEY STATION NO.                N.B.  NUMERALS ONLY"

1240  INPUT S
1250  PRINT "ENTER ELEVATION OF STATION"
1260  INPUT Z
1270  PRINT "ENTER HEIGHT OF INSTRUMENT ABOVE STATION"
1280  INPUT H0
1290  LET P1 = 0
1300  PRINT "DO YOU WISH PRINTOUT OF SURVEY STN.DATA (Y/N)";
1310  INPUT Q$
1320  IF Q$ = "Y" THEN 6250
1330  IF Q$ <  > "N" THEN 1300
```

```
14ØØ  COSUB 9ØØØ
141Ø  PRINT P$
142Ø  INPUT P
143Ø  PRINT V$
144Ø  PRINT S$
145Ø  INPUT V1,V2,V3
146Ø  LET VØ = V1 + V2 / 6Ø + V3 / 36ØØ
147Ø  PRINT U$
148Ø  INPUT U,M,L
149Ø  LET I = U - L
1495  IF VØ > 18Ø THEN VØ = 36Ø - VØ
15ØØ  LET A = VØ - 9Ø
151Ø  IF A < Ø THEN 16ØØ
152Ø  GOSUB 5ØØØ
153Ø  LET ZØ = Z + HØ - V - M
154Ø  GOTO 163Ø
16ØØ  LET A =  - 1 * A
161Ø  GOSUB 5ØØØ
162Ø  LET ZØ = Z + HØ + V - M
1625  REM  CONVERT DECIMALS FOR DISPLAY
163Ø  LET ZØ =  INT (ZØ * 1ØØØ + .5) / 1ØØØ
164Ø  LET H =  INT (H * 1ØØØ + .5) / 1ØØØ
1645  REM  CONVERT DECIMAL ANGLE TO DD.MMSS
165Ø D = VØ: GOSUB 8Ø5Ø:VØ = D
17ØØ  COSUB 9ØØØ
171Ø  PRINT "SURVEY POINT NO. ";P
172Ø  PRINT "========================================"
173Ø  PRINT "ZENITH ANGLE       = ";VØ
174Ø  PRINT
175Ø  PRINT "HORIZ. DIST.       = ";H
176Ø  PRINT "ELEV. SURVEY PT.   = ";ZØ
177Ø  PRINT
178Ø  PRINT "****************************************"
179Ø :
18ØØ  PRINT
181Ø  PRINT "DO YOU WISH PRINTOUT OF RESULTS (Y/N)"
182Ø  INPUT Q$
183Ø  IF Q$ = "N" THEN 2ØØØ
184Ø  IF Q$ < > "Y" THEN 181Ø
185Ø  GOSUB 65ØØ
2ØØØ  GOSUB 9ØØØ
2Ø1Ø  PRINT "DO YOU WISH TO ENTER MORE DATA FOR      SURVEY STN. NO. ";S;
     " (Y/N)"
2Ø2Ø  INPUT Q$
2Ø3Ø  IF Q$ = "Y" THEN 14ØØ
2Ø4Ø  IF Q$ < > "N" THEN 2ØØØ
2Ø5Ø  PRINT
2Ø6Ø  PRINT "DO YOU WISH TO ENTER DATA FOR ANOTHER   SURVEY STATION (Y/N)
     "
2Ø7Ø  INPUT Q$
2Ø8Ø  IF Q$ = "N" THEN 4ØØØ
2Ø9Ø  IF Q$ = "Y" THEN 211Ø
21ØØ  IF Q$ < > "N" THEN 2Ø6Ø
211Ø  IF P1 = Ø THEN 12ØØ
212Ø  PRINT D$;"PR#1": POKE 1784 + 1,8Ø
213Ø  PRINT "=========================================================
     ===================="
214Ø  PRINT D$;"PR#Ø"
215Ø  GOTO 12ØØ
4ØØØ  GOSUB 9ØØØ
```

```
4Ø1Ø  IF P1 = Ø THEN 4Ø5Ø
4Ø2Ø  PRINT D$;"PR#1": POKE 1784 + 1,8Ø
4Ø3Ø  PRINT "============================================================
      ===================="
4Ø4Ø  PRINT D$;"PR#Ø"
4Ø5Ø  VTAB (1Ø): HTAB (5)
4Ø6Ø  PRINT "END OF STADIA REDUCTION PROGRAM"
4Ø7Ø  PRINT
4Ø8Ø  PRINT "******************************************"
4Ø9Ø  PRINT
41ØØ  END
4985  REM  ******************************************
4995  REM  GOSUB ROUTINE FOR H,V,TACHY REDUCTIONS
5ØØØ  LET H = K * I *  COS (A / R) ^ 2 + C *  COS (A / R)
5Ø1Ø  LET V = .5 * K * I *  SIN (2 * A / R) + C *  SIN (A / R)
5Ø2Ø  RETURN
5985  REM  ******************************************
599Ø  REM  ROUTINE FOR LOCATION PRINTOUT
6ØØØ  PRINT D$;"PR#1": POKE 1784 + 1,8Ø
6Ø1Ø  PRINT "LOCATION OF SURVEY :- ";A$
6Ø2Ø  PRINT "************************************************************
      ********************"
6Ø3Ø  PRINT "OPERATOR'S NAME :-     ";B$
6Ø4Ø  PRINT "DATE OF SURVEY  :-     ";C$
6Ø5Ø  PRINT "INSTRUMENT USED :-     ";K$
6Ø6Ø  PRINT "INST. CONSTANTS K,C    ";K;",";C
6Ø7Ø  PRINT "************************************************************
      ********************"
6Ø8Ø  PRINT
6Ø9Ø  PRINT D$;"PR#Ø"
61ØØ  GOTO 12ØØ
625Ø  PRINT D$;"PR#1"
626Ø  PRINT
627Ø  PRINT "SURVEY STATION NO. ";S
628Ø  PRINT "************************************************************
      ********************"
629Ø  PRINT "ELEVATION OF STATION = ";Z
63ØØ  PRINT "HEIGHT OF INSTRUMENT = ";HØ
631Ø  PRINT "************************************************************
      ********************"
632Ø  PRINT
633Ø  PRINT D$;"PR#Ø"
634Ø  GOTO 14ØØ
6485  REM  ******************************************
649Ø  REM  STADIA RESULTS PRINTOUT
65ØØ  PRINT D$;"PR#1": POKE 1784 + 1,8Ø
651Ø  IF P1 = 1 THEN 656Ø
652Ø  PRINT "PT.NO.   ZENITH    - U -   - M -   - L -    HORIZ.DIST.    E
      LEV."
653Ø  PRINT "============================================================
      ===================="
654Ø  PRINT
655Ø  LET P1 = 1: REM  PRINT HEADING ONCE
656Ø :
657Ø  PRINT P;: POKE 36,9: PRINT VØ;: POKE 36,19: PRINT U;: POKE 36,27: PRINT
      M;: POKE 36,35: PRINT L;: POKE 36,44: PRINT H;: POKE 36,59: PRINT ZØ
665Ø  PRINT D$;"PR#Ø"
666Ø  RETURN
```

```
8043  REM  ***************************************************
8045  REM  ROUTINE FROM DECIMAL TO DD.MMSS
8050 D1 =  INT (D)
8055  REM  FIND TOTAL NO. OF SECONDS
8060 D2 = (D - D1) * 3600
8065  REM  FIND NO. OF MINUTES
8070 M1 =  INT (D2 / 60)
8075  REM  FIND NO. OF SECONDS
8080 S1 = D2 - 60 * M1 + .5
8090  IF S1 < 60 THEN 8130
8100 M1 = M1 + 1:S1 = 0
8110  IF M1 < 60 THEN 8130
8120 D1 = D1 + 1:M1 = 0
8130 D = D1 * 10000 + M1 * 100 + S1
8140 D =  INT (D) / 10000
8150  RETURN
8985  REM  ******************************************
8995  REM  ROUTINE TO CLEAR SCREEN
9000  HOME
9010  RETURN
```

3.3.5 Tacheometry reduction – computer printout

```
LOCATION OF SURVEY :- STRATHCLYDE UNIVERSITY - STEELHENGE
***************************************************************************
OPERATOR'S NAME :-     JOHN SMITH
DATE OF SURVEY  :-     05/05/83
INSTRUMENT USED :-     ZEISS ZENA T20A
INST. CONSTANTS K,C    100,0
***************************************************************************


SURVEY STATION NO. 20
***************************************************************************
ELEVATION OF STATION = 98.946
HEIGHT OF INSTRUMENT = 1.46
***************************************************************************

PT.NO.   ZENITH    - U -    - M -    - L -    HORIZ.DIST.    ELEV.
===========================================================================

1        88.562    .955     .833     .712     24.292         100.023
2        88.562    1.35     1.15     .95      39.986         99.[illegible]97
3        88.562    2.01     1.913    1.817    19.293         98.85
4        88.562    1.035    .965     .895     13.995         99.7
5        88.562    1.345    1.303    1.26     8.497          99.26
6        88.562    2.24     2.183    2.126    11.396         98.434
7        93.32     2.16     2.078    1.999    16.039         97.338
8        93.32     2.915    2.8      2.69     22.415         96.222
9        93.32     3.254    3.102    2.956    29.687         95.471
===========================================================================
```

3.4 3-D POSITIONS FROM TACHEOMETRY

In the previous tacheometric reduction program '<STADIA>', only the levels of the survey points and the distances from the survey stations were calculated. If the 3-D co-ordinates of the survey station are known, or can be found by sights onto two known reference stations whose 3-D positions are known, then the 3-D co-ordinates of the survey points can also be found. Additionally if a plotter is available, the points can be plotted simultaneously for record purposes.

3.4.1 3-D tacheometric reduction – subroutine index

	Line numbers	*Function*
(a)	10–320	Initialization and control
(b)	1000–1190	Entry of survey location, etc.
(c)	1200–1490	Entry of reference station data
(d)	1500–1690	Entry of survey station data
(e)	1700–2090	Entry and display of station data
(f)	2100–2700	Computation of *X*, *Y*, *Z* co-ordinates of survey station
(g)	2710–2790	Entry of *X*, *Y*, *Z* co-ordinates of known survey station
(h)	2800–2990	Screen display of station co-ordinates
(i)	3000–4660	Computation of *X*, *Y*, *Z* co-ordinates of survey points
(j)	4670–4720	Routine to run program again
(k)	5000–5070	Termination of program
(l)	5100–5250	Routine to enter reference station data
(m)	5300–5480	Routine to enter survey point data
(n)	5500–5520	Routine to calculate *H* and *V*
(o)	5600–5620	Routine to calculate angle of triangle
(p)	6000–6080	Routine to print location of survey, etc.
(q)	6250–6430	Routine to print station co-ordinates
(r)	6500–6630	Routine to print survey point data
(s)	8050–8150	Routine to convert decimal angles to DD.MMSS
(t)	9000–9010	Routine to clear screen
(u)	9100–9130	Routine to display error message

(a) Initialization and control

Line numbers 10–320

The program is initialized and space reserved for string variables for keyboard entry and regular screen display. Since the Apple uses radians for angle computations, *R* is set to 57.2957795 in line 320 and all entries for angles in degrees divided by *R* to convert to radians.

(b) Entry of survey location

Line numbers 1000–1190

Line 1000 clears the screen and presents a screen header with requests for site information for record purposes, location, operator, date, etc. After entry, line 1160 gives a printout, if required, in the subroutine at line 6000.

(c) Entry of reference station data

Line numbers 1200–1490

To simplify the computations, the reference station with the smaller *X* and/or *Y* co-ordinate is entered first, followed by its *X*, *Y*, *Z* co-ordinates. Checks are made in the program to see that if $X_1 = X_2$ then $Y_1 < Y_2$, and if $Y_1 = Y_2$ then $X_1 < X_2$. After entry, the baseline distance, B, between the two stations is calculated in line 1490 from the formula:

$$B = [(X_2 - X_1)^2 + (Y_2 - Y_1)^2]^{1/2} \tag{3.19}$$

(d) Entry of survey station data

Line numbers 1500–1690

After entry of the survey station number, if the X, Y, Z, co-ordinates are known from a previous survey, the program branches to line 2700, otherwise a request is made for the height of the theodolite above the station.

(e) Entry and display of station data

Line numbers 1700–2090

If the X, Y, Z co-ordinates of the survey station are unknown, the program proceeds to line 1700 where the data is entered for tacheometric observations on to the two reference stations. To simplify the entry of data, a subroutine is called from line 5100 in each case. After calculation of the distance to each station and the reduced level of the instrument from each station, the screen is cleared at line 1860 and the computed data displayed. A check is made at line 1960 to ensure that the two distances are greater than the baseline distance between the two stations. If not, an error message is displayed in lines 1970–1980. Assuming no error, the operator has a choice at lines 2000–2020 to select the correct calculated value of the instrument elevation from the lines 2070–2090.

(f) Computation of X, Y, Z co-ordinates of survey station

Line numbers 2100–2700

From the previous calculations, the distances between the two reference stations, R_1, R_2, and the survey station S are known, that is B, H_1 and H_2 as shown in Fig. 3.6. The angle A of the triangle R_1, S, R_2 can now be solved using the (S_1, S_2, S_3) triangle solution provided in the routine at line 5600. If only distances were known from S to H_1 and H_2 then there could be two solutions, one either side of the baseline. To find the correct solution, the operator is asked on which side of the baseline the survey point lies. If $X_1 = X_2$, that is a vertical baseline, the survey point will lie to east (E) or west (W) of the baseline, line 2140. In all other cases, $Y_1 = Y_2$, $Y_1 < Y_2$, $Y_1 > Y_2$, the survey point will lie north (N) or south (S) of the baseline, line 2240.

In each of the above instances, the co-ordinates of the survey point S are computed, depending on which side of the baseline the survey point lies. In the first case, $X_1 = X_2$, with a vertical baseline the solution is simple, and is given in lines 2180–2220. In the second case $Y_1 = Y_2$ the north solution is given in lines 2290–2300 with the south solution in lines 2490–2500. In the third and fourth

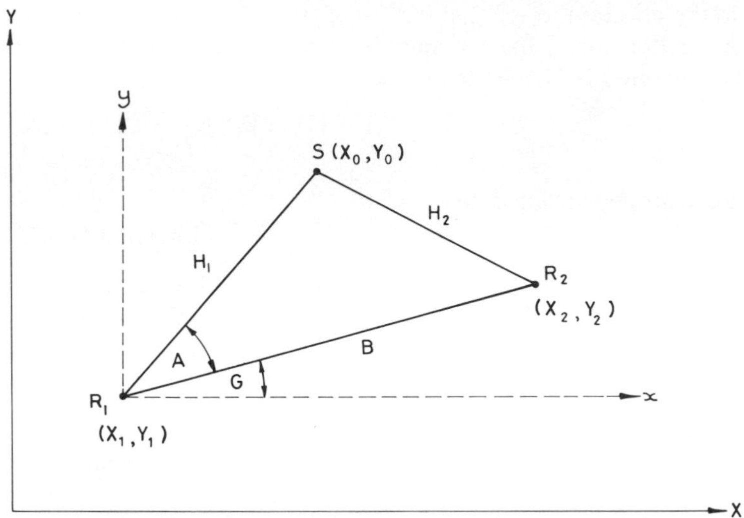

Fig. 3.6. Nomenclature for 3-D tacheometric reduction of observation station.

cases $Y_1 < Y_2$ and $Y_1 > Y_2$ it is essential to find the inclination or gradient of the baseline with the horizontal axis, that is angle G in Fig. 3.6. The co-ordinates of the point $S(X_0, Y_0)$ are then computed from the first reference point R_1 by using the combined angle of A and G as given in lines 2330–2650.

Each of the above calculation routines rejoins the main program at line 2660 where the result is determined to the third decimal place for display.

(g) Entry of *X, Y, Z* co-ordinates of known survey station

Line numbers 2710–2790

Where a previous survey has determined the X, Y, Z co-ordinates of the survey station, these can be entered here to skip over lines 1600–2700.

(h) Screen display of station co-ordinates

Line numbers 2800–2990

The results of the previous computations are then displayed with an option to obtain a printout using the routine at line 6250.

(i) Computation of *X, Y, Z* co-ordinates of survey points

Line numbers 3000–4660

The survey point computations start in this section. Initially one of the two reference stations has to be selected for zeroing the horizontal circle at line 3030. A check is inserted to ensure that one of the two previous reference stations R_1 and R_2 is chosen and then the X, Y co-ordinates of that station stored in (X_5, Y_5).

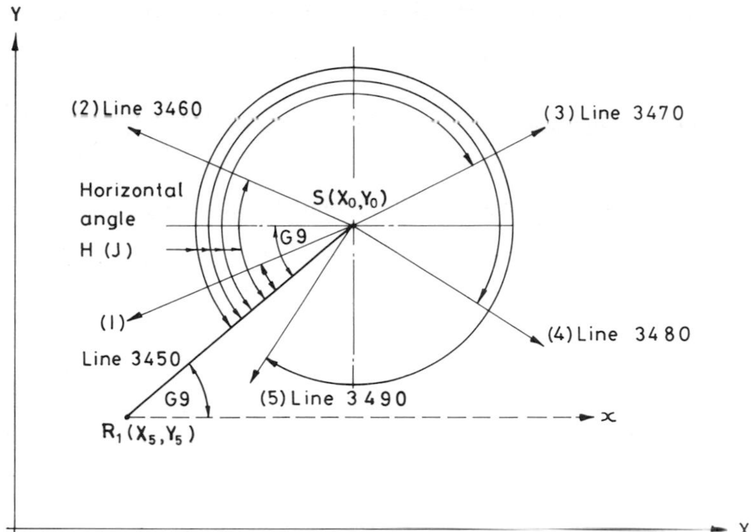

Fig. 3.7. Nomenclature of 3-D tacheometric reduction of points from observation station.

The X, Y co-ordinates of the survey station (X_0, Y_0) are then compared with the chosen reference station (X_5, Y_5). From this comparison the quadrant (see Fig. 3.1(b)) in which the new reference baseline falls is determined. The normal quadrants (Q) are 1 to 4. Depending on the value of Q, the gradient of the new baseline, $G9$ is calculated in lines 3160–3310. However, to cater for either vertical or horizontal baselines, a further four values of Q with designations from 5 to 8 are given in lines 3320–3360.

The total number of survey points observed from the survey station S are entered at line 3380, and the necessary array variables dimensioned in line 3400. A loop is then started in line 3410 to enter the survey points sequentially using the entry routine called from line 5300.

The main computations then commence at line 3440 where first the value of the horizontal angle $H(J)$ to the survey point is compared with the angle of the baseline $G9$, giving five solution areas, 1–5 in Fig. 3.7, lines 3450–3490. Then the appropriate solution for $X(J)$, $Y(J)$ is obtained knowing the previous value of Q for the baseline at lines 3500–3540, using an 'ON Q GOTO' solution. Each of these branch computations returns to the main program at line 4650 where a routine is provided for a printout of the results. The loop continues with the next survey point at line 4660.

(j) Routine to run program again

Line numbers 4670–4720

After completing the above loop for survey point entry, it is possible to run the program again. However, since it is not possible to re-dimension Apple variables again in the middle of a program, it is necessary to re-run from the beginning, as indicated at line 4720.

(k) Termination of program

Line numbers 5000–5070

If no further calculations are required, an end-of-program message is displayed to advise the operator that computations are complete.

(l) Routine to enter reference station data

Line numbers 5100–5250

This routine asks for tacheometric data to determine the difference in height and the distance between the reference station and the survey station. From the entries for zenith angle and the three staff readings, the routine at line 5500 is used to calculate H and V.

(m) Routine to enter survey point data

Line numbers 5300–5480

This routine asks for the clockwise horizontal angle from the reference station in addition to the zenith and staff readings. After computation, the horizontal angle is stored in $H(J)$, and the horizontal distance in $R(J)$ and the elevation in $Z(J)$ respectively, before returning to the main program.

(n) Routine to calculate H and V

Line numbers 5500–5520

The tacheometry calculations for H and V from Equations 3.16 and 3.17 are provided in this subroutine where W is the angle θ in Fig. 3.5. Note W is divided by R to convert the angle from degrees to radians before calculation.

(o) Routine to calculate angle of triangle

Line numbers 5600–5620

This routine was previously used in the triangle solution of three sides (S_1, S_2, S_3) in the <TRIANGLE> program where $B = S_1$, $H_1 = S_2$, $H_2 = S_3$ and $A = A_1$ where A is thc angle shown in Fig. 3.6.

(p) Routine to print location of survey, etc.

Line numbers 6000–6080

This routine is provided for record purposes, and is based on Applesoft DOS and may require alteration for different microcomputers and printers.

(q) Routine to print station co-ordinates

Line numbers 6250–6430

The same comments apply as above for (p).

(r) Routine to print survey point data

Line numbers 6500–6630

This printout routine, written in Applesoft DOS, assumes an 80-column printer to give a one-line display for each survey point. Prior to printing, the data is computed to the third decimal place and decimal angles converted to DD.MMSS, line numbers 6550–6590.

(s) Routine to convert decimal angles to DD.MMSS

Line numbers 8050–8150

This routine was described previously in Section 2.2.7.

(t) Routine to clear screen

Line numbers 9000–9010

To simplify portability, the Applesoft 'HOME' command to clear the screen has been placed in a subroutine to allow an easy change to another microcomputer.

(u) Routine to display error message

Line numbers 9100–9130

Should an error occur in the keyboard data entry, an error message is displayed asking for re-entry.

3.4.2 3-D tacheometry reduction – numeric variables

A	= Angle of triangle *B*, *H*1, *H*2
B	= Baseline distance between *R*1 and *R*2
*B*9	= New baseline distance between *S* and *R*9
C	= Additive constant
D, *D*1, *D*2	= Number of degrees in angle
E	= Storage for *Z* value of reference station
G, *G*9	= Gradient of baseline
*H*0	= Height of instrument above survey station
H	= Horizontal distance to survey point
*H*1	= Horizontal distance from *R*1 to *S*
*H*2	= Horizontal distance from *R*2 to *S*
*I*0, *I*1, *I*2	= Elevation of instrument
*I*9	= Selection of instrument elevation

Table—continued

J	= Loop counter for *N* survey points
K	= Multiplication constant
L	= Lower stadia hair reading
M	= Mid-stadia hair reading
*M*1	= Number of minutes in angle
N	= Number of survey points
*P*1	= Printer code
*P*7	= Screen display code
Q	= Quadrant for baseline *B*9
R	= 57.2957795, conversion degrees to radians
*R*1, *R*2, *R*9	= Reference station numbers
R(*N*)	= Horizontal range to survey point
S	= Survey station number
*S*1	= Number of seconds in angle
U	= Upper stadia hair reading
V	= Vertical height difference
*V*0, *V*1–*V*3, *V*9	= Zenith angle entries
W	= Angle of elevation or depression
*X*0, *X*1, *X*2, *X*5, *X*(*N*)	= *X* co-ordinates of stations
*Y*0, *Y*1, *Y*2, *Y*5, *Y*(*N*)	= *Y* co-ordinates of stations
*Z*0, *Z*1, *Z*2, *Z*(*N*)	= *Z* co-ordinates of stations

3.4.3 3-D tacheometry reduction – string variables

A$	Storage for location of survey
B$	Storage for operator's name
C$	Storage for date of survey
D$	= CHR$(4) i.e. CTRL-D
E$	= "SORRY, DATA ERROR . . . PLEASE RE-ENTER"
K$	Storage for instrument used
M$	= CHR$(13) + CHR$(4) i.e. allows program to re-run
P$	= "ENTER SURVEY POINT NUMBER, NUMERALS ONLY"
Q$	Question response (Y/N)
S$	= "IF IN DECIMAL DEGREES, FILL MINS & SECS WITH 0"
U$	= "ENTER UPPER(U), MID(M), LOWER(L) STADIA HAIR READINGS, SEPARATED BY COMMAS,"
V$	= "ENTER ZENITH ANGLE IN DEGS, MINS, SECS SEPARATED BY COMMAS"
W$	= "ENTER HORIZONTAL ANGLE IN CLOCKWISE DIRECTION FROM REF. STN."
X$	= "ENTER X(E) COORD. OF STN."
Y$	= "ENTER Y(N) COORD. OF STN."
Z$	= "ENTER Z COORD. OF STN."

3.4.4 3-D tacheometry reduction – BASIC program

```
10  REM  <TACHY.3-D> PROGRAM FOR APPLE II+  USES QUME PRINTER.
20  REM  TACHEOMETRY DATA REDUCTION PROGRAM FOR APPLE II+
30  REM  ***********************************************************
40  REM  BY DR. P.H. MILNE
50  REM  SURVEYING SECTION
60  REM  DEPARTMENT OF CIVIL ENGINEERING
70  REM  UNIVERSITY OF STATHCLYDE
80  REM  GLASGOW G4 ONG
90  REM  ***********************************************************
100  CLEAR
110  DIM A$(50),B$(25),C$(25),K$(25)
120  DIM E$(40),P$(50),S$(50),U$(80),V$(60),W$(70),X$(25),Y$(25),Z$(25)
130 E$ = "SORRY, DATA ERROR ---- PLEASE RE-ENTER."
140 P$ = "ENTER SURVEY POINT NUMBER, NUMERALS ONLY"
150 U$ = "ENTER UPPER(U), MID(M), LOWER(L) STADIA HAIR READINGS, SEPARATE
    D BY COMMAS,"
160 V$ = "ENTER ZENITH ANGLE IN DEGS,MINS,SECS    SEPARATED BY COMMAS,"
170 S$ = "IF IN DECIMAL DEGS, FILL MINS & SECS    WITH 0"
180 W$ = "ENTER HORIZONTAL ANGLE IN CLOCKWISE     DIRECTION FROM REF. STN
    ."
190 X$ = "ENTER X(E) COORD. OF STN."
200 Y$ = "ENTER Y(N) COORD. OF STN."
210 Z$ = "ENTER Z    COORD. OF STN."
220 D$ =  CHR$ (4): REM  CTRL-D
230 M$ =  CHR$ (13) +  CHR$ (4): REM  CTRL CHARACTERS TO RETURN TO <MENU>

300  REM  SET COUNTERS TO ZERO
310  LET H0 = 0:P1 = 0:P7 = 0
320  LET R = 57.2957795: REM  CONVERTS RADIANS TO DEGREES
1000  GOSUB 9000
1010  PRINT "UNIVERSITY OF STRATHCLYDE"
1020  PRINT "TACHEOMETRY DATA REDUCTION PROGRAM"
1030  PRINT "****************************************"
1040  PRINT
1050  PRINT "ENTER LOCATION OF WORK, ETC. WHEN       REQUESTED, AND PRESS
      <RETURN>"
1060  PRINT "ENTER LOCATION OF SURVEY :-"
1070  INPUT A$
1080  PRINT "ENTER OPERATOR'S NAME :-"
1090  INPUT B$
1100  PRINT "ENTER DATE OF SURVEY AS DD/MM/YY :-"
1110  INPUT C$
1120  PRINT "ENTER INSTRUMENT USED :-"
1130  INPUT K$
1140  PRINT "ENTER INSTRUMENT CONSTANTS, K,C         SEPARATED BY A COMMA
      ,"
1150  INPUT K,C
1160  PRINT "DO YOU WISH PRINTOUT OF LOCATION ETC.,  (Y/N) ";
1170  INPUT Q$
1180  IF Q$ = "Y" THEN 6000
1190  IF Q$ < > "N" THEN 1160
1200  GOSUB 9000
1210  PRINT "ENTER DETAILS OF REFERENCE STNS WHEN    REQUESTED, AND PRESS
      <RETURN>"
1220  PRINT
1230  PRINT "ENTER NO. OF REFERENCE STN. WITH SMALLER X AND/OR Y COORD.,
     NUMERALS ONLY :-"
```

```
1240  INPUT R1
1250  PRINT X$;R1;
1260  INPUT X1
1270  PRINT Y$;R1;
1280  INPUT Y1
1290  PRINT Z$;R1;
1300  INPUT Z1
1310  PRINT
1320  PRINT "ENTER NO. OF SECOND REFERENCE STN.      NUMERALS ONLY :-"
1330  INPUT R2
1340  IF R2 <  > R1 THEN 1370
1350  GOSUB 9100
1360  GOTO 1320
1370  PRINT X$;R2;
1380  INPUT X2
1390  IF X1 <  = X2 THEN 1420
1400  GOSUB 9100
1410  GOTO 1230
1420  PRINT Y$;R2;
1430  INPUT Y2
1435  IF X1 < X2 THEN 1470
1440  IF Y1 <  = Y2 THEN 1470
1450  GOSUB 9100
1460  GOTO 1230
1470  PRINT Z$;R2;
1480  INPUT Z2
1490  LET B =  SQR ((X2 - X1) ^ 2 + (Y2 - Y1) ^ 2)
1500  GOSUB 9000
1510  PRINT "ENTER SURVEY STATION INFORMATION WHEN   REQUESTED, AND PRESS
      <RETURN>"
1520  PRINT
1530  PRINT "ENTER SURVEY STATION NO., NUMERALS ONLY"
1540  INPUT S
1550  PRINT "ARE X,Y,Z COORDS OF SURVEY STN. ";S: PRINT "KNOWN (Y/N) ";
1560  INPUT Q$
1570  IF Q$ = "Y" THEN 2710
1580  IF Q$ <  > "N" THEN 1550
1590  REM
1600  GOSUB 9000
1610  PRINT "IS HEIGHT OF THEOD. ABOVE SURVEY STN. ";S: PRINT "KNOWN (Y/N
     ) ";
1620  INPUT Q$
1630  IF Q$ = "N" THEN 1680
1640  IF Q$ <  > "Y" THEN 1600
1650  PRINT "ENTER HEIGHT OF THEODOLITE :-";
1660  INPUT H0
1670  GOTO 1700
1680  LET H0 = 0
1690  REM
1700  GOSUB 9000
1710  PRINT "DATA FOR REFERENCE STN.";R1
1720  PRINT "****************************************"
1730  PRINT
1740  LET E = Z1
1750  GOSUB 5100
1760  LET H1 =  INT (H * 1000 + .5) / 1000
1770  LET I1 =  INT (I0 * 1000 + .5) / 1000
1780  GOSUB 9000
1790  PRINT "DATA FOR REFERENCE STN.";R2
```

```
1800  PRINT "****************************************"
1810  PRINT
1820  LET E = Z2
1830  GOSUB 5100
1840  LET H2 =  INT (H * 1000 + .5) / 1000
1850  LET I2 =  INT (I0 * 1000 + .5) / 1000
1860  GOSUB 9000
1870  PRINT "DATA FOR SURVEY STATION ";S
1880  PRINT "========================================"
1890  PRINT
1900  PRINT "DIST. TO REF. STN.";R1;" =";H1
1910  PRINT "DIST. TO REF. STN.";R2;" =";H2
1920  PRINT
1930  PRINT "INST.HT. FROM STN.";R1;" =";I1
1940  PRINT "INST.HT. FROM STN.";R2;" =";I2
1950  PRINT
1960  IF H1 + H2 > B THEN 2000
1970  GOSUB 9100
1980  PRINT "DISTANCES TO";R1;" AND";R2;" TOO SMALL !"
1990  GOTO 1710
2000  PRINT "WHICH VALUE OF INSTRUMENT HT. IS CORRECT"
2010  PRINT "PRESS <1> FOR ";R1;", <2> FOR ";R2: PRINT ", OR <3> FO
     AGE : ";
2020  INPUT I9
2030  ON I9 GOTO 2070,2080,2090
2040 :
2050 :
2060  IF I9 < 1 OR I9 > 3 THEN 1860
2070  LET I0 = I1: GOTO 2100
2080  LET I0 = I2: GOTO 2100
2090  LET I0 = ((I1 + I2) / 2) * 10 ^ 3:I0 =  INT (I0) / 10 ^ 3
2100  GOSUB 9000
2110  REM  CALCULATE ANGLE OF TRIANGLE WITH SIDES B,H1,H2
2120  GOSUB 5600
2130  IF X2 <  > X1 THEN 2240
2140  PRINT "DOES NEW STATION LIE TO EAST OR WEST    OF BASELINE (E
2150  INPUT Q$
2160  IF Q$ = "E" THEN 2210
2170  IF Q$ <  > "W" THEN 2140
2180  LET X0 = X1 - H1 *  SIN (A / R)
2190  LET Y0 = Y1 + H1 *  COS (A / R)
2200  GOTO 2660
2210  LET X0 = X1 + H1 *  SIN (A / R)
2220  LET Y0 = Y1 + H1 *  COS (A / R)
2230  GOTO 2660
2240  PRINT "DOES NEW STATION LIE TO NORTH OR SOUTH  OF BASELINE (N/S) ";

2250  INPUT Q$
2260  IF Q$ = "S" THEN 2480
2270  IF Q$ <  > "N" THEN 2240
2280  IF Y1 <  > Y2 THEN 2320
2290  LET X0 = X1 + H1 *  COS (A / R)
2300  LET Y0 = Y1 + H1 *  SIN (A / R)
2310  GOTO 2660
2320  IF Y1 > Y2 THEN 2380
2330  LET G = (Y2 - Y1) / (X2 - X1)
2340  LET G =  ATN (G) * R
2350  LET X0 = X1 + H1 *  COS ((A + G) / R)
2360  LET Y0 = Y1 + H1 *  SIN ((A + G) / R)
```

```
2370  GOTO 266Ø
2380  LET G = (Y1 - Y2) / (X2 - X1)
2390  LET G =  ATN (G) * R
2400 :
2410  IF G > (A) THEN 2450
2420  LET XØ = X1 + H1 *  COS ((A - G) / R)
2430  LET YØ = Y1 + H1 *  SIN ((A - G) / R)
2440  GOTO 2660
2450  LET XØ = X1 + H1 *  COS ((G - A) / R)
2460  LET YØ = Y1 - H1 *  SIN ((G - A) / R)
2470  GOTO 2660
2480  IF Y1 <  > Y2 THEN 2520
2490  LET XØ = X1 + H1 *  COS (A / R)
2500  LET YØ = Y1 - H1 *  SIN (A / R)
2510  GOTO 2660
2520  IF Y1 > Y2 THEN 2620
2530  LET G = (Y2 - Y1) / (X2 - X1)
2540  LET G =  ATN (G) * R
2550  IF G < (A) THEN 2590
2560  LET XØ = X1 + H1 *  COS ((G - A) / R)
2570  LET YØ   Y1 + H1 *  SIN ((G - A) / R)
2580  GOTO 2660
2590  LET XØ = X1 + H1 *  COS ((A - G) / R)
2600  LET YØ = Y1 - H1 *  SIN ((A - G) / R)
2610  GOTO 2660
2620  LET G = (Y1 - Y2) / (X2 - X1)
2630  LET G =  ATN (G) * R
2640  LET XØ = X1 + H1 *  COS ((A + G) / R)
2650  LET YØ = Y1 - H1 *  SIN ((A + G) / R)
2660  LET XØ =  INT (XØ * 1ØØØ + .5) / 1ØØØ
2670  LET YØ =  INT (YØ * 1ØØØ + .5) / 1ØØØ
2680  IF HØ = Ø THEN 27ØØ
2690  LET ZØ =  INT ((IØ - HØ) * 1ØØØ + .5) / 1ØØØ
2700  GOTO 2790
2710  PRINT X$;S;
2720  INPUT XØ
2730  PRINT Y$;S;
2740  INPUT YØ
2750  PRINT Z$;S;
2760  INPUT ZØ
2770  PRINT "ENTER HT. OF THEODOLITE ABOVE STN.";S
2780  INPUT HØ:IØ = ZØ + HØ:P7 = 1
2790  GOSUB 9ØØØ
2800  PRINT "RANGE-RANGE FIX ON STN. ";S
2810  PRINT "========================================"
2820  PRINT : IF P7 = 1 THEN 2850
2830  PRINT "DIST. TO REF.STN.";R1;" =";H1
2840  PRINT "DIST. TO REF.STN.";R2;" =";H2: PRINT
2850  PRINT "STN.   EASTING    NORTHING   LEVEL"
2860  PRINT "========================================"
2870  PRINT R1; TAB( 7);X1; TAB( 18);Y1; TAB( 29);Z1
2880  PRINT R2; TAB( 7);X2; TAB( 18);Y2; TAB( 29);Z2
2890  IF HØ = Ø THEN 2930
2900  PRINT S; TAB( 7);XØ; TAB( 18);YØ; TAB( 29);ZØ
2910  PRINT "HT. OF THEOD. AT ";S;" IS ";HØ
2920  GOTO 2950
2930  PRINT S; TAB( 7);XØ; TAB( 18);YØ
2940  PRINT "ELEV. OF THEOD. AT ";S;" IS ";IØ
2950  PRINT "****************************************"
2960  PRINT "DO YOU WISH PRINTOUT OF ABOVE DATA (Y/N)"
```

```
297Ø  INPUT Q$
298Ø  IF Q$ = "Y" THEN 625Ø
299Ø  IF Q$ < > "N" THEN 296Ø
3ØØØ  GOSUB 9ØØØ
3Ø1Ø  PRINT "SURVEY POINT COMPUTATIONS"
3Ø2Ø  PRINT "*******************************************"
3Ø3Ø  PRINT "WHICH REFERENCE STN. IS USED FOR ZEROING HORIZONTAL CIRCLE "
     ;R1;" OR ";R2
3Ø4Ø  INPUT R9
3Ø5Ø  IF R9 = R1 THEN 3Ø8Ø
3Ø6Ø  IF R9 = R2 THEN 3Ø9Ø
3Ø7Ø  IF R9 < > R1 AND R9 < > R2 THEN 3ØØØ
3Ø8Ø  LET X5 = X1:Y5 = Y1:B9 = SQR ((X5 - XØ) ^ 2 + (Y5 - YØ) ^ 2): GOTO
     31ØØ
3Ø9Ø  LET X5 = X2:Y5 = Y2:B9 = SQR ((X5 - XØ) ^ 2 + (Y5 - YØ) ^ 2)
31ØØ  IF XØ > X5 AND YØ > Y5 THEN 316Ø
311Ø  IF XØ > X5 AND YØ < Y5 THEN 32ØØ
312Ø  IF XØ < X5 AND YØ < Y5 THEN 324Ø
313Ø  IF XØ < X5 AND YØ > Y5 THEN 328Ø
314Ø  IF XØ = X5 THEN 332Ø
315Ø  IF YØ = Y5 THEN 335Ø
316Ø  LET Q = 1
317Ø  LET G9 = (YØ - Y5) / (XØ - X5)
318Ø  LET G9 = ATN (G9) * R
319Ø  GOTO 338Ø
32ØØ  LET Q = 2
321Ø  LET G9 = (XØ - X5) / (Y5 - YØ)
322Ø  LET G9 = ATN (G9) * R
323Ø  GOTO 338Ø
324Ø  LET Q = 3
325Ø  LET G9 = (Y5 - YØ) / (X5 - XØ)
326Ø  LET G9 = ATN (G9) * R
327Ø  GOTO 338Ø
328Ø  LET Q = 4
329Ø  LET G9 = (X5 - XØ) / (YØ - Y5)
33ØØ  LET G9 = ATN (G9) * R
331Ø  GOTO 338Ø
332Ø  IF YØ > Y5 THEN Q = 5
333Ø  IF YØ < Y5 THEN Q = 6
334Ø  GOTO 337Ø
335Ø  IF XØ < X5 THEN Q = 7
336Ø  IF XØ > X5 THEN Q = 8
337Ø G9 = Ø: REM  ONLY IF XØ=X5 OR YØ=Y5
338Ø  PRINT "ENTER NUMBER OF SURVEY POINTS OBSERVED  FROM STN. ";S
339Ø  INPUT N: LET N = N - 1
34ØØ  DIM H(N),P(N),R(N),X(N),Y(N),Z(N)
341Ø  FOR J = Ø TO N
342Ø  GOSUB 53ØØ
343Ø  GOSUB 9ØØØ
344Ø  PRINT "           CALCULATING ......"
345Ø  IF H(J) < G9 THEN 35ØØ
346Ø  IF H(J) < G9 + 9Ø THEN 351Ø
347Ø  IF H(J) < G9 + 18Ø THEN 352Ø
348Ø  IF H(J) < G9 + 27Ø THEN 353Ø
349Ø  GOTO 354Ø
35ØØ  ON Q GOTO 358Ø,361Ø,364Ø,367Ø
351Ø  ON Q GOTO 37ØØ,373Ø,376Ø,379Ø,382Ø,385Ø,388Ø,391Ø
352Ø  ON Q GOTO 394Ø,397Ø,4ØØØ,4Ø3Ø,4Ø6Ø,4Ø9Ø,412Ø,415Ø
353Ø  ON Q GOTO 418Ø,421Ø,424Ø,427Ø,43ØØ,433Ø,436Ø,439Ø
354Ø  ON Q GOTO 442Ø,445Ø,448Ø,451Ø,454Ø,457Ø,46ØØ,463Ø
```

```
3580  LET X(J) = X0 - R(J) * COS ((G9 - H(J)) / R)
3590  LET Y(J) = Y0 - R(J) * SIN ((G9 - H(J)) / R)
3600  GOTO 4650
3610  LET X(J) = X0 - R(J) * SIN ((G9 - H(J)) / R)
3620  LET Y(J) = Y0 + R(J) * COS ((G9 - H(J)) / R)
3630  GOTO 4650
3640  LET X(J) = X0 + R(J) * COS ((G9 - H(J)) / R)
3650  LET Y(J) = Y0 + R(J) * SIN ((G9 - H(J)) / R)
3660  GOTO 4650
3670  LET X(J) = X0 + R(J) * SIN ((G9 - H(J)) / R)
3680  LET Y(J) = Y0 - R(J) * COS ((G9 - H(J)) / R)
3690  GOTO 4650
3700  LET X(J) = X0 - R(J) * COS ((H(J) - G9) / R)
3710  LET Y(J) = Y0 + R(J) * SIN ((H(J) - G9) / R)
3720  GOTO 4650
3730  LET X(J) = X0 + R(J) * SIN ((H(J) - G9) / R)
3740  LET Y(J) = Y0 + R(J) * COS ((H(J) - G9) / R)
3750  GOTO 4650
3760  LET X(J) = X0 + R(J) * COS ((H(J) - G9) / R)
3770  LET Y(J) = Y0 - R(J) * SIN ((H(J) - G9) / R)
3780  GOTO 4650
3790  LET X(J) = X0 - R(J) * SIN ((H(J) - G9) / R)
3800  LET Y(J) = Y0 - R(J) * COS ((H(J) - G9) / R)
3810  GOTO 4650
3820  LET X(J) = X0 - R(J) * SIN (H(J) / R)
3830  LET Y(J) = Y0 - R(J) * COS (H(J) / R)
3840  GOTO 4650
3850  LET X(J) = X0 + R(J) * SIN (H(J) / R)
3860  LET Y(J) = Y0 + R(J) * COS (H(J) / R)
3870  GOTO 4650
3880  LET X(J) = X0 + R(J) * COS (H(J) / R)
3890  LET Y(J) = Y0 - R(J) * SIN (H(J) / R)
3900  GOTO 4650
3910  LET X(J) = X0 - R(J) * COS (H(J) / R)
3920  LET Y(J) = Y0 + R(J) * SIN (H(J) / R)
3930  GOTO 4650
3940  LET X(J) = X0 + R(J) * SIN ((H(J) - 90 - G9) / R)
3950  LET Y(J) = Y0 + R(J) * COS ((H(J) - 90 - G9) / R)
3960  GOTO 4650
3970  LET X(J) = X0 + R(J) * COS ((H(J) - 90 - G9) / R)
3980  LET Y(J) = Y0 - R(J) * SIN ((H(J) - 90 - G9) / R)
3990  GOTO 4650
4000  LET X(J) = X0 - R(J) * SIN ((H(J) - 90 - G9) / R)
4010  LET Y(J) = Y0 - R(J) * COS ((H(J) - 90 - G9) / R)
4020  GOTO 4650
4030  LET X(J) = X0 - R(J) * COS ((H(J) - 90 - G9) / R)
4040  LET Y(J) = Y0 + R(J) * SIN ((H(J) - 90 - G9) / R)
4050  GOTO 4650
4060  LET X(J) = X0 - R(J) * COS ((H(J) - 90) / R)
4070  LET Y(J) = Y0 + R(J) * SIN ((H(J) - 90) / R)
4080  GOTO 4650
4090  LET X(J) = X0 + R(J) * COS ((H(J) - 90) / R)
4100  LET Y(J) = Y0 - R(J) * SIN ((H(J) - 90) / R)
4110  GOTO 4650
4120  LET X(J) = X0 - R(J) * SIN ((H(J) - 90) / R)
4130  LET Y(J) = Y0 - R(J) * COS ((H(J) - 90) / R)
4140  GOTO 4650
4150  LET X(J) = X0 + R(J) * SIN ((H(J) - 90) / R)
4160  LET Y(J) = Y0 + R(J) * COS ((H(J) - 90) / R)
4170  GOTO 4650
```

```
4180  LET X(J) = X0 + R(J) * COS ((H(J) - 180 - G9) / R)
4190  LET Y(J) = Y0 - R(J) * SIN ((H(J) - 180 - G9) / R)
4200  GOTO 4650
4210  LET X(J) = X0 - R(J) * SIN ((H(J) - 180 - G9) / R)
4220  LET Y(J) = Y0 - R(J) * COS ((H(J) - 180 - G9) / R)
4230  GOTO 4650
4240  LET X(J) = X0 - R(J) * COS ((H(J) - 180 - G9) / R)
4250  LET Y(J) = Y0 + R(J) * SIN ((H(J) - 180 - G9) / R)
4260  GOTO 4650
4270  LET X(J) = X0 + R(J) * SIN ((H(J) - 180 - G9) / R)
4280  LET Y(J) = Y0 + R(J) * COS ((H(J) - 180 - G9) / R)
4290  GOTO 4650
4300  LET X(J) = X0 + R(J) * SIN ((H(J) - 180) / R)
4310  LET Y(J) = Y0 + R(J) * COS ((H(J) - 180) / R)
4320  GOTO 4650
4330  LET X(J) = X0 - R(J) * SIN ((H(J) - 180) / R)
4340  LET Y(J) = Y0 - R(J) * COS ((H(J) - 180) / R)
4350  GOTO 4650
4360  LET X(J) = X0 - R(J) * COS ((H(J) - 180) / R)
4370  LET Y(J) = Y0 + R(J) * SIN ((H(J) - 180) / R)
4380  GOTO 4650
4390  LET X(J) = X0 + R(J) * COS ((H(J) - 180) / R)
4400  LET Y(J) = Y0 - R(J) * SIN ((H(J) - 180) / R)
4410  GOTO 4650
4420  LET X(J) = X0 - R(J) * SIN ((H(J) - 270 - G9) / R)
4430  LET Y(J) = Y0 - R(J) * COS ((H(J) - 270 - G9) / R)
4440  GOTO 4650
4450  LET X(J) = X0 - R(J) * COS ((H(J) - 270 - G9) / R)
4460  LET Y(J) = Y0 + R(J) * SIN ((H(J) - 270 - G9) / R)
4470  GOTO 4650
4480  LET X(J) = X0 + R(J) * SIN ((H(J) - 270 - G9) / R)
4490  LET Y(J) = Y0 + R(J) * COS ((H(J) - 270 - G9) / R)
4500  GOTO 4650
4510  LET X(J) = X0 + R(J) * COS ((H(J) - 270 - G9) / R)
4520  LET Y(J) = Y0 - R(J) * SIN ((H(J) - 270 - G9) / R)
4530  GOTO 4650
4540  LET X(J) = X0 + R(J) * COS ((H(J) - 270) / R)
4550  LET Y(J) = Y0 - R(J) * SIN ((H(J) - 270) / R)
4560  GOTO 4650
4570  LET X(J) = X0 - R(J) * COS ((H(J) - 270) / R)
4580  LET Y(J) = Y0 + R(J) * SIN ((H(J) - 270) / R)
4590  GOTO 4650
4600  LET X(J) = X0 + R(J) * SIN ((H(J) - 270) / R)
4610  LET Y(J) = Y0 + R(J) * COS ((H(J) - 270) / R)
4620  GOTO 4650
4630  LET X(J) = X0 - R(J) * SIN ((H(J) - 270) / R)
4640  LET Y(J) = Y0 - R(J) * COS ((H(J) - 270) / R)
4650  GOSUB 6500
4660  NEXT J
4670  GOSUB 9000
4680  PRINT "DO YOU WISH TO RUN PROGRAM AGAIN (Y/N)"
4690  INPUT Q$
4700  IF Q$ = "N" THEN 5000
4710  IF Q$ < > "Y" THEN 4670
4720  PRINT M$"RUN TACHY.3-D"
4985  REM  *****************************************
4990  REM  END OF PROGRAM - RETURN TO MENU
5000  GOSUB 9000
5010  PRINT D$;"PR#1"
5020  PRINT "**********************************************************
      ********************"
```

```
5030  PRINT D$;"PR#Ø"
5040  VTAB (5): HTAB (5): PRINT "END OF TACHEOMETRY PROGRAM"
5050  PRINT : PRINT "******************************************"
5060  PRINT : PRINT : PRINT "    LOADING <MENU> FROM DISC ....."
5070  PRINT M$"RUN HELLO": REM  MENU PROGRAM
5093  REM  ************************************************
5095  REM  GOSUB ROUTINE FOR KEYBOARD ENTRY
5100  PRINT "*****   REFERENCE STATIONS INPUT   *****"
5110  PRINT V$;S$
5120  INPUT V1,V2,V3
5130  LET VØ = V1 + V2 / 6Ø + V3 / 36ØØ
5135  IF VØ > 18Ø THEN VØ = 36Ø - VØ
5140  PRINT U$
5150  INPUT U,M,L
5160  LET I = U - L
5170  LET W = VØ - 9Ø
5180  IF W < Ø THEN 522Ø
5190  GOSUB 55ØØ
5200  LET IØ = E + M + V
5210  GOTO 525Ø
5220  LET W =  - 1 * W
5230  GOSUB 55ØØ
5240  LET IØ = E + M - V
5250  RETURN
5293  REM  *************************************************
5295  REM  GOSUB ROUTINE FOR KEYBOARD ENTRY
5300  PRINT "*****   SURVEY POINT INPUT   *****"
5310  PRINT P$
5320  INPUT P(J)
5330  PRINT W$;S$
5340  INPUT H5,H6,H7
5345  LET H(J) = H5 + H6 / 6Ø + H7 / 36ØØ
5350  PRINT V$;S$
5360  INPUT V1,V2,V3
5370  LET V9 = V1 + V2 / 6Ø + V3 / 36ØØ
5375  IF V9 > 18Ø THEN V9 = 36Ø - V9
5380  PRINT U$
5390  INPUT U,M,L
5400  LET I = U - L:W = V9 - 9Ø
5410  IF W < Ø THEN 545Ø
5420  GOSUB 55ØØ
5430  LET R(J) = H:Z(J) = IØ - V - M
5440  GOTO 548Ø
5450  LET W =  - 1 * W
5460  GOSUB 55ØØ
5470  LET R(J) = H:Z(J) = IØ + V - M
5480  RETURN
5485  REM  **********************************************************
5490  REM  GOSUB ROUTINE FOR H,V
5500  LET H = K * I *  COS (W / R) ^ 2 + C *  COS (W / R)
5510  LET V = .5 * K * I *  SIN (2 * W / R) + C *  SIN (W / R)
5520  RETURN
5585  REM  **********************************************************
5590  REM  GOSUB ROUTINE FOR ANGLE OF TRIANGLE
5600  LET A = (B * B + H1 * H1 - H2 * H2) / (2 * H1 * B)
5610  LET A = 9Ø -  ATN (A /  SQR (1 - A * A)) * R
5620  RETURN
5985  REM  **********************************************************
5990  REM  PRINTOUT OF LOCATION, ETC
```

```
6ØØØ  PRINT D$;"PR#1": PRINT "LOCATION OF SURVEY :- ";A$
6Ø1Ø  PRINT "********************************************************
     ********************"
6Ø2Ø  PRINT "OPERATOR'S NAME :-     ";B$
6Ø3Ø  PRINT "DATE OF SURVEY  :-     ";C$
6Ø4Ø  PRINT "INSTRUMENT USED :-     ";K$
6Ø5Ø  PRINT "INSTRUMENT CONSTANTS =";K;",";C
6Ø6Ø  PRINT "********************************************************
     ********************"
6Ø7Ø  PRINT : PRINT D$;"PR#Ø"
6Ø8Ø  GOTO 12ØØ
6235  REM  *******************************************************
624Ø  REM  PRINTOUT OF STATION COORDS
625Ø  PRINT D$;"PR#1": POKE 1784 + 1,8Ø
626Ø  PRINT "RANGE - RANGE FIX ON STN. ";S
627Ø  PRINT "========================================================
     ====================": IF P7 = 1 THEN 63ØØ
628Ø  PRINT "DISTANCE TO REF. STN. ";R1;" = ";H1
629Ø  PRINT "DISTANCE TO REF. STN. ";R2;" = ";H2
63ØØ  PRINT
631Ø  PRINT "STN.NO. EASTING     NORTHING    LEVEL"
632Ø  PRINT R1;: POKE 36,8: PRINT X1;: POKE 36,2Ø: PRINT Y1;: POKE 36,32:
      PRINT Z1
633Ø  PRINT R2;: POKE 36,8: PRINT X2;: POKE 36,2Ø: PRINT Y2;: POKE 36,32:
      PRINT Z2
634Ø  PRINT
635Ø  IF HØ = Ø THEN 639Ø
636Ø  PRINT S;: POKE 36,8: PRINT XØ;: POKE 36,2Ø: PRINT YØ;: POKE 36,32: PRINT
     ZØ
637Ø  PRINT "HT. OF THEODOLITE ABOVE STN. ";S;" IS ";HØ
638Ø  GOTO 641Ø
639Ø  PRINT S;: POKE 36,8: PRINT XØ;: POKE 36,2Ø: PRINT YØ
64ØØ  PRINT "ELEV. OF THEODOLITE AT STN. ";S;" IS ";IØ
641Ø  PRINT "********************************************************
     ********************"
642Ø  PRINT : PRINT D$;"PR#Ø"
643Ø  GOTO 3ØØØ
6485  REM  *******************************************************
649Ø  REM  SURVEY POINT PRINTOUT
65ØØ  PRINT D$;"PR#1": POKE 1784 + 1,8Ø: IF P1 = 1 THEN 655Ø
651Ø  PRINT "PT.NO.  HORIZ.ANGLE  HORIZ.RANGE    EASTING        NORTHING
          LEVEL"
652Ø  PRINT "========================================================
     ===================="
653Ø  PRINT
654Ø  LET P1 = 1: REM  PRINT HEADING ONCE ONLY
6545  REM  CONVERT HORIZ. ANGLE TO DD.MMSS
655Ø D = H(J): GOSUB 8Ø5Ø:H(J) = D
656Ø R(J) =  INT (R(J) * 1ØØØ + .5) / 1ØØØ
657Ø X(J) =  INT (X(J) * 1ØØØ + .5) / 1ØØØ
658Ø Y(J) =  INT (Y(J) * 1ØØØ + .5) / 1ØØØ
659Ø Z(J) =  INT (Z(J) * 1ØØØ + .5) / 1ØØØ
66ØØ  PRINT P(J);: POKE 36,8: PRINT H(J);: POKE 36,21: PRINT R(J);: POKE
     36,36: PRINT X(J);: POKE 36,51: PRINT Y(J);: POKE 36,66: PRINT Z(J)
661Ø :
662Ø  PRINT D$;"PR#Ø"
663Ø  RETURN
8Ø43  REM  ***********************************************
8Ø45  REM  ROUTINE FROM DECIMAL TO DD.MMSS
8Ø5Ø D1 =  INT (D)
```

```
8055  REM  FIND TOTAL NO. OF SECONDS
8060 D2 = (D - D1) * 3600
8065  REM  FIND NO. OF MINUTES
8070 M1 =  INT (D2 / 60)
8075  REM  FIND NO. OF SECONDS
8080 S1 = D2 - 60 * M1 + .5
8090  IF S1 < 60 THEN 8130
8100 M1 = M1 + 1:S1 = 0
8110  IF M1 < 60 THEN 8130
8120 D1 = D1 + 1:M1 = 0
8130 D = D1 * 10000 + M1 * 100 + S1
8140 D =  INT (D) / 10000
8150  RETURN
8993  REM  **************************************************
8995  REM  GOSUB ROUTINE TO CLEAR SCREEN
9000  HOME
9010  RETURN
9085  REM  ******************************************
9090  REM  GOSUB ROUTINE FOR ERRORS
9100  PRINT
9110  PRINT E$
9120  PRINT
9130  RETURN
```

3.4.5 3-D tacheometry reduction – computer printout

```
LOCATION OF SURVEY :- STRATHCLYDE UNIVERSITY - STEELHENGE
***********************************************************************
OPERATOR'S NAME :-    JOHN SMITH
DATE OF SURVEY  :-    05/05/83
INSTRUMENT USED :-    ZEISS ZENA T20A
INSTRUMENT CONSTANTS =100,0
***********************************************************************

RANGE - RANGE FIX ON STN. 20
=======================================================================
DISTANCE TO REF. STN. 1 = 24.292
DISTANCE TO REF. STN. 2 = 39.986

STN.NO. EASTING      NORTHING     LEVEL
1       20           10           100.02
2       82.468       12           100

20      42.935       18.004       98.946
HT. OF THEODOLITE ABOVE STN. 20 IS 1.46
***********************************************************************

PT.NO.  HORIZ.ANGLE  HORIZ.RANGE    EASTING          NORTHING        LEVEL
=======================================================================

3       100.544      19.293         36.48            -.177           98.85
4       109.432      13.995         36.288           5.688           99.7
5       138.374      8.497          35.788           13.409          99.26
6       213.334      11.396         34.492           25.659          98.434
7       227.23       16.039         33.971           31.304          97.338
8       236.45       22.415         33.599           38.382          96.222
9       243.372      29.687         33.888           46.279          95.471
***********************************************************************
```

3.4.6 3-D tacheometry reduction – plotter routine

In addition to the printout obtained from the <TACHY.3-D> program, there are advantages in being able to simultaneously plot the survey points for record purposes.

Several A4 plotters are now available at reasonable cost, for example the HP-7470A two-pen plotter is only 60% of the earlier HP-7225A single-pen plotter, and can easily be linked to a microcomputer using an IEEE-488 interface. Some small changes are required to the previous program, as shown in Section 3.4.7.

3.4.7 3-D tacheometry plot – subroutine index

	Line numbers	*Function*
(aa)	2970–2990	Plot option after printout
(bb)	4655	Plotter survey point routine
(cc)	7000–7290	Plotter initialization and station plot
(dd)	7300–7340	Routine to mark station (#)
(ee)	7400–7420	Routine to label station
(ff)	7450–7460	Routine to plot level below station
(gg)	7500–7730	Routine to plot survey point (+)

(aa) Plot option after printout

Line numbers 2970–2990

These three lines of the <TACHY.3-D> program require to be changed to give the plot option in line 2980, which if required branches to the plotter subroutine at line 7000.

(bb) Plotter survey point routine

Line number 4655

This additional line is required to access the plotter subroutine at line 7500.

(cc) Plotter initialization and station plot

Line numbers 7000–7290

The plotter requires various control codes and strings and these are initialized in lines 7000–7010. Since the IEEE-488 interface tends to overheat in the main Apple board, it is housed in an expansion chassis, which is addressed in line 7020, and then the IEEE-488 card accessed in slot 4 in lines 7030–7040. The plotter is first scaled in line 7070 and a border or frame drawn in line 7080. The reference stations R_1, R_2 and the survey station S are then plotted and labelled using subroutines at lines 7300, 7400 and 7450.

(dd) Routine to mark station (#)

Line numbers 7300–7340

This routine moves to the point (X_9, Y_9) on the plotter and marks the station with a (#).

(ee) Routine to label station

Line numbers 7400–7420

This routine labels the station number contained in the string *PT$* above and to the right of the station.

(ff) Routine to plot level below station

Line numbers 7450–7460

This routine plots the level contained in the string *PZ$* below the station.

(gg) Routine to plot survey point (+)

Line numbers 7500–7730

The plotter is initialized at the outset and the plotter scaled as before. The survey point co-ordinates (*X*(*J*), *Y*(*J*)) are then stored in *X*8, *Y*8 which is plotted with a '+', line 7640, followed by the survey point number, line 7670 and the level, line 7700 as shown in Fig. 3.8.

3.4.8 3-D tacheometry plot – additional numeric variables

PP	= Plotter code
*X*8, *X*9	= Temporary storage of *X* co-ordinates
*Y*8, *Y*9	= Temporary storage of *Y* co-ordinates
*Z*8	= Temporary storage of *Z* co-ordinate

3.4.9 3-D tacheometry plot – additional string variables

EP$	= CHR$(3), control code for plotter
PC$	= "#", station point symbol
PL$	= Survey point number in string form for labelling
PP$	= "WT%" + *ZP$*, string to write data to plotter
PS$	= "+", survey point symbol
PT$	= Station number in string form for labelling
PZ$	= Station level in string form for labelling
ZP$	= CHR$(26), control code for plotter

3.4.10 3-D tacheometry plot – BASIC program addition

```
297Ø  INPUT Q$: IF Q$ = "Y" THEN 625Ø: IF Q$ <  > "N" THEN 296Ø
298Ø  PRINT "DO YOU WISH PLOT OF ABOVE DATA (Y/N)"
299Ø  INPUT Q$: IF Q$ = "Y" THEN 7ØØØ: IF Q$ <  > "N" THEN 298Ø
465Ø  GOSUB 65ØØ
4655  GOSUB 75ØØ: REM  PLOTTER POINT ROUTINE
466Ø  NEXT J
6985  REM  ###########################################################
699Ø  REM  PLOTTER ROUTINE FOR HP-747ØA
7ØØØ EP$ =  CHR$ (3):ZP$ =  CHR$ (26): REM  CTRL CODES FOR PLOTTER
7Ø1Ø PP$ = "WT%" + ZP$: REM  STRING TO WRITE TO PLOTTER
7Ø2Ø  POKE 49184,254: REM  SELECT EXPANSION CHASSIS
7Ø3Ø  PRINT D$;"PR#4": REM  IEEE-488 IN SLOT 4
7Ø4Ø  PRINT D$;"IN#4": REM  INITIALISE PLOTTER
7Ø5Ø  PRINT "SCØ": REM  SCREEN OFF
7Ø6Ø  PRINT PP$;"SC Ø,1ØØ,Ø,75;": REM  SCALE Xmin,Xmax,Ymin,Ymax
7Ø65  IF PP = 1 THEN 71ØØ
7Ø7Ø  PRINT PP$;"SP1;PU; PA Ø,Ø; PD;": REM  SELECT PEN 1 & GOTO ORIGIN
7Ø8Ø  PRINT PP$;"PA Ø,75,1ØØ,75,1ØØ,Ø,Ø,Ø;PU;": REM  FRAME
7Ø9Ø PP = 1: REM  ONLY DRAW FRAME ONCE
71ØØ X9 =  INT (X1):Y9 =  INT (Y1)
71Ø5  PRINT PP$;"SP2": REM  SELECT PEN 2
711Ø  GOSUB 73ØØ: REM  PLOT STN. #
712Ø PT$ =  STR$ (R1)
713Ø  GOSUB 74ØØ: REM  PLOT STN. NO.
714Ø PZ$ =  STR$ (Z1)
715Ø  GOSUB 745Ø: REM  PLOT LEVEL
716Ø X9 =  INT (X2):Y9 =  INT (Y2)
717Ø  GOSUB 73ØØ
718Ø PT$ =  STR$ (R2)
719Ø  GOSUB 74ØØ
72ØØ PZ$ =  STR$ (Z2)
721Ø  GOSUB 745Ø
722Ø X9 =  INT (XØ):Y9 =  INT (YØ)
723Ø  GOSUB 73ØØ
724Ø PT$ =  STR$ (S)
725Ø  GOSUB 74ØØ
7255  IF HØ = Ø THEN 728Ø
726Ø PZ$ =  STR$ (ZØ)
727Ø  GOSUB 745Ø
728Ø  POKE 49184,255: REM  DESELECT EXPANSION CHASSIS
729Ø  PRINT D$;"PR#Ø": PRINT D$;"IN#Ø": GOTO 3ØØØ
7295  REM  ####################################################
73ØØ  PRINT PP$;"PA";X9;",";Y9;";"
731Ø PC$ = "#": REM  STATION POINT SYMBOL
732Ø  PRINT PP$;"LB" + PC$ + EP$ + ";": REM  MARK STATION
733Ø  PRINT PP$;"PA";(X9 + 1);",";(Y9 + 1);";"
734Ø  RETURN
7395  REM  ####################################################
74ØØ  PRINT PP$;"LB STN." + PT$ + EP$ + ";"
741Ø  PRINT PP$;"PA";(X9 - 2);",";(Y9 - 1);";"
742Ø  RETURN
7445  REM  ####################################################
745Ø  PRINT PP$;"LB" + PZ$ + EP$ + ";"
746Ø  RETURN
7485  REM  ####################################################
749Ø  REM  PLOTTER ROUTINE FOR TACHY POINTS
75ØØ EP$ =  CHR$ (3):ZP$ =  CHR$ (26): REM  CTRL CODES FOR PLOTTER
```

```
7510 PP$ = "WT%" + ZP$: REM  WRITE STRING TO PLOTTER
7520  POKE 49184,254: REM  SELECT EXPANSION CHASSIS
7530  PRINT D$;"PR#4": REM  IEEE-488 IN SLOT 4
7540  PRINT D$;"IN#4": REM  INITIALISE PLOTTER
7550  PRINT "SCØ": REM  SCREEN OFF
7560  PRINT PP$;"SC Ø,1ØØ,Ø,75;": REM  SCALE XMIN,XMAX,YMIN,YMAX
7565  IF PP = 1 THEN 76ØØ
7570  PRINT PP$;"SP1;PU; PA Ø,Ø; PD;": REM  SELECT PEN 1 & GOTO ORIGIN
7580  PRINT PP$;"PA Ø,75,1ØØ,75,1ØØ,Ø,Ø,Ø;PU;": REM  FRAME PLOT
7590 PP = 1: REM  ONLY DRAW FRAME ONCE
76ØØ X8 =  INT (X(J))
7610 Y8 =  INT (Y(J))
7620  PRINT PP$;"SP1; PA";X8;",";Y8;";"
7630 PS$ = "+": REM  POINT PLOT SYMBOL
7640  PRINT PP$;"LB" + PS$ + EP$ + ";": REM  MARK POINT
7650  PRINT PP$;"PA";(X8 + 1);",";(Y8 + 1);";"
7660 PL$ =  STR$ (P(J)): REM  CHANGE POINT NO.
7670  PRINT PP$;"LB" + PL$ + EP$ + ";": REM  LABEL POINT NO.
7680  PRINT PP$;"PA";(X8 - 2);",";(Y8 - 1);";"
7685  LET Z8 = Z(J) * 1Ø ^ 2:Z8 =  INT (Z8) / 1Ø ^ 2
7690 PZ$ =  STR$ (Z8): REM  CHANGE LEVEL
77ØØ  PRINT PP$;"LB" + PZ$ + EP$ + ";": REM  PRINT LEVEL
7710  POKE 49184,255: REM  DESELECT EXPANSION CHASSIS
7720  PRINT D$;"PR#Ø": PRINT D$;"IN#Ø"
7730  RETURN
```

3.4.11 3-D tacheometry plot – data plot

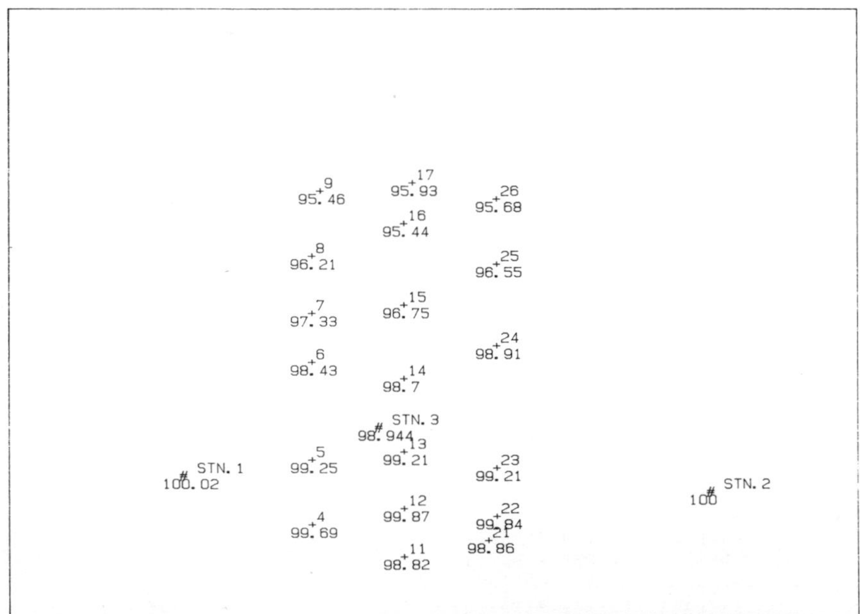

Fig. 3.8. Plot of 3-D tacheometric readings by microcomputer.

3.5 LOCATION OF STATIONS IN 2-D BY INTERSECTION

In the previous section, the location of a survey station was found by determining its distance and level from two known reference stations using tacheometry. Other co-ordinate geometry methods are also available for determining the location of an unknown station using a combination of distance, bearing and angle measurements from two known reference stations.

In the following program <INTERSECTIONS>, five different options are presented:

<1> Distance/Distance
<2> Bearing/Bearing
<3> Bearing/Distance
<4> Bearing/Angle
<5> Angle/Angle

The relationship between each of the stations, and their angles, bearings and distances are shown in Fig. 3.9.

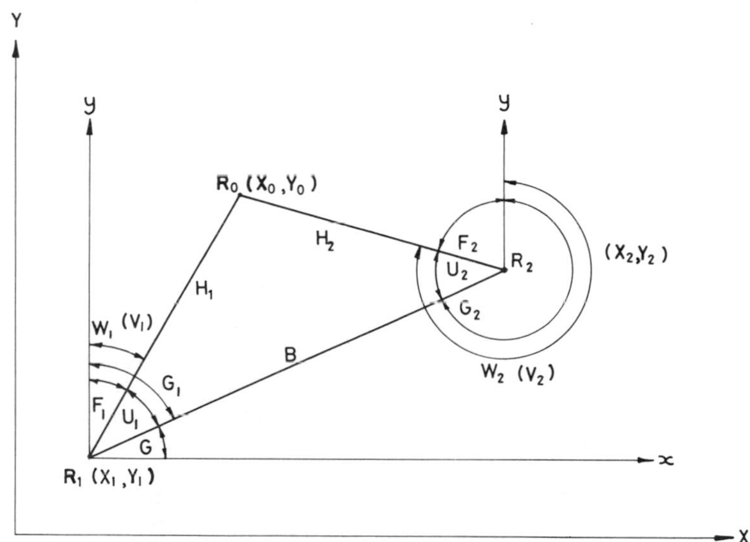

Fig. 3.9. Nomenclature for 2-D intersection solutions from two reference stations.

3.5.1 2-D intersection solutions – subroutine index

	Line numbers	*Function*
(a)	10–90	Initialization and control
(b)	100–190	Screen header display
(c)	200–240	Selection of angle format
(d)	300–430	Menu selection for intersections program

	Line numbers	*Function*
(e)	440–490	Option to change earlier reference stations
(f)	500–680	Entry of survey location, etc.
(g)	700–990	Entry of reference station data
(h)	1000–1980	<1> Distance/Distance Intersection
(i)	2000–2990	<2> Bearing/Bearing Intersection
(j)	3000–3990	<3> Bearing/Distance Intersection
(k)	4000–4550	<4> Bearing/Angle Intersection
(l)	4700–4740	Routine to check perpendicular
(m)	4750–4790	Routine to check perpendicular
(n)	4800–4890	Routine to find bearings
(o)	4900–4990	Routine to find bearings
(p)	5000–5290	<5> Angle/Angle Intersection
(q)	6000–6080	Routine to print survey location, etc.
(r)	6250–6740	Routine to print intersection solutions
(s)	6800–6940	Routine to display angles of triangle
(t)	7000–7080	Intersection solution with $X_1 = X_2$
(u)	7100–7180	Intersection solution with $Y_1 = Y_2$
(v)	7200–7380	Intersection solution with $Y_1 < Y_2$
(w)	7400–7570	Intersection solution with $Y_1 > Y_2$
(x)	7600–7740	Routine to find gradient of baseline
(y)	7750–7790	Triangle solution (A_1, S_1, A_3)
(z)	7800–7870	Triangle solution (S_1, S_2, S_3)
(aa)	7900–7970	Routines to find new station co-ordinates from R_1
(bb)	8000–8040	Routine to convert DD.MMSS angles to decimals
(cc)	8050–8140	Routine to convert decimal angles to DD.MMSS
(dd)	8150–8220	Routine to find reduced bearing from WCB
(ee)	8250–8320	Routine to find WCB from reduced bearing
(ff)	8350–8390	Routine to find angle at R_1
(gg)	8400–8440	Routine to find angle at R_2
(hh)	8460–8480	Routine to store WCB
(ii)	8500–8520	Routine to store reduced bearings
(jj)	8550–8640	Routine for screen display of results
(kk)	8650–8750	Routine to determine WCB and reduced bearings
(ll)	8800–8880	Routine to determine reduced bearings, etc.
(mm)	8900–8940	Routine to enter survey station
(nn)	9000–9010	Routine to clear screen
(oo)	9020–9090	Routines to find new station co-ordinates from R_2
(pp)	9100–9130	Routine to display error message
(qq)	9200–9290	Error trace routine
(rr)	9300–9360	Triangle solution ($S1$, $S2$, $A2$) – first
(ss)	9400–9460	Triangle solution ($S1$, $S2$, $A2$) – second
(tt)	9500–9620	Termination of program

(a) Initialization and control

Line numbers 10–90

All numeric variables and required string variables are initialized and two 'DEF FN' functions used in lines 80 and 90 to define arccosine and arcsine as discussed previously in Section 2.2.8.

(b) Screen header display

Line numbers 100–190

This is the screen display used for civil engineering students at Strathclyde University. The operator can alter this as required to suit another organization.

(c) Selection of angle format

Line numbers 200–240

The operator has a choice in line 200 of entering angles in decimal or DD.MMSS format, where the *D*9 flag is set to 1 for decimals and 2 for DD.MMSS in line 210.

(d) Menu selection for intersection program

Line numbers 300–430

A choice of five separate programs is offered in lines 320–380. On the first occasion the menu is presented ($N = 0$), the program branches to line 500 to allow the survey location, etc. to be recorded.

(e) Option to change earlier reference stations

Line numbers 440–490

This option is only presented on the second and subsequent displays of the menu, allowing several intersection programs to be run using the same reference stations. If required, the reference stations can be changed by branching to line 700.

(f) Entry of survey location, etc.

Line numbers 500–680

This routine is only accessed the first time the program is run to allow a record to be kept of site location, operator, date, etc. After entry, line 650 gives an option of a printout if required in the subroutine at line 6000.

(g) Entry of reference station data

Line numbers 700–990

As in the previous tacheometry programs, the reference station with the smaller X and/or Y co-ordinate is entered first. After entry, the baseline length B between the two stations is calculated from Equation 3.19 in line 980.

(h) <1> Distance/Distance Intersection

Line numbers 1000–1980

At the outset, the operator is asked to select the final output of the results, either in whole circle bearings (WCB) or reduced bearings, thus setting the flag *W*9 to 1 or 2 in line 1050. The reference number of the survey station is entered in line 1090. The distances from the two reference stations to the survey station, H_1 and H_2 respectively, as shown in Fig. 3.9, are entered in line numbers 1140 and 1210. A check is made in line 1310 to ensure that there is a solution where the sum of H_1 and H_2 is greater than B the baseline and, if not, an error message is displayed.

Knowing the three sides B, H_1 and H_2 of the triangle R_2, R_1, R_0 in Fig. 3.9 allows the triangle to be solved for angles U_1 and U_2 using the ($S1$, $S2$, $S3$) triangle solution given in an earlier program. The subroutine for this triangle solution is at line 7800.

It is now necessary to determine the angle of inclination or gradient of the baseline B with the X-axis, shown as angle G in Fig. 3.9. There can be four orientations of the baseline:

(i) $X_1 = X_2$, i.e. baseline vertical, $J = 1$
(ii) $Y_1 = Y_2$, i.e. baseline horizontal, $J = 2$
(iii) $Y_1 < Y_2$, i.e. baseline positive gradient, $J = 3$
(iv) $Y_1 > Y_2$, i.e. baseline negative gradient, $J = 4$

For each orientation, knowing the value of G, the whole circle bearings of the lines from R_1 to R_2 and R_2 to R_1 are calculated and stored in G_1 and G_2 respectively using the subroutine at line 7600. Depending upon the value of J the program then branches at line 1410 to subroutines to compute the co-ordinates of the two solutions, one either side of the baseline.

The first solution to this intersection program is displayed in lines 1500–1730 with a printout option at line 1740. The second solution is then displayed in lines 1790–1930 using the same display header for the first solution from lines 1500–1590. Again a printout option is given in line 1940 before returning to the menu. In each solution the angles of the triangle are calculated together with the whole circle bearings or reduced bearings of the lines using various subroutines as listed.

(i) <2> Bearing/Bearing Intersection

Line numbers 2000–2990

As in (h) above, either whole circle bearings or reduced bearings are selected at the beginning, followed by the entry for the survey station.

The WCB option starts at line 2100 and the bearings are entered in decimals as W_1 and W_2 at line numbers 2120 and 2260 respectively. If the WCBs are in DD.MMSS then the bearings are entered as Z_1 and Z_2 at line numbers 2160 and 2300 with subsequent conversion to W_1 and W_2 in decimals for calculation using the subroutine at line 8000. Once the WCBs in decimals (now designated V_1, V_2) have been calculated, the bearing quadrants (see Fig. 3.1) and the reduced bearings are computed using the subroutine at line 8150.

The reduced bearing option starts at line 2400 where the bearings are entered in decimals as F_1 and F_2 at line numbers 2420 and 2560 respectively. If the reduced bearings are in DD.MMSS then the bearings are entered as P_1 and P_2 at line numbers 2460 and 2600 with subsequent conversion to F_1 and F_2 in decimals for calculation using the routine at line 8000. Once the reduced bearings in decimals have been calculated, the quadrant and WCB of the bearings (now designated V_1, V_2) are determined using the subroutine at line 8250.

Both the WCB and reduced bearing calculations merge at line 2700, where the first calculation using the subroutine at line 7600 determines the gradient (G) of the baseline between the reference stations and the WCBs of the baseline from each reference station, G_1 and G_2 respectively, as shown in Fig. 3.9.

Knowing G_1 and G_2 together with the WCBs V_1 and V_2, the angles U_1 and U_2 of the triangle R_0, R_1, R_2 (Fig. 3.9) can be determined using the subroutines at line numbers 8350 and 8400. As the baseline B is also known, the other sides H_1 and H_2 can be found using the <TRIANGLE> solution for ($A1$, $S1$, $A3$) in the subroutine at line 7750.

From a knowledge of the co-ordinates of the first reference station, the reduced bearing of the line to the survey station and its distance, the co-ordinates of the survey station are found from line 2760. These values are then rounded to the third decimal figure for the screen display starting at line 2800, with an option to obtain a printout at line 2950.

(j) <3> Bearing/Distance Intersection

Line numbers 3000–3990

As before, either WCB or reduced bearings are selected at the beginning, followed by the entry for the survey station.

If the bearing is from the first reference station (R_1) the program follows through lines 3080–3390, and if from the second reference station (R_2), lines 3400–3690. In each case the bearing, either WCB or reduced bearing is entered in either decimals or DD.MMSS and the angle converted to decimals before determining the quadrant (Fig. 3.1). After entering the horizontal distance from the other reference station, the program checks that the distance is greater than the perpendicular distance to the line to give a solution using the subroutines at line numbers 4700 and 4750.

The triangle R_0, R_1, R_2 in Fig. 3.9 can now be solved using the <TRIANGLE> solution ($S1$, $S2$, $A2$). If the baseline B is greater than the distance measurement, H_1 or H_2, lines 3320 and 3620, there will be two solutions in a similar manner to Fig. 3.3. The first triangle solution subroutine is at line 9300 and the second at line 9400.

Once the other side of the triangle has been found, depending on the quadrant in which the recorded bearing lies, the co-ordinates of the point R_0 are found using lines 3350 or 3660 for both solutions if applicable.

Both bearing/distance programs merge at line 3700 with a screen display of the results to each solution, with a printout option if required.

(k) <4> Bearing/Angle Intersection

Line numbers 4000–4550

As before, either WCB or reduced bearings are selected at the beginning. Use is made of the bearing entry in the previous intersection program <3> by branching from line 4040 to 3040 returning with decimal angles and bearing quadrants as indicated in line 4045.

The gradient of the baseline and the WCBs G_1 and G_2 (Fig. 3.9) from each end of the baseline are then found using the subroutine at line 4050. If the bearing is from reference station R_1, then the angle U_1 is found using the subroutine at line 8350 and then angle U_2 entered from the keyboard. In the triangle R_0, R_1, R_2, both angles U_1 and U_2 are known at either end of the baseline B, line 4150, giving the triangle solution ($A1$, $S1$, $A3$) using the subroutine at line 7750 to give H_1 and H_2. From this data the co-ordinates of the survey station R_0 can be determined in line 4180. A similar procedure is followed if the bearing is from reference station R_2 where the solution commences at line 4220 having branched from 4060. Once U_1 and U_2 have been calculated the program jumps back to re-use the lines 4140–4210, both solutions returning to line 4300 for the screen display in the same format as before.

(l) Routine to check perpendicular

Line numbers 4700–4740

From a knowledge of the angle U_1 and the baseline B in triangle R_0, R_1, R_2 (Fig. 3.9) the perpendicular distance from the second reference point R_2 to the bearing from R_1 can be calculated from:

$$H_9 = B \sin U_1 \tag{3.21}$$

If the entry for H_2 is less than H_9 there can be no solution and an error message is displayed in line 4720 with an opportunity to return and re-enter.

(m) Routine to check perpendicular

Line numbers 4750–4790

This routine is similar to (l) above where U_2 replaces U_1 in Equation 3.21.

(n) Routine to find bearings

Line numbers 4800–4890

This routine is designed to find the WCBs and reduced bearings from reference station 2 when only bearings from reference station 1 are known.

(o) Routine to find bearings

Line numbers 4900–4990

This routine is similar to (n) above with the reference stations reversed.

(p) <5> Angle/Angle Intersection

Line numbers 5000–5290

As before, either WCB or reduced bearings are selected at the beginning, followed by the entry for the survey station. The two angles U_1 and U_2, Fig. 3.9 are then entered in line 5080 and 5140 either in decimals or DD.MMSS. If the latter, the DD.MMSS angles are converted to decimals to use the triangle solution ($A1$, $S1$, $A3$) since U_1, B and U_2 are known, and the other two sides H_1, H_2 are solved by the subroutine at line 7750. The subroutine at line 7600 is then used to find the gradient of the baseline and the bearings from each end of the baseline.

Knowing the value of J for the baseline, two solutions are obtained either side of the baseline, from the subroutines called in line number 5240. Since the screen display for program <1> with two solutions is similar to this program <5>, it is possible to use the previous program lines from 1520–1980.

(q) Routine to print survey location, etc.

Line numbers 6000–6080

This routine is provided for record purposes, and is based on Applesoft DOS and may require alteration for different microcomputers and printers.

(r) Routine to print intersection solutions

Line numbers 6250–6740

The format for the printout has been standardized so that it can be used by each of the programs 1–5 selected from the menu, giving initially the type of intersection solution, lines 6300–6340, and then whether a first or second solution, lines 6460–6470. The chosen format, WCB or reduced bearing is also checked and if the DD.MMSS angle format was used, the computed angles and bearings in decimals are converted back to DD.MMSS with the subroutine at line 8050.

(s) Routine to display angles of triangle

Line numbers 6800–6940

This additional routine has been used in conjunction with the screen display routine (jj) at line numbers 8550–8640 to give the angles U_1 and U_2 whether or not they were entered in either of the angle intersection solutions <4> and <5>.

(t) Intersection solution with $X_1 = X_2$

Line numbers 7000–7080

This is the case of a vertical baseline ($J = 1$), Fig. 3.10(a). The first solution lies to the left or west of the line and is stored in (X_9, Y_9). The second solution lies to the right or east of the line with co-ordinates (X_0, Y_0).

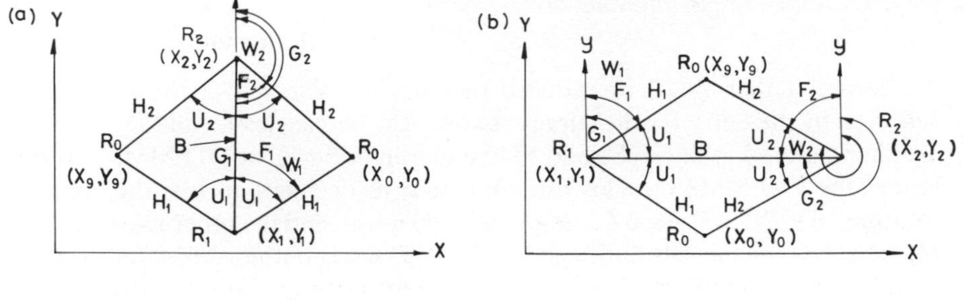

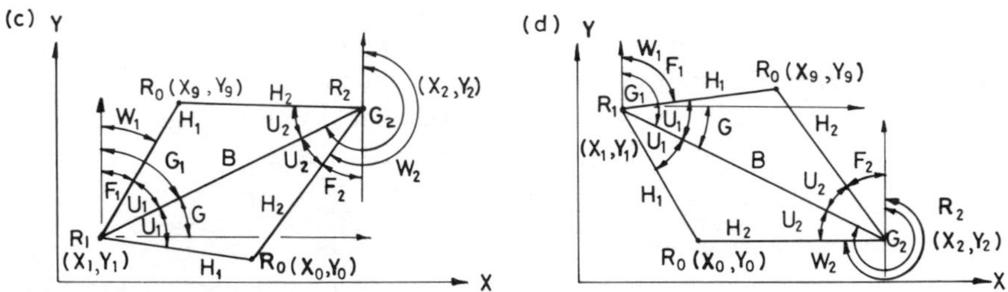

Fig. 3.10. Nomenclature for four (a)–(d) types of reference baseline in the 2-D intersection solutions.

(u) Intersection solution with $Y_1 = Y_2$

Line numbers 7100–7180

This is the case of a horizontal baseline ($J = 2$), Fig. 3.10(b). The first solution lies above or north of the baseline and is stored in (X_9, Y_9). The second solution lies below or south of the line with co-ordinates (X_0, Y_0).

(v) Intersection solution with $Y_1 < Y_2$

Line numbers 7200–7380

This is the case of a baseline with a positive gradient from left to right ($J = 3$), Fig. 3.10(c). The first and second solution locations were described in (u) above.

(w) Intersection solution with $Y_1 > Y_2$

Line numbers 7400–7570

This is the case of a baseline with a negative gradient from left to right ($J = 4$), Fig. 3.10(c). The first and second solution locations were described in (u) above.

(x) Routine to find gradient of baseline

Line numbers 7600–7740

By comparing the values of X_1 and X_2 or Y_1 and Y_2 the gradient of the baseline can be categorized and given a J value (see Sections (s)–(v)). If the baseline is vertical or horizontal the computation of the bearings from each station of the other G_1 and G_2 are simple. If however $J = 3$ or 4, that is, an inclined baseline, the gradient G of the line has to be found as discussed previously, Fig. 3.6, before G_1 and G_2 can be calculated.

(y) Triangle solution (*A*1, *S*1, *A*3)

Line numbers 7750–7790

The solution of this triangle configuration, Fig. 3.2, was discussed previously in Section <2> of the <TRIANGLE> program.

(z) Triangle solution (*S*1, *S*2, *S*3)

Line numbers 7800–7870

The solution of this triangle configuration, Fig. 3.2, was previously discussed in Section <1> of the <TRIANGLE> program.

(aa) Routine to find new station co-ordinates from R_1

Line numbers 7900–7970

Each of these four small routines, one for each of the four quadrants, Fig. 3.1(b), calculates the co-ordinates of the new survey station from R_1 knowing X_1, Y_1, H_1 and F_1, the latter being the reduced bearing relative to the N–S meridian.

(bb) Routine to convert DD.MMSS angles to decimals

Line numbers 8000–8040

This routine was discussed previously in Section 2.2.6.

(cc) Routine to convert decimal angles to DD.MMSS

Line numbers 8050–8140

This routine was discussed previously in Section 2.2.6.

(dd) Routine to find reduced bearing from WCB

Line numbers 8150–8220

If the WCB format of angle entry is chosen, the reduced bearings are calculated and the quadrant noted (Figs 3.1(a) and (b)) for future calculations.

(ee) Routine to find WCB from reduced bearing

Line numbers 8250–8320

If the reduced bearing format for angle entry is chosen, the WCBs are calculated and the quadrant noted (Figs 3.1(a) and (b)) for future calculations.

(ff) Routine to find angle at R_1

Line numbers 8350–8390

This routine will find the angle U_1 knowing the values of V_1 and G_1 shown in Fig. 3.9.

(gg) Routine to find angle at R_2

Line numbers 8400–8440

This routine will find the angle U_2 knowing the values of V_2 and G_2 shown in Fig. 3.9.

(hh) Routine to store WCB

Line numbers 8460–8480

At the final display and printout the angles Z_8 and Z_9 are used to store either W_1 and W_2 or Z_1 and Z_2.

(ii) Routine to store reduced bearings

Line numbers 8500–8520

At the final display and printout, the angles Z_8 and Z_9 are used to store either F_1 and F_2 or P_1 and P_2.

(jj) Routine for screen display of results

Line numbers 8550–8640

To simplify presentation of the screen display each time, this routine was written in a standard format.

(kk) Routine to determine WCB and reduced bearings

Line numbers 8650–8750

From a knowledge of quadrant and reduced bearing (Fig. 3.1(b)) the WCB is found.

(ll) Routine to determine reduced bearings, etc.

Line numbers 8800–8880

From a knowledge of G_2 and U_2 the WCB and reduced bearings from R_2 are found, Fig. 3.9.

(mm) Routine to enter survey station

Line numbers 8900–8940

This routine is provided to check that the input R_0 is not the same as either R_1 or R_2 and, if so, then an error message is displayed.

(nn) Routine to clear screen

Line numbers 9000–9010

To simplify portability the Applesoft 'HOME' command to clear the screen has been placed in a subroutine to allow an easy change to another microcomputer.

(oo) Routines to find new station co-ordinates from R_2

Line numbers 9020–9090

These routines are similar to those for station R_1 discussed in (aa) above.

(pp) Routine to display error message

Line numbers 9100–9130

If a wrong keyboard entry has been made, the error is trapped and this error message displayed and re-entry requested.

(qq) Error trace routine

Line numbers 9200–9290

This error trace routine was discussed earlier in Section 2.2.9.

(rr) Triangle solution ($S1$, $S2$, $A2$) – first

Line numbers 9300–9360

The solution of this triangle configuration, Fig. 3.3, was discussed previously in Section <5> of the <TRIANGLE> program.

(ss) Triangle solution ($S1$, $S2$, $A2$) – second

Line numbers 9400–9460

The second solution for this triangle configuration Fig. 3.3 where $S_2 > S_1$, that is $B > H_1$ or H_2, was discussed previously in Section <5> of the <TRIANGLE> program.

(tt) Termination of program

Line numbers 9500–9620

When the <QUIT> selection is made, either after the initial screen header or from the program menu, an end-of-program message is displayed to advise the operator that computations are complete.

3.5.2 2-D intersection solutions – numeric variables

*A*1, *A*2, *A*3	= Angles of triangle for triangle solutions
B	= Baseline distance between (*X*1, *Y*1) and (*X*2, *Y*2)
C	= Counter in error trace routine
*C*1, *C*2	= Angles subtended by reference stations at survey station
*C*9	= Defined function for arccosine
D	= Angle in degrees
*D*0, *D*5	= Temporary storage for DD.MMSS to decimals
*D*1	= Integer value of *D* degrees
*D*2	= Total number of seconds in fraction of *D* degrees
*D*9	= Menu selection for decimals <1> or DD.MMSS <2>
E	= Error number
*F*0, *F*5, *F*9	= Temporary storage of reduced bearings in decimals
*F*1, *F*2	= Reduced bearings in decimals
G, *G*1, *G*2	= Whole circle bearings of baselines
*H*1, *H*2	= Horizontal distances from reference stations to survey station
J	= 1 to 4 depending on gradient of baseline
K	= Counter for first and second solutions
M, *M*0, *M*1	= Minutes of angle in subroutine
P	= 1/2 perimeter of triangle
*P*1, *P*2, *P*5	= Reduced bearings in DD.MMSS
*P*7	= Printer code
Q, *Q*0, *Q*1, *Q*2	
*Q*5, *Q*9	= Storage of quadrant codes 1 to 4
*R*0, *R*1, *R*2	= Survey and reference station numbers
R	= 57.2957795 for degrees
S, *S*0	= Seconds of angle in subroutine
*S*1–*S*3	= Sides of triangle for triangle solutions
*S*9	= Defined function for arcsine
T	= Menu selection for intersection solution routine
*T*1–*T*3	= Temporary angle storage in defined function routines for arccosine and arcsine
*U*0–*U*2, *U*9	= Angles between baseline and survey station bearings
V–*V*2	= Calculated whole circle bearings in decimals
W, *W*5	= Temporary storage of whole circle bearings in decimals
*W*1, *W*2	= Whole circle bearings in decimals
*W*9	= Menu selection for whole circle bearings <1> or reduced bearings <2>
*X*0–*X*2	= *X*(*E*) co-ordinate of survey and reference stations
*X*5, *X*9	= Temporary storage of *X*(*E*) co-ordinates
*Y*0–*Y*2	= *Y*(*N*) co-ordinate of survey and reference stations
*Y*5, *Y*9	= Temporary storage of *Y*(*N*) co-ordinates
*Z*1, *Z*2	= Whole circle bearings in DD.MMSS
*Z*5	= Temporary storage of whole circle bearing in DD.MMSS
*Z*8, *Z*9	= Temporary storage of bearings for display and printout

3.5.3 2-D intersection solutions – string variables

A$	Storage for location of survey
B$	Storage for operator's name

Table—continued

C$	Storage for date of survey
D$	= CHR$(4), DOS command, CTRL-D
*E*0$	= "SORRY, DATA ERROR . . . PLEASE RE-ENTER"
E$, *E*1$, *E*2$	= Reduced bearing indicators "E" or "W"
K$	Storage for instrument used
N$, *N*1$, *N*2$	= Reduced bearing indicators "N" or "S"
P$	= "DO YOU WISH PRINTOUT OF ABOVE DATA (Y/N)"
Q$	= Question response (Y/N)
R$	= "ENTER REDUCED BEARING"
*R*0$	= "STN. RD.BR. DIST. NORTHING EASTING"
S$	= "SELECT <1> WHOLE CIRCLE BEARINGS, OR <2> REDUCED BEARINGS"
W$	= "ENTER WHOLE CIRCLE BEARING"
*W*0$	= "STN. W.C.B. DIST. NORTHING EASTING"
X$	= "ENTER X(E) COORD. OF STN."
Y$	= "ENTER Y(N) COORD. OF STN."

3.5.4 2-D intersection solutions – BASIC program

```
10  REM  <INTERSECTIONS> PROGRAM FOR APPLE II+  USES QUME PRINTER
20 D$ =  CHR$ (4): ONERR  GOTO 9200
30 A = 0:B = 0:E = 0:G = 0:J = 0:K = 0:N = 0:S = 0:U = 0:V = 0:W = 0
40 R0$ = "STN. RD.BR.  DIST.    EASTING    NORTHING":W0$ = "STN. W.C.B.  DI
     ST.  EASTING   NORTHING"
50 E0$ = "SORRY, DATA ERROR ... PLEASE RE-ENTER."
55 P$ = "DO YOU WISH PRINTOUT OF ABOVE DATA (Y/N)"
60 R$ = "ENTER REDUCED BEARING ":W$ = "ENTER WHOLE CIRCLE BEARING "
65 S$ = "SELECT <1> WHOLE CIRCLE BEARINGS,        OR      <2> REDUCED BEARI
     NGS."
70 X$ = "ENTER X(E) COORD. OF STN. ":Y$ = "ENTER Y(N) COORD. OF STN. "
80  DEF  FN C9(C9) = (1.5707964 -  ATN (C9 /  SQR (1 - C9 * C9))) * R: REM
     ARCCOSINE
90  DEF  FN S9(S9) =  ATN (S9 /  SQR (1 - S9 * S9)) * R: REM  ARCSINE
95  REM  ***************************************************
100  GOSUB 9000
110  PRINT "****************************************"
120  PRINT "*                                      *"
130  PRINT "*     UNIVERSITY  OF  STRATHCLYDE      *"
140  PRINT "*  DEPARTMENT  OF  CIVIL  ENGINEERING  *"
150  PRINT "*                                      *"
160  PRINT "*          SURVEYING  SECTION          *"
170  PRINT "*       INTERSECTION  SOLUTIONS        *"
180  PRINT "****************************************"
190  PRINT
200  PRINT "ARE ANGLES IN DECIMALS <1>, DD.MMSS <2> OR DO YOU WISH TO QUI
     T <3>."
210  INPUT D9
220  IF D9 < 1 OR D9 > 3 THEN 200
230  ON D9 GOTO 240,240,9500
240 R = 57.2957795
300  GOSUB 9000:K = 0:K2 = 0
310  PRINT
320  PRINT "SELECT INTERSECTION PROGRAM FROM MENU :-"
330  PRINT
340  PRINT "    <1> DISTANCE / DISTANCE"
350  PRINT "    <2> BEARING  / BEARING"
```

```
360  PRINT "      <3> BEARING  / DISTANCE"
370  PRINT "      <4> BEARING  / ANCLE"
380  PRINT "      <5> ANGLE    / ANGLE"
390  PRINT "      <6> QUIT"
400  INPUT T
410  IF T < 1 OR T > 6 THEN 300
420  IF T = 6 THEN 9500
430  IF N = 0 THEN 500
440  PRINT "DO YOU WISH TO CHANGE REFERENCE STATIONS";R1;" AND ";R2;" (Y/
     N).";
450  INPUT Q$
460  IF Q$ = "Y" THEN 700
470  IF Q$ < > "N" THEN 440
480 N = N + 1: REM  COUNTER
490  ON T GOTO 1000,2000,3000,4000,5000
500  GOSUB 9000
510  PRINT "UNIVERSITY OF STRATHCLYDE"
520  PRINT "INTERSECTIONS PROGRAM"
530  PRINT "****************************************"
540  PRINT
550  PRINT "ENTER LOCATION OF WORK, ETC. WHEN        REQUESTED, AND PRESS
     <RETURN>"
560  PRINT "ENTER LOCATION OF SURVEY :-"
570  INPUT A$
580  PRINT "ENTER OPERATOR'S NAME :-"
590  INPUT B$
600  PRINT "ENTER DATE OF SURVEY AS DD/MM/YY :-"
610  INPUT C$
620  PRINT "ENTER INSTRUMENT USED :-"
630  INPUT K$
640  PRINT
650  PRINT "DO YOU WISH PRINTOUT OF LOCATION ETC.,  (Y/N) ";
660  INPUT Q$
670  IF Q$ = "Y" THEN 6000
680  IF Q$ < > "N" THEN 650
690 :
700  GOSUB 9000
710  PRINT "ENTER DETAILS OF REFERENCE STNS. WHEN   REQUESTED, AND PRESS
     <RETURN> "
720  PRINT
730  PRINT "ENTER NO. OF REFERENCE STN. WITH SMALLER X AND/OR Y COORD., N
     UMERALS ONLY :- "
740  INPUT R1
750  PRINT X$;R1;
760  INPUT X1
770  PRINT Y$;R1;
780  INPUT Y1
810  PRINT
820  PRINT "ENTER NO. OF SECOND REFERENCE STN.      NUMERALS ONLY :-"
830  INPUT R2
840  IF R1 < > R2 THEN 870
850  GOSUB 9100
860  GOTO 820
870  PRINT X$;R2;
880  INPUT X2
890  IF X1 < = X2 THEN 920
900  GOSUB 9100
910  GOTO 730
920  PRINT Y$;R2;
```

```
 93Ø  INPUT Y2
 94Ø  IF X1 < X2 THEN 98Ø
 95Ø  IF Y1 < Y2 THEN 98Ø
 96Ø  GOSUB 91ØØ
 97Ø  GOTO 73Ø
 975  REM  CALCULATE BASELINE LENGTH B
 98Ø B =  SQR ((X2 - X1) ^ 2 + (Y2 - Y1) ^ 2)
 99Ø  GOTO 48Ø: REM  RETURN TO MENU SELECTION
 995  REM  **************************************************
1ØØØ  GOSUB 9ØØØ
1Ø1Ø  PRINT "<1> DISTANCE/DISTANCE PROGRAM :-"
1Ø2Ø  PRINT "****************************************"
1Ø3Ø  PRINT
1Ø4Ø  PRINT S$
1Ø5Ø  INPUT W9
1Ø6Ø  IF W9 < Ø OR W9 > 2 THEN 1ØØØ
1Ø7Ø :
1Ø8Ø  PRINT
1Ø9Ø  GOSUB 89ØØ
11ØØ  PRINT "DATA FOR REFERENCE STN. ";R1
111Ø  PRINT "******************************"
112Ø  PRINT
113Ø  PRINT "ENTER DISTANCE FROM ";R1;" TO ";RØ
114Ø  INPUT H1
115Ø  IF H1 <  = Ø THEN 113Ø
116Ø  PRINT
117Ø  PRINT "DATA FOR REFERENCE STN. ";R2
118Ø  PRINT "******************************"
119Ø  PRINT
12ØØ  PRINT "ENTER DISTANCE FROM ";R2;" TO ";RØ
121Ø  INPUT H2
122Ø  IF H2 <  = Ø THEN 12ØØ
123Ø  PRINT
124Ø  GOSUB 9ØØØ
125Ø  PRINT "DATA FOR SURVEY STATION ";RØ
126Ø  PRINT "========================================"
127Ø  PRINT
128Ø  PRINT "DIST. TO REF.STN. ";R1;" = ";H1
129Ø  PRINT "DIST. TO REF.STN. ";R2;" = ";H2
13ØØ  PRINT
13Ø5  REM  CHECK H1 + H2 > B ELSE ERROR
131Ø  IF H1 + H2 > B THEN 135Ø
132Ø  GOSUB 91ØØ
133Ø  PRINT "DISTANCES TO ";R1;" AND ";R2;" TOO SMALL"
134Ø  GOTO 11ØØ
135Ø  GOSUB 9ØØØ
136Ø  PRINT "CALCULATING ......"
1365  REM  CALCULATE ANGLES OF TRIANGLE
137Ø S1 = B:S2 = H1:S3 = H2
138Ø  GOSUB 78ØØ
139Ø U1 = A1:U2 = A3:C1 = A2
1395  REM  FIND GRADIENT OF BASELINE B
14ØØ  GOSUB 76ØØ
14Ø5  REM  RETURN WITH J AND G1 FOR R1
141Ø  ON J GOSUB 7ØØØ,71ØØ,72ØØ,74ØØ
1415  REM  RETURN WITH FIRST SOLUTION (X9,Y9) AND SECOND SOLUTION (XØ,YØ)

142Ø K = 2: REM  TWO SOLUTIONS
1495  REM  SCREEN DISPLAY OF RESULTS
```

```
15ØØ  GOSUB 9ØØØ
151Ø  PRINT "<1> DISTANCE/DISTANCE FIX ON STN. ";RØ
152Ø  PRINT "****************************************"
153Ø  PRINT
154Ø  IF W9 = 2 THEN 157Ø
155Ø  PRINT WØ$: REM  WHOLE CIRCLE HEADING
156Ø  GOTO 158Ø
157Ø  PRINT RØ$: REM  RED. BEARING HEADING
158Ø  PRINT "========================================"
159Ø  PRINT : IF K = 3 THEN 18ØØ
16ØØ  PRINT "FIRST SOLUTION :-"
161Ø  PRINT R1; TAB( 22);X1; TAB( 32);Y1
1615  REM  HAVE (X9,Y9) FIND F1,P1,W1,Z1,N1$,E1$
162Ø F5 = F9:Q5 = Q9
163Ø  GOSUB 865Ø
164Ø F1 = F5:P1 = P5:W1 = W5:Z1 = Z5:N1$ = N$:E1$ = E$
165Ø K = 2: REM  FIRST SOLUTION
1655  REM  NOW FIND F2,P2,W2,Z2,N2$,E2$
166Ø  GOSUB 88ØØ
167Ø  IF W9 = 2 THEN 17ØØ
168Ø  GOSUB 846Ø
169Ø  GOTO 171Ø
17ØØ  GOSUB 85ØØ
17Ø5  REM  STORE (X9,Y9) IN (X5,Y5) FOR DISPLAY
171Ø X5 =  INT (X9 * 1ØØØ + .5) / 1ØØØ:Y5 =  INT (Y9 * 1ØØØ + .5) / 1ØØØ
172Ø  GOSUB 855Ø
173Ø  PRINT
174Ø  PRINT P$
175Ø  INPUT Q$
176Ø  IF Q$ = "Y" THEN 625Ø
177Ø  IF Q$ <  > "N" THEN 174Ø
178Ø K = 3: REM  PRINTOUT NOT REQUIRED, GOTO DISPLAY OF SECOND SOLUTION
179Ø  GOTO 152Ø
18ØØ  PRINT "SECOND SOLUTION :-"
181Ø  PRINT R1; TAB( 22);X1; TAB( 32);Y1
1815  REM  HAVE (XØ,YØ) FIND F1,P1,W1,Z1,N1$,E1$
182Ø F5 = FØ:Q5 = QØ
183Ø  GOSUB 865Ø
184Ø F1 = F5:P1 = P5:W1 = W5:Z1 = Z5:N1$ = N$:E1$ = E$
185Ø K = 3
186Ø  GOSUB 88ØØ
187Ø  IF W9 = 2 THEN 19ØØ
188Ø  GOSUB 846Ø
189Ø  GOTO 191Ø
19ØØ  GOSUB 85ØØ
19Ø5  REM  STORE (XØ,YØ) IN (X5,Y5) FOR DISPLAY
191Ø X5 =  INT (XØ * 1ØØØ + .5) / 1ØØØ:Y5 =  INT (YØ * 1ØØØ + .5) / 1ØØØ
192Ø  GOSUB 855Ø
193Ø  PRINT
194Ø  PRINT P$
195Ø  INPUT Q$
196Ø  IF Q$ = "Y" THEN 625Ø
197Ø  IF Q$ <  > "N" THEN 194Ø
198Ø  GOTO 3ØØ
1995  REM  ****************************************************
2ØØØ  GOSUB 9ØØØ
2Ø1Ø  PRINT "<2> BEARING/BEARING PROGRAM :-"
2Ø2Ø  PRINT "****************************************"
2Ø3Ø  PRINT
```

```
2040  PRINT "SELECT <1> WHOLE CIRCLE BEARINGS,"
2050  PRINT "OR     <2> REDUCED BEARINGS."
2060  INPUT W9
2070  IF W9 < 0 OR W9 > 2 THEN 2000
2080  PRINT : GOSUB 8900: REM  R0 ENTRY
2090  IF W9 = 2 THEN 2400
2095  REM  W9=1, WHOLE CIRCLE BEARING ENTRY
2100  IF D9 = 2 THEN 2150
2105  REM  I.E. W.C.B. ANGLE IN DECIMALS
2110  PRINT W$;R1;" TO ";R0
2120  INPUT W1
2130  IF W1 < 0 OR W1 > 360 THEN 2110
2140  GOTO 2210
2145  REM  I.E. W.C.B. ANGLE IN DD.MMSS
2150  PRINT W$;R1;" TO ";R0
2160  INPUT Z1
2170  IF Z1 < 0 OR Z1 > 360 THEN 2150
2175  REM  CONVERT Z1 TO W1 IN DECIMALS
2180 D5 = Z1
2190  GOSUB 8000
2195  REM  STORE DECIMAL VALUE IN W1
2200 W1 = D5
2205  REM  FIND Q1 AND F1 FROM W1
2210 W = W1
2220  GOSUB 8150
2225  REM  STORE Q,F ETC IN Q1,F1 ETC
2230 Q1 = Q:F1 = F:V1 = W:N1$ = N$:E1$ = E$
2235  REM  NOW COLLECT SECOND W.C.B.
2240  IF D9 = 2 THEN 2290
2245  REM  I.E. W.C.B. ANGLE IN DECIMALS
2250  PRINT W$;R2;" TO ";R0
2260  INPUT W2
2270  IF W2 < 0 OR W2 > 360 THEN 2250
2280  GOTO 2350
2285  REM  I.E. W.C.B. ANGLE IN DD.MMSS
2290  PRINT W$;R2;" TO ";R0
2300  INPUT Z2
2310  IF Z2 < 0 OR Z2 > 360 THEN 2290
2315  REM  CONVERT Z2 TO W2 IN DECIMALS
2320 D5 = Z2
2330  GOSUB 8000
2335  REM  STORE DECIMAL VALUE IN W2
2340 W2 = D5
2345  REM  FIND Q2,F2 ETC FROM W2
2350 W = W2
2360  GOSUB 8150
2365  REM  STORE Q,F ETC IN Q2,F2 ETC
2370 Q2 = Q:F2 = F:V2 = W:N2$ = N$:E2$ = E$
2375  REM  NOW HAVE ALL W.C.B. INFORMATION SO GOTO SOLUTION
2380  GOTO 2700
2390  PRINT
2395  REM  W9=2, REDUCED BEARINGS ENTRY
2400  IF D9 = 2 THEN 2450
2405  REM  I.E. REDUCED BEARINGS IN DECIMALS
2410  PRINT R$;R1;" TO ";R0;" (N$,F1,E$)"
2420  INPUT N1$,F1,E1$
2430  IF F1 < 0 OR F1 > 90 THEN 2410
2440  GOTO 2510
2445  REM  I.E. REDUCED BEARINGS IN DD.MMSS
```

```
245Ø  PRINT R$;R1;" TO ";RØ;" (N$,P1,E$)"
246Ø  INPUT N1$,P1,E1$
247Ø  IF P1 < Ø OR P1 > 9Ø THEN 245Ø
2475  REM  CONVERT P1 TO F1 IN DECIMALS
248Ø D5 = P1
249Ø  GOSUB 8ØØØ
2495  REM  STORE DECIMAL ANGLE IN F1
25ØØ F1 = D5
25Ø5  REM  FIND Q1,V1 FROM F1 ETC
251Ø F = F1:N$ = N1$:E$ = E1$
252Ø  GOSUB 825Ø
2525  REM  STORE Q,V IN Q1,V1
253Ø Q1 = Q:V1 = V
2535  REM  NOW COLLECT SECOND BEARING
254Ø  IF D9 = 2 THEN 259Ø
2545  REM  I.E. BEARING IN DECIMALS
255Ø  PRINT R$;R2;" TO ";RØ;" (N$,F2,E$)"
256Ø  INPUT N2$,F2,E2$
257Ø  IF F2 < Ø OR F2 > 9Ø THEN 255Ø
258Ø  GOTO 265Ø
2585  REM  I.E. BEARING IN DD.MMSS
259Ø  PRINT R$;R2;" TO ";RØ;" (N$,P2,E$)"
26ØØ  INPUT N2$,P2,E2$
261Ø  IF P2 < Ø OR P2 > 9Ø THEN 259Ø
2615  REM  CONVERT P2 TO F2 IN DECIMALS
262Ø D5 = P2
263Ø  GOSUB 8ØØØ
2635  REM  STORE DECIMAL ANGLE IN F2
264Ø F2 = D5
2645  REM  FIND Q2,V2 FROM F2
265Ø F = F2:N$ = N2$:E$ = E2$
266Ø  GOSUB 825Ø
2665  REM  STORE Q,V IN Q2,V2
267Ø Q2 = Q:V2 = V
269Ø :
2695  REM  START OF SOLUTION - FIRST FIND GRADIENT OF BASELINE
27ØØ  GOSUB 76ØØ
27Ø5  REM  RETURN WITH G1,G2 NOW FIND U1,U2
271Ø  GOSUB 835Ø
272Ø  GOSUB 84ØØ
2725  REM  NOW KNOW U1,B,U2 IN TRIANGLE, SOLVE FOR H1,H2,C1
273Ø S1 = B:A1 = U1:A3 = U2
274Ø  GOSUB 775Ø
2745  REM  CONVERT BACK TO H1,H2 ETC
275Ø H1 = S2:H2 = S3:C1 = A2
2755  REM  NOW KNOW X1,Y1,H1,F1,Q1 FIND (XØ,YØ)
276Ø  ON Q1 GOSUB 79ØØ,792Ø,794Ø,796Ø
277Ø H1 =  INT (H1 * 1ØØØ + .5) / 1ØØØ:H2 =  INT (H2 * 1ØØØ + .5) / 1ØØØ
278Ø XØ =  INT (XØ * 1ØØØ + .5) / 1ØØØ:YØ =  INT (YØ * 1ØØØ + .5) / 1ØØØ
279Ø K = 1: REM  ONLY ONE SOLUTION
2795  REM  SCREEN DISPLAY OF RESULT
28ØØ  GOSUB 9ØØØ
281Ø  PRINT "<2> BEARING/BEARING FIX ON STN. ";RØ
282Ø  PRINT "****************************************"
283Ø  PRINT
284Ø  IF W9 = 2 THEN 287Ø
285Ø  PRINT WØ$
286Ø  GOTO 288Ø
287Ø  PRINT RØ$
```

```
2880  PRINT "========================================"
2890  PRINT R1; TAB( 22);X1; TAB( 32);Y1
2900  IF W9 = 2 THEN 2920
2910  GOSUB 8460: GOTO 2930
2920  GOSUB 8500
2930 X5 = X0:Y5 = Y0: GOSUB 8550
2940  PRINT
2950  PRINT P$
2960  INPUT Q$
2970  IF Q$ = "Y" THEN 6250
2980  IF Q$ <  > "N" THEN 2940
2990  GOTO 300
2995  REM  ***************************************************
3000  GOSUB 9000: PRINT "<3> BEARING/DISTANCE PROGRAM :-"
3010  PRINT "***************************************": PRINT
3020 H = 0: PRINT S$
3030  INPUT W9: IF W9 < 0 OR W9 > 2 THEN 3000
3040  PRINT : GOSUB 8900: REM  R0 ENTRY
3050  PRINT "BEARING FROM WHICH STN. ";R1;" OR ";R2
3060  INPUT R9: IF R9 = R2 THEN 3400
3070  IF R9 <  > R1 THEN 3050
3080  IF W9 = 2 THEN 3190
3085  REM  I.E. W9=1, W.C.B. ENTRY
3090  IF D9 = 2 THEN 3130
3095  REM  I.E. W.C.B. ANGLE IN DECIMALS FROM R1
3100  PRINT W$;R1;" TO ";R0
3110  INPUT W1: IF W1 < 0 OR W1 > 360 THEN 3100
3120  GOTO 3160
3125  REM  I.E. W.C.B. ANGLE IN DD.MMSS
3130  PRINT W$;R1;" TO ";R0
3140  INPUT Z1: IF Z1 < 0 OR Z1 > 360 THEN 3130
3145  REM  CONVERT Z1 TO W1 IN DECIMALS
3150 D5 = Z1: GOSUB 8000:W1 = D5
3155  REM  FIND Q1,F1 FROM W1
3160 W = W1: GOSUB 8150
3165  REM  STORE Q,F ETC. IN Q1,F1 ETC.
3170 Q1 = Q:F1 = F:V1 = W:N1$ = N$:E1$ = E$: IF H = 1 THEN  RETURN
3175  REM  IF H=1 ROUTINE USED BY SOLUTION <4> SO RETURN
3180  GOTO 3280
3185  REM  W9=2, RED. BEARING ENTRY
3190  IF D9 = 2 THEN 3230
3195  REM  I.E. RED. BEARING IN DECIMALS
3200  PRINT R$;R1;" TO ";R0;" (N$,F1,E$)"
3210  INPUT N1$,F1,E1$: IF F1 < 0 OR F1 > 90 THEN 3200
3220  GOTO 3260
3225  REM  I.E. RED. BEARING IN DD.MMSS
3230  PRINT R$;R1;" TO ";R0;" (N$,P1,E$)"
3240  INPUT N1$,P1,E1$: IF P1 < 0 OR P1 > 90 THEN 3230
3245  REM  CONVERT P1 TO F1 IN DECIMALS
3250 D5 = P1: GOSUB 8000:F1 = D5
3255  REM  FIND Q1,V1 FROM F1, ETC.
3260 F = F1:N$ = N1$:E$ = E1$
3265  REM  FIND Q,V AND STORE IN Q1,V1
3270  GOSUB 8250:Q1 = Q:V1 = V: IF H = 1 THEN  RETURN
3275  REM  NOW COLLECT DISTANCE H2
3280  PRINT "ENTER DISTANCE FROM ";R2;" TO ";R0
3290  INPUT H2: IF H2 < 0 THEN 3280
3295  REM  FIND GRADIENT G1,G2,J AND U1
3300  GOSUB 7600: GOSUB 8350: GOTO 4700
```

```
3305  REM  IF H9<H2 SOLVE TRIANGLE B,H2,U1 WITH (S1,S2,A2) SOLUTION
3310 S1 = H2:S2 = B:A2 = U1
3315  REM  IF S2 > S1 TWO SOLUTIONS K=1,2
3320 K = 1: IF S2 > S1 THEN K = 2
3330  GOSUB 9300
3340 H1 = S3:U2 = A1:U6 = A3
3345  REM  FIND (X0,Y0) FROM Q1
3350  ON Q1 GOSUB 7900,7920,7940,7960
3355  REM  RETURN WITH (X0,Y0) FOR DISPLAY
3360 X0 =  INT (X0 * 1000 + .5) / 1000:Y0 =  INT (Y0 * 1000 + .5) / 1000:
     H1 =  INT (H1 * 1000 + .5) / 1000: GOTO 3700
3365  REM  IF S2 > S1 THEN SECOND SOLUTION, K=3
3370 S1 = H2:S2 = B:A2 = U1
3375  REM  SOLVE TRIANGLE (S1,S2,A2)
3380  GOSUB 9400
3390 H1 = S3:U2 = A1:U6 = A3: GOTO 3350
3395  REM  SOLUTION FOR BEARING FROM R2
3400  IF W9 = 2 THEN 3500
3405  REM  W9=1, WHOLE CIRCLE BEARING
3410  IF D9 = 2 THEN 3450
3415  REM  I.E. W.C.B. IN DECIMALS
3420  PRINT W$;R2;" TO ";R0
3430  INPUT W2: IF W2 < 0 OR W2 > 360 THEN 3420
3440  GOTO 3480
3445  REM  I.E. W.C.B. ANGLE IN DD.MMSS
3450  PRINT W$;R2;" TO ";R0
3460  INPUT Z2: IF Z2 < 0 OR Z2 > 360 THEN 3450
3465  REM  CONVERT Z2 TO W2 IN DECIMALS
3470 D5 = Z2: GOSUB 8000:W2 = D5
3475  REM  FIND Q2,F2 FROM W2
3480 W = W2: GOSUB 8150:Q2 = Q:F2 = F:V2 = W:N2$ = N$:E2$ = E$: IF H = 1 THEN
      RETURN
3490  GOTO 3590
3495  REM  W9=2, RED. BEARING ENTRY
3500  IF D9 = 2 THEN 3540
3505  REM  RED. BEARING IN DECIMALS
3510  PRINT R$;R2;" TO ";R0;" (N$,F2,E$)"
3520  INPUT N2$,F2,E2$: IF F2 < 0 OR F2 > 90 THEN 3510
3530  GOTO 3570
3535  REM  I.E. RED. BEARING IN DD.MMSS
3540  PRINT R$;R2;" TO ";R0;" (N$,P2,E$)"
3550  INPUT N2$,P2,E2$: IF P2 < 0 OR P2 > 90 THEN 3540
3555  REM  CONVERT P2 TO F2 IN DECIMALS
3560 D5 = P2: GOSUB 8000:F2 = D5
3565  REM  FIND Q2,V2 FROM F2, ETC
3570 F = F2:N$ = N2$:E$ = E2$
3575  REM  STORE Q,V IN Q2,V2
3580  GOSUB 8250:Q2 = Q:V2 = V: IF H = 1 THEN  RETURN
3585  REM  NOW COLLECT DISTANCE H1
3590  PRINT "ENTER DISTANCE FROM ";R1;" TO ";R0
3600  INPUT H1: IF H1 < 0 THEN 3590
3605  REM  FIND GRADIENT G1,G2,J AND U2
3610  GOSUB 7600: GOSUB 8400: GOTO 4750
3615  REM  IF H9<H1 SOLVE TRIANGLE B,H1,U2 WITH (S1,S2,A2) SOLUTION
3620 S1 = H1:S2 = B:A2 = U2
3625  REM  IF S2 > S1 TWO SOLUTIONS K2=1,2
3630 K2 = 1: IF S2 > S1 THEN K2 = 2
3640  GOSUB 9300
3650 H2 = S3:U1 = A1:U6 = A3
```

```
3655  REM  FIND (XØ,YØ) FROM Q2
366Ø  ON Q2 GOSUB 9Ø2Ø,9Ø4Ø,9Ø6Ø,9Ø8Ø
3665  REM  RETURN WITH (XØ,YØ) FOR DISPLAY
367Ø XØ =  INT (XØ * 1ØØØ + .5) / 1ØØØ:YØ =  INT (YØ * 1ØØØ + .5) / 1ØØØ:
    H2 =  INT (H2 * 1ØØØ + .5) / 1ØØØ: GOTO 37ØØ
3675  REM  IF S2 > S1 THEN SECOND SOLUTION, K=3
368Ø S1 = H1:S2 = B:A2 = U2: GOSUB 94ØØ
369Ø H2 = S3:U1 = A1:U6 = A3: GOTO 366Ø
37ØØ  GOSUB 9ØØØ
371Ø  PRINT "<3> BEARING/DISTANCE FIX ON STN. ";RØ
372Ø  PRINT "****************************************"
373Ø  PRINT
374Ø  IF W9 = 2 THEN 377Ø
375Ø  PRINT WØ$
376Ø  GOTO 378Ø
377Ø  PRINT RØ$
378Ø  PRINT "========================================"
379Ø  PRINT
38ØØ  IF K = 1 OR K2 = 1 THEN 384Ø
381Ø  IF K = 3 OR K2 = 3 THEN 383Ø
3815  REM  IF K=2 OR K2=2 THEN TWO SOLUTIONS
382Ø  PRINT "FIRST SOLUTION :-": GOTO 384Ø
383Ø  PRINT "SECOND SOLUTION :-"
384Ø  PRINT R1; TAB( 22);X1; TAB( 32);Y1
3845  REM  FIND SECOND BEARING
385Ø  GOSUB 43ØØ: IF W9 = 2 THEN 387Ø
3855  REM  PREPARE ANGLES FOR DISPLAY
386Ø  GOSUB 846Ø: GOTO 388Ø
387Ø  GOSUB 85ØØ
388Ø X5 = XØ:Y5 = YØ: GOSUB 855Ø
389Ø  PRINT
39ØØ  PRINT P$
391Ø  INPUT Q$: IF Q$ = "Y" THEN 625Ø
392Ø  IF Q$ <  > "N" THEN 39ØØ
393Ø  IF K = 2 OR K2 = 2 THEN 399Ø
3935  IF K = 3 OR K2 = 3 THEN 398Ø
394Ø  IF D9 = 1 THEN 396Ø
3945  REM  CHANGE U1,U2 BACK TO DECIMALS
395Ø D5 = U1: GOSUB 8ØØØ:U1 = D5
3955 D5 = U2: GOSUB 8ØØØ:U2 = D5
396Ø  IF R9 = R1 AND K = 3 THEN 337Ø
397Ø  IF R9 = R2 AND K2 = 3 THEN 368Ø
3975  REM  RETURN TO MENU
398Ø  GOTO 3ØØ
399Ø K = 3:K2 = 3: GOTO 394Ø
3995  REM  **************************************************
4ØØØ  GOSUB 9ØØØ: PRINT "<4> BEARING/ANGLE PROGRAM :-"
4Ø1Ø  PRINT "****************************************"
4Ø2Ø  PRINT S$
4Ø3Ø  INPUT W9: IF W9 < Ø OR W9 > 2 THEN 4ØØØ
4Ø35  REM  USE ENTRY FOR <3> WITH H=1
4Ø4Ø H = 1: GOSUB 3Ø4Ø:K = 1:K2 = 1
4Ø45  REM  RETURNS FROM LINES 318Ø,327Ø,349Ø OR 358Ø WITH Q1,F1,V1,N1$,E1
    $ OR Q2,F2,V2,N2$,E2$
4Ø5Ø  GOSUB 76ØØ: REM  FIND J,G1,G2
4Ø55  REM  FIND IF R9=R1 OR R2
4Ø6Ø  IF R9 = R2 THEN 422Ø
4Ø65  REM  I.E. BEARING FROM R1,FIND U1
4Ø7Ø  GOSUB 835Ø
```

```
4080  PRINT "ENTER ANGLE BETWEEN ";R2;"-";R1;" AND ";R2;"-";RØ
4090  IF D9 = 2 THEN 412Ø
4095  REM  I.E. D9=1, ANGLE IN DECIMALS
41ØØ  INPUT U2: IF U2 < Ø OR U2 > 18Ø THEN 4Ø8Ø
411Ø  GOTO 414Ø
4115  REM  I.E. D9=2, ANGLE IN DD.MMSS
412Ø  INPUT UØ: IF UØ < Ø OR UØ > 18Ø THEN 4Ø8Ø
4125  REM  CONVERT UØ TO U2 IN DECIMALS
413Ø D5 = UØ: GOSUB 8ØØØ:U2 = D5
4135  REM  ARRIVE WITH B,U1,U2 FOR TRIANGLE SOLUTION
414Ø  GOSUB 9ØØØ: PRINT "CALCULATING ..."
415Ø A1 = U1:S1 = B:A3 = U2: GOSUB 775Ø
4155  REM  RETURN WITH S2,S3,A2
416Ø H1 = S2:H2 = S3:C1 = A2
4165  REM  FIND (XØ,YØ) FROM Q1,Q2
417Ø  IF R9 = R2 THEN 419Ø
4175  REM  NOW KNOW X1,Y1,H1,F1,Q1 FIND (XØ,YØ)
418Ø  ON Q1 GOSUB 79ØØ,792Ø,794Ø,796Ø: GOTO 42ØØ
419Ø  ON Q2 GOSUB 9Ø2Ø,9Ø4Ø,9Ø6Ø,9Ø8Ø
42ØØ XØ =  INT (XØ * 1ØØØ + .5) / 1ØØØ:YØ =  INT (YØ * 1ØØØ + .5) / 1ØØØ:
     H1 =  INT (H1 * 1ØØØ + .5) / 1ØØØ:H2 =  INT (H2 * 1ØØØ + .5) / 1ØØØ
421Ø  GOTO 43ØØ
4215  REM  BEARING FROM R2, FIND U2
422Ø  GOSUB 84ØØ
423Ø  PRINT "ENTER ANGLE BETWEEN ";R1;"-";R2;" AND ";R1;"-"RØ
424Ø  IF D9 = 2 THEN 427Ø
4245  REM  I.E. D9=1, ANGLE IN DECIMALS
425Ø  INPUT U1: IF U1 < Ø OR U1 > 18Ø THEN 423Ø
426Ø  GOTO 429Ø
4265  REM  I.E. D9=2, ANGLE IN DD.MMSS
427Ø  INPUT UØ: IF UØ < Ø OR UØ > 18Ø THEN 423Ø
4275  REM  CONVERT UØ TO U1 IN DECIMALS
428Ø D5 = UØ: GOSUB 8ØØØ:U1 = D5
4285  REM  ARRIVE WITH B,U1,U2 FOR TRIANGLE SOLUTION
429Ø  GOTO 414Ø
4295  REM  PREPARE DATA FOR DISPLAY
43ØØ  IF R9 = R2 THEN 432Ø
43Ø5  REM  R9=R1, BEARING FROM R1, KNOW W1 FIND W2
431Ø  GOSUB 48ØØ: GOTO 433Ø
4315  REM  R9=R2, BEARING FROM R2, KNOW W2 FIND W1
432Ø  GOSUB 49ØØ
4325  REM  LINES 43ØØ-434Ø USED BY SOLUTION <3> FOR K,K2 = 1,2,3
433Ø  IF T = 4 THEN 435Ø
434Ø  RETURN
4345  REM  DISPLAY OF RESULTS
435Ø  GOSUB 9ØØØ
436Ø  PRINT "<4> BEARING/ANGLE FIX ON STN. ";RØ
437Ø  PRINT "****************************************"
438Ø  PRINT
439Ø  IF W9 = 2 THEN 442Ø
44ØØ  PRINT WØ$
441Ø  GOTO 443Ø
442Ø  PRINT RØ$
443Ø  PRINT "========================================"
444Ø  PRINT
445Ø  PRINT R1; TAB( 22);X1; TAB( 32);Y1
4455  IF W9 = 2 THEN 447Ø
446Ø  GOSUB 846Ø: GOTO 448Ø
4465  REM  I.E. R9=R1, BEARING FROM R1, ANGLE FROM R2
```

```
447Ø  GOSUB 85ØØ
4475  REM  ANGLE UØ IN DD.MMSS CONVERT TO U2
448Ø X5 = XØ:Y5 = YØ
449Ø  GOSUB 855Ø
45ØØ  PRINT
45Ø5  REM  I.E. R9=R2, BEARING FROM R2, ANGLE FROM R1
451Ø  PRINT P$
4515  REM  ANGLE UØ IN DD.MMSS CONVERT TO U1
452Ø  INPUT Q$
453Ø  IF Q$ = "Y" THEN 625Ø
454Ø  IF Q$ < > "N" THEN 451Ø
4545  REM  RETURN TO MENU
455Ø  GOTO 3ØØ
4693  REM  ****************************************************
4695  REM  ROUTINE TO CHECK TRIANGLE
47ØØ H9 = B *  SIN (U1 / R)
471Ø  IF H2 > = H9 THEN 331Ø
4715  REM  IF H2 < H9 NO SOLUTION
472Ø  PRINT "NO SOLUTION :-"
473Ø  GOSUB 91ØØ
4735  REM  RETURN AND RE-ENTER H2
474Ø  GOTO 328Ø
4743  REM  ****************************************************
4745  REM  ROUTINE TO CHECK TRIANGLE
475Ø H9 = B *  SIN (U2 / R)
476Ø  IF H1 > = H9 THEN 362Ø
4765  REM  IF H1 < H9 NO SOLUTION
477Ø  PRINT "NO SOLUTION :-"
478Ø  GOSUB 91ØØ
4785  REM  RETURN AND RE-ENTER H1
479Ø  GOTO 359Ø
4793  REM  ****************************************************
4795  REM  ROUTINE TO FIND W2,Z2,F2,P2 FROM W1,G1,G2,U2
48ØØ  IF W1 < G1 OR W1 > G1 + 18Ø THEN 482Ø
48Ø5  REM  I.E. W1 > G1 AND < G1+18Ø
481Ø W2 = G2 - U2: GOTO 483Ø
482Ø W2 = G2 + U2: IF W2 > 36Ø THEN W2 = W2 - 36Ø
4825  REM  CHECK IF W.C.B. OR RED. BEARING
483Ø  IF W9 = 2 THEN 486Ø
4835  REM  I.E. W9=1, WHOLE CIRCLE BEARING
484Ø  IF D9 = 1 THEN  RETURN
4845  REM  I.E. W.C.B. IN DD.MMSS
485Ø D = W2: GOSUB 8Ø5Ø:Z2 = D: RETURN
4855  REM  I.E. W9=2, RED. BEARINGS
486Ø W = W2: GOSUB 815Ø
487Ø Q2 = Q:F2 = F:N2$ = N$:E2$ = E$
488Ø  IF D9 = 1 THEN  RETURN
4885  REM  I.E. RED. BEARING IN DD.MMSS
489Ø D = F2: GOSUB 8Ø5Ø:P2 = D: RETURN
4893  REM  ****************************************************
4895  REM  ROUTINE TO FIND W1,Z1,F1,P1 FROM W2,G1,G2,U1
49ØØ  IF W2 < G2 - 18Ø OR W2 > G2 THEN 492Ø
49Ø5  REM  I.E. W2 > G2-18Ø OR < G2
491Ø W1 = G1 + U1: GOTO 493Ø
492Ø W1 = G1 - U1: IF W1 < Ø THEN W1 = 36Ø + W1
4925  REM  CHECK IF W.C.B. OR RED. BEARING
493Ø  IF W9 = 2 THEN 496Ø
4935  REM  I.E. W9=1, WHOLE CIRCLE BEARING
494Ø  IF D9 = 1 THEN  RETURN
```

```
4945  REM  I.E. D9=2, W.C.B. IN DD.MMSS
495Ø D = W1: GOSUB 8Ø5Ø:Z1 = D: RETURN
4955  REM  I.E. W9=2, RED. BEARING
496Ø W = W1: GOSUB 815Ø
497Ø Q1 = Q:F1 = F:N1$ = N$:E1$ = E$
498Ø  IF D9 = 1 THEN  RETURN
4985  REM  I.E. D9=2, RED. BEARING IN DD.MMSS
499Ø D = F1: GOSUB 8Ø5Ø:P1 = D: RETURN
4993  REM  ******************************************************
5ØØØ  GOSUB 9ØØØ: PRINT "<5> ANGLE/ANGLE PROGRAM :-"
5Ø1Ø  PRINT "***************************************"
5Ø2Ø  PRINT S$
5Ø3Ø  INPUT W9
5Ø4Ø  IF W9 < Ø OR W9 > 2 THEN 5ØØØ
5Ø5Ø  PRINT
5Ø6Ø  GOSUB 89ØØ: REM  RØ ENTRY
5Ø7Ø  PRINT
5Ø8Ø  PRINT "ENTER ANGLE BETWEEN ";R1;"-";R2;" AND ";R1;"-";RØ
5Ø9Ø  IF D9 = 2 THEN 512Ø
5Ø95  REM  I.E. D9=1, ANGLE IN DECIMALS
51ØØ  INPUT U1: IF U1 < Ø OR U1 > 18Ø THEN 5Ø8Ø
511Ø  GOTO 514Ø
5115  REM  I.E. D9=2, ANGLE IN DD.MMSS
512Ø  INPUT UØ: IF UØ < Ø OR UØ > 18Ø THEN 5Ø8Ø
5125  REM  CONVERT UØ TO U1 IN DECIMALS
513Ø D5 = UØ: GOSUB 8ØØØ:U1 = D5
5135  REM  NOW COLLECT U2
514Ø  PRINT "ENTER ANGLE BETWEEN ";R2;"-";R1;" AND ";R2;"-";RØ
515Ø  IF D9 = 2 THEN 518Ø
5155  REM  I.E. D9=1, ANGLE IN DECIMALS
516Ø  INPUT U2: IF U2 < Ø OR U2 > 18Ø THEN 514Ø
517Ø  GOTO 52ØØ
5175  REM  I.E. D9=2, ANGLE IN DD.MMSS
518Ø  INPUT U9: IF U9 < Ø OR U9 > 18Ø THEN 514Ø
5185  REM  CONVERT U9 TO U2 IN DECIMALS
519Ø D5 = U9: GOSUB 8ØØØ:U2 = D5
5195  REM  ARRIVE WITH B,U1,U2 FOR TRIANGLE SOLUTION (A1,S1,A3)
52ØØ A1 = U1:S1 = B:A3 = U2
521Ø  GOSUB 775Ø
522Ø H1 = S2:H2 = S3:C1 = A2
5225  REM  NOW FIND J,G1,G2
523Ø  GOSUB 76ØØ
5235  REM  FIND SOLUTIONS FROM J
524Ø  ON J GOSUB 7ØØØ,71ØØ,72ØØ,74ØØ
5245  REM  RETURN WITH TWO SOLUTIONS, FIRST(LEFT) AT (X9,Y9), SECOND(RIGH
     T) AT (XØ,YØ)
525Ø H1 =  INT (H1 * 1ØØØ + .5) / 1ØØØ:H2 =  INT (H2 * 1ØØØ + .5) / 1ØØØ
526Ø K = 2: REM  FIRST SOLUTION
5265  REM  USE DISPLAY FOR PROGRAM <1>
527Ø  GOSUB 9ØØØ
528Ø  PRINT "<5> ANGLE/ANGLE FIX ON STN. ";RØ
529Ø  GOTO 152Ø
5985  REM  ************************************************************
599Ø  REM  PRINTOUT OF LOCATION, ETC
6ØØØ  PRINT D$;"PR#1": PRINT "LOCATION OF SURVEY :- ";A$
6Ø1Ø  PRINT "********************************************************
     *********************"
6Ø2Ø  PRINT "OPERATOR'S NAME :- ";B$
6Ø3Ø  PRINT "DATE OF SURVEY  :- ";C$
```

```
6Ø4Ø  PRINT "INSTRUMENT USED :- ";K$
6Ø6Ø  PRINT "**************************************************************
     ********************"
6Ø7Ø  PRINT : PRINT D$;"PR#Ø"
6Ø8Ø  GOTO 7ØØ
6243  REM  ************************************************
6245  REM  PRINTOUT OF COORDINATES ON QUME
625Ø  PRINT D$;"PR#1": POKE 1784 + 1,8Ø
626Ø  PRINT :P7 = 1: REM  PRINTER CODE
627Ø  PRINT "INTERSECTION PROGRAM RESULTS :-"
629Ø  ON T GOTO 63ØØ,631Ø,632Ø,633Ø,634Ø
63ØØ  PRINT "<1> DISTANCE/DISTANCE FIX ON STN. ";RØ: GOTO 635Ø
631Ø  PRINT "<2> BEARING/BEARING FIX ON STN. ";RØ: GOTO 635Ø
632Ø  PRINT "<3> BEARING/DISTANCE FIX ON STN. ";RØ: GOTO 635Ø
633Ø  PRINT "<4> BEARING/ANGLE FIX ON STN. ";RØ: GOTO 635Ø
634Ø  PRINT "<5> ANGLE/ANGLE FIX ON STN. ";RØ:
635Ø  PRINT "************************************************"
638Ø  IF W9 = 2 THEN 641Ø
639Ø  PRINT WØ$: REM  W.C.B.
64ØØ  GOTO 642Ø
641Ø  PRINT RØ$: REM  RED. BEARING
642Ø  PRINT "====================================================="
644Ø  IF K = 1 OR K2 = 1 THEN 649Ø
645Ø  IF K = 3 OR K2 = 3 THEN 647Ø
6455  REM  I.E. K=2, TWO SOLUTIONS
646Ø K = 3:K2 = 3: PRINT "FIRST SOLUTION :-": GOTO 649Ø
647Ø K = 2:K2 = 2: PRINT "SECOND SOLUTION :-"
649Ø  PRINT R1;: POKE 36,22: PRINT X1;: POKE 36,32: PRINT Y1
65Ø5  REM  W9=1 FOR W.C.B., W9=2 FOR RED.BEARING
651Ø  IF W9 = 1 THEN 653Ø
6515  REM  W9=2, RED. BEARING
652Ø  POKE 36,4: PRINT N1$;Z8;E1$;: POKE 36,14: PRINT H1: GOTO 654Ø
653Ø  POKE 36,4: PRINT Z8;: POKE 36,14: PRINT H1
654Ø :
655Ø  PRINT RØ;: POKE 36,22: PRINT X5;: POKE 36,32: PRINT Y5
656Ø :
657Ø  IF W9 = 1 THEN 659Ø
658Ø  POKE 36,4: PRINT N2$;Z9;E2$;: POKE 36,14: PRINT H2: GOTO 66ØØ
659Ø  POKE 36,4: PRINT Z9;: POKE 36,14: PRINT H2
66ØØ :
661Ø  PRINT R2;: POKE 36,22: PRINT X2;: POKE 36,32: PRINT Y2
663Ø  PRINT "====================================================="
665Ø  POKE 36,4: PRINT U1;: POKE 36,14: PRINT "= ANGLE ";R1;"-";R2;" AND
     ";R1;"-";RØ
667Ø  POKE 36,4: PRINT U2;: POKE 36,14: PRINT "= ANGLE ";R2;"-";R1;" AND
     ";R2;"-";RØ
669Ø  PRINT "*****************************************************"
67ØØ :
67Ø5  REM  RETURN TO SCREEN
671Ø  PRINT D$;"PR#Ø"
6715  REM  CHECK IF SECOND SOLUTION
672Ø  IF K = 1 OR K2 = 1 THEN 3ØØ
6725  REM  IF K,K2=1,2 RETURN TO MENU
673Ø  IF K = 2 OR K2 = 2 THEN 3ØØ
6735  REM  I.E. K,K2=3, SECOND SOLUTION
674Ø  ON T GOTO 15ØØ,3ØØ,394Ø,3ØØ,178Ø
6793  REM  ************************************************
6795  REM  PREPARE U1,U2 FOR DISPLAY
68ØØ  IF D9 = 2 THEN 683Ø
```

```
6805  REM  D9=1, ANGLES IN DECIMALS
6810 U1 =  INT (U1 * 1000 + .5) / 1000:U2 =  INT (U2 * 1000 + .5) / 1000
6820  GOTO 6900
6825  REM  I.E. D9=2, CONVERT ANGLES TO DD.MMSS
6830  IF K = 1 OR K = 2 THEN 6870
6835  REM  FIRST TIME CHANGE ANGLE
6840  IF T = 3 AND K = 3 THEN 6870
6845  REM  SECOND TIME ONLY CHANGE ANGLE IF T=3 AND K=3
6850  IF T = 1 AND K = 3 THEN 6900
6860  IF T = 5 AND K = 3 THEN 6900
6870 D = U1: GOSUB 8050:U1 = D
6880 D = U2: GOSUB 8050:U2 = D
6890 :
6900  PRINT  TAB( 4);U1; TAB( 14);" = ANGLE ";R1;"-";R2;" AND ";R1;"-";R0

6910  PRINT
6920  PRINT  TAB( 4);U2; TAB( 14);" = ANGLE ";R2;"-";R1;" AND ";R2;"-";R0

6930  PRINT "****************************************"
6940  RETURN
6993  REM  ***************************************************
6995  REM  J=1, VERTICAL BASELINE
7000  IF U1 > 90 THEN 7050
7010 F1 = U1: GOSUB 7960: REM  Q1=4
7015  REM  STORE FIRST SOLUTION IN X9 ETC
7020 X9 = X0:Y9 = Y0:F9 = F1:Q9 = 4
7025  REM  GET SECOND SOLUTION, Q1=1
7030  GOSUB 7900
7035  REM  SECOND SOLUTION USES X0 ETC
7040 F0 = F1:Q0 = 1: RETURN
7045  REM  U1 > 90 POINT BELOW Y1
7050 F1 = 180 - U1: GOSUB 7940: REM  Q1=3
7055  REM  STORE FIRST SOLUTION IN X9 ETC
7060 X9 = X0:Y9 = Y0:F9 = F1:Q9 = 3
7065  REM  GET SECOND SOLUTION, Q1=2
7070  GOSUB 7920
7075  REM  SECOND SOLUTION USES X0 ETC
7080 F0 = F1:Q0 = 3: RETURN
7093  REM  ***************************************************
7095  REM  J=2, BASELINE HORIZONTAL, G1=90
7100  IF U1 > 90 THEN 7150
7110 F1 = G1 - U1: GOSUB 7900: REM  Q1=1
7115  REM  STORE FIRST SOLUTION IN X9 ETC
7120 X9 = X0:Y9 = Y0:F9 = F1:Q9 = 1
7125  REM  GET SECOND SOLUTION, Q1=2
7130  GOSUB 7920
7135  REM  SECOND SOLUTION USES X0 ETC
7140 F0 = F1:Q0 = 2: RETURN
7145  REM  U1 > 90, POINT TO LEFT OF X1
7150 F1 = U1 - G1: GOSUB 7960: REM  Q1=4
7155  REM  STORE FIRST SOLUTION IN X9 ETC
7160 X9 = X0:Y9 = Y0:F9 = F1:Q9 = 4
7165  REM  GET SECOND SOLUTION, Q1=3
7170  GOSUB 7940
7175  REM  SECOND SOLUTION USES X0 ETC
7180 F0 = F1:Q0 = 3: RETURN
7193  REM  ***************************************************
7195  REM  J=3, BASELINE POSITIVE GRADIENT FROM LEFT TO RIGHT
7200 F1 = G1 - U1
```

```
7210  IF F1 >  = 0 THEN 7260
7220 F1 = F1 * ( - 1)
7230  IF F1 <  = 90 THEN 7280
7240 F1 = 180 - F1:Q9 = 3: GOSUB 7940
7250  GOTO 7290
7260 Q9 = 1: GOSUB 7900
7270  GOTO 7290
7280 Q9 = 4: GOSUB 7960
7285  REM  STORE FIRST SOLUTION IN X9 ETC
7290 X9 = X0:Y9 = Y0:F9 = F1
7300 F1 = G1 + U1
7310  IF F1 <  = 90 THEN 7350
7320  IF F1 <  = 180 THEN 7370
7330 F1 = F1 - 180:Q0 = 3: GOSUB 7940
7340  GOTO 7380
7350 Q0 = 1: GOSUB 7900
7360  GOTO 7380
7370 F1 = 180 - F1:Q0 = 2: GOSUB 7920
7375  REM  SECOND SOLUTION USES X0 ETC
7380 F0 = F1: RETURN
7393  REM  ****************************************************
7395  REM  J=4, BASELINE NEGATIVE GRADIENT FROM LEFT TO RIGHT
7400 F1 = G1 - U1
7410  IF F1 < 0 THEN 7450
7420  IF F1 <  = 90 THEN 7470
7430 F1 = 180 - F1:Q9 = 2: GOSUB 7920
7440  GOTO 7480
7450 F1 = F1 * ( - 1):Q9 = 4: GOSUB 7960
7460  GOTO 7480
7470 Q9 = 1: GOSUB 7900
7475  REM  STORE FIRST SOLUTION IN X9 ETC
7480 X9 = X0:Y9 = Y0:F9 = F1
7490 F1 = G1 + U1
7500  IF F1 <  = 180 THEN 7540
7510  IF F1 <  = 270 THEN 7560
7520 F1 = 360 - F1:Q0 = 4: GOSUB 7960
7530  GOTO 7570
7540 F1 = 180 - F1:Q0 = 2: GOSUB 7920
7550  GOTO 7570
7560 F1 = F1 - 180:Q0 = 3: GOSUB 7940
7565  REM  SECOND SOLUTION USES X0 ETC
7570 F0 = F1: RETURN
7593  REM  ****************************************************
7595  REM  ROUTINE TO FIND J,G1 AND G2, GRADIENT OF BASELINE
7600  IF X1 = X2 THEN 7650
7610  IF Y1 = Y2 THEN 7670
7620  IF Y1 < Y2 THEN 7690
7630  IF Y1 > Y2 THEN 7720
7640 :
7645  REM  J=1, BASELINE VERTICAL
7650 J = 1:G1 = 0:G2 = 180
7660  RETURN
7665  REM  J=2, BASELINE HORIZONTAL
7670 J = 2:G1 = 90:G2 = 270
7680  RETURN
7685  REM  J=3, BASELINE +VE GRADIENT
7690 G =  ATN ((Y2 - Y1) / (X2 - X1)) * R
7700 J = 3:G1 = 90 - G:G2 = 270 - G
7710  RETURN
```

```
7715  REM  J=4, BASELINE -VE GRADIENT
772Ø G =  ATN ((Y1 - Y2) / (X2 - X1)) * R
773Ø J = 4:G1 = 9Ø + G:G2 = 27Ø + G
774Ø  RETURN
7743  REM  ***************************************************
7745  REM  <TRIANGLE> SOLUTION (A1,S1,A3)
775Ø T2 =  -  COS ((A3 + A1) / R)
776Ø A2 =  FN C9(T2)
777Ø S2 = S1 *  SIN (A3 / R) /  SIN (A2 / R)
778Ø S3 = S1 *  COS (A3 / R) + S2 *  COS (A2 / R)
779Ø  RETURN
7793  REM  ***************************************************
7795  REM  <TRIANGLE> SOLUTION (S1,S2,S3)
78ØØ P = (S1 + S2 + S3) / 2
781Ø T3 =  SQR (P * (P - S2) / (S1 * S3))
782Ø A3 = 2 *  FN C9(T3)
783Ø T2 =  SQR (P * (P - S1) / (S2 * S3))
784Ø A2 = 2 *  FN C9(T2)
785Ø T1 =  -  COS ((A3 + A2) / R)
786Ø A1 =  FN C9(T1)
787Ø  RETURN
7893  REM  ***************************************************
7895  REM  ROUTINES TO FIND POINT COORDS
79ØØ XØ = X1 + H1 *  SIN (F1 / R):YØ = Y1 + H1 *  COS (F1 / R)
791Ø  RETURN
792Ø XØ = X1 + H1 *  SIN (F1 / R):YØ = Y1 - H1 *  COS (F1 / R)
793Ø  RETURN
794Ø XØ = X1 - H1 *  SIN (F1 / R):YØ = Y1 - H1 *  COS (F1 / R)
795Ø  RETURN
796Ø XØ = X1 - H1 *  SIN (F1 / R):YØ = Y1 + H1 *  COS (F1 / R)
797Ø  RETURN
7993  REM  ***************************************************
7995  REM  ROUTINE DD.MMSS TO DECIMALS
8ØØØ DØ =  INT (D5)
8ØØ5  REM  FIND NO. OF MINUTES,MØ
8Ø1Ø M1 = D5 * 1ØØ - DØ * 1ØØ:MØ =  INT (M1 + .1)
8Ø15  REM  FIND NO. OF SECONDS,SØ
8Ø2Ø SØ =  INT (M1 * 1ØØ - MØ * 1ØØ + .5)
8Ø25  REM  COLLECT DEGS,MINS,SECS
8Ø3Ø D5 = DØ + MØ / 6Ø + SØ / 36ØØ
8Ø4Ø  RETURN
8Ø43  REM  ***************************************************
8Ø45  REM  ROUTINE DECIMAL TO DD.MMSS
8Ø5Ø D1 =  INT (D)
8Ø55  REM  FIND TOTAL NO. OF SECONDS
8Ø6Ø D2 = (D - D1) * 36ØØ
8Ø65  REM  FIND NO. OF MINUTES
8Ø7Ø M =  INT (D2 / 6Ø)
8Ø75  REM  FIND NO. OF SECONDS
8Ø8Ø S = D2 - 6Ø * M + .5
8Ø9Ø  IF S < 6Ø THEN 813Ø
81ØØ M = M + 1:S = Ø
811Ø  IF M < 6Ø THEN 813Ø
812Ø D1 = D1 + 1:M = Ø
813Ø D = D1 * 1ØØØØ + M * 1ØØ + S
814Ø D =  INT (D) / 1ØØØØ: RETURN
8143  REM  ***************************************************
8145  REM  ROUTINE TO FIND QUADRANT,Q AND REDUCED BEARING,F FROM W.C.B.
815Ø  IF W <  = 9Ø THEN 819Ø
```

```
8160  IF W > 90 AND W <  = 180 THEN 8200
8170  IF W > 180 AND W <  = 270 THEN 8210
8180  IF W > 270 THEN 8220
8185  REM  FIRST QUADRANT
8190 Q = 1:F = W:N$ = "N":E$ = "E": RETURN
8195  REM  SECOND QUADRANT
8200 Q = 2:F = 180 - W:N$ = "S":E$ = "E": RETURN
8205  REM  THIRD QUADRANT
8210 Q = 3:F = W - 180:N$ = "S":E$ = "W": RETURN
8215  REM  FOURTH QUADRANT
8220 Q = 4:F = 360 - W:N$ = "N":E$ = "W": RETURN
8243  REM  **************************************************
8245  REM  ROUTINE TO FIND QUADRANT,Q AND W.C.B. FROM REDUCED BEARING
8250  IF N$ = "N" AND E$ = "E" THEN 8290
8260  IF N$ = "S" AND E$ = "E" THEN 8300
8270  IF N$ = "S" AND E$ = "W" THEN 8310
8280  IF N$ = "N" AND E$ = "W" THEN 8320
8285  REM  FIRST QUADRANT
8290 Q = 1:V = F: RETURN
8295  REM  SECOND QUADRANT
8300 Q = 2:V = 180 - F: RETURN
8305  REM  THIRD QUADRANT
8310 Q = 3:V = 180 + F: RETURN
8315  REM  FOURTH QUADRANT
8320 Q = 4:V = 360 - F: RETURN
8343  REM  **************************************************
8345  REM  ROUTINE TO FIND U1 FROM V1,G1
8350  IF V1 > G1 + 180 THEN 8380
8360  IF V1 > G1 THEN 8390
8370 U1 = G1 - V1: RETURN
8380 U1 = 360 - V1 + G1: RETURN
8390 U1 = V1 - G1: RETURN
8393  REM  **************************************************
8395  REM  ROUTINE TO FIND U2 FROM V2,G2
8400  IF V2 > G2 THEN 8430
8410  IF V2 > G2 - 180 THEN 8440
8420 U2 = V2 + 360 - G2: RETURN
8430 U2 = V2 - G2: RETURN
8440 U2 = G2 - V2: RETURN
8453  REM  **************************************************
8455  REM  ROUTINE FOR W.C.B. ANGLES Z8,Z9
8460  IF D9 = 2 THEN 8480
8470 Z8 =  INT (W1 * 1000 + .5) / 1000:Z9 =  INT (W2 * 1000 + .5) / 1000:
     RETURN
8480 Z8 = Z1:Z9 = Z2: RETURN
8493  REM  **************************************************
8495  REM  ROUTINE FOR RD.BR. ANGLES Z8,Z9
8500  IF D9 = 2 THEN 8520
8510 Z8 =  INT (F1 * 1000 + .5) / 1000:Z9 =  INT (F2 * 1000 + .5) / 1000:
     RETURN
8520 Z8 = P1:Z9 = P2: RETURN
8543  REM  **************************************************
8545  REM  SCREEN DISPLAY Z8,Z9
8550  IF W9 = 1 THEN 8570
8560  PRINT  TAB( 4);N1$;Z8;E1$; TAB( 14);H1: GOTO 8580
8570  PRINT  TAB( 4);Z8; TAB( 14);H1
8580  PRINT R0; TAB( 22);X5; TAB( 32);Y5
8590  IF W9 = 1 THEN 8610
8600  PRINT  TAB( 4);N2$;Z9;E2$; TAB( 14);H2: GOTO 8620
```

```
8610  PRINT  TAB( 4);Z9; TAB( 14);H2
8620  PRINT R2; TAB( 22);X2; TAB( 32);Y2
8630  PRINT "========================================"
8640  GOSUB 6800: RETURN
8643  REM  ****************************************************
8645  REM  ROUTINE TO DETERMINE W.C.B. AND RD.BR. FROM Q5 AND F5
8650  IF Q5 = 1 THEN W5 = F5
8660  IF Q5 = 2 THEN W5 = 180 - F5
8670  IF Q5 = 3 THEN W5 = 180 + F5
8680  IF Q5 = 4 THEN W5 = 360 - F5
8685  REM  CONVERT W5 TO DD.MMSS
8690 D = W5: GOSUB 8050:Z5 = D
8695  REM  NOW FIND RED. BEARINGS
8700  IF Q5 = 1 THEN N$ = "N": IF Q5 = 1 THEN E$ = "E"
8710  IF Q5 = 2 THEN N$ = "S": IF Q5 = 2 THEN E$ = "E"
8720  IF Q5 = 3 THEN N$ = "S": IF Q5 = 3 THEN E$ = "W"
8730  IF Q5 = 4 THEN N$ = "N": IF Q5 = 4 THEN E$ = "W"
8735  REM  CONVERT F5 TO DD.MMSS
8740 D = F5: GOSUB 8050:P5 = D
8745  REM  RETURN WITH F5,P5,W5,Z5,N$,E$
8750  RETURN
8793  REM  ****************************************************
8795  REM  GOSUB ROUTINE FOR F2,P2,W2,Z2,N2$,E2$ FROM G2 AND U2
8800  IF K = 3 THEN 8830
8810 W2 = G2 + U2
8820  IF W2 >  = 360 THEN W2 = W2 - 360
8830  IF K = 3 THEN W2 = G2 - U2
8835  REM  CONVERT W2 TO DD.MMSS
8840 D = W2: GOSUB 8050:Z2 = D
8845  REM  FIND RED. BEARINGS
8850 W = W2: GOSUB 8150
8860 Q2 = Q:F2 = F:N2$ = N$:E2$ = E$
8865  REM  CONVERT F2 TO DD.MMSS
8870 D = F2: GOSUB 8050:P2 = D
8880  RETURN
8893  REM  ****************************************************
8895  REM  ROUTINE TO ENTER STATION R0
8900  PRINT "ENTER SURVEY STATION NO., NUMERALS ONLY"
8910  INPUT R0
8920  IF R0 <  > R1 AND R0 <  > R2 THEN  RETURN
8930  GOSUB 9100
8940  GOTO 8900
8985  REM  ******************************************
8990  REM  GOSUB ROUTINE TO CLEAR SCREEN
9000  HOME
9010  RETURN
9013  REM  ****************************************************
9015  REM  ROUTINES TO FIND (X0,Y0) FROM Q2
9020 X0 = X2 + H2 *  SIN (F2 / R):Y0 = Y2 + H2 *  COS (F2 / R)
9030  RETURN
9040 X0 = X2 + H2 *  SIN (F2 / R):Y0 = Y2 - H2 *  COS (F2 / R)
9050  RETURN
9060 X0 = X2 - H2 *  SIN (F2 / R):Y0 = Y2 - H2 *  COS (F2 / R)
9070  RETURN
9080 X0 = X2 - H2 *  SIN (F2 / R):Y0 = Y2 + H2 *  COS (F2 / R)
9090  RETURN
9093  REM  ****************************************************
9095  REM  GOSUB ROUTINE FOR ERRORS
9100  PRINT
```

```
9110  PRINT EØ$
9120  PRINT
9130  RETURN
9193  REM  ***************************************************
9195  REM  ERROR TRACE ROUTINE
9200  PRINT  CHR$ (7): REM  CTRL-B (BELL)
9210 E =  PEEK (222): REM  GET ERROR NO.
9220  IF C = 1 THEN 9260
9230  INVERSE
9240  PRINT "ERROR NO. ";E;" FOUND"
9250 C = 1: TRACE : RESUME
9260  PRINT "ERROR ON SECOND LINE NO. "
9270  NORMAL
9280  NOTRACE
9290  STOP
9293  REM  **************************************************
9295  REM  TRIANGLE SOLUTION NO.<5>-FIRST
9300 T3 = S2 / S1 *  SIN (A2 / R)
9310 A3 =  FN S9(T3)
9320 T1 =  -  COS ((A3 + A2) / R)
9330 A1 =  FN C9(T1)
9340 S3 = S1 *  COS (A3 / R) + S2 *  COS (A2 / R)
9345  REM  STORE DECIMAL VALUES OF A3 AND A2 IN D3,D4 FOR SECOND SOLUTION
9350 D3 = A3:D4 = A2
9360  RETURN
9393  REM  **************************************************
9395  REM  TRIANGLE SOLUTION NO.<6>-SECOND
9400 T3 =  -  COS (D3 / R)
9410 A3 =  FN C9(T3)
9415  REM  RECALL DECIMAL VALUE A2
9420 A2 = D4
9430 T1 =  -  COS ((A3 + A2) / R)
9440 A1 =  FN C9(T1)
9450 S3 = S1 *  COS (A3 / R) + S2 *  COS (A2 / R)
9460  RETURN
9493  REM  ***************************************************
9495  REM  END OF PROGRAM
9500  GOSUB 9000
9510  PRINT
9520  PRINT "END INTERSECTION PROGRAM"
9530  PRINT
9540  PRINT "*************************"
9550  REM  PROGRAM PREPARED BY
9560  REM  DR. P.H. MILNE
9570  REM  DEPT. OF CIVIL ENGINEERING
9580  REM  UNIVERSITY OF STRATHCLYDE
9590  REM  GLASGOW G4 ONG
9600  REM  SCOTLAND
9610  REM  ***************************
9620  END
```

3.5.5 2-D intersection solutions – computer printout

```
LOCATION OF SURVEY :- STRATHCLYDE UNIVERSITY - STEELHENCE
**********************************************************************
OPERATOR'S NAME :- JOHN SMITH
DATE OF SURVEY  :- 12/05/83
INSTRUMENT USED :- ZEISS ZENA T20A + CD6
**********************************************************************
```

```
INTERSECTION PROGRAM RESULTS :-
<1> DISTANCE/DISTANCE FIX ON STN. 3
**************************************************
STN. W.C.B.  DIST.    EASTING   NORTHING
===================================================
FIRST SOLUTION :-
1                     100       100
    38.3834   180.376
3                     212.638   240.883
    296.2707  121.542
2                     321.456   186.743
===================================================
    29.5802   = ANGLE 1-2 AND 1-3
    47.5031   = ANGLE 2-1 AND 2-3
***************************************************

INTERSECTION PROGRAM RESULTS :-
<1> DISTANCE/DISTANCE FIX ON STN. 3
**************************************************
STN. W.C.B.  DIST.    EASTING   NORTHING
===================================================
SECOND SOLUTION :-
1                     100       100
    98.3438   180.376
3                     278.359   73.099
    201.0625  121.542
2                     321.456   186.743
===================================================
    29.5802   = ANGLE 1-2 AND 1-3
    47.5031   = ANGLE 2-1 AND 2-3
***************************************************

INTERSECTION PROGRAM RESULTS :-
<2> BEARING/BEARING FIX ON STN. 3
**************************************************
STN. W.C.B.  DIST.    EASTING   NORTHING
===================================================
1                     100       100
    38.3834   180.376
3                     212.638   240.884
    296.2707  121.542
2                     321.456   186.743
===================================================
    29.5802   = ANGLE 1-2 AND 1-3
    47.5031   = ANGLE 2-1 AND 2-3
***************************************************

INTERSECTION PROGRAM RESULTS :-
<3> BEARING/DISTANCE FIX ON STN. 3
**************************************************
STN. W.C.B.  DIST.    EASTING   NORTHING
===================================================
FIRST SOLUTION :-
1                     100       100
    38.3834   231.707
3                     244.692   280.976
    320.4959  121.542
2                     321.456   186.743
```

```
====================================================
    29.5802    = ANGLE 1-2 AND 1-3
    72.1323    = ANGLE 2-1 AND 2-3
****************************************************

INTERSECTION PROGRAM RESULTS :-
<3> BEARING/DISTANCE FIX ON STN. 3
**************************************************
STN. W.C.B.  DIST.    EASTING   NORTHING
====================================================
SECOND SOLUTION :-
1                        100        100
    38.3834   180.377
3                        212.639    240.884
    296.2709  121.542
2                        321.456    186.743
====================================================
    29.5802    = ANGLE 1-2 AND 1-3
    47.5033    = ANGLE 2-1 AND 2-3
****************************************************

INTERSECTION PROGRAM RESULTS :-
<3> BEARING/DISTANCE FIX ON STN. 3
**************************************************
STN. W.C.B.  DIST.    EASTING   NORTHING
====================================================
FIRST SOLUTION :-
1                        100        100
    14.1541   180.376
3                        144.435    274.817
    296.2707  197.721
2                        321.456    186.743
====================================================
    54.2055    = ANGLE 1-2 AND 1-3
    47.5031    = ANGLE 2-1 AND 2-3
****************************************************

INTERSECTION PROGRAM RESULTS :-
<3> BEARING/DISTANCE FIX ON STN. 3
**************************************************
STN. W.C.B.  DIST.    EASTING   NORTHING
====================================================
SECOND SOLUTION :-
1                        100        100
    38.3833   180.376
3                        212.637    240.884
    296.2707  121.543
2                        321.456    186.743
====================================================
    29.5803    = ANGLE 1-2 AND 1-3
    47.5031    = ANGLE 2-1 AND 2-3
****************************************************

INTERSECTION PROGRAM RESULTS :-
<4> BEARING/ANGLE FIX ON STN. 3
**************************************************
STN. W.C.B.  DIST.    EASTING   NORTHING
====================================================
1                        100        100
    38.3834   180.376
```

```
3                      212.638   240.884
    296.2707  121.542
2                      321.456   186.743
=====================================================
    29.5802   = ANGLE 1-2 AND 1-3
    47.5031   = ANGLE 2-1 AND 2-3
*****************************************************

INTERSECTION PROGRAM RESULTS :-
<5> ANGLE/ANGLE FIX ON STN. 3
***************************************************
STN. W.C.B.  DIST.    EASTING    NORTHING
=====================================================
FIRST SOLUTION :-
1                      100       100
    38.3834   180.376
3                      212.638   240.884
    296.2707  121.542
2                      321.456   186.743
=====================================================
    29.5802   = ANGLE 1-2 AND 1-3
    47.5031   = ANGLE 2-1 AND 2-3
*****************************************************

INTERSECTION PROGRAM RESULTS :-
<5> ANGLE/ANGLE FIX ON STN. 3
***************************************************
STN. W.C.B.  DIST.    EASTING    NORTHING
=====================================================
SECOND SOLUTION :-
1                      100       100
    98.3438   180.376
3                      278.359   73.098
    201.0625  121.542
2                      321.456   186.743
=====================================================
    29.5802   = ANGLE 1-2 AND 1-3
    47.5031   = ANGLE 2-1 AND 2-3
*****************************************************
```

3.6 LOCATION OF STATIONS IN 2-D BY RESECTION

In the previous program <INTERSECTION>, to find the co-ordinates of a new survey station, there were sometimes two solutions to the problem depending on the nature of the data entered, either bearings, angles or distances from two known reference stations.

There are, however, occasions when it is necessary to establish in the field a survey station which has not previously been observed from any other stations. It is therefore necessary to determine the station's co-ordinates from three known stations, and this technique is sometimes referred to as the 'three point problem', or more commonly resection. Either the three distances, H_1, H_2, H_3, from the survey station to the three known reference stations (Fig. 3.11), must be measured or the angles U_5 and U_6 subtended by the two reference baselines at the survey station, have to be measured.

If three distance measurements are made to the new survey station, a unique solution will be obtained. However, when angle measurements are made, should the new survey station lie on the circumscribing circle through the three reference stations and the survey station, the three-point problem will be insoluble. Should this happen, further observations, for example distances, will be required from two of the reference stations to give a unique solution.

3.6.1 2-D resection solutions – subroutine index

	Line numbers	*Function*
(a)	10–90	Initialization and control
(b)	100–190	Screen header display
(c)	200–240	Selection of angle format
(d)	300–430	Menu selection for resection solution
(e)	440–490	Option to change earlier reference stations
(f)	500–680	Entry of survey location, etc.
(g)	700–910	Entry of reference station data
(h)	1000–1590	<1> Distance/Distance Resection
(i)	1600–1990	Screen display of results
(j)	2000–2370	<2> Angle/Angle Resection
(k)	6000–6080	Routine to print survey location, etc.
(l)	6250–6590	Routine to print resection solutions
(m)	7800–7870	Triangle solution (*S*1, *S*2, *S*3)
(n)	8000–8040	Routine to convert DD.MMSS to decimals
(o)	8050–8140	Routine to convert decimals to DD.MMSS
(p)	9000–9010	Routine to clear screen
(q)	9100–9130	Routine to display error message
(r)	9200–9290	Error trace routine
(s)	9500–9620	Termination of program

(a) Initialization and control

Line numbers 10–90

All numeric variables and required string variables are initialized and two 'DEF FN' functions used in lines 80 and 90 to define arccosine and arcsine as discussed previously in Section 2.2.8.

(b) Screen header display

Line numbers 100–190

This is the screen display used for civil engineering students at Strathclyde University. The operator can alter this as required to suit another organization.

(c) Selection of angle format

Line numbers 200–240

The operator has a choice in line 200 of entering angles in decimal or DD.MMSS format, where the *D*9 flag is set to 1 for decimals and 2 for DD.MMSS in line 210.

(d) Menu selection for resection solution

Line numbers 300–430

A choice of two programs is offered in lines 340–350, either distance/distance measurements or angle measurements to three known reference stations. On the first occasion, the menu is presented ($N = 0$), the program branches to line 500 to allow the survey location, etc., to be recorded.

(e) Option to change earlier reference stations

Line numbers 440–490

This option is only presented on the second and subsequent display of the menu, allowing several resection programs to be run using the same reference stations. If required, the reference stations can be changed by branching to line 700.

(f) Entry of survey location, etc.

Line numbers 500–680

This routine is only accessed the first time the program is run to allow a record to be kept of site location, operator, date, etc. After keyboard entry, line 650 gives an option of a printout if required in the subroutine at line 6000.

(g) Entry of reference station data

Line numbers 700–910

The numbers and co-ordinates of each of the three reference stations are entered in rotation, either clockwise or anticlockwise, so it is immaterial where the new survey station lies with respect to the known reference stations.

(h) <1> Distance/Distance Resection

Line numbers 1000–1590

The distances from the three reference stations, R_1, R_2, R_3, to the survey station R_0 are entered in line numbers 1130, 1200, 1330. In each case after entry, a check is made to ensure that the sum of the distances between a pair of stations is greater than the baseline distance between the reference stations. These distances can be either slant range or horizontal distances, as checked in line 1400. If slant range, the horizontal distances are calculated in line 1440 from a knowledge of the level of the stations, thus reducing the solution to two dimensions (2-D).

The distance/distance calculations start on line 1450 where a screen message is displayed to advise the operator that computations have commenced. From a knowledge of the horizontal distances, H_1, H_2, H_3, from the survey station R_0 to the three reference stations, R_1, R_2, R_3 as shown in Fig. 3.11, three equations can be written considering R_0 as the intersection point of three circles centred at the three reference stations. These three equations are as follows:

$$(X_0 - X_1)^2 + (Y_0 - Y_1)^2 = H_1^2 \tag{3.22}$$

$$(X_0 - X_2)^2 + (Y_0 - Y_2)^2 = H_2^2 \tag{3.23}$$

$$(X_0 - X_3)^2 + (Y_0 - Y_3)^2 = H_3^2 \tag{3.24}$$

To obtain a solution for X_0 and Y_0 it is necessary to eliminate the X^2 and Y^2 terms by subtracting pairs of circle equations, from Equations 3.22–3.24 to give the equations of the radical axes:

$$(X_2 - X_1)X_0 + (Y_2 - Y_1)Y_0 = (H_2^2 - H_1^2)/2 \tag{3.25}$$

$$(X_3 - X_2)X_0 + (Y_3 - Y_2)Y_0 = (H_3^2 - H_2^2)/2 \tag{3.26}$$

$$(X_1 - X_2)X_0 + (Y_1 - Y_3)Y_0 = (H_1^2 - H_3^2)/2 \tag{3.27}$$

Solving any two of these three simultaneous equations will give the desired co-ordinates for X_0 and Y_0:

$$\begin{aligned} X_0 = \{ & [(X_1^2 - X_2^2) + (Y_1^2 - Y_2^2) - (H_1^2 - H_2^2)](Y_2 - Y_3) \\ & - [(X_2^2 - X_3^2) + (Y_2^2 - Y_3^2) - (H_2^2 - H_3^2)](Y_1 - Y_2)\}/ \\ & 2[(X_1 - X_2)(Y_2 - Y_3) - (X_2 - X_3)(Y_1 - Y_2)] \end{aligned} \tag{3.28}$$

$$\begin{aligned} Y_0 = \{ & [(X_1^2 - X_2^2) + (Y_1^2 - Y_2^2) - (H_1^2 - H_2^2)](X_2 - X_3) \\ & - [(X_2^2 - X_3^2) + (Y_2^2 - Y_3^2) - (H_2^2 - H_3^2)](X_1 - X_2)\}/ \\ & 2[(Y_1 - Y_2)(X_2 - X_3) - (Y_2 - Y_3)(X_1 - X_2)] \end{aligned} \tag{3.29}$$

These two Equations 3.28 and 3.29 give a unique solution for the survey point, $R_0(X_0, Y_0)$ with no ambiguity wherever the point R_0 lies with respect to the reference stations R_1, R_2 and R_3. To simplify the BASIC programming in entering the Equations 3.28 and 3.29, several array variables A_1–A_4, B_1–B_4 and C_1–C_2 are declared in line numbers 1460–1500, with the solutions to X_0 and Y_0 in line numbers 1510 and 1520.

To be able to compare different solutions to the same survey point, the same screen display is used for both distance/distance ($T = 1$) and angle/angle ($T = 2$) programs. Since the latter program requires angle inputs, the angles U_1–U_6 in Fig. 3.11 are therefore computed in lines 1540–1590, using the triangle solution (S_1, S_2, S_3) discussed in Section 3.2, which is included as a subroutine at line 7800.

(i) Screen display of results

Line numbers 1600–1990

Rather than use 'PRINT USING' statements for the screen display, which are machine dependent, each of the calculated distances is presented to the third decimal figure and angles presented as decimal or DD.MMSS as selected at the

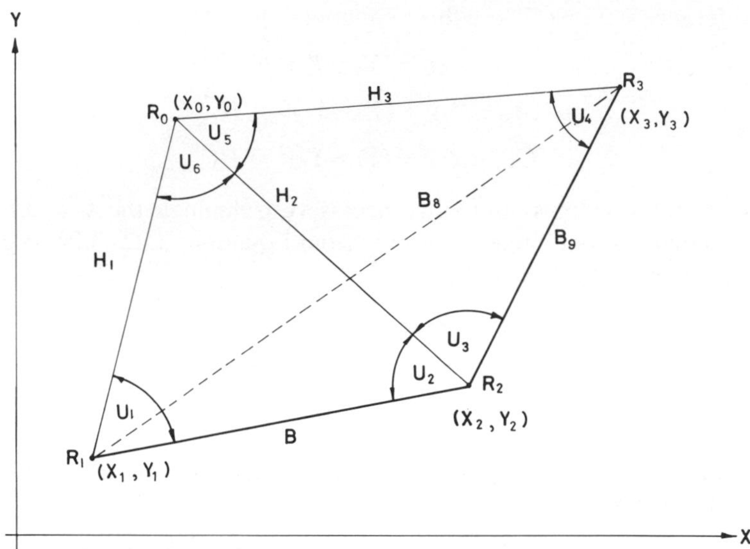

Fig. 3.11. Nomenclature for 2-D resection solutions from distances to three reference stations.

outset (line 200). In the case of the angles, loops are used to fill the array variable $U(5)$ with the six angles U_1–U_6.

The screen is cleared at line 1730 and the type of solution displayed in either line 1750 or 1760. A loop is used in lines 1800–1820 to display the co-ordinates of each of the reference stations and the survey point. This is followed by a display of all the distances and angles in Fig. 3.11. The option of a hard copy printout is offered in line 1960 using a subroutine at line 6250. On completion, the program returns to the menu selection at line 300.

(j) <2> Angle/Angle Resection

Line numbers 2000–2370

The angle/angle resection case is more complicated than the previous distance/distance case since the survey point can be in three zones as indicated in Fig. 3.12. That is, the survey station can (a) lie on the opposite side of station R_2 from the baseline R_1–R_3 (Fig. 3.12(a)); (b) lie between the station R_2 and the baseline R_1–R_3 (Fig. 3.12(b)), or (c) lie on the opposite side of the baseline R_1–R_3 to station R_2. The latter is the more normal case for resection, as shown in Fig. 3.11.

The relationship of R_0 to the baseline R_1–R_3 and station R_2 is requested in lines 2060–2080 and the flag I8 set to 1 if case (a), and 2 if case (b) or (c). There are only two real solution cases; the first when the angle R_1–R_2–R_3 in quadrilateral R_0, R_1, R_2, R_3 is greater than 180° (I8 = 1) and the second when angle R_1–R_2–R_3 in the quadrilateral R_0, R_1, R_2, R_3 is less than 180° (I8 = 2). The angle R_1–R_2–R_3 (E_4, designated $E4$ in line 2230) in triangle R_1, R_2, R_3 is calculated

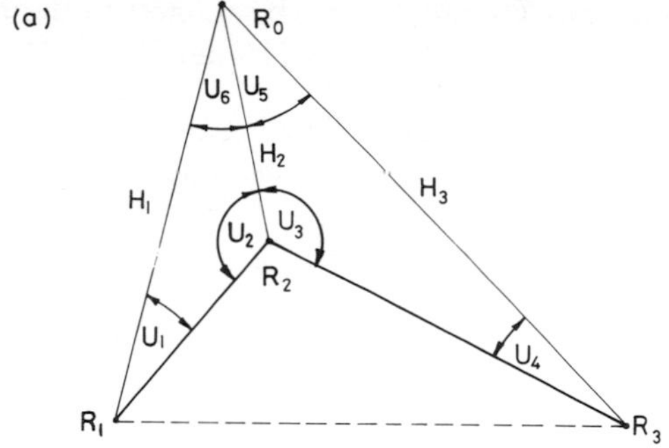

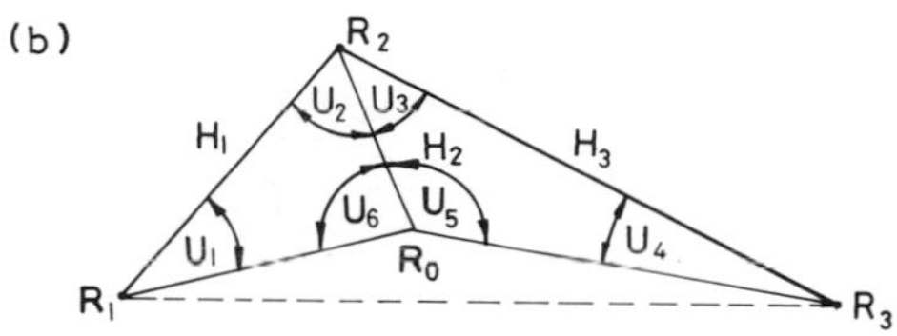

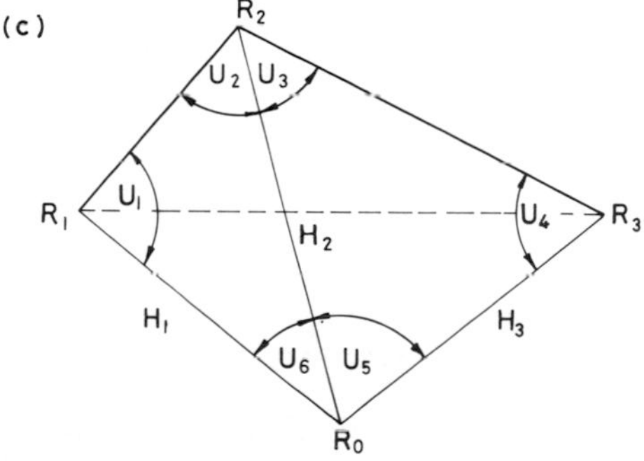

Fig. 3.12. Nomenclature for 2-D resection solutions from angles to three reference stations showing three possible situations (a)–(c).

using the (S_1, S_2, S_3) triangle solution. If I8 = 1, i.e. Fig. 3.12(a), then the angle is greater than 180° and the complement of E_4 is computed. This angle E_4 now represents the angle made up by the two angles U_2 and U_3, therefore:

$$E_4 = U_2 + U_3 \tag{3.30}$$

Let E_5 represent the other two unknown angles U_1 and U_4 so that:

$$E_5 = U_1 + U_4 \tag{3.31}$$

$$E_5 = 360 - E_4 - U_5 - U_6 \tag{3.32}$$

Now the individual angles U_1 and U_4 can be related to the common distance R_0–R_2 (i.e. H_2) from the sine rule in Fig. 3.11 where:

$$H_2/\sin U_1 = B/\sin U_6 \tag{3.33}$$

and

$$H_2/\sin U_4 = B_9/\sin U_5 \tag{3.34}$$

Equating H_2 from these two Equations, 3.33 and 3.34 gives:

$$B\sin U_1/\sin U_6 = B_9\sin U_4/\sin U_5 \tag{3.35}$$

which can be resolved into:

$$\sin U_4 = (B\sin U_5/B_9\sin U_6)\sin U_1 \tag{3.36}$$

where from Equation 3.31 we have:

$$\sin U_4 = \sin(E_5 - U_1) \tag{3.37}$$

i.e.

$$\sin U_4 = \sin E_5\cos U_1 - \cos E_5\sin U_1 \tag{3.38}$$

Equating $\sin U_4$ in the Equations 3.36 and 3.38 gives:

$$(B\sin U_5/B_9\sin U_6)\sin U_1 = \sin E_5\cos U_1 - \cos E_5\sin U_1 \tag{3.39}$$

which when reorganized becomes:

$$\sin U_1[(B\sin U_5/B_9\sin U_6) + \cos E_5] = \sin E_5\cos U_1 \tag{3.40}$$

which reduces to:

$$\tan U_1 = \sin E_5/[(B\sin U_5/B_9\sin U_6) + \cos E_5] \tag{3.41}$$

and hence U_1 from the arctangent formula as solved in line 2300. Once U_1 has been found, U_4 can be obtained from Equation 3.31. With two of the angles known in each of the triangles R_0, R_1, R_2 and R_0, R_3, R_2, the other two angles U_2 and U_3 can be solved in lines 2320–2330. Since all the angles and the baselines R_1–R_2 and R_2–R_3 are known, the distances H_1, H_2, H_3 (Fig. 3.11) can now be found using the sine rule in lines 2340–2360. Once the three distances are known, the same distance/distance solution used earlier in lines 1460–1520 can be used to solve for (X_0, Y_0) for the survey point R_0. The program then continues to use the same screen display as before in lines 1600–1990.

(k) Routine to print survey location, etc.

Line numbers 6000–6080

This routine is provided for record purposes, and is based on Applesoft DOS and may require alteration for different microcomputers and printers.

(l) Routine to print resection solutions

Line numbers 6250–6590

The format for the printout has been standardized so that it can be used by both the distance/distance and angle/angle solutions. As mentioned above, the routine uses Applesoft DOS and may require to be altered for other microcomputers and printers.

(m) Triangle solution ($S1$, $S2$, $S3$)

Line numbers 7800–7870

The solution of this triangle configuration, Fig. 3.2, was discussed previously in Section <1> of the <TRIANGLE> program.

(n) Routine to convert DD.MMSS to decimals

Line numbers 8000–8040

This routine was discussed previously in Section 2.2.6.

(o) Routine to convert decimals to DD.MMSS

Line numbers 8050–8140

This routine was discussed previously in Section 2.2.6.

(p) Routine to clear screen

Line numbers 9000–9010

To simplify portability, the Applesoft 'HOME' command to clear the screen has been placed in a subroutine to allow an easy change to another microcomputer.

(q) Routine to display error message

Line numbers 9100–9130

If a wrong keyboard entry has been made, the error is trapped and this error message displayed and re-entry requested.

(r) Error trace routine

Line numbers 9200–9290

This error trace routine was discussed earlier in Section 2.2.9.

(s) Termination of program

Line numbers 9500–9620

When the '<QUIT>' selection is made either after the initial screen header or from the program menu, an end-of-program message is displayed to advise the operator that computations are complete.

3.6.2 2-D resection solutions – numeric variables

*A*1–*A*4	= Three circle array variables
*B*1–*B*4	= Three circle array variables
B, *B*8, *B*9	= Baseline distances between stations
C	= Counter in error trace routine
*C*1–*C*2	= Three circle array variables
*C*9	= Defined function for arccosine
D	= Angle in degrees
*D*0, *D*5	= Temporary storage for DD.MMSS to decimals
*D*1	= Integer value of *D* degrees
*D*2	= Total number of seconds in fraction of *D* degrees
*D*9	= Menu selection for decimals <1> or DD.MMSS <2>
E	= Error number
*E*4–*E*6, *E*9	= Angles used in angle/angle solution
*H*1–*H*3	= Horizontal distances from survey station to reference stations
I	= Loop variable
*I*8	= Relationship of *R*0 to *R*1, *R*2 and *R*3
*I*9	= Selection of slant <1> or horizontal <2> distances
J	= Loop variable
*K*5	= Multiplying constant in angle/angle solution
M, *M*0, *M*1	= Minutes of angle in subroutine
P	= 1/2 perimeter of triangle
R	= 57.2957795 to convert degrees to radians
*R*0–*R*3	= Reference numbers of stations
R(*I*)	= Reference numbers of stations
S, *S*0	= Seconds of angle in subroutine
*S*1–*S*3	= Sides of triangle for triangle solutions
*S*9	= Defined function for arcsine
T	= Type of solution selected
*T*1–*T*3	= Angles of triangle for triangle solutions
*U*1–*U*6	= Angles of two triangles in survey
U(*I*)	= Angles of two triangles in survey
*X*0–*X*3	= *X* co-ordinates of stations
X(*I*)	= *X* co-ordinates of stations
*Y*0–*Y*3	= *Y* co-ordinates of stations
Y(*I*)	= *Y* co-ordinates of stations
*Z*0–*Z*3	= *Z* co-ordinates of stations
Z(*I*)	= *Z* co-ordinates of stations

3.6.3 2-D resection solutions – string variables

A$	Storage for location of survey
B$	Storage for operator's name
C$	Storage for date of survey
D$	= CHR$(4), DOS command
*E*0$	= "SORRY DATA ERROR . . . PLEASE RE-ENTER"
K$	Storage for instrument used
P$	= "DO YOU WISH PRINTOUT OF ABOVE DATA (Y/N)"

Table—continued

Q$ = Question response (Y/N)
X$ = "ENTER X(E) COORD. OF STN."
Y$ = "ENTER Y(N) COORD. OF STN."
Z$ = "ENTER LEVEL OF STN."

3.6.4 2-D resection solutions – BASIC program

```
10  REM  <RESECTION> PROGRAM FOR APPLE II+  USES QUME PRINTER
20 D$ =  CHR$ (4): ONERR  GOTO 9200
30 A = 0:B = 0:E = 0:G = 0:J = 0:K = 0:N = 0:S = 0:U = 0:V = 0:W = 0
50 E0$ = "SORRY, DATA ERROR ... PLEASE RE-ENTER."
55 P$ = "DO YOU WISH PRINTOUT OF ABOVE DATA (Y/N)"
70 X$ = "ENTER X(E) COORD. OF STN. ":Y$ = "ENTER Y(N) COORD. OF STN. "
75 Z$ = "ENTER LEVEL OF STN. "
80  DEF  FN C9(C9) = (1.5707964 -  ATN (C9 /  SQR (1 - C9 * C9))) * R: REM
    ARCCOSINE
90  DEF  FN S9(S9) =  ATN (S9 /  SQR (1 - S9 * S9)) * R: REM  ARCSINE
95  REM  ************************************************
100  GOSUB 9000
110  PRINT "****************************************"
120  PRINT "*                                      *"
130  PRINT "*     UNIVERSITY  OF  STRATHCLYDE      *"
140  PRINT "*  DEPARTMENT  OF  CIVIL  ENGINEERING  *"
150  PRINT "*                                      *"
160  PRINT "*          SURVEYING  SECTION          *"
170  PRINT "*         RESECTION  SOLUTIONS         *"
180  PRINT "****************************************"
190  PRINT
200  PRINT "ARE ANGLES IN DECIMALS <1>, DD.MMSS <2> OR DO YOU WISH TO QUI
    T <3>."
210  INPUT D9
220  IF D9 < 1 OR D9 > 3 THEN 200
230  ON D9 GOTO 240,240,9500
240 R = 57.2957795
250  IF D9 = 1 THEN H$ = "DD.DEC"
260  IF D9 = 2 THEN H$ = "DD.MMSS"
300  GOSUB 9000
310  PRINT
320  PRINT "SELECT RESECTION PROGRAM FROM MENU :-"
330  PRINT
340  PRINT "    <1> DISTANCE / DISTANCE"
350  PRINT "    <2> ANGLE    / ANGLE"
360  PRINT "    <3> QUIT"
400  INPUT T
410  IF T < 1 OR T > 3 THEN 300
420  IF T = 3 THEN 9500
430  IF N = 0 THEN 500
440  PRINT "DO YOU WISH TO CHANGE REFERENCE STATIONS (Y/N) ";
450  INPUT Q$
460  IF Q$ = "Y" THEN 700
470  IF Q$ <  > "N" THEN 440
480 N = N + 1: REM  COUNTER
490  ON T GOTO 1000,2000
500  GOSUB 9000
510  PRINT "UNIVERSITY OF STRATHCLYDE"
520  PRINT "RESECTION PROGRAM"
530  PRINT "****************************************"
```

```
540  PRINT
550  PRINT "ENTER LOCATION OF WORK, ETC. WHEN       REQUESTED, AND PRESS
     <RETURN>"
560  PRINT "ENTER LOCATION OF SURVEY :-"
570  INPUT A$
580  PRINT "ENTER OPERATOR'S NAME :-"
590  INPUT B$
600  PRINT "ENTER DATE OF SURVEY AS DD/MM/YY :-"
610  INPUT C$
620  PRINT "ENTER INSTRUMENT USED :-"
630  INPUT K$
640  PRINT
650  PRINT "DO YOU WISH PRINTOUT OF LOCATION ETC.,  (Y/N) ";
660  INPUT Q$
670  IF Q$ = "Y" THEN 6000
680  IF Q$ < > "N" THEN 650
690 :
700  GOSUB 9000
710  PRINT "ENTER DETAILS OF REFERENCE STNS. WHEN   REQUESTED, AND PRESS
     <RETURN> "
720  PRINT
730  PRINT "ENTER NUMBERS OF REFERENCE STATIONS IN  ROTATION, NUMERALS ON
     LY :-"
740  FOR I = 0 TO 2
750  PRINT "REFERENCE STN. NO. ";
760  INPUT R(I)
770  PRINT X$;R(I);
780  INPUT X(I)
790  PRINT Y$;R(I);
800  INPUT Y(I)
810  PRINT Z$;R(I);
820  INPUT Z(I)
830  PRINT : NEXT I
835  REM  STORE STATIONS,ETC.
840 R1 = R(0):R2 = R(1):R3 = R(2)
850 X1 = X(0):X2 = X(1):X3 = X(2)
860 Y1 = Y(0):Y2 = Y(1):Y3 = Y(2)
870 Z1 = Z(0):Z2 = Z(1):Z3 = Z(2)
875  REM  CALCULATE BASELINE LENGTHS
880 B =  SQR ((X2 - X1) ^ 2 + (Y2 - Y1) ^ 2)
890 B8 =  SQR ((X3 - X1) ^ 2 + (Y3 - Y1) ^ 2)
900 B9 =  SQR ((X3 - X2) ^ 2 + (Y3 - Y2) ^ 2)
910  GOTO 480
995  REM  ***************************************************
1000  GOSUB 9000
1010  PRINT "<1> DISTANCE/DISTANCE PROGRAM :-"
1020  PRINT "****************************************"
1030  PRINT
1040  GOSUB 8900
1050  PRINT
1100  PRINT "DATA FOR REFERENCE STN. ";R1
1110  PRINT "******************************"
1120  PRINT
1130  PRINT "ENTER DISTANCE FROM ";R1;" TO ";R0
1140  INPUT H1
1150  IF H1 <  = 0 THEN 1130
1160  PRINT
1170  PRINT "DATA FOR REFERENCE STN. ";R2
1180  PRINT "******************************"
```

```
1190  PRINT
1200  PRINT "ENTER DISTANCE FROM ";R2;" TO ";RØ
1210  INPUT H2
1220  IF H2 <  = Ø THEN 12ØØ
1230  PRINT
1240  IF H1 + H2 > B THEN 128Ø
1250  GOSUB 91ØØ
1260  PRINT "DISTANCES TO ";R1;" AND ";R2;" TOO SMALL"
1270  GOTO 11ØØ
1280  PRINT
1290  PRINT "DATA FOR REFERENCE STN. ";R3
1300  PRINT "********************************"
1310  PRINT
1320  PRINT "ENTER DISTANCE FROM ";R3;" TO ";RØ
1330  INPUT H3
1340  IF H3 < Ø THEN 132Ø
1350  PRINT
1360  IF H2 + H3 > B9 THEN 14ØØ
1370  GOSUB 91ØØ
1380  PRINT "DISTANCES TO ";R2;" AND ";R3;" TOO SMALL"
1390  GOTO 117Ø
1395  REM  ARRIVE HERE WITH H1,H2,H3
1400  PRINT "ARE DISTANCES SLANT<1> OR HORIZONTAL<2>."
1410  INPUT I9
1420  IF I9 < 1 OR I9 > 2 THEN 14ØØ
1430  IF I9 = 2 THEN 145Ø
1435  REM  CALC. HORIZ.DIST FROM SLANT RANGE
1440 H1 =  SQR (H1 ^ 2 - (Z1 - ZØ) ^ 2):H2 =  SQR (H2 ^ 2 - (Z2 - ZØ) ^ 2
     ):H3 =  SQR (H3 ^ 2 - (Z3 - ZØ) ^ 2)
1445  REM  HORIZONTAL SOLUTION STARTS HERE
1450  GOSUB 9ØØØ: PRINT "CALCULATING .... "
1455  REM  ALLOCATE ARRAY VARIABLES
1460 A1 = X1 ^ 2 - X2 ^ 2:A2 = X2 ^ 2 - X3 ^ 2
1470 A3 = X1 - X2:A4 = X2 - X3
1480 B1 = Y1 ^ 2 - Y2 ^ 2:B2 = Y2 ^ 2 - Y3 ^ 2
1490 B3 = Y1 - Y2:B4 = Y2 - Y3
1500 C1 = H1 ^ 2 - H2 ^ 2:C2 = H2 ^ 2 - H3 ^ 2
1505  REM  THREE CIRCLES SOLUTION
1510 XØ = ((A1 + B1 - C1) * B4 - (A2 + B2 - C2) * B3) / (2 * (A3 * B4 - A
     4 * B3))
1520 YØ = ((A1 + B1 - C1) * A4 - (A2 + B2 - C2) * A3) / (2 * (B3 * A4 - B
     4 * A3))
1525  REM  THREE CIRCLES SOLUTION USED BY BOTH T=1 AND T=2
1530  IF T = 2 THEN 16ØØ
1535  REM  I.E. T=1, DIST/DIST SOLN., FIND ANGLES
1540 S1 = H1:S2 = B:S3 = H2
1545  REM  USE TRIANGLE SOLUTION (S1,S2,S3)
1550  GOSUB 78ØØ
1560 U1 = T1:U2 = T2:U6 = T3
1565  REM  SECOND TRIANGLE
1570 S1 = H2:S2 = B9:S3 = H3
1575  REM  USE TRIANGLE SOLUTION (S1,S2,S3)
1580  GOSUB 78ØØ
1590 U3 = T1:U4 = T2:U5 = T3
1595  REM  COMPUTATIONS COMPLETE, PREPARE DATA FOR SCREEN DISPLAY
1600 XØ =  INT (XØ * 1ØØØ + .5) / 1ØØØ:YØ =  INT (YØ * 1ØØØ + .5) / 1ØØØ
1610 X(3) = XØ:Y(3) = YØ
1620 B =  INT (B * 1ØØØ + .5) / 1ØØØ:B8 =  INT (B8 * 1ØØØ + .5) / 1ØØØ:B9
     =  INT (B9 * 1ØØØ + .5) / 1ØØØ
```

```
1630 H1 =  INT (H1 * 1000 + .5) / 1000:H2 =  INT (H2 * 1000 + .5) / 1000:
     H3 =  INT (H3 * 1000 + .5) / 1000
1635  REM  PREPARE ANGLES, CHECK D9 FOR DECIMALS<1> OR DD.MMSS<2>
1640 U(0) = U1:U(1) = U2:U(2) = U3:U(3) = U4:U(4) = U5:U(5) = U6
1650  IF D9 = 1 THEN 1700
1660  FOR J = 0 TO 5
1665  REM  CONVERT DECIMALS TO DD.MMSS
1670 D = U(J): GOSUB 8050:U(J) = D
1680  NEXT J
1690  GOTO 1730
1695  REM  PREPARE DECIMAL ANGLES FOR DISPLAY
1700  FOR J = 0 TO 5
1710 U(J) =  INT (U(J) * 1000 + .5) / 1000
1720  NEXT J
1730  GOSUB 9000
1740  PRINT "<RESECTION> PROGRAM": IF T = 2 THEN 1760
1750  PRINT "<1> DIST/DIST SOLUTION :-": GOTO 1770
1760  PRINT "<2> ANGLE/ANGLE SOLUTION :-"
1770  PRINT "**************************************"
1780  PRINT "STN.NO. EASTING    NORTHING   LEVEL"
1790  PRINT "======================================"
1800  FOR I = 0 TO 3
1810  PRINT R(I); TAB( 8);X(I); TAB( 19);Y(I); TAB( 30);Z(I)
1820  NEXT I
1830  PRINT "--------------------------------------"
1840  PRINT "DISTANCE  ";R1;"-";R2;" = ";B
1850  PRINT "........  ";R1;"-";R3;" = ";B8
1860  PRINT "........  ";R2;"-";R3;" = ";B9
1870  PRINT "........  ";R1;"-";R0;" = ";H1
1880  PRINT "........  ";R2;"-";R0;" = ";H2
1890  PRINT "........  ";R3;"-";R0;" = ";H3
1900  PRINT "ANGLE  ";R1;"-";R0;"-";R2;" = ";U(5)
1910  PRINT ".....  ";R2;"-";R0;"-";R3;" = ";U(4)
1920  PRINT ".....  ";R2;"-";R1;"-";R0;" = ";U(0)
1930  PRINT ".....  ";R1;"-";R2;"-";R0;" = ";U(1)
1940  PRINT ".....  ";R3;"-";R2;"-";R0;" = ";U(2)
1950  PRINT ".....  ";R2;"-";R3;"-";R0;" = ";U(3)
1960  PRINT P$;: INPUT Q$
1970  IF Q$ = "Y" THEN 6250
1980  IF Q$ < > "N" THEN 1960
1990  GOTO 300
1995  REM  ************************************************
2000  GOSUB 9000
2010 :
2020  PRINT "<2> ANGLE/ANGLE RESECTION"
2030  PRINT "**************************************"
2040  GOSUB 8900
2050  PRINT
2060  PRINT "ENTER RELATION OF ";R0;" TO BASELINE ";R1;"-";R3
2070  PRINT "   <1> OPPOSITE SIDE OF STN. ";R2
2080  PRINT "   <2> BETWEEN BASELINE AND STN. ";R2
2085  PRINT "        OR ON OPPOSITE SIDE OF BASELINE          TO STN. ";R2
2090  INPUT I8: IF I8 < 1 OR I8 > 2 THEN 2060
2100  PRINT "ENTER ANGLE ";R1;"-";R0;"-";R2
2110  INPUT U6
2120  IF U6 < 0 OR U6 > 180 THEN 2100
2130  PRINT
2140  PRINT "ENTER ANGLE ";R2;"-";R0;"-";R3
2150  INPUT U5
```

```
2160  IF U5 < 0 OR U5 > 180 THEN 2140
2170  GOSUB 9000: PRINT "CALCULATING ...."
2180  IF D9 = 1 THEN 2210
2185  REM  I.E. D9=2, ANGLES IN DD.MMSS, CONVERT TO DECIMALS
2190 D5 = U6: GOSUB 8000:U6 = D5
2200 D5 = U5: GOSUB 8000:U5 = D5
2205  REM  FIND ANGLE E4, R1-R2-R3
2210 S1 = B:S2 = B9:S3 = B8
2215  REM  FIND E4 FROM TRIANGLE SOLUTION (S1,S2,S3)
2220  GOSUB 7800
2225  REM  IF R0 ON SAME SIDE OF BASELINE R1-R3 AS R2 THEN COMPLIMENT OF E4
2230 E4 = T1: IF I8 = 1 THEN E4 = 360 - E4
2235  REM  LET E5=U1+U4
2240 E5 = 360 - E4 - U5 - U6
2245  REM  STORE E5 IN E6 FOR FUTURE USE
2250 E6 = E5
2255  REM  CALCULATE FACTOR K5
2260 K5 = B *  SIN (U5 / R) / (B9 *  SIN (U6 / R))
2285  REM  FIND U1
2290 U1 =  ATN ( SIN (E5 / R) / (K5 +  COS (E5 / R))) * R
2300  IF U1 < 0 THEN U1 = 180 + U1
2305  REM  E6=U1+U4
2310 U4 = E6 - U1
2315  REM  FIND REST OF ANGLES
2320 U2 = 180 - U1 - U6
2330 U3 = 180 - U4 - U5
2335  REM  USE SINE RULE TO FIND H1,H2,H3
2340 H1 = B *  SIN (U2 / R) /  SIN (U6 / R)
2350 H2 = B *  SIN (U1 / R) /  SIN (U6 / R)
2360 H3 = B9 *  SIN (U3 / R) /  SIN (U5 / R)
2365  REM  USE THREE CIRCLES PROGRAM IN <1> FOR SOLUTION
2370  GOTO 1460
5985  REM  ***************************************************************
5990  REM  PRINTOUT OF LOCATION, ETC
6000  PRINT D$;"PR#1": PRINT "LOCATION OF SURVEY :- ";A$
6010  PRINT "*******************************************************************************"
6020  PRINT "OPERATOR'S NAME :- ";B$
6030  PRINT "DATE OF SURVEY  :- ";C$
6040  PRINT "INSTRUMENT USED :- ";K$
6060  PRINT "*******************************************************************************"
6070  PRINT : PRINT D$;"PR#0"
6080  GOTO 700
6243  REM  **************************************************
6245  REM  PRINTOUT OF COORDINATES ON QUME
6250  PRINT D$;"PR#1": POKE 1784 + 1,80
6260  PRINT :P7 = 1: REM  PRINTER CODE
6270  PRINT "RESECTION PROGRAM RESULTS :-"
6290  IF T = 2 THEN 6320
6300  PRINT "<1> DISTANCE/DISTANCE SOLUTION :-"
6310  GOTO 6330
6320  PRINT "<2> ANGLE/ANGLE SOLUTION :-"
6330  PRINT "****************************************"
6350  PRINT "STN.NO. EASTING    NORTHING   LEVEL"
6360  PRINT "========================================"
6370  FOR I = 0 TO 3
```

```
6380  PRINT R(I);: POKE 36,8: PRINT X(I);: POKE 36,19: PRINT Y(I);: POKE
    36,3Ø: PRINT Z(I)
639Ø  NEXT I
64ØØ  PRINT "--------------------------------------"
642Ø  PRINT "DISTANCE  ";R1;"-";R2;: POKE 36,2Ø: PRINT " = ";B
643Ø  PRINT "........  ";R1;"-";R3;: POKE 36,2Ø: PRINT " = ";B8
644Ø  PRINT "........  ";R2;"-";R3;: POKE 36,2Ø: PRINT " = ";B9
645Ø  PRINT "........  ";R1;"-";RØ;: POKE 36,2Ø: PRINT " = ";H1
646Ø  PRINT "........  ";R2;"-";RØ;: POKE 36,2Ø: PRINT " = ";H2
647Ø  PRINT "........  ";R3;"-";RØ;: POKE 36,2Ø: PRINT " = ";H3
648Ø  PRINT "--------------------------------------"
649Ø  POKE 36,23: PRINT H$
65ØØ  PRINT "ANGLE  ";R1;"-";RØ;"-";R2;: POKE 36,2Ø: PRINT " = ";U(5)
651Ø  PRINT ".....  ";R2;"-";RØ;"-";R3;: POKE 36,2Ø: PRINT " = ";U(4)
652Ø  PRINT ".....  ";R2;"-";R1;"-";RØ;: POKE 36,2Ø: PRINT " = ";U(Ø)
653Ø  PRINT ".....  ";R1;"-";R2;"-";RØ;: POKE 36,2Ø: PRINT " = ";U(1)
654Ø  PRINT ".....  ";R3;"-";R2;"-";RØ;: POKE 36,2Ø: PRINT " = ";U(2)
655Ø  PRINT ".....  ";R2;"-";R3;"-";RØ;: POKE 36,2Ø: PRINT " = ";U(3)
656Ø  PRINT "****************************************"
657Ø  PRINT
6575  REM  RETURN TO SCREEN
658Ø  PRINT D$;"PR#Ø"
659Ø  GOTO 3ØØ
7793  REM  ****************************************************
7795  REM  <TRIANGLE> SOLUTION (S1,S2,S3)
78ØØ P = (S1 + S2 + S3) / 2
781Ø T3 =  SQR (P * (P - S2) / (S1 * S3))
782Ø T3 = 2 *  FN C9(T3)
783Ø T2 =  SQR (P * (P - S1) / (S2 * S3))
784Ø T2 = 2 *  FN C9(T2)
785Ø T1 =  -  COS ((T3 + T2) / R)
786Ø T1 =  FN C9(T1)
787Ø  RETURN
7993  REM  ****************************************************
7995  REM  ROUTINE DD.MMSS TO DECIMALS
8ØØØ DØ =  INT (D5)
8ØØ5  REM  FIND NO. OF MINUTES,MØ
8Ø1Ø M1 = D5 * 1ØØ - DØ * 1ØØ:MØ =  INT (M1 + .1)
8Ø15  REM  FIND NO. OF SECONDS,SØ
8Ø2Ø SØ =  INT (M1 * 1ØØ - MØ * 1ØØ + .5)
8Ø25  REM  COLLECT DEGS,MINS,SECS
8Ø3Ø D5 = DØ + MØ / 6Ø + SØ / 36ØØ
8Ø4Ø  RETURN
8Ø43  REM  ****************************************************
8Ø45  REM  ROUTINE DECIMAL TO DD.MMSS
8Ø5Ø D1 =  INT (D)
8Ø55  REM  FIND TOTAL NO. OF SECONDS
8Ø6Ø D2 = (D - D1) * 36ØØ
8Ø65  REM  FIND NO. OF MINUTES
8Ø7Ø M =  INT (D2 / 6Ø)
8Ø75  REM  FIND NO. OF SECONDS
8Ø8Ø S = D2 - 6Ø * M + .5
8Ø9Ø  IF S < 6Ø THEN 813Ø
81ØØ M = M + 1:S = Ø
811Ø  IF M < 6Ø THEN 813Ø
812Ø D1 = D1 + 1:M = Ø
813Ø D = D1 * 1ØØØØ + M * 1ØØ + S
814Ø D =  INT (D) / 1ØØØØ: RETURN
```

```
8895  REM  ROUTINE TO ENTER STATION RØ
8900  PRINT "ENTER SURVEY STATION NO., NUMERALS ONLY"
8910  INPUT RØ
8920  IF RØ = R1 THEN 8900
8930  IF RØ = R2 THEN 8900
8940  IF RØ = R3 THEN 8900
8950  PRINT Z$;RØ
8960  INPUT ZØ
8970 R(3) = RØ:Z(3) = ZØ
8980  RETURN
8985  REM  ******************************************
8990  REM  GOSUB ROUTINE TO CLEAR SCREEN
9000  HOME
9010  RETURN
9093  REM  ***************************************************
9095  REM  GOSUB ROUTINE FOR ERRORS
9100  PRINT
9110  PRINT EØ$
9120  PRINT
9130  RETURN
9193  REM  ***************************************************
9195  REM  ERROR TRACE ROUTINE
9200  PRINT  CHR$ (7): REM  CTRL-B (BELL)
9210 E =  PEEK (222): REM  GET ERROR NO.
9220  IF C = 1 THEN 9260
9230  INVERSE
9240  PRINT "ERROR NO. ";E;" FOUND"
9250 C = 1: TRACE : RESUME
9260  PRINT "ERROR ON SECOND LINE NO. "
9270  NORMAL
9280  NOTRACE
9290  STOP
9493  REM  ***************************************************
9495  REM  END OF PROGRAM
9500  GOSUB 9000
9510  PRINT
9520  PRINT "END RESECTION PROGRAM"
9530  PRINT
9540  PRINT "*************************"
9550  REM  PROGRAM PREPARED BY
9560  REM  DR. P.H. MILNE
9570  REM  DEPT. OF CIVIL ENGINEERING
9580  REM  UNIVERSITY OF STRATHCLYDE
9590  REM  GLASGOW G4 ONG
9600  REM  SCOTLAND
9610  REM  ***************************
9620  END
```

3.6.5 2-D resection solutions – computer printout

```
LOCATION OF SURVEY :- GLENMORE PARK
**************************************************************************
OPERATOR'S NAME :- JOHN SMITH
DATE OF SURVEY  :- 17/Ø5/83
INSTRUMENT USED :- AGA GEODIMETER 14Ø
**************************************************************************
```

```
RESECTION PROGRAM RESULTS :-
<1> DISTANCE/DISTANCE SOLUTION :-
****************************************
STN.NO. EASTING     NORTHING    LEVEL
========================================
1        8235.3      4253         100
2        7000        7214.3       125
3        6470.6      6976.2       120
4        7278.143    8378.426     150
----------------------------------------
DISTANCE  1-2           = 3208.623
........  1-3           = 3244.994
........  2-3           = 580.479
........  1-4           = 4235.007
........  2-4           = 1196.892
........  3-4           = 1618.135
----------------------------------------
                          DD.MMSS
ANGLE   1-4-2           = 26.3
.....   2-4-3           = 16.3
.....   2-1-4           = 9.3452
.....   1-2-4           = 143.5508
.....   3-2-4           = 127.3913
.....   2-3-4           = 35.5047
****************************************


RESECTION PROGRAM RESULTS :-
<2> ANGLE/ANGLE SOLUTION :-
****************************************
STN.NO. EASTING     NORTHING    LEVEL
========================================
1        8235.3      4253         100
2        7000        7214.3       125
3        6470.6      6976.2       120
4        7278.143    8378.426     150
----------------------------------------
DISTANCE  1-2           = 3208.623
........  1-3           = 3244.994
........  2-3           = 580.479
........  1-4           = 4235.008
........  2-4           = 1196.894
........  3-4           = 1618.136
----------------------------------------
                          DD.MMSS
ANGLE   1-4-2           = 26.3
.....   2-4-3           = 16.3
.....   2-1-4           = 9.3452
.....   1-2-4           = 143.5508
.....   3-2-4           = 127.3913
.....   2-3-4           = 35.5047
****************************************
```

3.7 EDM SLOPE REDUCTION

Some of the more modern EDM (Electromagnetic Distance Measurement) instruments like the Aga Geodimeter 120, 140; Hewlett-Packard 3820A, Wild

Fig. 3.13. Aga Geodimeter 140 total station used in conjunction with Geodat data recorder. (Photograph by courtesy of AGA Geotronics.)

DI3S, DI4, and Zeiss Elta series, will all automatically reduce the EDM slant range between the instrument and the reflector to the horizontal. However, many of the earlier instruments (Aga 110, Kern DM501, Tellurometer CD6, etc.) give only the slant range, and it is also necessary to measure either the zenith angle by theodolite, or the difference in station elevations by levelling to enable the horizontal distance to be computed.

In some of the earlier add-on EDM instruments like the Tellurometer CD6, it is necessary not only to measure the EDM and reflector heights, but also to measure the theodolite and target height due to separated optics. In more modern EDM instruments the theodolite and EDM optics are coincident, and so only one measurement is required. The following program <EDM.SLOPE> is directed towards the earlier type of EDM instrument. The program can however be adapted easily by removing the extraneous measurements in the formulae.

The user is given two program options for short range or long range EDM. In the first case, no allowance is made for the curvature and refraction of the EDM path, whereas in the second case, full allowance is made with corrections to the zenith angle and distance.

In some of the more advanced combined electronic theodolite and EDM instruments, provision is made for the transfer of data direct to a data recorder. Some of these, like the Geodat for the Aga Geodimeters (Figs 1.7 and 3.13) and the MEM (data memory) for the Zeiss Elta Series, Fig. 3.14, are specifically designed for the task, with, in the latter case, a special transcription device, the

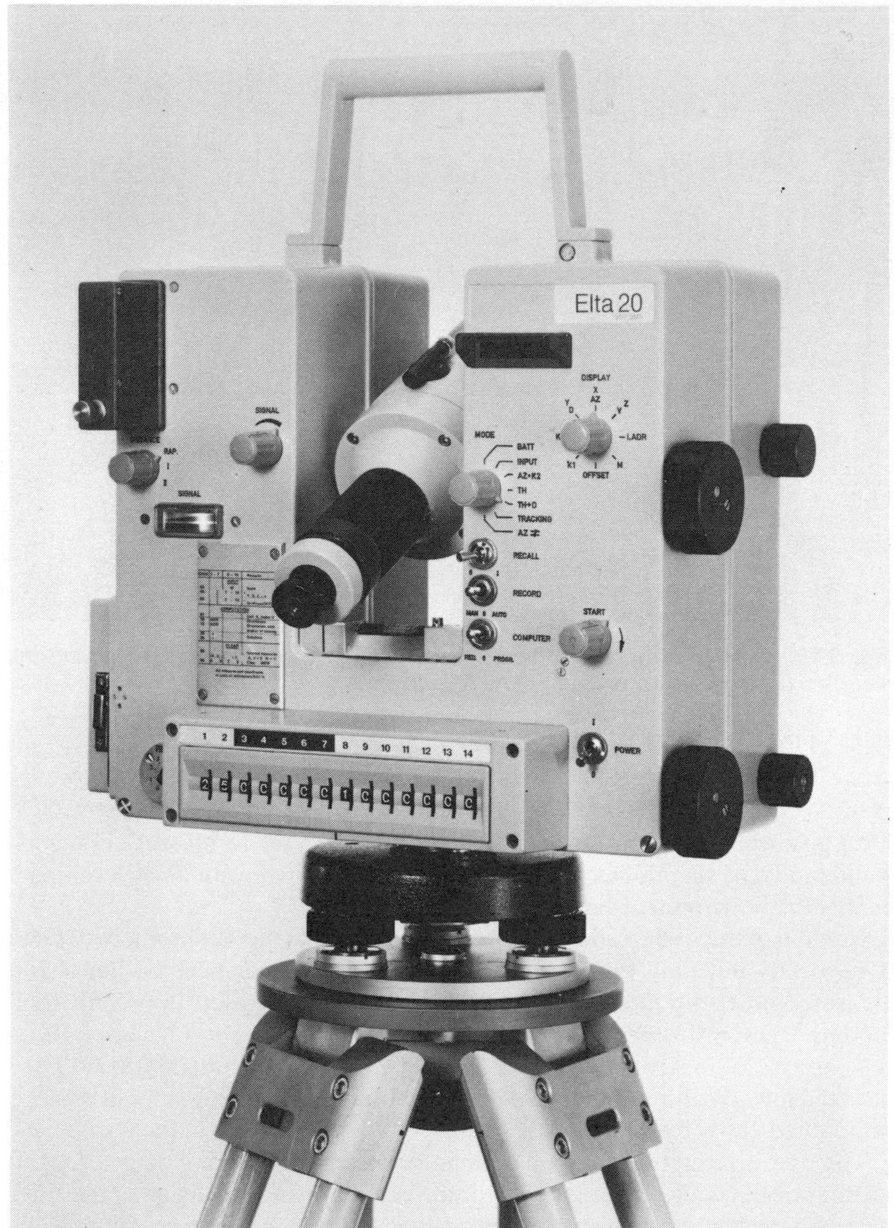

Fig. 3.14. Zeiss Elta 20 electronic tacheometer with a digital-reading, one second theodolite and MEM recorder. (Photograph by courtesy of Carl Zeiss (Oberkochen) Ltd.)

Fig. 3.15. Zeiss DAC 100 transcription device for transferring data from the MEM recorder to, say, an HP-85. (Photograph by courtesy of Carl Zeiss (Oberkochen) Ltd.)

Fig. 3.16. A Kern E1 Total Station theodolite and DM 502 EDM is shown being used in conjunction with a Hewlett-Packard HP-41CV for data recording. (Photograph by courtesy of Survey & General Instrument Co. Ltd.)

DAC 100. Both data recorders can be connected to an HP-85 for downloading the data, Fig. 3.15.

In the case of the Kern E1 Total Station theodolite and EDM, either a Kern R48 data recorder or a Hewlett-Packard HP-41CV may be used to record all the observations, Fig. 3.16. The data from the HP-41CV may be downloaded using an HP-IL interface into either an HP-85 (Fig. 1.6) or an HP-9816 (Fig. 1.10).

3.7.1 EDM slope reduction – subroutine index

	Line numbers	*Function*
(a)	10–90	Initialization and control
(b)	100–190	Screen header display
(c)	200–240	Selection of angle format
(d)	250–290	Menu selection for short/long range EDM
(e)	300–370	Menu selection for feet/metre units
(f)	500–680	Entry of survey location, etc.
(g)	700–910	Entry of survey station data
(h)	1000–1240	Entry of survey point data

	Line numbers	*Function*
(i)	1250–1390	<1> Short range EDM calculations
(j)	1400–1520	<2> Long range EDM calculations
(k)	1540–1900	Screen display of results
(l)	6000–6080	Routine to print survey location, etc.
(m)	6250–6350	Routine to print survey station data
(n)	6500–6540	Routine to print EDM slope solution
(o)	8000–8040	Routine to convert DD.MMSS to decimals
(p)	8050–8140	Routine to convert decimals to DD.MMSS
(q)	9000–9010	Routine to clear screen
(r)	9100–9130	Routine to display error message
(s)	9200–9290	Error trace routine
(t)	9500–9620	Termination of program

(a) Initialization and control

Line numbers 10–90

All numeric variables and required string variables are initialized.

(b) Screen header display

Line numbers 100–190

This is the screen display used for civil engineering students at Strathclyde University. The operator can alter this as required to suit another organization.

(c) Selection of angle format

Line numbers 200–240

The operator has a choice in line 200 of entering angles in decimal or DD.MMSS format, where the *D*9 flag is set to 1 for decimals and 2 for DD.MMSS in line 210.

(d) Menu selection for short/long-range EDM

Line numbers 250–290

The operator has the option of selecting whether the distances measured are great enough to warrant the curvature and refraction of the EDM distance measured to be allowed for in the computations.

(e) Menu selection for feet/metre units

Line numbers 300–370

Some EDM instruments have an option of recording distance in imperial units (feet) or metric units (metres). To cater for both types of observations, the units used are entered at line 320 and the corresponding radius of the earth computed in the subsequent lines.

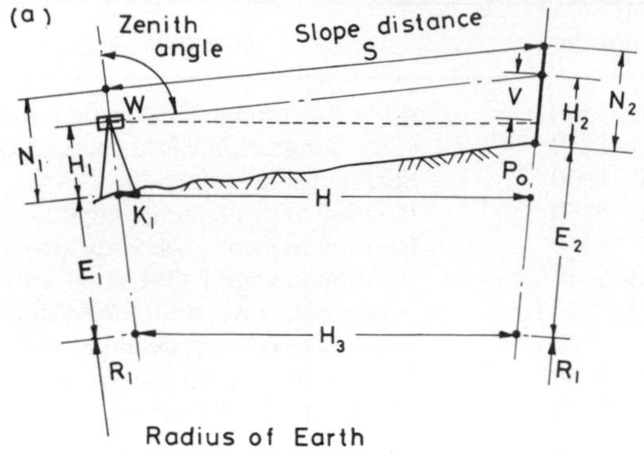

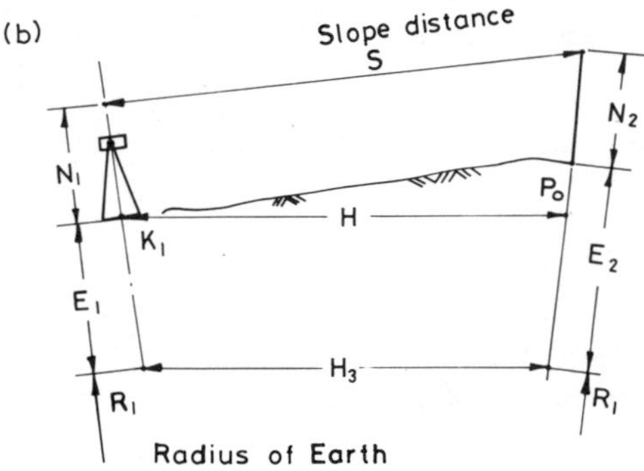

Fig. 3.17. Nomenclature for EDM slope distance reduction to the horizontal: (a) Zenith angle known; (b) Target elevation known.

(f) Entry of survey location, etc.

Line numbers 500–680

This routine is only accessed the first time the program is run to allow a record to be kept of site location, operator, date, etc. After keyboard entry, line 650 gives an option of a printout if required in the subroutine at line 6000.

(g) Entry of survey station data

Line numbers 700–910

This routine is accessed initially at the start of the program and later only if data for a second or third survey station is to be entered.

The essential details required are survey station number (K_1), elevation of station (E_1), height of theodolite above station (H_1) and, if necessary, height of EDM unit above station (N_1), all as shown in Fig. 3.17. An option to obtain a printout of the survey station data is given in line 880, using the printer routine at line 6250.

(h) Entry of survey point data

Line numbers 1000–1240

The routine is accessed each time data is to be entered for a survey point. The essential details required are survey point number (P_0), the corrected slope distance (S), the height of the target above the point (H_2), and if necessary the height of the reflector or prism above the point (N_2), all as shown in Fig. 3.17. In addition, it is essential to know either the zenith angle (W) or the survey point elevation (E_2).

(i) <1> Short-range EDM calculations

Line numbers 1250–1390

This routine for EDM calculations assumes that only short ranges have been measured and that there has been no curvature of the ray path. After atmospheric corrections for temperature and pressure have been applied to the observed slope distance, the corrected slope distance S is used to compute the horizontal distance and target elevation if required from the zenith angle. The nomenclature for the EDM slope reduction program is shown on Fig. 3.17 where four heights require to be measured in the field:

(i) The theodolite height, H_1
(ii) The EDM instrument height, N_1
(iii) The target height, H_2
(iv) The reflector (prism) height, N_2.

The zenith angle W is also required if the elevation of the target survey point is unknown, Fig. 3.17(a). Since curvature and refraction are neglected at short ranges, the horizontal distance and vertical height difference between the survey station and survey point are calculated in line 1300 from:

$$H = S\cos W_0 \quad (3.42)$$

and

$$V = S\sin W_0 \quad (3.43)$$

where W_0 is the angle of elevation or depression of the target sight above the horizontal, i.e., if the zenith angle W is less than 90° then $W_0 = 90 - W$, and if W is greater than 90° then $W_0 = W - 90$. Therefore the target elevation can be determined in lines 1310, 1350 from:

$$E_2 = E_1 + H_1 \pm V - H_2 \quad (3.44)$$

where the positive/negative choice depends upon whether the zenith angle is elevated or depressed from the horizontal. If the zenith angle W is not known, so that Equation 3.42 cannot be used to calculate H, it is necessary to know the elevation of the target point E_2. The vertical height difference between the two points is calculated in line 1370 from:

$$V = E_2 + H_2 - E_1 - H_1 \quad (3.45)$$

and the horizontal distance determined in line 1380 from:

$$H = S - V^2/2S \tag{3.46}$$

This horizontal distance H is at the elevation of the initial survey station.

(j) <2> Long-range EDM calculations

Line numbers 1400–1520

The nomenclature for long-range EDM is the same as for short-range EDM as shown in Fig. 3.17 and described above in the previous Section (i). The zenith angle as previously is W, but before it can be used in the calculations, the angle must be corrected for the earth's curvature (C), refraction (R_2), where the coefficient of refraction is taken to be 0.071, and also the difference in height of the instrument and the target (N_9) as given in lines 1410–1450 using the following equations:

$$C = \sin^{-1}(S\sin W/2R_1) \tag{3.47}$$

$$R_2 = 19 \times 10^{-8} S\sin W \tag{3.48}$$

$$N_9 = N_1 - N_2 - H_1 + H_2 \tag{3.49}$$

The corrected zenith angle W_9 is then given by:

$$W_9 = W - C + R_2 + \sin^{-1}[N_9 \sin(W - 2C + R_2)/S] \tag{3.50}$$

The horizontal distance (H) between the survey station and survey point, and the vertical height difference (V) between the instrument and the target can then be calculated in lines 1460–1470 from:

$$H = S\sin(W_9 - C)/\cos C \tag{3.51}$$

$$V = S\cos W_9/\cos C \tag{3.52}$$

The elevation of the survey point can then be found in line 1480 from:

$$E_2 = E_1 + N_1 + V - N_2 \tag{3.53}$$

If the elevation of the survey point is known at the outset, then the calculations are simplified, and the horizontal distance is computed in lines 1500–1520 from:

$$H = \left[\frac{S^2 - (E_2 + N_2 - E_1 - N_1)^2}{(R_1 + E_1 + N_1)(R_1 + E_2 + N_2)}\right]^{1/2} (R_1 + E_1) \tag{3.54}$$

(k) Screen display of results

Line numbers 1540–1900

The horizontal distance H, calculated above, is at the elevation of the survey station, Fig. 3.17. If it is desired to plot the data on a national grid drawing, the horizontal distance, H_3, at sea level is required (Fig. 3.17) and is calculated at line 1540 from:

$$H_3 = HR_1/(R_1 + E_1) \tag{3.55}$$

where R_1 is radius of earth, and E_1 is elevation of survey station. Then the various computed variables H_3, H and E_2 are tidied up to three decimal places for display in lines 1600–1750. An option to obtain a computer printout is given at line 1760 and provided in a routine at line 6500.

Further options are given in lines 1810 and 1860 to enter more data for the same survey station or to enter data for another station.

(l) Routine to print survey location, etc.

Line numbers 6000–6080

This routine is provided for record purposes, and is based on Applesoft DOS and may require alteration for different microcomputers and printers.

(m) Routine to print survey station data

Line numbers 6250–6350

The same comments apply to this routine, as to (l) above.

(n) Routine to print EDM slope solutions

Line numbers 6500–6540

This routine uses the same format as the previous screen display in lines 1610–1750. The same comments apply to this printout as to (l) above.

(o) Routine to convert DD.MMSS to decimals

Line numbers 8000–8040

This routine was discussed previously in Section 2.2.6.

(p) Routine to convert decimals to DD.MMSS

Line numbers 8050–8140

This routine was discussed previously in Section 2.2.6.

(q) Routine to clear screen

Line numbers 9000–9010

To simplify portability, the Applesoft 'HOME' command to clear the screen has been placed in a subroutine to allow an easy change to another microcomputer.

(r) Routine to display error message

Line numbers 9100–9130

If a wrong keyboard entry has been made, the error is trapped and this error message displayed and re-entry requested.

(s) Error trace routine

Line numbers 9200–9290

This error trace routine was discussed earlier in Section 2.2.9.

(t) Termination of program

Line numbers 9500–9620

When the '<QUIT>' selection is made either after the initial screen header or from the program menu, an end-of-program message is displayed to advise the operator that computations are complete.

3.7.2 EDM slope reduction – numeric variables

C	= Zenith correction due to the earth's curvature
*C*0	= Counter in error trace routine
*C*9	= Temporary angle storage
D, *D*0, *D*1, *D*2, *D*5	= Degrees of angle
*D*9	= Flag for Decimal <1> or DD.MMSS <2>
E	= Error number
*E*1	= Elevation of survey station
*E*2	= Elevation of survey point
H, *H*3	= Horizontal distances
*H*1	= Height of theodolite
*H*2	= Height of target
*H*5, *H*6	= Temporary storage of variables
K	= Flag for zenith angle <1> or target elevation <2>
*K*1	= Survey station number
L	= Line number for error
M, *M*0, *M*1	= Minutes of angles
*N*1	= Height of EDM at station *K*1
*N*2	= Height of reflector at point *P*0
*N*9	= *N*1 − *N*2 − *H*1 + *H*2
*P*0	= Survey point number
*P*7	= Print code number
R	= 57.2957795 for conversion to radians
*R*0	= Units, feet <1> or metres <2>
*R*1	= Radius of earth
*R*2	= Refraction correction to zenith angle
S	= Slope distance
*S*0, *S*1	= Seconds of angle
T	= Short range EDM <1> or long range EDM <2>
V	= Vertical height difference
W	= Zenith angle
*W*0	= Angle of elevation or depression
*W*8	= Temporary storage of variables
*W*9	= Zenith angle adjusted for curvature and refraction

3.7.3 EDM slope reduction – string variables

A$	Storage for location of survey
B$	Storage for operator's name
C$	Storage for date of survey
D$	= CHR$(4), i.e. CTRL-D
*E*0$	= "SORRY, DATA ERROR . . . PLEASE RE-ENTER"
K$	Storage for instrument used
P$	= "DO YOU WISH PRINTOUT OF ABOVE DATA (Y/N)"
Q$	= Question response (Y/N)
V$	= "DD.DEC" if *D*9 = 1, "DD.MMSS" if *D*9 = 2

3.7.4 EDM slope reduction – BASIC program

```
10  REM  <EDM.SLOPE> PROGRAM FOR APPLE II+  USES QUME PRINTER
20 D$ =  CHR$ (4): ONERR  GOTO 9200
30 A = Ø:B = Ø:E = Ø:G = Ø:J = Ø:K = Ø:N = Ø:S = Ø:U = Ø:V = Ø:W = Ø
40 P$ = "DO YOU WISH PRINTOUT OF ABOVE DATA (Y/N)"
50 EØ$ = "SORRY, DATA ERROR ... PLEASE RE-ENTER."
95  REM  ***************************************************
100  GOSUB 9000
110  PRINT "****************************************"
120  PRINT "*                                      *"
130  PRINT "*     UNIVERSITY  OF  STRATHCLYDE      *"
140  PRINT "*  DEPARTMENT  OF  CIVIL  ENGINEERING  *"
150  PRINT "*                                      *"
160  PRINT "*          SURVEYING  SECTION          *"
170  PRINT "*          EDM SLOPE  SOLUTIONS        *"
180  PRINT "****************************************"
190  PRINT
200  PRINT "ARE ANGLES IN DECIMALS <1>, DD.MMSS <2> OR DO YOU WISH TO QUI
     T <3>."
210  INPUT D9
220  IF D9 < 1 OR D9 > 3 THEN 200
230  ON D9 GOTO 240,240,9500
240 R = 57.2957795
250  PRINT
260  PRINT "SELECT FROM TWO EDM PROGRAMS:-"
270  PRINT "<1> SHORT RANGE (NO CURVATURE EFFECTS),"
280  PRINT "<2> LONG RANGE (CURVATURE CORRECTION)."
290  INPUT T: IF T < 1 OR T > 2 THEN 250
300  PRINT
310  PRINT "SELECT UNITS <1> IMPERIAL (FEET), OR                 <2> METR
     IC (METRES)."
320  INPUT RØ
330  IF RØ < 1 OR RØ > 2 THEN 310
335  REM  R1 IS RADIUS OF EARTH
340  IF RØ = 1 THEN R1 = 20906 * 1000
350  IF RØ = 2 THEN R1 = 20906 * 1000 / 3.280839895
360  IF D9 = 1 THEN V$ = "DD.DEC"
370  IF D9 = 2 THEN V$ = "DD.MMSS"
380  IF T = 1 THEN W$ = "(SHORT)"
390  IF T = 2 THEN W$ = "(LONG)"
495  REM  KEYBOARD ENTRY OF LOCATION ETC.
500  GOSUB 9000
```

```
510  PRINT "UNIVERSITY OF STRATHCLYDE"
520  PRINT "EDM.SLOPE PROGRAM"
530  PRINT "****************************************"
540  PRINT
550  PRINT "ENTER LOCATION OF WORK, ETC. WHEN        REQUESTED, AND PRESS
     <RETURN>"
560  PRINT "ENTER LOCATION OF SURVEY :-"
570  INPUT A$
580  PRINT "ENTER OPERATOR'S NAME :-"
590  INPUT B$
600  PRINT "ENTER DATE OF SURVEY AS DD/MM/YY :-"
610  INPUT C$
620  PRINT "ENTER INSTRUMENT USED :-"
630  INPUT K$
640  PRINT
650  PRINT "DO YOU WISH PRINTOUT OF LOCATION ETC.,  (Y/N) ";
660  INPUT Q$
670  IF Q$ = "Y" THEN 6000
680  IF Q$ < > "N" THEN 650
690 :
700  GOSUB 9000
710  PRINT "ENTER DETAILS OF SURVEY STATION WHEN    REQUESTED AND PRESS <
     RETURN>."
720  PRINT
730  PRINT "ENTER NO. OF SURVEY STN., NUMERALS ONLY"
740  INPUT K1
750  IF K1 < 0 THEN 730
760  PRINT
770  PRINT "ENTER ELEVATION OF SURVEY STN. ";K1
780  INPUT E1
790  PRINT
800  PRINT "ENTER HEIGHT OF THEODOLITE AT STN. ";K1
810  INPUT H1
820  IF H1 < 0 THEN 800
830  PRINT
840  PRINT "ENTER HEIGHT OF EDM AT STN. ";K1
850  INPUT N1
860  IF N1 < 0 THEN 840
870  PRINT
880  PRINT P$
890  INPUT Q$
900  IF Q$ = "Y" THEN 6250
910  IF Q$ < > "N" THEN 880
995  REM ***********************************************
1000  GOSUB 9000
1010  PRINT "ENTER SURVEY POINT NO., NUMERALS ONLY"
1020  INPUT P0
1030  PRINT
1040  PRINT "ENTER CORRECTED SLOPE DISTANCE"
1050  INPUT S
1060  IF S < 0 THEN 1040
1070  PRINT
1080  PRINT "ENTER HEIGHT OF TARGET"
1090  INPUT H2
1100  IF H2 < 0 THEN 1080
1110  PRINT
1120  PRINT "ENTER HEIGHT OF REFLECTOR"
1130  INPUT N2
1140  IF N2 < 0 THEN 1120
```

```
1145  REM  CHECK IF ZENITH ANGLE KNOWN
115Ø  PRINT : PRINT "ENTER <1> IF ZENITH ANGLE KNOWN          OR    <2> IF
      TARGET ELEVATION KNOWN."
116Ø  INPUT K: IF K < 1 OR K > 2 THEN 115Ø
117Ø  IF K = 2 THEN 123Ø
118Ø  PRINT : PRINT "ENTER ZENITH ANGLE IN ";V$
119Ø  INPUT W: IF D9 = 1 THEN 121Ø
12ØØ D5 = W: GOSUB 8ØØØ:W = D5
121Ø  IF W > 18Ø THEN W = 36Ø - W
122Ø  GOTO 125Ø
1225  REM  I.E. K=2, TARGET ELEVATION KNOWN
123Ø  PRINT : PRINT "ENTER ELEVATION OF SURVEY POINT ";PØ
124Ø  INPUT E2
125Ø  GOSUB 9ØØØ: PRINT "CALCULATING .... "
126Ø  IF T = 2 THEN 14ØØ
1265  REM  CHECK IF TARGET ELEVATION KNOWN (K=2)
127Ø  IF K = 2 THEN 137Ø
128Ø  IF W > 9Ø THEN 133Ø
1285  REM  I.E. W < 9Ø, ELEVATED ANGLE
129Ø WØ = 9Ø - W
13ØØ V = S *  SIN (WØ / R):H = S *  COS (WØ / R)
131Ø E2 = E1 + H1 + V - H2
132Ø  GOTO 154Ø
1325  REM  W > 9Ø, DEPRESSED ANGLE
133Ø WØ = W - 9Ø
134Ø V = S *  SIN (WØ / R):H = S *  COS (WØ / R)
135Ø E2 = E1 + H1 - V - H2
136Ø  GOTO 154Ø
1365  REM  K=2, TARGET ELEVATION KNOWN
137Ø V = E2 + H2 - E1 - H1
138Ø H = S - V * V / (2 * S)
139Ø  GOTO 154Ø
1395  REM  T=2, LONG RANGE EDM
14ØØ  IF K = 2 THEN 15ØØ
141Ø C9 = S *  SIN (W / R) / (2 * R1):C =  ATN (C9 /  SQR (1 - C9 * C9))
142Ø R2 = .ØØØØØØ19 * S *  SIN (W / R)
143Ø N9 = N1 - N2 - H1 + H2
144Ø W8 = N9 *  SIN ((W - 2 * C + R2) / R) / S
145Ø W9 = W - C + R2 +  ATN (W8 /  SQR (1 - W8 * W8))
146Ø H = S *  SIN ((W9 - C) / R) /  COS (C / R)
147Ø V = S *  COS (W9 / R) /  COS (C / R)
148Ø E2 = E1 + N1 + V - N2
149Ø  GOTO 154Ø
1495  REM  K=2, TARGET ELEVATION KNOWN
15ØØ H5 = S * S - (E2 + N2 - E1 - N1) ^ 2
151Ø H6 = (R1 + E1 + N1) * (R1 + E2 + N2)
152Ø H =  SQR (H5 / H6) * (R1 + E1)
153Ø :
1535  REM  PREPARE DATA FOR DISPLAY
154Ø H3 = H * R1 / (R1 + E1)
155Ø H3 =  INT (H3 * 1ØØØ + .5) / 1ØØØ
156Ø H =  INT (H * 1ØØØ + .5) / 1ØØØ
157Ø E2 =  INT (E2 * 1ØØØ + .5) / 1ØØØ
158Ø  IF D9 = 1 THEN 16ØØ
159Ø D = W: GOSUB 8Ø5Ø:W = D
16ØØ  GOSUB 9ØØØ:P7 = Ø
161Ø  PRINT "EDM SLOPE REDUCTION PROGRAM :-"
162Ø  PRINT "****************************************"
163Ø :
```

```
164Ø  PRINT "SURVEY POINT NO. ";PØ;" : ";W$
165Ø  PRINT "========================================"
166Ø  PRINT "SLOPE DISTANCE            = ";S
167Ø  PRINT "HEIGHT OF TARGET          = ";H2
168Ø  PRINT "HEIGHT OF REFLECTOR       = ";N2
169Ø  IF K = 2 THEN 171Ø
170Ø  PRINT "ZENITH ANGLE              = ";W
171Ø  PRINT "ELEVATION SURVEY POINT    = ";E2
172Ø  PRINT
173Ø  PRINT "HORIZONTAL DISTANCE       = ";H
174Ø  PRINT "HORIZ.DIST. AT SEA LEVEL = ";H3
175Ø  PRINT "****************************************": IF P7 = 1 THEN  RETURN

176Ø  PRINT P$
177Ø  INPUT Q$
178Ø  IF Q$ = "Y" THEN 65ØØ
179Ø  IF Q$ < > "N" THEN 176Ø
18ØØ  GOSUB 9ØØØ
181Ø  PRINT "DO YOU WISH TO ENTER MORE DATA FOR      SURVEY STN. ";K1;" (
     Y/N) ";
182Ø  INPUT Q$
183Ø  IF Q$ = "Y" THEN 1ØØØ
184Ø  IF Q$ < > "N" THEN 18ØØ
185Ø  PRINT
186Ø  PRINT "DO YOU WISH TO ENTER DATA FOR ANOTHER   SURVEY STN. (Y/N)"
187Ø  INPUT Q$
188Ø  IF Q$ = "Y" THEN 7ØØ
189Ø  IF Q$ < > "N" THEN 186Ø
19ØØ  GOTO 1ØØ: REM  RETURN TO HEADER
5985  REM  ************************************************************
599Ø  REM  PRINTOUT OF LOCATION, ETC
6ØØØ  PRINT D$;"PR#1": PRINT "LOCATION OF SURVEY :- ";A$
6Ø1Ø  PRINT "************************************************************
     ********************"
6Ø2Ø  PRINT "OPERATOR'S NAME :- ";B$
6Ø3Ø  PRINT "DATE OF SURVEY  :- ";C$
6Ø4Ø  PRINT "INSTRUMENT USED :- ";K$
6Ø6Ø  PRINT "************************************************************
     ********************"
6Ø7Ø  PRINT D$;"PR#Ø"
6Ø8Ø  GOTO 7ØØ
6243  REM  ****************************************************
6245  REM  PRINTOUT OF STATION DATA
625Ø  PRINT D$;"PR#1": POKE 1784 + 1,8Ø
626Ø  PRINT :P7 = 1: REM  PRINTER CODE
627Ø  PRINT "EDM.SLOPE SURVEY STATION ";K1
628Ø  PRINT "****************************************"
629Ø  PRINT "ELEVATION OF STATION = ";E1
63ØØ  PRINT "HEIGHT OF THEODOLITE = ";H1
631Ø  PRINT "HEIGHT OF EDM UNIT   = ";N1
632Ø  PRINT "========================================"
633Ø :
634Ø  PRINT D$;"PR#Ø"
635Ø  GOTO 1ØØØ
6493  REM  ****************************************************
6495  REM  PRINTOUT OF EDM SOLUTION
65ØØ  PRINT D$;"PR#1": POKE 1784 + 1,8Ø
651Ø  PRINT :P7 = 1: REM  PRINTER CODE
6515  REM  USE DISPLAY AT LINES 161Ø-175Ø FOR PRINTER WITH P7=1
```

```
6520  GOSUB 1610
6530  PRINT D$;"PR#Ø"
6540  GOTO 18ØØ
7993  REM  ***************************************************
7995  REM  ROUTINE DD.MMSS TO DECIMALS
8ØØØ DØ =  INT (D5)
8ØØ5  REM  FIND NO. OF MINUTES,MØ
8Ø1Ø M1 = (D5 - DØ) * 1ØØ:MØ =  INT (M1 + .1)
8Ø15  REM  FIND NO. OF SECONDS,SØ
8Ø2Ø SØ =  INT (M1 * 1ØØ - MØ * 1ØØ + .5)
8Ø25  REM  COLLECT DEGS,MINS,SECS
8Ø3Ø D5 = DØ + MØ / 6Ø + SØ / 36ØØ
8Ø4Ø  RETURN
8Ø43  REM  ***************************************************
8Ø45  REM  ROUTINE DECIMAL TO DD.MMSS
8Ø5Ø D1 =  INT (D)
8Ø55  REM  FIND TOTAL NO. OF SECONDS
8Ø6Ø D2 = (D - D1) * 36ØØ
8Ø65  REM  FIND NO. OF MINUTES
8Ø7Ø M =  INT (D2 / 6Ø)
8Ø75  REM  FIND NO. OF SECONDS
8Ø8Ø S1 = D2 - 6Ø * M + .5
8Ø9Ø  IF S1 < 6Ø THEN 813Ø
81ØØ M = M + 1:S1 = Ø
811Ø  IF M < 6Ø THEN 813Ø
812Ø D1 = D1 + 1:M = Ø
813Ø D = D1 * 1ØØØØ + M * 1ØØ + S1
814Ø D =  INT (D) / 1ØØØØ: RETURN
8985  REM  ******************************************
899Ø  REM  GOSUB ROUTINE TO CLEAR SCREEN
9ØØØ  HOME
9Ø1Ø  RETURN
9Ø93  REM  ***************************************************
9Ø95  REM  GOSUB ROUTINE FOR ERRORS
91ØØ  PRINT
911Ø  PRINT EØ$
912Ø  PRINT
913Ø  RETURN
9193  REM  ***************************************************
9195  REM  ERROR TRACE ROUTINE
92ØØ  PRINT  CHR$ (7): REM  CTRL-B (BELL)
921Ø E =  PEEK (222): REM  GET ERROR NO.
922Ø  IF CØ = 1 THEN 926Ø
923Ø  INVERSE
924Ø  PRINT "ERROR NO. ";E;" FOUND"
925Ø CØ = 1: TRACE : RESUME
926Ø  PRINT "ERROR ON SECOND LINE NO. "
927Ø  NORMAL
928Ø  NOTRACE
929Ø  STOP
9493  REM  ***************************************************
9495  REM  END OF PROGRAM
95ØØ  GOSUB 9ØØØ
951Ø  PRINT
952Ø  PRINT "END RESECTION PROGRAM"
953Ø  PRINT
954Ø  PRINT "************************"
955Ø  REM  PROGRAM PREPARED BY
956Ø  REM  DR. P.H. MILNE
```

```
9570 REM DEPT. OF CIVIL ENGINEERING
9580 REM UNIVERSITY OF STRATHCLYDE
9590 REM GLASGOW G4 ONG
9600 REM SCOTLAND
9610 REM ***************************
9620 END
```

3.7.5 EDM slope reduction – computer printout

```
LOCATION OF SURVEY :- STRATHCLYDE UNIVERSITY - STEPPS
************************************************************************
OPERATOR'S NAME :- JOHN SMITH
DATE OF SURVEY  :- 24/04/83
INSTRUMENT USED :- CD6 + T15
************************************************************************

EDM.SLOPE SURVEY STATION 4
****************************************
ELEVATION OF STATION = 88.06
HEIGHT OF THEODOLITE = 1.6
HEIGHT OF EDM UNIT   = 1.8
========================================

EDM SLOPE REDUCTION PROGRAM :-
****************************************
SURVEY POINT NO. 3 : (SHORT)
========================================
SLOPE DISTANCE            = 197.307
HEIGHT OF TARGET          = 1.53
HEIGHT OF REFLECTOR       = 1.71
ZENITH ANGLE              = 89.313
ELEVATION SURVEY POINT    = 89.766

HORIZONTAL DISTANCE       = 197.3
HORIZ.DIST. AT SEA LEVEL = 197.297
****************************************

EDM SLOPE REDUCTION PROGRAM :-
****************************************
SURVEY POINT NO. 5 : (SHORT)
========================================
SLOPE DISTANCE            = 206.871
HEIGHT OF TARGET          = 1.55
HEIGHT OF REFLECTOR       = 1.72
ZENITH ANGLE              = 89.121
ELEVATION SURVEY POINT    = 90.988

HORIZONTAL DISTANCE       = 206.851
HORIZ.DIST. AT SEA LEVEL = 206.848
****************************************

EDM SLOPE REDUCTION PROGRAM :-
****************************************
SURVEY POINT NO. 6 : (SHORT)
========================================
SLOPE DISTANCE            = 483.458
HEIGHT OF TARGET          = 1.44
HEIGHT OF REFLECTOR       = 1.64
```

```
ZENITH ANGLE               = 89.44
ELEVATION SURVEY POINT     = 90.47

HORIZONTAL DISTANCE        = 483.453
HORIZ.DIST. AT SEA LEVEL = 483.446
****************************************

EDM SLOPE REDUCTION PROGRAM :-
****************************************
SURVEY POINT NO. 7 : (SHORT)
========================================
SLOPE DISTANCE             = 566.86
HEIGHT OF TARGET           = 1.07
HEIGHT OF REFLECTOR        = 1.24
ELEVATION SURVEY POINT     = 102.905

HORIZONTAL DISTANCE        = 566.679
HORIZ.DIST. AT SEA LEVEL = 566.671
****************************************

EDM SLOPE REDUCTION PROGRAM :-
****************************************
SURVEY POINT NO. 8 : (SHORT)
========================================
SLOPE DISTANCE             = 541.397
HEIGHT OF TARGET           = 1.155
HEIGHT OF REFLECTOR        = 1.33
ELEVATION SURVEY POINT     = 97.305

HORIZONTAL DISTANCE        = 541.325
HORIZ.DIST. AT SEA LEVEL = 541.318
****************************************

EDM SLOPE REDUCTION PROGRAM :-
****************************************
SURVEY POINT NO. 7 : (LONG)
========================================
SLOPE DISTANCE             = 566.86
HEIGHT OF TARGET           = 1.07
HEIGHT OF REFLECTOR        = 1.24
ELEVATION SURVEY POINT     = 102.905

HORIZONTAL DISTANCE        = 566.679
HORIZ.DIST. AT SEA LEVEL = 566.671
****************************************

EDM SLOPE REDUCTION PROGRAM :-
****************************************
SURVEY POINT NO. 13 : (LONG)
========================================
SLOPE DISTANCE             = 1543.287
HEIGHT OF TARGET           = 1.56
HEIGHT OF REFLECTOR        = 1.72
ZENITH ANGLE               = 88.542
ELEVATION SURVEY POINT     = 117.612

HORIZONTAL DISTANCE        = 1543.005
HORIZ.DIST. AT SEA LEVEL = 1542.984
****************************************
```

3.8 3-D POSITIONS FROM EDM

The previous program <EDM.SLOPE> calculated the horizontal distance knowing slope distance together with the zenith angle, or the target elevation. One of the advantages of modern EDM instruments where the EDM and theodolite are combined in one unit, is that they can be used for tacheometric-type surveys. Thus if the co-ordinates of the station are known, or if two observations can be made to two known reference stations, then the 3-D position of the survey points can be found. The previous program <TACHY.3-D> has been adapted to utilize the advantages of EDM slope readings.

3.8.1 Alterations to <TACHY.3-D>

The previous program <TACHY.3-D> only requires fifteen line number changes for use as an <EDM.3-D> program, of which two are remarks (REM) and four are deletions.

The program uses the EDM short range equations presented in Equations 3.42 and 3.43, which are accessed via a GOSUB routine at line 5500. The subroutine index and previous program are given in Section 3.4 and only alterations are noted here.

3.8.2 3-D Positions from EDM – numeric variables

*E*1	=	Slope distance
M	=	Height of target/reflector

3.8.3 3-D Positions from EDM – string variables

U$	=	“ENTER SLOPE DISTANCE AND TARGET HEIGHT SEPARATED BY COMMAS”

3.8.4 3-D Positions from EDM – BASIC program

```
10  REM  <EDM.3-D> PROGRAM FOR APPLE II+  USES QUME PRINTER
20  REM  EDM.TACHY DATA REDUCTION PROGRAM FOR APPLE II+
150 U$ = "ENTER SLOPE DISTANCE AND TARGET HEIGHT  SEPARATED BY COMMAS"
1020  PRINT "EDM.TACHY DATA REDUCTION PROGRAM"
1140 :
1150 :
4720  PRINT M$"RUN EDM.3-D"
5040  VTAB (5): HTAB (5): PRINT "END OF EDM PROGRAM"
5150  INPUT E1,M
5160 :
5390  INPUT E1,M
5400  LET W = V9 - 90
```

```
5500  LET H = El *  COS (W / R)
5510  LET V = El *  SIN (W / R)
6050 :
```

3.8.5 3-D Positions from EDM – computer printout

```
LOCATION OF SURVEY :- GLENMORE PARK
*************************************************************************
OPERATOR'S NAME :-     JOHN SMITH
DATE OF SURVEY  :-     20/05/83
INSTRUMENT USED :-     CD6 + T15
*************************************************************************

RANGE - RANGE FIX ON STN. 3
=========================================================================

STN.NO. EASTING      NORTHING     LEVEL
1       100          100          132.67
2       473.265      194.328      126.34
3       215.87       264.35       122.94
HT. OF THEODOLITE ABOVE STN. 3 IS 1.45
*************************************************************************

PT.NO.  HORIZ.ANGLE  HORIZ.RANGE    EASTING        NORTHING       LEVEL
=========================================================================
11      32.44        103.399        120.055        225.479        125.457
12      12.252       95.418         145.401        200.018        123.588
13      352.203      88.558         174.943        185.817        122.305
14      344.124      75.848         190.682        192.807        121.101
15      336.561      68.378         201.512        197.496        122.469
*************************************************************************
```

3.9 TRAVERSE SURVEYS AND ADJUSTMENT

In land surveying, the co-ordinate position of points in a horizontal plane is required for many reasons, the three main ones being:

(i) Control of topographic surveying,
(ii) Control for construction surveying and setting out,
(iii) Control for aerial surveying.

Previous sections in this chapter have dealt with the determination of the co-ordinates of stations either by tacheometry, intersection or resection, with direct measurements, either distance and/or angle, to adjacent known reference stations. A traverse survey allows a large number of stations to be linked in sequence by measuring the angles between successive stations and the distance between the stations. A traverse survey therefore provides a network of control stations over an area, for subsequent use, either for topographic surveying or engineering surveying where details are required for the preparation of site plans prior to the design and setting out of an engineering project.

The three principal types of traverses are:

(i) Closed,
(ii) Link,
(iii) Open,

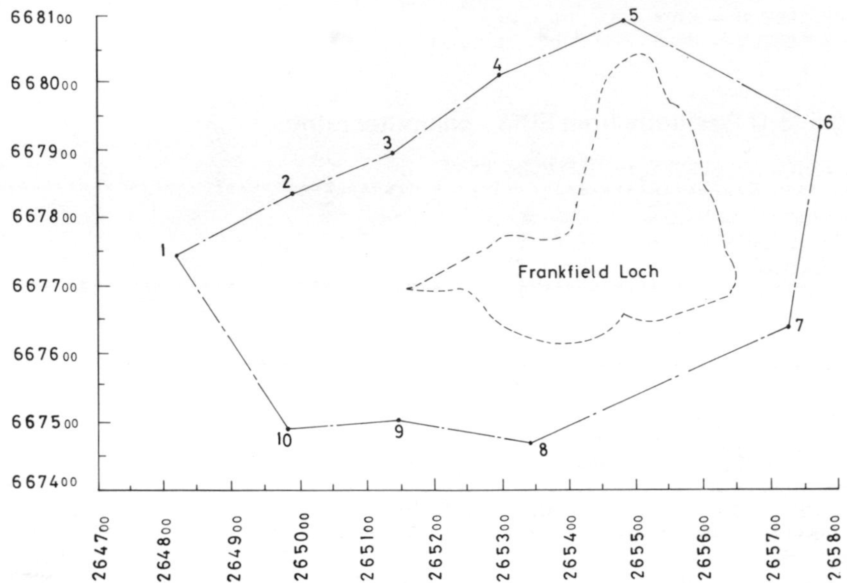

Fig. 3.18. Closed traverse of stations 1–10 carried out by undergraduate civil engineering students of Strathclyde University at Stepps Playing Fields near Glasgow.

where only the first two types of traverses are capable of adjustment. In the case of a closed traverse (Fig. 3.18), the traverse starts and finishes at Station 1, where the lengths measured form a polygon, sometimes called a closed-route traverse. To calculate the relative co-ordinates of the other stations in the traverse, the co-ordinates of Station 1 are required together with the whole circle bearing (WCB) of the first length (1–2). A link traverse need not close back on to the initial station, but can close to another station whose co-ordinates are known, for example Station 6 in Fig. 3.18.

In both the closed and link traverses there is an external check on the observations since the traverses start and finish at known points. It is thus possible in closed and link traverses to determine the misclosure, that is the difference between the computed and known co-ordinates of the final station, thus allowing traverse adjustment.

If an operator has a choice of running a closed or a link traverse, the former should always be chosen, because it is then possible to obtain a check on the angle measurements. Any angular misclosure is found by comparing the sum of the observed angles with one of the following theoretical values given in Equations 3.56 and 3.57:

$$\text{Sum of internal angles} = (2n-4)\times 90° \qquad (3.56)$$

$$\text{Sum of external angles} = (2n+4)\times 90° \qquad (3.57)$$

where n is the number of angles or sides of the polygon. If the misclosure is acceptable, the error is then equally distributed round the angles to satisfy

Equations 3.56 and 3.57. Such a check is not possible with either link or open traverses, so great care has to be taken in the angle measurement process, and the angles read at least four times in both face left and face right positions to determine the mean angle.

The computer program <TRAVERSE> included in this section allows a closed or link traverse to be adjusted either by the Bowditch or Transit method. These are the two most common adjustment methods used by students in land surveying or engineering surveying. It should however be pointed out that these techniques were developed before the advent of EDM which provides much greater accuracy than that obtainable previously by chaining or banding. The Bowditch method is deduced from least squares and is based on the assumptions that the errors in the measurements of the lengths, are directly proportional to the square roots of the lengths measured, and the errors in the bearings of the lines are inversely proportional to the square roots of the lengths of the lines. The Transit method has however no mathematical basis, and accordingly will invariably give different adjustments when compared with the Bowditch adjustments. Research in various academic centres into the best method of traverse adjustment using EDM measurements is currently underway to provide a more exact method than the two methods described here.

In both the Bowditch and Transit methods, the initial stages of computation are the same:

(i) Abstraction of mean angles and distances from field observations;
(ii) (a) Adjustment of mean observed angles (Equations 3.56 or 3.57);
(b) Adjustment of mean measured distances to corrected horizontal distances;
(iii) Calculation of partial co-ordinates, and
(iv) Computation of initial co-ordinates using unadjusted partial co-ordinates.

From the above calculations, the misclosure at the closing station can then be found.

In the Bowditch method of adjustment, the following corrections are applied:

correction to partial departures =

$$\frac{\text{total error in departures}}{\text{total length of traverse}} \times \text{length of side} \tag{3.58}$$

correction to partial latitudes =

$$\frac{\text{total error in latitudes}}{\text{total length of traverse}} \times \text{length of side} \tag{3.59}$$

In the Transit method of adjustment, the following corrections are applied:

correction to partial departures =

$$\frac{\text{total error in departures}}{\text{absolute sum of departures}} \times \text{absolute departure of line} \tag{3.60}$$

correction to partial latitudes =

$$\frac{\text{total error in latitudes}}{\text{absolute sum of latitudes}} \times \text{absolute latitude of line} \qquad (3.61)$$

These corrections when applied to the partial departures and latitudes should ensure closure at the final station. Care should be taken when calculating the Transit corrections in Equations 3.60 and 3.61 that it is the absolute values which are used rather than algebraic values.

3.9.1 Traverse surveys – subroutine index

	Line numbers	*Function*
(a)	10–90	Initialization and control
(b)	100–190	Screen header display
(c)	200–240	Selection of angle format
(d)	300–430	Menu selection for type of traverse
(e)	500–700	Entry of survey location, etc.
(f)	1000–1290	<1> Closed Traverse routine
(g)	1300–1410	Station co-ordinate calculations
(h)	1420–1770	Screen display of results
(i)	1800–1830	Traverse adjustment selection
(j)	1840–1920	Bowditch method of adjustment
(k)	1930–1980	Option to calculate new distances
(l)	2000–2160	<2> Link Traverse routine
(m)	2200–2450	Transit method of adjustment
(n)	2500–2710	Routine to calculate new distances
(o)	3000–3020	<3> Open Traverse routine
(p)	6000–6080	Routine to print survey location, etc.
(q)	6250–6430	Routine to print traverse solution
(r)	6500–6680	Routine to print partial eastings and northings
(s)	6750–6890	Routine to print new distances
(t)	8000–8040	Routine to convert DD.MMSS to decimals
(u)	8050–8140	Routine to convert decimals to DD.MMSS
(v)	9000–9010	Routine to clear screen
(w)	9100–9130	Routine to display error message
(x)	9200–9290	Error trace routine
(y)	9500–9620	Termination of program

(a) Initialization and control

Line numbers 10–90

All numeric variables and required string variables are initialized, except those that depend on the number of traverse stations which are dimensioned in lines 1060, 2200 and 2500.

(b) Screen header display

Line numbers 100–190

This is the screen display used for civil engineering students at Strathclyde University. The operator can alter this as required to suit another organization.

(c) Selection of angle format

Line numbers 200–240

The operator has a choice in line 200 of entering angles in decimal or DD.MMSS format, where the $D9$ flag is set to 1 for decimals and 2 for DD.MMSS in line 210.

(d) Menu selection for type of traverse

Line numbers 300–430

A choice of three traverse programs are presented in lines 340–360:

<1> CLOSED <2>LINK <3> OPEN

and the string $H\$$ set to the appropriate menu selection.

(e) Entry of survey location, etc.

Line numbers 500–700

This routine is included to allow a record to be kept of site location, operator, date, etc. After keyboard entry, the option of a printout at line 6000 is provided.

(f) <1> Closed traverse routine

Line numbers 1000–1290

The keyboard entries for this routine are also used by the link and open traverse. After entering the total number of traverse stations, the program requests the stations in sequence, line numbers 1070–1110, so it is not necessary to choose consecutive numbers to label the stations. These numbers are stored in the array $S(N-1)$ where N is the total number of stations. The co-ordinates of the initial station are then entered in lines 1120–1130. If a link traverse ($T = 2$) is being entered, the program branches from line 1140 to 2030 to collect the co-ordinates of the final station.

The whole circle bearings (WCBs) and horizontal distances are then entered consecutively in a loop between lines 1160 and 1230. If the angles are entered in the DD.MMSS format they are automatically converted to decimals. The total length of the traverse is then computed in another loop between lines 1240 and 1290.

(g) Station co-ordinate calculations

Line numbers 1300–1410

The computation of the co-ordinates of the stations commences in a loop between lines 1300 and 1390. From the WCBs, $W(K)$ the correct quadrant (Fig. 3.1) is determined in line 1310 and the respective reduced bearing (B), together with the departure, $D(K)$, and latitude, $L(K)$, computed in lines 1340–1370. The co-ordinates of the next station in the traverse are then computed in line 1380. On completion of the loop, the misclosure at the final station is calculated in line 1410 for the closed traverse ($T = 1$) and line 2100 for the link traverse ($T = 2$). In addition, the closure distance ($C0$) is computed.

(h) Screen display of results

Line numbers 1420–1770

Initially all computed co-ordinates are prepared for display to the third decimal figure and the angles transferred back to the DD.MMSS angle format if required. The results of the co-ordinate computations are displayed in lines 1500–1590 with a printout option provided at line 6250. This is followed by a display of the partial eastings and northings together with the misclosure and closing error in lines 1650–1730, with a printout option provided at line 6500. This routine is also used for the display of adjusted co-ordinates.

(i) Traverse adjustment selection

Line numbers 1800–1830

This option is only presented for closed or link traverses and gives the choice of traverse adjustment either by the Bowditch ($U = 1$) or Transit ($U = 2$) method.

(j) Bowditch method of adjustments

Line numbers 1840–1920

The Bowditch corrections to the partial departures and latitudes were given in Equations 3.58 and 3.59. The corrections to departures are computed in lines 1850 where $H(K)$ is the line distance, $T1(N-1)$ is total length of traverse and $E(N+1)$ is total error or misclosure in departures. The corrections to the latitudes are computed similarly in line 1860 where $N(N+1)$ is the total error or misclosure in latitudes.

The adjusted co-ordinates of each subsequent station are then computed in lines 1870–1880. Once the co-ordinates of the final station are obtained, the misclosure is recalculated for the closed traverse ($T = 1$) at line 1410 and for the link traverse ($T = 2$) at line 2100, with a subsequent re-display of the adjusted co-ordinates in the routine at lines 1420–1770.

(k) Option to calculate new distances

Line numbers 1930–1980

After the adjustment of the co-ordinates of the traverse stations, an option is given to calculate the distances between the adjusted stations for comparison with the measured lengths.

(l) <2> Link traverse routine

Line numbers 2000–2160

After presenting confirmation of the link traverse option in line 2010, the program branches to the previous keyboard entry routine for the closed traverse at line 1020, returning from line 1140 to collect the co-ordinates of the final station in the link traverse (lines 2040–2070). The program then returns to line 1160 to collect the WCBs and distances. After computation of all the station co-ordinates, the program returns to compute the misclosure in lines 2100–2110. The program then reverts to the screen display of results provided at line 1420.

(m) Transit method of adjustment

Line numbers 2200–2450

The Transit corrections to the partial departures and latitudes were given in Equations 3.60 and 3.61. Initially the absolute sum of the departures $D2(N-1)$ is computed in the loop between lines 2210–2290 and the absolute sum of the latitudes, $L2(N-1)$, is computed in the loop between lines 2300–2350.

The corrections to the partial departures and latitudes are computed in the loop between lines 2360–2410, where the total error or misclosure in departures is $E(N+1)$ and in latitudes is $N(N+1)$. It should be noted that it is not the algebraic values of the departures and latitudes of the lines which are used in Equations 3.60 and 3.61, but the absolute values, 'ABS(D(K))' and 'ABS(L(K))'.

The adjusted co-ordinates of each subsequent station are then computed in lines 2390–2400. Once the co-ordinates of the final station are obtained, the misclosure is recalculated for the closed traverse ($T = 1$) at line 1410 and for the link traverse ($T = 2$) at line 2100, with a subsequent re-display of the adjusted co-ordinates in the routine at lines 1420–1770.

(n) Routine to calculate new distances

Line numbers 2500–2710

The new or adjusted distances are computed from a knowledge of the station co-ordinates at either end of each line and the result stored in array $R(N-1)$ for comparison with the observed distances in array $H(N-1)$. The total length of the traverse using the adjusted co-ordinates is calculated and stored in $T2(N-1)$. After display, an option to obtain a printout of the new distances is available at line 6750.

(o) <3> Open traverse routine

Line numbers 3000–3020

After presenting confirmation of the open traverse option in line 3010, the program branches to the previous keyboard entry routine for the closed traverse at line 1020 in a similar manner to that discussed previously in (l).

(p) Routine to print survey location, etc.

Line numbers 6000–6080

This routine is provided for record purposes and is based on Applesoft DOS and may require alteration for different microcomputers and printers.

(q) Routine to print traverse solution

Line numbers 6250–6430

The format has been standardized so that it can be used by any of the traverse types, closed, link and open, no matter whether the adjustment has been carried out by the Bowditch or the Transit method. As mentioned above, this routine uses Applesoft DOS.

(r) Routine to print partial eastings and northings

Line numbers 6500–6680

This routine is provided to record the partial eastings and northings, the misclosure, the closing distance and closing error for the traverse. This routine, like the two previous ones, uses Applesoft DOS, and can be used for both unadjusted and adjusted traverses.

(s) Routine to print new distances

Line numbers 6750–6890

After a closed or link traverse has been adjusted, this routine prints out the horizontal distances between the stations, computed from the adjusted coordinates. This routine, like the three preceding ones, uses Applesoft DOS.

(t) Routine to convert DD.MMSS to decimals

Line numbers 8000–8040

This routine was discussed previously in Section 2.2.6.

(u) Routine to convert decimals to DD.MMSS

Line numbers 8050–8140

This routine was discussed previously in Section 2.2.6.

(v) Routine to clear screen

Line numbers 9000–9010

To simplify portability, the Applesoft 'HOME' command to clear the screen has been placed in a subroutine to allow an easy change to another microcomputer.

(w) Routine to display error message

Line numbers 9100–9130

If a wrong keyboard entry has been made, the error is trapped and this error message displayed and re-entry requested.

(x) Error trace routine

Line numbers 9200–9290

This error trace routine was discussed previously in Section 2.2.9.

(y) Termination of program

Line numbers 9500–9620

When the '<QUIT>' selection is made either after the initial screen header or from the program menu, an end-of-program message is displayed to advise the operator that computations are complete.

3.9.2 Traverse surveys – numeric variables

B	= Reduced bearing of WCB
C	= Counter in error trace routine
$C0$	= Closure distance in traverse
$D, D0, D1, D5$	= Degrees of angle
$D2$	= Total number of seconds in angle
$D9$	= Flag for Decimals <1> or DD.MMSS <2>
$D(0)–D(N-1)$	= Station departures array
$D1(0)–D1(N-1)$	= Temporary storage of station departures array
$D2(N-1)$	= Absolute sum of departures
E	= Error number
$E(0)–E(N)$	= Station eastings array
$E(N+1)$	= Closure error in eastings
$E(N+2)$	= Eastings of closure station in link traverse
$E1(0)–E1(N-1)$	= Partial eastings array
$H(0)–H(N-1)$	= Horizontal distance array
I	= Integer values used in loops
J	= Integer values used in loops
K	= Integer values used in loops
L	= Line number for error
$L(0)–L(N-1)$	= Station latitudes array
$L1(0)–L1(N-1)$	= Temporary storage of station latitudes array
$L2(N-1)$	= Absolute sum of latitudes
$M, M0, M1$	= Minutes of angles
N	= Number of stations
$N(0)–N(N)$	= Station northings array
$N(N+1)$	= Closure error in northings
$N(N+2)$	= Northing of closure station in link traverse
$N1(0)–N1(N-1)$	= Partial northings array
$P1$	= Counter to check on DD.MMSS angle conversion
Q	= Quadrant solution
$Q(0)–Q(N-1)$	= Quadrants array for bearings
R	= 57.2957795 for angle conversion to radians
$R(0)–R(N-1)$	= New horizontal distance array
$S, S0$	= Seconds of angle
$S(N)$	= Station numbers array
T	= Traverse type, selection from menu

Table—continued

$T(N-1)$	= Horizontal distances array
$T1(N-1)$	= Total horizontal distance array
$T2(N-1)$	= Total new horizontal distance array
U	= Traverse adjustment <1> Bowditch or <2> Transit
$W(N-1)$	= Whole circle bearings array

3.9.3 Traverse surveys – string variables

A$	Storage for location of survey
B$	Storage for operator's name
C$	Storage for date of survey
D$	= CHR$(4), DOS command, CTRL-D
E0$	= "SORRY DATA ERROR . . . PLEASE RE-ENTER"
H$	= "<1> CLOSED" when $T = 1$, "<2> LINK" when $T = 2$ and "<3> OPEN" when $T = 3$
K$	Storage for instrument used
P$	= "DO YOU WISH PRINTOUT OF ABOVE DATA (Y/N)"
Q$	= Question response (Y/N)
T$	= "BOWDITCH ADJUSTMENT" if $U = 1$, "TRANSIT ADJUSTMENT" if $U = 2$
X$	= "ENTER X(E) COORD.OF STN. "
Y$	= "ENTER Y(N) COORD.OF STN."

3.9.4 Traverse surveys – BASIC program

```
10  REM  <TRAVERSE> PROGRAM FOR APPLE II+  USES QUME PRINTER
20 D$ =  CHR$ (4): ONERR  GOTO 9200
30 E = 0:I = 0:J = 0:K = 0:N = 0:P1 = 0:T = 0:C = 0
50 E0$ = "SORRY, DATA ERROR ... PLEASE RE-ENTER."
55 P$ = "DO YOU WISH PRINTOUT OF ABOVE DATA (Y/N)"
60 T$ = " "
70 X$ = "ENTER X(E) COORD. OF STN. ":Y$ = "ENTER Y(N) COORD. OF STN. "
90 :
95  REM  **************************************************
100  GOSUB 9000
110  PRINT "****************************************"
120  PRINT "*                                      *"
130  PRINT "*      UNIVERSITY  OF  STRATHCLYDE     *"
140  PRINT "*  DEPARTMENT  OF  CIVIL  ENGINEERING  *"
150  PRINT "*                                      *"
160  PRINT "*           SURVEYING  SECTION         *"
170  PRINT "*          TRAVERSE   SOLUTIONS        *"
180  PRINT "****************************************"
190  PRINT
200  PRINT "ARE ANGLES IN DECIMALS <1>, DD.MMSS <2> OR DO YOU WISH TO QUI
     T <3>."
210  INPUT D9
220  IF D9 < 1 OR D9 > 3 THEN 200
230  ON D9 GOTO 240,240,9500
240 R = 57.2957795
300  GOSUB 9000
```

```
310  PRINT
320  PRINT "SELECT TRAVERSE PROGRAM FROM MENU :-"
330  PRINT
340  PRINT "     <1> CLOSED"
350  PRINT "     <2> LINK  "
360  PRINT "     <3> OPEN  "
370  PRINT "     <4> QUIT  "
380  INPUT T
390  IF T < 1 OR T > 4 THEN 300
400  IF T = 4 THEN 9500
410  IF T = 1 THEN H$ = "<1> CLOSED "
420  IF T = 2 THEN H$ = "<2> LINK "
430  IF T = 3 THEN H$ = "<3> OPEN "
500  GOSUB 9000
510  PRINT "UNIVERSITY OF STRATHCLYDE"
520  PRINT "TRAVERSE PROGRAM"
530  PRINT "***************************************"
540  PRINT
550  PRINT "ENTER LOCATION OF WORK, ETC. WHEN       REQUESTED, AND PRESS
     <RETURN>"
560  PRINT "ENTER LOCATION OF SURVEY :-"
570  INPUT A$
580  PRINT "ENTER OPERATOR'S NAME :-"
590  INPUT B$
600  PRINT "ENTER DATE OF SURVEY AS DD/MM/YY :-"
610  INPUT C$
620  PRINT "ENTER INSTRUMENT USED :-"
630  INPUT K$
640  PRINT
650  PRINT "DO YOU WISH PRINTOUT OF LOCATION ETC.,  (Y/N) ";
660  INPUT Q$
670  IF Q$ = "Y" THEN 6000
680  IF Q$ < > "N" THEN 650
690 :
700  ON T GOTO 1000,2000,3000
995  REM  ************************************************
1000  GOSUB 9000
1010  PRINT "<1> CLOSED TRAVERSE PROGRAM"
1020  PRINT "***************************************"
1030  PRINT
1040  PRINT "ENTER NUMBER OF TRAVERSE STATIONS"
1050  INPUT N: IF T = 2 OR T = 3 THEN N = N - 1
1060  DIM S(N),W(N - 1),H(N - 1),D(N - 1),L(N - 1),E(N + 2),N(N + 2),Q(N -
     1),T(N - 1),T1(N - 1),E1(N - 1),N1(N - 1)
1070  PRINT "ENTER THE STATION NUMBERS IN SEQUENCE,  INCLUDING THE CLOSIN
     G STATION"
1080  FOR J = 0 TO N
1090  PRINT "ENTER STATION NO. ";
1100  INPUT S(J)
1110  NEXT J
1120  PRINT X$;S(0);: INPUT E(0)
1130  PRINT Y$;S(0);: INPUT N(0)
1140  IF T = 2 THEN 2030
1150 :
1155  REM  NOW ENTER W.C.B. AND DISTANCE
1160  FOR K = 0 TO N - 1
1170  PRINT "ENTER W.C.B. FROM ";S(K);" TO ";S(K + 1)
1180  INPUT W(K)
1190  IF D9 = 1 THEN 1210
```

```
1195  REM  I.E. D9=2, CONVERT DD.MMSS TO DECIMALS
1200 D5 = W(K): GOSUB 8000:W(K) = D5
1210  PRINT "ENTER HORIZ. DIST. FROM ";S(K);" TO ";S(K + 1)
1220  INPUT H(K)
1230  NEXT K
1235  REM  CALCULATE TOTAL LENGTH OF TRAVERSE
1240  FOR I = Ø TO N - 1
1250 T(I) = H(I)
1260  NEXT I
1265  REM  COLLECT TOTAL LENGTH IN T1(N-1)
1270  FOR J = 1 TO N - 1:T1(Ø) = T(Ø)
1280 T1(J) = T1(J - 1) + T(J)
1290  NEXT J
1295  REM  CALCULATE CO-ORDS OF STATIONS
1300  FOR K = Ø TO N - 1
1310 Q =  INT (W(K) / 90 + 1)
1320 Q(K) = Q
1330  ON Q GOTO 1340,1350,1360,1370
1340 B = W(K):D(K) = H(K) *  SIN (B / R):L(K) = H(K) *  COS (B / R): GOTO
     1380
1350 B = 180 - W(K):D(K) = H(K) *  SIN (B / R):L(K) =  - H(K) *  COS (B /
     R): GOTO 1380
1360 B = W(K) - 180:D(K) =  - H(K) *  SIN (B / R):L(K) =  - H(K) *  COS (
     B / R): GOTO 1380
1370 B = 360 - W(K):D(K) =  - H(K) *  SIN (B / R):L(K) = H(K) *  COS (B /
     R)
1380 E(K + 1) = E(K) + D(K):N(K + 1) = N(K) + L(K)
1390  NEXT K: IF T = 2 THEN 2100
1400  IF T = 3 THEN 1430
1405  REM  FIND ERROR AT CLOSURE
1410 E(N + 1) = E(Ø) - E(N):N(N + 1) = N(Ø) - N(N):CØ =  INT (( SQR ((E(N
     + 1)) ^ 2 + (N(N + 1)) ^ 2)) * 1000 + .5) / 1000
1415  REM  PREPARE DATA FOR DISPLAY
1420 E(N + 1) =  INT (E(N + 1) * 1000 + .5) / 1000:N(N + 1) =  INT (N(N +
     1) * 1000 + .5) / 1000
1430  IF D9 = 1 OR P1 = 1 THEN 1460
1435  REM  CONVERT ANGLES BACK TO DD.MMSS
1440  FOR I = Ø TO N - 1
1450 D = W(I): GOSUB 8050:W(I) = D: NEXT I
1460  FOR J = 1 TO N
1470 E(J) =  INT (E(J) * 1000 + .5) / 1000:N(J) =  INT (N(J) * 1000 + .5)
     / 1000
1480 D(J - 1) =  INT ((D(J - 1) + E1(J - 1)) * 1000 + .5) / 1000:L(J - 1)
     =  INT ((L(J - 1) + N1(J - 1)) * 1000 + .5) / 1000
1490  NEXT J
1495  REM  SCREEN DISPLAY OF RESULTS
1500  GOSUB 9000
1510  PRINT H$;"RESULTS";T$
1520  PRINT "***************************************"
1530  PRINT "PT. W.C.B.  DISTANCE  EASTING  NORTHING"
1540  PRINT "======================================="
1550  PRINT S(Ø); TAB( 21);E(Ø); TAB( 31);N(Ø)
1560  FOR I = Ø TO N - 1
1570  PRINT S(I + 1); TAB( 3);W(I); TAB( 12);H(I); TAB( 21);E(I + 1); TAB(
      31);N(I + 1)
1580  NEXT I: IF T = 2 THEN 2150
1590  PRINT "======================================="
1600  PRINT P$
1610  INPUT Q$
```

```
1620  IF Q$ = "Y" THEN 625Ø
163Ø  IF Q$ < > "N" THEN 16ØØ
164Ø  PRINT
165Ø  PRINT "PT.-PT."; TAB( 16);"PART.EAST."; TAB( 28);"PART.NORTH."
166Ø  FOR J = Ø TO N - 1
167Ø  PRINT S(J);"-";S(J + 1); TAB( 16);D(J); TAB( 28);L(J)
168Ø  NEXT J: PRINT : IF T = 3 THEN 171Ø
169Ø  PRINT "DIFFERENCE IN EASTINGS    = ";E(N + 1): PRINT "DIFFERENCE IN
    NORTHINGS  = ";N(N + 1)
17ØØ  PRINT "CLOSURE DISTANCE          = ";CØ
171Ø  PRINT "TOTAL LENGTH OF TRAVERSE = ";T1(N - 1): IF T = 3 THEN 173Ø
1715  IF CØ = Ø THEN 173Ø
172Ø  PRINT "TRAVERSE CLOSING ERROR   = 1/"; INT (T1(N - 1) / CØ + .5)
173Ø  PRINT "****************************************"
174Ø  PRINT P$: INPUT Q$
175Ø  IF Q$ = "Y" THEN 65ØØ
176Ø  IF Q$ < > "N" THEN 173Ø
177Ø  IF T = 3 THEN 95ØØ
1775  REM  ADJUSTMENT OPTION ONLY PRESENTED ON FIRST OCCASION (P1=Ø)
178Ø  IF P1 = 1 THEN 194Ø
18ØØ  PRINT "DO YOU WISH TO ADJUST TRAVERSE USING    <1> BOWDITCH, <2> TR
    ANSIT, <3> NEITHER."
181Ø  INPUT U
182Ø  IF U < 1 OR U > 3 THEN 18ØØ
183Ø  ON U GOTO 184Ø,22ØØ,193Ø
1835  REM  BOWDITCH METHOD OF ADJUSTMENT
184Ø  FOR K = Ø TO N - 1
185Ø E1(K) = (H(K) / T1(N - 1)) * E(N + 1)
186Ø N1(K) = (H(K) / T1(N - 1)) * N(N + 1)
187Ø E(K + 1) = E(K) + D(K) + E1(K)
188Ø N(K + 1) = N(K) + L(K) + N1(K)
189Ø  NEXT K
1895  REM  RETURN TO PREVIOUS ROUTINE TO CHECK ERROR AND DISPLAY RESULTS
19ØØ T$ = " BOWDITCH ADJUSTMENT":P1 = 1: REM  NOTE CORRECTIONS APPLIED TO
     DD.MMSS ANGLES
191Ø  IF T = 1 THEN 141Ø
192Ø  IF T = 2 THEN 21ØØ
193Ø  IF P1 = Ø THEN 95ØØ
194Ø  PRINT "DO YOU WISH TO CALCULATE DISTANCES (Y/N)"
195Ø  INPUT Q$
196Ø  IF Q$ = "Y" THEN 25ØØ
197Ø  IF Q$ < > "N" THEN 194Ø
198Ø  GOTO 95ØØ
1995  REM  **************************************************
2ØØØ  GOSUB 9ØØØ
2Ø1Ø  PRINT "<2> LINK TRAVERSE PROGRAM"
2Ø2Ø  GOTO 1Ø2Ø
2Ø3Ø  PRINT
2Ø4Ø  PRINT X$;S(N)
2Ø5Ø  INPUT E(N + 2)
2Ø6Ø  PRINT Y$;S(N)
2Ø7Ø  INPUT N(N + 2)
2Ø8Ø  GOTO 116Ø
2Ø9Ø :
21ØØ E(N + 1) = E(N + 2) - E(N):N(N + 1) = N(N + 2) - N(N)
211Ø CØ =  INT (( SQR ((E(N + 1)) ^ 2 + (N(N + 1)) ^ 2)) * 1ØØØ + .5) / 1
     ØØØ
212Ø  GOTO 142Ø
215Ø  PRINT S(N); TAB( 21);E(N + 2); TAB( 31);N(N + 2)
```

```
2160  GOTO 1590
2193  REM  ***********************************************
2195  REM  TRANSIT METHOD OF ADJUSTMENT
2200  DIM D1(N - 1),D2(N - 1),L1(N - 1),L2(N - 1)
2205  REM  STORE DEPARTURES D(K) IN D1(K) AND LATITUDES L(K) IN L1(K)
2210  FOR K = Ø TO N - 1
2220 D1(K) = D(K):L1(K) = L(K)
2230  NEXT K
2235  REM  CALCULATE TOTAL DEPARTURES IN D2(N-1)
2240  IF D1(Ø) < 1 THEN D1(Ø) = ( - 1) * D1(Ø)
2250 D2(Ø) = D1(Ø)
2260  FOR I = 1 TO N - 1
2270  IF D1(I) < 1 THEN D1(I) = ( - 1) * D1(I)
2280 D2(I) = D2(I - 1) + D1(I)
2290  NEXT I
2295  REM  CALCULATE TOTAL LATITUDES IN L2(N-1)
2300  IF L1(Ø) < 1 THEN L1(Ø) = ( - 1) * L1(Ø)
2310 L2(Ø) = L1(Ø)
2320  FOR J = 1 TO N - 1
2330  IF L1(J) < 1 THEN L1(J) = ( - 1) * L1(J)
2340 L2(J) = L2(J - 1) + L1(J)
2350  NEXT J
2355  REM  CALCULATE TRANSIT CORRECTIONS
2360  FOR K = Ø TO N - 1
2370 E1(K) = (E(N + 1) / D2(N - 1)) *  ABS (D(K))
2380 N1(K) = (N(N + 1) / L2(N - 1)) *  ABS (L(K))
2390 E(K + 1) = E(K) + D(K) + E1(K)
2400 N(K + 1) = N(K) + L(K) + N1(K)
2410  NEXT K
2415  REM  RETURN TO PREVIOUS ROUTINES TO CHECK ERROR AND DISPLAY RESULTS

2420 P1 = 1: REM  NOTE THAT DD.MMSS CORRECTION APPLIED
2430 T$ = " TRANSIT ADJUSTMENT"
2435  REM  CHECK WHETHER CLOSED OR LINK TRAVERSE
2440  IF T = 1 THEN 1410
2445  REM  I.E. T=2, LINK TRAVERSE
2450  GOTO 2100
2493  REM  ************************************************
2495  REM  CALCULATE DISTANCES FROM NEW COORDINATES
2500  DIM R(N - 1),T2(N - 1)
2510  FOR I = Ø TO N - 1
2520 R(I) =  SQR ((E(I + 1) - E(I)) ^ 2 + (N(I + 1) - N(I)) ^ 2)
2530 R(I) =  INT (R(I) * 1000 + .5) / 1000: NEXT I
2540  FOR J = 1 TO N - 1
2550 T2(Ø) = R(Ø)
2560 T2(J) = T2(J - 1) + R(J)
2570  NEXT J
2600  GOSUB 9000
2610  PRINT "PT.-PT."; TAB( 10);"NEW DISTANCE"
2620  PRINT "===================================="
2630  FOR K = Ø TO N - 1
2640  PRINT S(K);"-";S(K + 1); TAB( 10);R(K)
2650  NEXT K
2660  PRINT "****************************************"
2670  PRINT P$
2680  INPUT Q$
2690  IF Q$ = "Y" THEN 6750
2700  IF Q$ < > "N" THEN 2670
2710  GOTO 9500
```

```
2995  REM  ***************************************************
3ØØØ  GOSUB 9ØØØ
3Ø1Ø  PRINT "<3> OPEN TRAVERSE PROGRAM"
3Ø2Ø  GOTO 1Ø2Ø
5985  REM  ************************************************************
599Ø  REM  PRINTOUT OF LOCATION, ETC
6ØØØ  PRINT D$;"PR#1": PRINT "LOCATION OF SURVEY :- ";A$
6Ø1Ø  PRINT "**************************************************************
      ********************"
6Ø2Ø  PRINT "OPERATOR'S NAME :- ";B$
6Ø3Ø  PRINT "DATE OF SURVEY  :- ";C$
6Ø4Ø  PRINT "INSTRUMENT USED :- ";K$
6Ø6Ø  PRINT "**************************************************************
      ********************"
6Ø7Ø  PRINT : PRINT D$;"PR#Ø"
6Ø8Ø  GOTO 7ØØ
6243  REM  *****************************************************
6245  REM  PRINTOUT OF COORDINATES ON QUME
625Ø  PRINT D$;"PR#1": POKE 1784 + 1,8Ø
626Ø P7 = 1: REM  PRINTER CODE
627Ø  PRINT "TRAVERSE PROGRAM RESULTS :-"
629Ø  PRINT H$;"TRAVERSE SOLUTION :";T$
63ØØ  PRINT "**************************************************************
      "
632Ø  PRINT "STN.";: POKE 36,7: PRINT "W.C.B.";: POKE 36,17: PRINT "DISTA
      NCE";: POKE 36,29: PRINT "EASTING";: POKE 36,41: PRINT "NORTHING"
633Ø  PRINT "==============================================================
      "
634Ø  PRINT S(Ø);: POKE 36,28: PRINT E(Ø);: POKE 36,4Ø: PRINT N(Ø)
635Ø  FOR I = Ø TO N - 1
636Ø  PRINT S(I + 1);: POKE 36,6: PRINT W(I);: POKE 36,16: PRINT H(I);: POKE
      36,28: PRINT E(I + 1);: POKE 36,4Ø: PRINT N(I + 1)
637Ø  NEXT I
638Ø  IF T = 1 OR T = 3 THEN 64ØØ
639Ø  PRINT S(N);: POKE 36,28: PRINT E(N + 2);: POKE 36,4Ø: PRINT N(N + 2
      )
64ØØ  PRINT "**************************************************************
      "
641Ø  PRINT
6415  REM  RETURN TO SCREEN
642Ø  PRINT D$;"PR#Ø"
643Ø  GOTO 164Ø
6493  REM  ***************************************************
6495  REM  PRINTOUT OF PARTIAL EASTINGS AND NORTHINGS
65ØØ  PRINT D$;"PR#1": POKE 1784 + 1,8Ø
652Ø  PRINT "PARTIAL EASTINGS AND NORTHINGS"
653Ø  PRINT "**************************************************************
      "
655Ø  PRINT "STN-STN";: POKE 36,1Ø: PRINT "PART.EASTINGS";: POKE 36,25: PRINT
      "PART.NORTHINGS"
656Ø  PRINT "========================================"
657Ø  FOR J = Ø TO N - 1
658Ø  PRINT S(J);: POKE 36,4: PRINT "-";: POKE 36,5: PRINT S(J + 1);: POKE
      36,1Ø: PRINT D(J);: POKE 36,25: PRINT L(J)
659Ø  NEXT J: PRINT
66ØØ  PRINT "DIFF. IN EASTINGS       = ";E(N + 1)
661Ø  PRINT "DIFF. IN NORTHINGS      = ";N(N + 1)
662Ø  PRINT "CLOSURE DISTANCE        = ";CØ
663Ø  PRINT "TOTAL LENGTH TRAVERSED  = ";T1(N - 1)
```

```
6635  IF C0 = 0 THEN 6650
6640  PRINT "TRAVERSE CLOSING ERROR  = 1/"; INT ((T1(N - 1) / C0) + .5)
6650  PRINT "*************************************************************
     ": PRINT
6655  REM  RETURN TO SCREEN
6660  PRINT D$;"PR#0"
6670  IF T = 3 THEN 100
6680  GOTO 1800
6743  REM  *************************************************
6745  REM  PRINTOUT OF NEW DISTANCES
6750  PRINT D$;"PR#1": POKE 1784 + 1,80
6760  PRINT "NEW DISTANCES FROM ADJUSTED CO-ORDINATES"
6770  PRINT "*************************************************************
     "
6790  PRINT "STN-STN";: POKE 36,10: PRINT "NEW DISTANCE"
6800  PRINT "========================================="
6810  FOR J = 0 TO N - 1
6820  PRINT S(J);: POKE 36,4: PRINT "-";: POKE 36,5: PRINT S(J + 1);: POKE
     36,10: PRINT R(J)
6830  NEXT J
6840  PRINT
6850  PRINT "TOTAL DISTANCE TRAVERSED = ";T2(N - 1)
6860  PRINT "*****************************************"
6870  PRINT
6875  REM  RETURN TO SCREEN
6880  PRINT D$;"PR#0"
6890  GOTO 9500
7993  REM  ***************************************************
7995  REM  ROUTINE DD.MMSS TO DECIMALS
8000 D0 =  INT (D5)
8005  REM  FIND NO. OF MINUTES,M0
8010 M1 = (D5 - D0) * 100:M0 =  INT (M1 + .1)
8015  REM  FIND NO. OF SECONDS,S0
8020 S0 =  INT (M1 * 100 - M0 * 100 + .5)
8025  REM  COLLECT DEGS,MINS,SECS
8030 D5 = D0 + M0 / 60 + S0 / 3600
8040  RETURN
8043  REM  ***************************************************
8045  REM  ROUTINE DECIMAL TO DD.MMSS
8050 D1 =  INT (D)
8055  REM  FIND TOTAL NO. OF SECONDS
8060 D2 = (D - D1) * 3600
8065  REM  FIND NO. OF MINUTES
8070 M =  INT (D2 / 60)
8075  REM  FIND NO. OF SECONDS
8080 S = D2 - 60 * M + .5
8090  IF S < 60 THEN 8130
8100 M = M + 1:S = 0
8110  IF M < 60 THEN 8130
8120 D1 = D1 + 1:M = 0
8130 D = D1 * 10000 + M * 100 + S
8140 D =  INT (D) / 10000: RETURN
8985  REM  ******************************************
8990  REM  GOSUB ROUTINE TO CLEAR SCREEN
9000  HOME
9010  RETURN
9093  REM  ***************************************************
9095  REM  GOSUB ROUTINE FOR ERRORS
9100  PRINT
```

```
9110  PRINT E0$
9120  PRINT
9130  RETURN
9193  REM  ****************************************************
9195  REM  ERROR TRACE ROUTINE
9200  PRINT  CHR$ (7): REM  CTRL-B (BELL)
9210 E =  PEEK (222): REM  GET ERROR NO.
9220  IF C = 1 THEN 9260
9230  INVERSE
9240  PRINT "ERROR NO. ";E;" FOUND"
9250 C = 1: TRACE : RESUME
9260  PRINT "ERROR ON SECOND LINE NO. "
9270  NORMAL
9280  NOTRACE
9290  STOP
9493  REM  ****************************************************
9495  REM  END OF PROGRAM
9500  GOSUB 9000
9510  PRINT
9520  PRINT "END TRAVERSE PROGRAM"
9530  PRINT
9540  PRINT "**************************"
9550  REM  PROGRAM PREPARED BY
9560  REM  DR. P.H. MILNE
9570  REM  DEPT. OF CIVIL ENGINEERING
9580  REM  UNIVERSITY OF STRATHCLYDE
9590  REM  GLASGOW G4 ONG
9600  REM  SCOTLAND
9610  REM  ***************************
9620  END
```

3.9.5 Traverse surveys – computer printout

```
LOCATION OF SURVEY :- STRATHCLYDE UNIVERSITY - STEPPS
***************************************************************************
OPERATOR'S NAME :- 2ND YEAR GROUP A
DATE OF SURVEY  :- 1/2 JUNE 1983
INSTRUMENT USED :- ZEISS ZENA T20A + CD6
***************************************************************************

TRAVERSE PROGRAM RESULTS :-
<1> CLOSED TRAVERSE SOLUTION :
***********************************************************
```

STN.	W.C.B.	DISTANCE	EASTING	NORTHING
1			264817.996	667743.484
2	62.0904	194.095	264989.611	667834.154
3	67.5111	156.248	265134.332	667893.057
4	55.1614	197.298	265296.481	668005.458
5	65.4152	206.848	265485	668090.586
6	119.334	331.705	265773.527	667926.939
7	189.4312	298.868	265723.068	667632.361
8	246.3042	417.312	265340.334	667466.036
9	279.1648	194.585	265148.296	667497.415
10	266.4001	169.072	264979.51	667487.585
1	327.4619	302.666	264818.101	667743.62

```
***********************************************************
```

```
PARTIAL EASTINGS AND NORTHINGS
*************************************************************
STN-STN   PART.EASTINGS   PART.NORTHINGS
========================================
1  -2     171.615         90.67
2  -3     144.72          58.903
3  -4     162.15          112.401
4  -5     188.519         85.128
5  -6     288.527         -163.647
6  -7     -50.459         -294.578
7  -8     -382.734        -166.325
8  -9     -192.038        31.379
9  -10    -168.786        -9.83
10 -1     -161.409        256.035

DIFF. IN EASTINGS          = -.105
DIFF. IN NORTHINGS         = -.136
CLOSURE DISTANCE           = .171
TOTAL LENGTH TRAVERSED     = 2468.697
TRAVERSE CLOSING ERROR     = 1/14437
*************************************************************

TRAVERSE PROGRAM RESULTS :-
<1> CLOSED TRAVERSE SOLUTION : BOWDITCH ADJUSTMENT
*************************************************************
STN.   W.C.B.     DISTANCE    EASTING      NORTHING
=============================================================
1                             264817.996   667743.484
2      62.0904    194.095     264989.603   667834.143
3      67.5111    156.248     265134.316   667893.038
4      55.1614    197.298     265296.458   668005.428
5      65.4152    206.848     265484.968   668090.544
6      119.334    331.705     265773.481   667926.879
7      189.4312   298.868     265723.009   667632.285
8      246.3042   417.312     265340.257   667465.937
9      279.1648   194.585     265148.211   667497.305
10     266.4001   169.072     264979.418   667487.465
1      327.4619   302.666     264817.996   667743.484
*************************************************************

PARTIAL EASTINGS AND NORTHINGS
*************************************************************
STN-STN   PART.EASTINGS   PART.NORTHINGS
========================================
1  -2     171.607         90.659
2  -3     144.713         58.894
3  -4     162.142         112.39
4  -5     188.51          85.117
5  -6     288.513         -163.665
6  -7     -50.472         -294.594
7  -8     -382.752        -166.348
8  -9     -192.046        31.368
9  -10    -168.793        -9.839
10 -1     -161.422        256.018

DIFF. IN EASTINGS          = 0
DIFF. IN NORTHINGS         = 0
CLOSURE DISTANCE           = 0
TOTAL LENGTH TRAVERSED     = 2468.697
*************************************************************
```

```
NEW DISTANCES FROM ADJUSTED CO-ORDINATES
*************************************************************
STN-STN   NEW DISTANCE
======================================
1   -2    194.083
2   -3    156.239
3   -4    197.285
4   -5    206.835
5   -6    331.702
6   -7    298.886
7   -8    417.338
8   -9    194.591
9   -10   169.08
10  -1    302.659

TOTAL DISTANCE TRAVERSED = 2468.698
****************************************
```

```
TRAVERSE PROGRAM RESULTS :-
<1> CLOSED TRAVERSE SOLUTION : TRANSIT ADJUSTMENT
*************************************************************
STN.   W.C.B.     DISTANCE     EASTING      NORTHING
=============================================================
1                              264817.996   667743.484
2      62.0904    194.095      264989.602   667834.144
3      67.5111    156.248      265134.314   667893.041
4      55.1614    197.298      265296.455   668005.43
5      65.4152    206.848      265484.963   668090.549
6      119.334    331.705      265773.475   667926.884
7      189.4312   298.868      265723.013   667632.275
8      246.3042   417.312      265340.258   667465.932
9      279.1648   194.585      265148.209   667497.307
10     266.4001   169.072      264979.414   667487.476
1      327.4619   302.666      264817.996   667743.484
*************************************************************
```

```
PARTIAL EASTINGS AND NORTHINGS
*************************************************************
STN-STN   PART.EASTINGS  PART.NORTHINGS
======================================
1   -2    171.606        90.66
2   -3    144.712        58.897
3   -4    162.141        112.389
4   -5    188.509        85.119
5   -6    288.511        -163.665
6   -7    -50.462        -294.61
7   -8    -382.755       -166.343
8   -9    -192.049       31.376
9   -10   -168.795       -9.831
10  -1    -161.418       256.008

DIFF. IN EASTINGS        = 0
DIFF. IN NORTHINGS       = 0
CLOSURE DISTANCE         = 0
TOTAL LENGTH TRAVERSED   = 2468.697
*************************************************************
```

```
NEW DISTANCES FROM ADJUSTED CO-ORDINATES
************************************************************
STN-STN   NEW DISTANCE
========================================
1    -2    194.082
2    -3    156.238
3    -4    197.284
4    -5    206.835
5    -6    331.701
6    -7    298.9
7    -8    417.338
8    -9    194.595
9    -10   169.081
10   -1    302.648

TOTAL DISTANCE TRAVERSED = 2468.702
****************************************


LOCATION OF SURVEY :- STRATHCLYDE UNIVERSITY - STEPPS
*************************************************************************
OPERATOR'S NAME :- 2ND YEAR GROUP B
DATE OF SURVEY  :- 5/6 JUNE 1983
INSTRUMENT USED :- ZEISS ZENA T20A + CD6
*************************************************************************


TRAVERSE PROGRAM RESULTS :-
<2> LINK TRAVERSE SOLUTION :
************************************************************
STN.   W.C.B.     DISTANCE    EASTING      NORTHING
============================================================
1                             264817.996  667743.484
2      62.0904    194.095     264989.611  667834.154
3      67.5111    156.248     265134.332  667893.057
4      55.1614    197.298     265296.481  668005.458
5      65.4152    206.848     265485      668090.586
6      119.334    331.705     265773.527  667926.939
6                             265773.481  667926.879
************************************************************


PARTIAL EASTINGS AND NORTHINGS
************************************************************
STN-STN   PART.EASTINGS   PART.NORTHINGS
========================================
1    -2    171.615         90.67
2    -3    144.72          58.903
3    -4    162.15          112.401
4    -5    188.519         85.128
5    -6    288.527         -163.647

DIFF. IN EASTINGS        = -.046
DIFF. IN NORTHINGS       = -.059
CLOSURE DISTANCE         = .075
TOTAL LENGTH TRAVERSED   = 1086.194
TRAVERSE CLOSING ERROR   = 1/14483
************************************************************
```

```
TRAVERSE PROGRAM RESULTS :-
<2> LINK TRAVERSE SOLUTION : BOWDITCH ADJUSTMENT
***************************************************************
STN.   W.C.B.     DISTANCE     EASTING      NORTHING
===============================================================
1                              264817.996   667743.484
2      62.Ø9Ø4    194.Ø95      264989.6Ø3   667834.143
3      67.5111    156.248      265134.316   667893.Ø38
4      55.1614    197.298      265296.458   668ØØ5.428
5      65.4152    2Ø6.848      265484.968   668Ø9Ø.545
6      119.334    331.7Ø5      265773.481   667926.879
6                              265773.481   667926.879
***************************************************************
```

```
PARTIAL EASTINGS AND NORTHINGS
***************************************************************
STN-STN    PART.EASTINGS   PART.NORTHINGS
=========================================
1   -2     171.6Ø7         9Ø.659
2   -3     144.713         58.895
3   -4     162.142         112.39
4   -5     188.51          85.117
5   -6     288.513         -163.665

DIFF. IN EASTINGS        = Ø
DIFF. IN NORTHINGS       = Ø
CLOSURE DISTANCE         = Ø
TOTAL LENGTH TRAVERSED   = 1Ø86.194
***************************************************************
```

3.10 CO-ORDINATE TRANSFORMATION

Initial site surveys and traverses are often referred to local or assumed co-ordinates. However, if the co-ordinates of local traverse stations require to be related, for example, to national grid, it is necessary to transform the co-ordinates from the first system to the second system.

Co-ordinates may be transformed from one system to another by scaling, rotation and translation with respect to the first system. If the transformation parameters, that is scale factor, rotation angle and translation, are known, then they can be directly entered together with the co-ordinates of one point in both systems. The most common use for co-ordinate transformation is when several co-ordinates need to be transformed using the co-ordinates of two or three known points in each system. A minimum of two points is required. If more than two points are known then the transformation parameters are computed by a least squares method using all the entered points.

The following computer program <TRANSFORMATION> gives the operator the option of transforming points using either of the above methods.

3.10.1 Co-ordinate transformation – subroutine index

	Line numbers	Function
(a)	10–90	Initialization and control
(b)	100–190	Screen header display
(c)	200–240	Selection of angle format
(d)	300–390	Menu selection for points/rotation
(e)	500–700	Entry of survey location, etc.
(f)	1000–1760	<1> Co-ordinate transformation – points
(g)	2000–2360	<2> Co-ordinate transformation – rotation
(h)	6000–6080	Routine to print survey location, etc.
(i)	6250–6480	Routine to print input co-ordinates
(j)	6500–6650	Routine to print transformed co-ordinates
(k)	8000–8040	Routine to convert DD.MMSS to decimals
(l)	8050–8140	Routine to convert decimals to DD.MMSS
(m)	9000–9010	Routine to clear screen
(n)	9100–9130	Routine to display error message
(o)	9200–9290	Error trace routine
(p)	9500–9620	Termination of program

(a) Initialization and control

Line numbers 10–90

All numeric variables and required string variables are initialized.

(b) Screen header display

Line numbers 100–190

This is the screen header display for civil engineering students at Strathclyde University. The operator can alter this as required to suit another organization.

(c) Selection of angle format

Line numbers 200–240

The operator has a choice in line 200 of entering angles in decimal or DD.MMSS format, where the *D*9 flag is set to 1 for decimals and 2 for DD.MMSS in line 210.

(d) Menu selection for points/rotation

Line numbers 300–390

The operator has the choice of selecting one of two options:

<1> Points in both systems,
<2> Rotation angle and point,

where in the first option <1> more than two points in each system are required. In the second option <2> the transformation parameters: scale factor, rotation

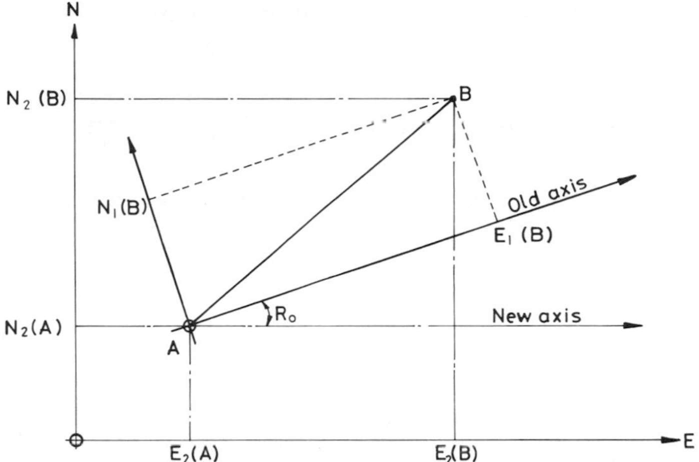

Fig. 3.19. Nomenclature for co-ordinate transformation either by rotation or by matching co-ordinates.

angle and translation, are required, plus one point with co-ordinates in both systems, Fig. 3.19.

(e) Entry of survey location, etc.

Line numbers 500–700

This routine is provided to keep a record of site location, operator, date, etc. After entry, line 650 gives an option of a printout if required in the subroutine at line 6000.

(f) <1> Co-ordinate transformation – points

Line numbers 1000–1760

At the outset, the number of points (N) whose co-ordinates are known in both systems is entered in line 1040. If the entry for N is less than two, there can be no solution, so the error is trapped and a message is displayed in lines 1050–1090. Depending on the value of N, the required numeric variables are dimensioned in line 1100 and a loop set up from line 1130 to line 1290 to enter the X and Y co-ordinates of all the N points in each co-ordinate system. The variables T_1 to T_7 are also calculated within this loop. These variables are then used to evaluate the multiplying constants K_1 and K_2 used in determining the easting and northing translations (E_T and N_T) in lines 1320–1330, where:

$$E_T = [\Sigma E_2 - K_1 \Sigma E_1 + K_2 \Sigma N_1]/N \tag{3.62}$$

$$N_T = [\Sigma N_2 - K_1 \Sigma N_1 - K_2 \Sigma E_1]/N \tag{3.63}$$

The subscripts 1 and 2 for the eastings (E) and northings (N) are used to designate either the first or second co-ordinate system. The two multiplying constants, K_1 and K_2, are evaluated in lines 1300–1310 from:

$$K_1 = \frac{[\Sigma(E_1E_2+N_1N_2)-(\Sigma E_1\Sigma E_2+\Sigma N_1\Sigma N_2)/N]}{[\Sigma(E_1^2+N_1^2)-\{(\Sigma E_1)^2+(\Sigma N_1)^2\}/N]} \tag{3.64}$$

$$K_2 = \frac{[\Sigma(E_1N_2-N_1E_2)-(\Sigma E_1\Sigma N_2-\Sigma N_1\Sigma E_2)/N]}{[\Sigma(E_1^2+N_1^2)-\{(\Sigma E_1)^2+(\Sigma N_1)^2\}/N]} \tag{3.65}$$

where $T_1 = \Sigma(E_1E_2+N_1N_2)$, line 1240

$T_2 = \Sigma E_1$, $T_3 = \Sigma E_2$, line 1250

$T_4 = \Sigma N_1$, $T_5 = \Sigma N_2$, line 1260

$T_6 = (\Sigma E_1)^2+(\Sigma N_1)^2$, line 1270

$T_7 = \Sigma(E_1N_2-N_1E_2)$, line 1280

Once the easting and northing translations have been calculated from Equations 3.62 and 3.63, the screen is cleared and the input points in both systems displayed, and the operator is given an opportunity to obtain a printout for record purposes at line 6250.

The number of points (T) to be transformed from the first to the second system is entered in the section commencing at line 1500 and the necessary numeric variables dimensioned in line 1550 where the easting and northing subscript 8 now refers to the first system and 9 to the second system. The eastings and northings of the transformed points in the second system are determined in lines 1620–1630 from:

$$E_9 = K_1E_8-K_2N_8+E_T \tag{3.66}$$

$$N_9 = K_1N_8+K_2E_8+N_T \tag{3.67}$$

Once all the entered points have been transformed, an option is given in line 1730 to obtain a printout for record purposes at line 6500.

(g) <2> Co-ordinate transformation – rotation

Line numbers 2000–2360

The three transformation parameters, rotation angle (R_0), scale factor (S_F) and translation are entered in lines 2030–2150, where the rotation can be clockwise or anticlockwise. On completion of entry, the point whose co-ordinates were entered in both systems is displayed using the previous screen display routine at lines 1350–1480, with a printout option as before.

After display and printout the program returns to line 2160 to enter the number of points (T) to be transformed, and the required numeric variables are dimensioned in line 2180. The subscripts for eastings and northings, 8 and 9 are used to refer to the first and second co-ordinate systems. The easting and northing translations, Fig. 3.19, are determined in lines 2250–2260 from:

$$E_T = E_8 - E_1 \quad (3.68)$$

$$N_T = N_8 - N_1 \quad (3.69)$$

The rotated azimuth angle (W) and the scaled horizontal distance (H) are then computed using E_T and N_T from Equations 3.68 and 3.69 in lines 2270–2280, where:

$$W = R_0 + \tan^{-1}(E_T/N_T) \quad (3.70)$$

$$H = S_F[(E_T)^2 + (N_T)^2]^{1/2} \quad (3.71)$$

Use is now made of the previous screen display at lines 1640–1710 to display the co-ordinates resulting from the transformation. Once all the entered points have been transformed, an option is given in line 2330 to obtain a printout for record purposes at line 6500.

(h) Routine to print survey location, etc.

Line numbers 6000–6080

This routine is provided for record purposes, and is based on Applesoft DOS and may require alteration for different microcomputers and printers.

(i) Routine to print input co-ordinates

Line numbers 6250–6480

The same comments apply to this routine, as to (h) above. This routine is used by both of the transformation options. If the second option ($CT = 2$) is chosen, where the transformation parameters are known, these are included in the printout at line 6280.

(j) Routine to print transformed co-ordinates

Line numbers 6500–6650

The same comments apply to this routine, as to that of (h) above. In this routine, both the points entered in the first system and the transformed co-ordinates of the second system are printed.

(k) Routine to convert DD.MMSS to decimals

Line numbers 8000–8040

This routine was discussed previously in Section 2.2.6.

(l) Routine to convert decimals to DD.MMSS

Line numbers 8050–8140

This routine was discussed previously in Section 2.2.6.

(m) Routine to clear screen

Line numbers 9000–9010

To simplify portability, the Applesoft 'HOME' command to clear the screen has been placed in a subroutine to allow an easy change to another microcomputer.

(n) Routine to display error message

Line numbers 9100–9130

If a wrong keyboard entry has been made, the error is trapped and this error message displayed and re-entry requested.

(o) Error trace routine

Line numbers 9200–9290

This error trace routine was discussed earlier in Section 2.2.9.

(p) Termination of program

Line numbers 9500–9620

When either the '<QUIT>' selection is made after the initial screen header, or the co-ordinate transformations are complete, an end-of-program message is displayed to advise the operator that computations are complete.

3.10.2 Co-ordinate transformation – numeric variables

C	= Counter in error trace routine
CT	= Transformation type, selection from menu
D, $D0$, $D1$, $D5$	= Degrees of angle
$D2$	= Total number of seconds in angle
$D9$	= Flag for decimals <1> or DD.MMSS <2>
E	= Error number
ET	= Easting translation
$E1(N-1)$	= Eastings array, first system, input points
$E2(N-1)$	= Eastings array, second system, input points
$E8(T-1)$	= Eastings array, first system, transformed points
$E9(T-1)$	= Eastings array, second system, transformed points
H	= Scaled horizontal distance
I	= Integer values used in loops
J	= Integer values used in loops
K	= Integer values used in loops
$K1$, $K2$	= Multiplying constants used in translation
M, $M0$, $M1$	= Minutes in angles
N	= Number of points in both systems
NT	= Northing translation
$N1(N-1)$	= Northings array, first system, input points
$N2(N-1)$	= Northings array, second system, input points
$N8(T-1)$	= Northings array, first system, transformed points

Table—continued

$N9(T-1)$ = Northings array, second system, transformed points
R = 57.2957795 for angle conversion to radians
$R0$ = Angle of rotation
S, $S0$ = Seconds of angle
SF = Scale factor
$S(N-1)$ = Point number array, input points
T = Number of transformed points
$T1-T7$ = Summation variables
$T(T-1)$ = Point number array, transformed points
W = Rotated azimuth angle

3.10.3 Co-ordinate transformation – string variables

A$ Storage for location of survey
B$ Storage for operator's name
C$ Storage for date of survey
*D$* = CHR$(4), i.e. CTRL-D
E0$ = "SORRY, DATA ERROR . . . PLEASE RE-ENTER"
H1$ = " FIRST SYSTEM SECOND SYSTEM"
H2$ = "PT. =============== ================="
H3$ = "NO. EASTING NORTHING EASTING NORTHING"
P$ = "DO YOU WISH PRINTOUT OF DATA (Y/N)"
Q$ = Question response (Y/N)
X$ = "ENTER X(E) COORD. OF POINT"
Y$ = "ENTER Y(N) COORD. OF POINT"

3.10.4 Co-ordinate transformation – BASIC program

```
1Ø  REM  <TRANSFORMATION> PROGRAM FOR APPLE II+  USES QUME PRINTER
2Ø D$ =  CHR$ (4): ONERR  GOTO 92ØØ
3Ø E = Ø:I = Ø:J = Ø:K = Ø:N = Ø:P1 = Ø:T = Ø:C = Ø
4Ø T1 = Ø:T2 = Ø:T3 = Ø:T4 = Ø:T5 = Ø:T6 = Ø:T7 = Ø
5Ø EØ$ = "SORRY, DATA ERROR ... PLEASE RE-ENTER."
55 P$ = "DO YOU WISH PRINTOUT OF ABOVE DATA (Y/N)"
7Ø X$ = "ENTER X(E) COORD. OF POINT ":Y$ = "ENTER Y(N) COORD. OF POINT "
8Ø H1$ = "        FIRST SYSTEM       SECOND SYSTEM  "
85 H2$ = "PT.   ================  ================="
9Ø H3$ = "NO.   EASTING NORTHING  EASTING NORTHING "
95  REM  ***************************************************
1ØØ  GOSUB 9ØØØ
11Ø  PRINT "****************************************"
12Ø  PRINT "*                                      *"
13Ø  PRINT "*     UNIVERSITY  OF  STRATHCLYDE      *"
14Ø  PRINT "*  DEPARTMENT  OF  CIVIL  ENGINEERING  *"
15Ø  PRINT "*                                      *"
16Ø  PRINT "*           SURVEYING  SECTION         *"
17Ø  PRINT "* CO-ORDINATE TRANSFORMATION SOLUTIONS *"
18Ø  PRINT "****************************************"
19Ø  PRINT
2ØØ  PRINT "ARE ANGLES IN DECIMALS <1>, DD.MMSS <2> OR DO YOU WISH TO QUI
     T <3>."
21Ø  INPUT D9
```

```
220  IF D9 < 1 OR D9 > 3 THEN 200
230  ON D9 GOTO 240,240,9500
240 R = 57.2957795
300  GOSUB 9000
310  PRINT
320  PRINT "SELECT TRANSFORMATION PROGRAM FROM MENU"
330  PRINT
340  PRINT "    <1> POINTS IN BOTH SYSTEMS"
350  PRINT "    <2> ROTATION ANGLE AND POINT"
360  PRINT "    <3> QUIT "
370  INPUT CT
380  IF CT < 1 OR CT > 3 THEN 300
390  IF CT = 3 THEN 9500
500  GOSUB 9000
510  PRINT "UNIVERSITY OF STRATHCLYDE"
520  PRINT "CO-ORDINATE TRANSFORMATION PROGRAM :-"
530  PRINT "****************************************"
540  PRINT
550  PRINT "ENTER LOCATION OF WORK, ETC. WHEN       REQUESTED, AND PRESS
     <RETURN>"
560  PRINT "ENTER LOCATION OF SURVEY :-"
570  INPUT A$
580  PRINT "ENTER OPERATOR'S NAME :-"
590  INPUT B$
600  PRINT "ENTER DATE OF SURVEY AS DD/MM/YY :-"
610  INPUT C$
640  PRINT
650  PRINT "DO YOU WISH PRINTOUT OF LOCATION ETC.,  (Y/N) ";
660  INPUT Q$
670  IF Q$ = "Y" THEN 6000
680  IF Q$ < > "N" THEN 650
690 :
700  ON CT GOTO 1000,2000
995  REM ***************************************************
1000  GOSUB 9000
1010  PRINT "<1> CO-ORDINATE TRANSFORMATION - POINTS"
1020  PRINT "****************************************"
1030  PRINT "ENTER NUMBER OF POINTS IN BOTH SYSTEMS"
1040  INPUT N
1050  IF N =  > 2 THEN 1100
1060  PRINT  CHR$ (7): REM  CTRL-B (BELL)
1070  GOSUB 9100
1080  PRINT "TWO OR MORE POINTS ARE REQUIRED IN EACH SYSTEM !"
1090  GOTO 1020
1100  DIM E1(N - 1),E2(N - 1),N1(N - 1),N2(N - 1),S(N - 1)
1110  GOSUB 9000
1120  PRINT
1130  FOR I = 0 TO N - 1
1140  PRINT "FIRST SYSTEM - ENTER A POINT NUMBER"
1150  PRINT "****************************************"
1160  INPUT S(I)
1170  PRINT X$;S(I);: INPUT E1(I)
1180  PRINT Y$;S(I);: INPUT N1(I)
1190  PRINT
1200  PRINT "SECOND SYSTEM - ENTER COORDS OF ";S(I)
1210  PRINT "****************************************"
1220  PRINT X$;S(I);: INPUT E2(I)
1230  PRINT Y$;S(I);: INPUT N2(I)
```

```
1235  REM  CALCULATE VARIABLES T1-T7
124Ø T1 = T1 + (E1(I) * E2(I) + N1(I) * N2(I))
125Ø T2 = T2 + E1(I):T3 = T3 + E2(I)
126Ø T4 = T4 + N1(I):T5 = T5 + N2(I)
127Ø T6 = T6 + ((E1(I)) ^ 2 + (N1(I)) ^ 2)
128Ø T7 = T7 + (E1(I) * N2(I) - N1(I) * E2(I))
129Ø  PRINT : NEXT I
1295  REM  CALCULATE MULTIPLYING CONSTANTS
13ØØ K1 = (T1 - (T2 * T3 + T4 * T5) / N) / (T6 - (T2 ^ 2 + T4 ^ 2) / N)
131Ø K2 = (T7 - (T2 * T5 - T3 * T4) / N) / (T6 - (T2 ^ 2 + T4 ^ 2) / N)
132Ø ET = (T3 - K1 * T2 + K2 * T4) / N
133Ø NT = (T5 - K1 * T4 - K2 * T2) / N
134Ø  GOSUB 9ØØØ
135Ø  PRINT "CO-ORDINATE TRANSFORMATION - INPUT PTS."
136Ø  PRINT "****************************************"
137Ø  PRINT H1$
138Ø  PRINT H2$
139Ø  PRINT H3$
14ØØ  PRINT "========================================"
141Ø  FOR J = Ø TO N - 1
142Ø  PRINT S(J); TAB( 5);E1(J); TAB( 14);N1(J); TAB( 23);E2(J); TAB( 32)
     ;N2(J)
143Ø  NEXT J
144Ø  PRINT "****************************************"
145Ø  PRINT P$;: INPUT Q$
146Ø  IF Q$ = "Y" THEN 625Ø
147Ø  IF Q$ < > "N" THEN 145Ø
148Ø  IF CT = 2 THEN 216Ø
149Ø :
15ØØ  GOSUB 9ØØØ
151Ø  PRINT "TRANSFORMATION FROM FIRST SYSTEM"
152Ø  PRINT "****************************************"
153Ø  PRINT "ENTER NUMBER OF POINTS TO BE TRANSFORMED"
154Ø  INPUT T
155Ø  DIM E8(T - 1),E9(T - 1),N8(T - 1),N9(T - 1),T(T - 1)
156Ø  PRINT "ENTER POINT NUMBERS IN SEQUENCE"
157Ø  FOR K = Ø TO T - 1
158Ø  PRINT "ENTER POINT NUMBER ";
159Ø  INPUT T(K)
16ØØ  PRINT X$;T(K);: INPUT E8(K)
161Ø  PRINT Y$;T(K);: INPUT N8(K)
162Ø E9(K) = K1 * E8(K) - K2 * N8(K) + ET:E9(K) =  INT ((E9(K)) * 1ØØØ +
     .5) / 1ØØØ
163Ø N9(K) = K1 * N8(K) + K2 * E8(K) + NT:N9(K) =  INT ((N9(K)) * 1ØØØ +
     .5) / 1ØØØ
164Ø  PRINT "****************************************"
165Ø  PRINT "CO-ORDINATE TRANSFORMATION - OUTPUT"
166Ø  PRINT "****************************************"
167Ø  PRINT H1$: PRINT H2$: PRINT H3$
168Ø  PRINT "========================================"
169Ø  PRINT T(K); TAB( 5);E8(K); TAB( 14);N8(K); TAB( 23);E9(K); TAB( 32)
     ;N9(K)
17ØØ  PRINT "****************************************"
171Ø  PRINT : IF CT = 2 THEN 232Ø
172Ø  NEXT K
173Ø  PRINT : PRINT P$;: INPUT Q$
174Ø  IF Q$ = "Y" THEN 65ØØ
175Ø  IF Q$ < > "N" THEN 173Ø
```

```
1755  REM  END OF PROGRAM SINCE CANNOT RE-DIMENSION VARIABLES
176Ø  GOTO 95ØØ
1995  REM  ***************************************************
2ØØØ  GOSUB 9ØØØ
2Ø1Ø  PRINT "<2> CO-ORDINATE TRANSFORMATION - ": PRINT "ROTATION ANGLE AN
     D POINT"
2Ø2Ø  PRINT "****************************************"
2Ø3Ø  PRINT "ENTER ROTATION ANGLE (+ FOR CLOCKWISE,  - FOR ANTI-CLOCKWISE
     )"
2Ø4Ø  INPUT RØ
2Ø5Ø  IF D9 = 1 THEN 2Ø7Ø
2Ø55  REM  I.E. D9=2, CONVERT DD.MMSS TO DECIMALS
2Ø6Ø D5 = RØ: GOSUB 8ØØØ:RØ = D5
2Ø7Ø  PRINT : PRINT "ENTER SCALE FACTOR"
2Ø8Ø  INPUT SF
2Ø9Ø  PRINT :N = 1: PRINT "ENTER NUMBER OF A POINT IN FIRST SYSTEM"
21ØØ  INPUT S(Ø)
211Ø  PRINT : PRINT X$;S(Ø);: INPUT E1(Ø)
212Ø  PRINT Y$;S(Ø);: INPUT N1(Ø)
213Ø  PRINT : PRINT "ENTER SAME POINT IN SECOND SYSTEM"
214Ø  PRINT : PRINT X$;S(Ø);: INPUT E2(Ø)
215Ø  PRINT Y$;S(Ø);: INPUT N2(Ø): GOSUB 135Ø
216Ø  PRINT : PRINT "ENTER NUMBER OF POINTS TO BE TRANSFORMED"
217Ø  INPUT T
218Ø  DIM E8(T - 1),E9(T - 1),N8(T - 1),N9(T - 1),T(T - 1)
219Ø  PRINT "ENTER POINT NUMBERS IN SEQUENCE"
22ØØ  FOR K = Ø TO T - 1
221Ø  PRINT "ENTER POINT NUMBER ";
222Ø  INPUT T(K)
223Ø  PRINT X$;T(K);: INPUT E8(K)
224Ø  PRINT Y$;T(K);: INPUT N8(K)
225Ø ET = E8(K) - E1(Ø)
226Ø NT = N8(K) - N1(Ø)
227Ø W = ( ATN (ET / NT)) * R + RØ
228Ø H = SF *  SQR (ET ^ 2 + NT ^ 2)
229Ø E9(K) = E2(Ø) + H *  SIN (W / R)
23ØØ N9(K) = N2(Ø) + H *  COS (W / R)
231Ø  GOTO 164Ø
232Ø  PRINT : NEXT K
233Ø  PRINT : PRINT P$;: INPUT Q$
234Ø  IF Q$ = "Y" THEN 65ØØ
235Ø  IF Q$ <  > "N" THEN 233Ø
2355  REM  END OF PROGRAM SINCE CANNOT RE-DIMENSION VARIABLES
236Ø  GOTO 95ØØ
5985  REM  ************************************************************
     **
599Ø  REM  PRINTOUT OF LOCATION, ETC
6ØØØ  PRINT D$;"PR#1": PRINT "LOCATION OF SURVEY :- ";A$
6Ø1Ø  PRINT "**********************************************************
     ********************"
6Ø2Ø  PRINT "OPERATOR'S NAME :- ";B$
6Ø3Ø  PRINT "DATE OF SURVEY  :- ";C$
6Ø6Ø  PRINT "**********************************************************
     ********************"
6Ø7Ø  PRINT : PRINT D$;"PR#Ø"
6Ø8Ø  GOTO 7ØØ
6243  REM  ***************************************************
6245  REM  PRINTOUT OF COORDINATES ON QUME
625Ø  PRINT D$;"PR#1": POKE 1784 + 1,8Ø
```

```
6260 P7 = 1: REM  PRINTER CODE
6270  PRINT "CO-ORDINATE TRANSFORMATION - INPUT POINTS"
6280  IF CT = 2 THEN 6450
6290  PRINT "************************************************************
     "
6300  POKE 36,12: PRINT "FIRST SYSTEM";: POKE 36,32: PRINT "SECOND SYSTEM
     "
6310  PRINT "POINT";: POKE 36,10: PRINT "=================";: POKE 36,30
     : PRINT "================="
6320  PRINT "NO.";: POKE 36,10: PRINT "EASTING";: POKE 36,20: PRINT "NORT
     HING";: POKE 36,30: PRINT "EASTING";: POKE 36,40: PRINT "NORTHING"
6330  PRINT "============================================================
     "
6340  FOR I = Ø TO N - 1
6350  PRINT S(I);: POKE 36,1Ø: PRINT E1(I);: POKE 36,2Ø: PRINT N1(I);: POKE
     36,3Ø: PRINT E2(I);: POKE 36,4Ø: PRINT N2(I)
6360  NEXT I
6370  PRINT "************************************************************
     "
6380  PRINT
6385  REM  RETURN TO SCREEN
6390  PRINT D$;"PR#Ø"
6400  IF CT = 1 THEN 15ØØ
6410  IF CT = 2 THEN 216Ø
6443  REM  ****************************************
6445  REM  EXTRA PRINTOUT IF CT=2
6450  IF D9 = 1 THEN 647Ø
6460 D = RØ: GOSUB 8Ø5Ø:RØ = D
6470  PRINT "ROTATION ANGLE = ";RØ
6480  PRINT "SCALE FACTOR   = ";SF
6490  GOTO 629Ø
6493  REM  *************************************************
6495  REM  PRINTOUT OF TRANSFORMATION POINTS
6500  PRINT D$;"PR#1": POKE 1784 + 1,8Ø
6510  PRINT
6520  PRINT "CO-ORDINATE TRANSFORMATION - OUTPUT POINTS"
6540  PRINT "************************************************************
     "
6550  POKE 36,12: PRINT "FIRST SYSTEM";: POKE 36,32: PRINT "SECOND SYSTEM
     "
6560  PRINT "POINT";: POKE 36,1Ø: PRINT "=================";: POKE 36,3Ø
     : PRINT "================="
6570  PRINT "NO.";: POKE 36,1Ø: PRINT "EASTING";: POKE 36,2Ø: PRINT "NORT
     HING";: POKE 36,3Ø: PRINT "EASTING";: POKE 36,4Ø: PRINT "NORTHING"
6580  PRINT "============================================================
     "
6590  FOR I = Ø TO T - 1
6600  PRINT T(I);: POKE 36,1Ø: PRINT E8(I);: POKE 36,2Ø: PRINT N8(I);: POKE
     36,3Ø: PRINT E9(I);: POKE 36,4Ø: PRINT N9(I)
6610  NEXT I
6620  PRINT "************************************************************
     "
6630  PRINT
6635  REM  RETURN TO SCREEN
6640  PRINT D$;"PR#Ø"
6650  GOTO 95ØØ
7993  REM  ***************************************************
7995  REM  ROUTINE DD.MMSS TO DECIMALS
8000 DØ =  INT (D5)
```

```
8005  REM  FIND NO. OF MINUTES,MØ
8010 M1 = (D5 - DØ) * 100:MØ =  INT (M1 + .1)
8015  REM  FIND NO. OF SECONDS,SØ
8020 SØ =  INT (M1 * 100 - MØ * 100 + .5)
8025  REM  COLLECT DEGS,MINS,SECS
8030 D5 = DØ + MØ / 60 + SØ / 3600
8040  RETURN
8043  REM  ****************************************************
8045  REM  ROUTINE DECIMAL TO DD.MMSS
8050 D1 =  INT (D)
8055  REM  FIND TOTAL NO. OF SECONDS
8060 D2 = (D - D1) * 3600
8065  REM  FIND NO. OF MINUTES
8070 M =  INT (D2 / 60)
8075  REM  FIND NO. OF SECONDS
8080 S = D2 - 60 * M + .5
8090  IF S < 60 THEN 8130
8100 M = M + 1:S = 0
8110  IF M < 60 THEN 8130
8120 D1 = D1 + 1:M = 0
8130 D = D1 * 10000 + M * 100 + S
8140 D =  INT (D) / 10000: RETURN
8985  REM  ***************************************************
8990  REM  GOSUB ROUTINE TO CLEAR SCREEN
9000  HOME
9010  RETURN
9093  REM  ****************************************************
9095  REM  GOSUB ROUTINE FOR ERRORS
9100  PRINT
9110  PRINT EØ$
9120  PRINT
9130  RETURN
9193  REM  ****************************************************
9195  REM  ERROR TRACE ROUTINE
9200  PRINT  CHR$ (7): REM  CTRL-B (BELL)
9210 E =  PEEK (222): REM  GET ERROR NO.
9220  IF C = 1 THEN 9260
9230  INVERSE
9240  PRINT "ERROR NO. ";E;" FOUND"
9250 C = 1: TRACE : RESUME
9260  PRINT "ERROR ON SECOND LINE NO. "
9270  NORMAL
9280  NOTRACE
9290  STOP
9493  REM  ****************************************************
9495  REM  END OF PROGRAM
9500  GOSUB 9000
9510  PRINT
9520  PRINT "END TRANSFORMATION PROGRAM"
9530  PRINT
9540  PRINT "**************************"
9550  REM  PROGRAM PREPARED BY
9560  REM  DR. P.H. MILNE
9570  REM  DEPT. OF CIVIL ENGINEERING
9580  REM  UNIVERSITY OF STRATHCLYDE
9590  REM  GLASGOW G4 ONG
9600  REM  SCOTLAND
9610  REM  ****************************
9620  END
```

3.10.5 Co-ordinate transformation – computer printout

```
LOCATION OF SURVEY :- GLENMORE PARK
****************************************************************************
OPERATOR'S NAME :- JOHN SMITH
DATE OF SURVEY  :- 23/05/83
****************************************************************************
CO-ORDINATE TRANSFORMATION - INPUT POINTS
***************************************************************
            FIRST SYSTEM          SECOND SYSTEM
POINT       ==================    ==================
NO.         EASTING   NORTHING    EASTING   NORTHING
===============================================================
1           561.673   224.54      515.353   165.977
2           468.71    356.577     429.427   302.698
***************************************************************


CO-ORDINATE TRANSFORMATION - OUTPUT POINTS
***************************************************************
            FIRST SYSTEM          SECOND SYSTEM
POINT       ==================    ==================
NO.         EASTING   NORTHING    EASTING   NORTHING
===============================================================
3           307.327   232.414     261.767   187.151
4           400       150         350       100
***************************************************************


CO-ORDINATE TRANSFORMATION - INPUT POINTS
ROTATION ANGLE = 3
SCALE FACTOR   = 1
***************************************************************
            FIRST SYSTEM          SECOND SYSTEM
POINT       ==================    ==================
NO.         EASTING   NORTHING    EASTING   NORTHING
===============================================================
4           400       150         350       100
***************************************************************


CO-ORDINATE TRANSFORMATION - OUTPUT POINTS
***************************************************************
            FIRST SYSTEM          SECOND SYSTEM
POINT       ==================    ==================
NO.         EASTING   NORTHING    EASTING   NORTHING
===============================================================
1           561.673   224.54      515.353   165.977
2           468.71    356.577     429.427   302.698
3           307.327   232.414     261.767   187.151
***************************************************************
```

4

Areas, volumes and mass-haul diagrams

4.1 INTRODUCTION

One of the essential tasks of the land surveyor is to measure and calculate accurately areas of land and volumes of material. The previous chapter dealt with precise station fixing as used in theodolite and traverse surveying. This chapter introduces some of the more important and most often used techniques in the calculation of areas and volumes. In carrying out the field observations prior to the calculations, it is essential that the surveyor obtains all the necessary data to ensure as accurate a computation as possible. It is also most important that the contract documents for a specific project are examined in detail to check the specified methods of calculation, e.g., volumes by end areas or prismoidal rule.

4.2 CALCULATION OF PLAN AREAS

Area calculations normally fall into three categories; they are either straight sided, irregular, or a combination of both. Two common methods for evaluating areas are the use of a planimeter for mechanical integration or plotting on gridded paper and counting squares. Both rely on the accurate plotting of the area to be measured prior to the area measurement. However for large areas, these two techniques can be very tedious, and computer oriented techniques can therefore have considerable advantages.

If the area is bounded by straight lines, it can generally be resolved into a matrix of triangles and rectangles. For triangular areas, the program <TRIANGLE> discussed in Section 3.2 can be utilized where the area can be found from one of three equations:

(1) $$\text{Area} = [P(P-a)(P-b)(P-c)]^{1/2} \tag{4.1}$$

where a, b, c are sides of triangle and

$$P = (a+b+c)/2 \tag{4.2}$$

(2) Area = (base of triangle × height of triangle)/2 (4.3)

(3) Area = $ab\sin C$ (4.4)

where C is angle between side lengths a and b.

In the earlier computer program <TRIANGLE>, the sides were labelled S_1, S_2, S_3 and the angles A_1, A_2, A_3 (Fig. 3.2). Using this BASIC substitution gives for Equation 4.4 the following:

$$A = (S_1 S_3 \sin A_3)/2 \quad (4.5)$$

as used in line 6410 of the <TRIANGLE> program.

However for large plan areas where the co-ordinates of the points are known, there are advantages in using the co-ordinates themselves. For smaller areas, for example the boundary area between a traverse line and a fence, where offset measurements have been taken at right angles to the traverse line, either the Trapezium Rule or Simpson's Rule can be used depending upon whether the boundary is straight or curved. All three of these methods are included in the following BASIC program.

4.2.1 Area solutions – subroutine index

	Line numbers	*Function*
(a)	10–90	Initialization and control
(b)	100–190	Screen header display
(c)	200–280	Program selection from menu
(d)	500–700	Entry of area location, etc.
(e)	1000–1470	<1> Area from co-ordinates
(f)	2000–2190	<2> Trapezoidal Rule
(g)	3000–3250	<3> Simpson's Rule
(h)	6000–6080	Routine to print area location, etc.
(i)	6250–6420	Routine to print solutions to <1>
(j)	6500–6650	Routine to print solutions to <2> and <3>
(k)	9000–9010	Routine to clear screen
(l)	9100–9130	Routine to display error message
(m)	9200–9290	Error trace routine
(n)	9500–9620	Termination of program

(a) Initialization and control

Line numbers 10–90

All numeric variables and required string variables are initialized.

(b) Screen header display

Line numbers 100–190

This is the screen display used for civil engineering students at Strathclyde University. The operator can alter this as required to suit another organization.

(c) Program menu selection

Line numbers 200–280

A choice of three area programs is offered in lines 220–240:

<1> AREA FROM CO-ORDINATES
<2> TRAPEZOIDAL RULE
<3> SIMPSON'S RULE

(d) Entry of area location

Line numbers 500–700

This routine is presented for record purposes to annotate the printout with title of area, operator's name and date. The program then branches to the selected program from the menu.

(e) <1> Area from co-ordinates

Line numbers 1000–1470

The number of stations to be used in the figure is entered in line 1040, and a check made that it is three or more, or else an error message is displayed. Depending on the number (N) of stations, the required numeric arrays are dimensioned in line 100.

The operator is then requested to enter the stations and their co-ordinates, X or easting and Y or northing in clockwise rotation, Fig. 4.1. The total area A for N stations is then given by:

$$A = [(N_1E_2 + N_2E_3 + N_3E_4 + \ldots + N_{N-1}E_N + N_NE_1) - (E_1N_2 + E_2N_3 + E_3N_4 + \ldots + E_{N-1}N_N + E_NN_1)]/2 \tag{4.6}$$

where the subscripts 1, 2, . . . N refer to the station numbers. This total is calculated in the lines 1220–1280, and then all the stations, their co-ordinates and the total area displayed in lines 1300–1430, with an option to obtain a hard copy printout at line 1440. As it is not possible to redimension variables in Applesoft, the program terminates after this option. If it was desired to run numerous calculations end-on, the array variables could be dimensioned in lines 10–90 choosing a reasonable number, 11, say, to reserve space for up to 12 stations.

(f) <2> Trapezoidal Rule

Line numbers 2000–2190

This is a useful method of calculating the area, for example, between a traverse line and a boundary line, where the boundary has been measured by offsets taken an equal distance apart at right angles to the traverse line, Fig. 4.2. It should be noted that it is assumed that the distance apart of the offsets, L, is short

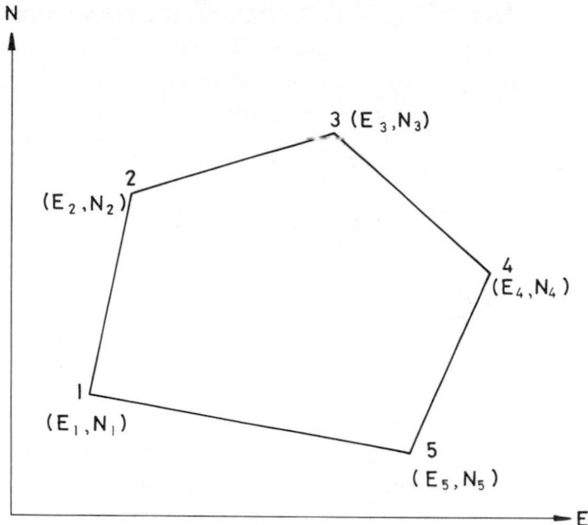

Fig. 4.1. Nomenclature for areas from co-ordinates.

enough for the length of the boundary between the offsets to be taken as straight lines, thus dividing the area into a series of trapezoids. The offsets, H, are entered in a loop, lines 2080–2120, where N is the number of offsets. The total area is then given by:

$$A = L[(H_1 + H_N)/2 + H_2 + H_3 + \ldots + H_{N-1}] \tag{4.7}$$

This formula can be used irrespective of whether there is an even or odd number of offsets. However if the boundary is very irregular between the offsets, the next method is to be preferred.

(g) <3> Simpson's Rule

Line numbers 3000–3250

This method assumes that the irregular boundary is composed of a series of parabolic arcs, and the area is calculated for three offsets in turn, that is over two strips. Thus it is essential to ensure there is an *odd* number of offsets to give an

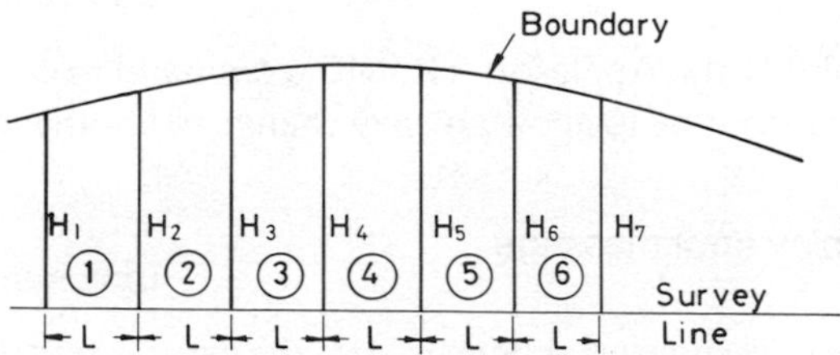

Fig. 4.2. Nomenclature for Simpson's Rule in area calculations.

even number of equal strips, Fig. 4.2. As before, the number of offsets is N and the distance apart L. A check is made in line 3050 that an odd number of offsets is entered. If not, an error message is displayed and re-entry requested. If the first three offsets are considered in Fig. 4.2 the area is given by:

$$A_{1-3} = (H_1 + 4H_2 + H_3)L/3 \tag{4.8}$$

For the general case of N offsets (N odd), the area will be:

$$\begin{aligned} A_{1-N} = [&\text{sum of first and last offsets} \\ &+ 2(\text{sum of remaining odd offsets}) \\ &+ 4(\text{sum of even offsets})]L/3 \end{aligned} \tag{4.9}$$

A loop is used starting at line 3150 to collect the odd offsets in T_1 and the even offsets in T_2 to calculate the area A from Equation 4.9 in line 3190. After the result has been displayed, an option is given for a printout.

(h) Routine to print area, location, etc.

Line numbers 6000–6080

This routine is provided for record purposes to annotate the printout, and is based on Applesoft DOS. It may require alteration for different microcomputers and printers.

(i) Routine to print solution to <1>

Line numbers 6250–6420

This routine prints out the stations and their co-ordinates in addition to the area. It can be used with either metric or imperial units.

(j) Routine to print solution to <2> and <3>

Line numbers 6500–6650

The type of solution – Trapezoidal or Simpson's Rule – is printed first, followed by the offsets and the total area. It can be used with either metric or imperial units.

(k) Routine to clear screen

Line numbers 9000–9010

To simplify portability the Applesoft 'HOME' command to clear the screen has been placed in a subroutine to allow an easy change to another microcomputer.

(l) Routine to display error message

Line numbers 9100–9130

If a wrong keyboard response has been made, the error is trapped and this error message displayed and re-entry requested.

(m) Error trace routine

Line numbers 9200–9290

This error trace routine was discussed earlier in Section 2.2.9.

(n) Termination of program

Line numbers 9500–9620

This program can be terminated either from the initial screen header, or on completion of the calculations. An end-of-program message is displayed to advise the operator that computations are complete.

4.2.2 Area calculations – numeric variables

A	= Area
C	= Counter in error trace routine
E	= Error number
$E(N-1)$	= Eastings array of stations
$H(N-1)$	= Distance array for offsets
I	= Integer loop variable
J	= Integer loop variable
K	= Integer loop variable
L	= Distance between offsets
N	= Number of stations/offsets
$N(N-1)$	= Northings array of stations
$S(N-1)$	= Number array of stations
T	= Selection of program from menu
$T1$, $T2$	= Temporary storage variables for calculating area

4.2.3 Area calculations – string variables

A$	Storage for location of area
B$	Storage for operator's name
C$	Storage for date of calculation
D$	= CHR$(4), DOS command, CTRL-D
E0$	= "SORRY, DATA ERROR . . . PLEASE RE-ENTER"
P$	= "DO YOU WISH PRINTOUT OF ABOVE DATA (Y/N)"
Q$	= Question response (Y/N)
T$	= Either "<2> TRAPEZOIDAL RULE" or "<3> SIMPSON'S RULE"
X$	= "ENTER X(E) COORD. OF POINT"
Y$	= "ENTER Y(N) COORD. OF POINT"

4.2.4 Area calculations – BASIC program

```
10  REM  <AREAS> PROGRAM FOR APPLE II+  USES QUME PRINTER
20 D$ =  CHR$ (4): ONERR  GOTO 9200
30 E = 0:I = 0:J = 0:K = 0:N = 0:P1 = 0:T = 0:C = 0
```

```
40 A = 0:T1 = 0:T2 = 0
50 E0$ = "SORRY, DATA ERROR ... PLEASE RE-ENTER."
55 P$ = "DO YOU WISH PRINTOUT OF ABOVE DATA (Y/N)"
70 X$ = "ENTER X(E) COORD. OF POINT ":Y$ = "ENTER Y(N) COORD. OF POINT "
95  REM *****************************************************
100  GOSUB 9000
110  PRINT "****************************************"
120  PRINT "*                                      *"
130  PRINT "*     UNIVERSITY  OF  STRATHCLYDE      *"
140  PRINT "*  DEPARTMENT  OF  CIVIL  ENGINEERING  *"
150  PRINT "*                                      *"
160  PRINT "*            SURVEYING  SECTION        *"
170  PRINT "*            PROGRAMS FOR AREAS        *"
180  PRINT "****************************************"
190  PRINT
200  PRINT "SELECT AREA PROGRAM FROM MENU :-"
210  PRINT
220  PRINT "    <1> AREAS FROM CO-ORDINATES"
230  PRINT "    <2> TRAPEZOIDAL RULE        "
240  PRINT "    <3> SIMPSON'S RULE          "
250  PRINT "    <4> QUIT                    "
260  INPUT T
270  IF T < 1 OR T > 4 THEN 200
280  IF T = 4 THEN 9500
500  GOSUB 9000
510  PRINT "UNIVERSITY OF STRATHCLYDE"
520  PRINT "PROGRAMS FOR AREAS"
530  PRINT "****************************************"
540  PRINT
550  PRINT "ENTER LOCATION OF WORK, ETC. WHEN       REQUESTED, AND PRESS
     <RETURN>"
560  PRINT "ENTER LOCATION OF AREA :-"
570  INPUT A$
580  PRINT "ENTER OPERATOR'S NAME :-"
590  INPUT B$
600  PRINT "ENTER DATE OF CALCULATION AS DD/MM/YY :-"
610  INPUT C$
640  PRINT
650  PRINT "DO YOU WISH PRINTOUT OF LOCATION ETC.,  (Y/N) ";
660  INPUT Q$
670  IF Q$ = "Y" THEN 6000
680  IF Q$ < > "N" THEN 650
690 :
700  ON T GOTO 1000,2000,3000
995  REM ****************************************************
1000  GOSUB 9000
1010  PRINT "<1> AREA FROM CO-ORDINATES PROGRAM"
1020  PRINT "****************************************"
1030  PRINT "ENTER NUMBER OF STATIONS IN FIGURE ";
1040  INPUT N
1050  IF N = > 3 THEN 1100
1060  PRINT  CHR$ (7): REM  CTRL-B (BELL)
1070  GOSUB 9100
1080  PRINT "THREE OR MORE POINTS ARE REQUIRED !"
1090  GOTO 1020
1100  DIM E(N - 1),N(N - 1),S(N - 1)
1110  GOSUB 9000
1120  PRINT
```

```
1130  PRINT "ENTER STN. NUMBERS IN CLOCKWISE ROTATION"
1140  PRINT "****************************************"
1150  FOR I = Ø TO N - 1
1160  PRINT "ENTER STATION NUMBER ";: INPUT S(I)
1170  PRINT X$;S(I);: INPUT E(I)
1180  PRINT Y$;S(I);: INPUT N(I)
1190  PRINT
1200  NEXT I
1205  REM  NOW CALCULATE AREA FROM VARIABLES T1 AND T2
1210  FOR J = Ø TO N - 2
1220 T1 = T1 + (N(J)) * (E(J + 1))
1230 T2 = T2 + (E(J)) * (N(J + 1))
1240  NEXT J
1250 T1 = T1 + (N(N - 1)) * (E(Ø))
1260 T2 = T2 + (E(N - 1)) * (N(Ø))
1270 A = (T1 - T2) / 2
1280 A =  INT (A * 1000 + .5) / 1000
1290 :
1300  GOSUB 9000
1310  PRINT "<1> AREA FROM CO-ORDINATES :-"
1320  PRINT "****************************************"
1330  PRINT "STN.NO."; TAB( 10);"EASTING"; TAB( 25);"NORTHING"
1340  PRINT "========================================"
1350  FOR I = Ø TO N - 1
1360  PRINT S(I); TAB( 10);E(I); TAB( 25);N(I)
1370  NEXT I
1380  PRINT "========================================"
1390  PRINT
1400  PRINT "TOTAL AREA = ";A;" SQ. UNITS"
1410  PRINT
1420  PRINT "****************************************"
1430  PRINT
1440  PRINT P$;: INPUT Q$
1450  IF Q$ = "Y" THEN 6250
1460  IF Q$ <  > "N" THEN 1440
1465  REM  END OF PROGRAM SINCE CANNOT RE-DIMENSION VARIABLES
1470  GOTO 9500
1993  REM  ****************************************
2000  GOSUB 9000
2010  PRINT "<2> TRAPEZOIDAL RULE :-"
2020  PRINT "****************************************"
2030  PRINT
2040  PRINT "ENTER NUMBER OF OFFSETS ";: INPUT N
2050  PRINT "ENTER DISTANCE BETWEEN OFFSETS ";: INPUT L
2060  DIM H(N - 1):T$ = "<2> TRAPEZOIDAL RULE :-"
2070  PRINT
2080  FOR I = Ø TO N - 1
2090  PRINT "ENTER OFFSET NO. ";(I + 1);" = ";
2100  INPUT H(I)
2110 T1 = T1 + H(I)
2120  NEXT I
2130 A = L * (T1 - (H(Ø) + H(N - 1)) / 2)
2140 A =  INT (A * 1000 + .5) / 1000
2150  PRINT : PRINT "TOTAL AREA = ";A;" SQ. UNITS"
2160  PRINT : PRINT P$;: INPUT Q$
2170  IF Q$ = "Y" THEN 6500
2180  IF Q$ <  > "N" THEN 2160
2190  GOTO 9500
```

```
2993  REM  ******************************************
3ØØØ  GOSUB 9ØØØ
3Ø1Ø  PRINT "<3> SIMPSON'S RULE :-"
3Ø2Ø  PRINT "****************************************"
3Ø3Ø  PRINT
3Ø4Ø  PRINT "ENTER NUMBER OF OFFSETS": PRINT "N.B. MUST BE ODD NUMBER ";:
      INPUT N
3Ø5Ø  IF (N - 1) / 2 =  INT ((N - 1) / 2) THEN 3Ø8Ø
3Ø6Ø  GOSUB 91ØØ
3Ø7Ø  GOTO 3Ø3Ø
3Ø8Ø  PRINT "ENTER DISTANCE BETWEEN OFFSETS ";: INPUT L
3Ø9Ø  DIM H(N - 1):T$ = "<3> SIMPSON'S RULE :-"
31ØØ  PRINT
311Ø  FOR I = Ø TO N - 1
312Ø  PRINT "ENTER OFFSET NO. ";(I + 1);" = ";
313Ø  INPUT H(I)
314Ø  NEXT I
315Ø  FOR J = Ø TO N - 1 STEP 2
3155  REM  COLLECT ODD OFFSETS IN T1
316Ø T1 = T1 + H(J): NEXT J
3165  REM  COLLECT EVEN OFFSETS IN T2
317Ø  FOR K = 1 TO N - 2 STEP 2
318Ø T2 = T2 + H(K): NEXT K
3185  REM  CALCULATE AREA
319Ø A = (4 * T2 + 2 * T1 - H(Ø) - H(N - 1)) * L / 3
32ØØ A =  INT (A * 1ØØØ + .5) / 1ØØØ
321Ø  PRINT : PRINT "TOTAL AREA = ";A;" SQ. UNITS"
322Ø  PRINT : PRINT P$;: INPUT Q$
323Ø  IF Q$ = "Y" THEN 65ØØ
324Ø  IF Q$ <  > "N" THEN 322Ø
325Ø  GOTO 95ØØ
5985  REM  ************************************************************
     **
599Ø  REM  PRINTOUT OF LOCATION, ETC
6ØØØ  PRINT D$;"PR#1": PRINT "LOCATION OF AREA :- ";A$
6Ø1Ø  PRINT "****************************************"
6Ø2Ø  PRINT "OPERATOR'S NAME :- ";B$
6Ø3Ø  PRINT "DATE OF CALCULATION :- ";C$
6Ø6Ø  PRINT "****************************************"
6Ø7Ø  PRINT : PRINT D$;"PR#Ø"
6Ø8Ø  GOTO 7ØØ
6243  REM  ************************************************
6245  REM  PRINTOUT OF COORDINATES ON QUME
625Ø  PRINT D$;"PR#1": POKE 1784 + 1,8Ø
626Ø  PRINT :P7 = 1: REM  PRINTER CODE
627Ø  PRINT "<1> AREA FROM CO-ORDINATES :-"
629Ø  PRINT "****************************************"
63ØØ  PRINT "STN.NO.";: POKE 36,1Ø: PRINT "EASTING";: POKE 36,25: PRINT "
     NORTHING"
631Ø  PRINT "========================================"
632Ø  FOR J = Ø TO N - 1
633Ø  PRINT S(J);: POKE 36,1Ø: PRINT E(J);: POKE 36,25: PRINT N(J)
634Ø  NEXT J
635Ø  PRINT "========================================"
637Ø  PRINT "TOTAL AREA = ";A;" SQ. UNITS"
639Ø  PRINT "****************************************"
64ØØ  PRINT
64Ø5  REM  RETURN TO SCREEN AT END OF PROGRAM
```

```
6410  PRINT D$;"PR#0"
6420  GOTO 9500
6493  REM  ****************************************
6495  REM  PRINTOUT FOR T=2 AND 3
6500  PRINT D$;"PR#1": POKE 1784 + 1,80
6510  PRINT T$;" L = ";L
6520 :
6530  PRINT "****************************************"
6540  PRINT "OFFSET NO.    DISTANCE"
6550  PRINT "======================"
6560  FOR J = 0 TO N - 1
6570  PRINT (J + 1);: POKE 36,15: PRINT H(J)
6580  NEXT J
6590  PRINT "========================================"
6600  PRINT "TOTAL AREA = ";A;" SQ. UNITS"
6620  PRINT "****************************************"
6630  PRINT
6635  REM  RETURN TO SCREEN
6640  PRINT D$;"PR#0"
6650  GOTO 9500
8985  REM  ****************************************
8990  REM  GOSUB ROUTINE TO CLEAR SCREEN
9000  HOME
9010  RETURN
9093  REM  ************************************************
9095  REM  GOSUB ROUTINE FOR ERRORS
9100  PRINT
9110  PRINT E0$
9120  PRINT
9130  RETURN
9193  REM  ************************************************
9195  REM  ERROR TRACE ROUTINE
9200  PRINT  CHR$ (7): REM  CTRL-B (BELL)
9210 E =  PEEK (222): REM  GET ERROR NO.
9220  IF C = 1 THEN 9260
9230  INVERSE
9240  PRINT "ERROR NO. ";E;" FOUND"
9250 C = 1: TRACE : RESUME
9260  PRINT "ERROR ON SECOND LINE NO. "
9270  NORMAL
9280  NOTRACE
9290  STOP
9493  REM  ************************************************
9495  REM  END OF PROGRAM
9500  GOSUB 9000
9510  PRINT
9520  PRINT "END AREAS PROGRAM"
9530  PRINT
9540  PRINT "****************************************"
9550  REM  PROGRAM PREPARED BY
9560  REM  DR. P.H. MILNE
9570  REM  DEPT. OF CIVIL ENGINEERING
9580  REM  UNIVERSITY OF STRATHCLYDE
9590  REM  GLASGOW G4 0NG
9600  REM  SCOTLAND
9610  REM  **************************
9620  END
```

4.2.5 Area calculations – computer printout

```
LOCATION OF AREA :- STRATHCLYDE PARK
****************************************
OPERATOR'S NAME :- JOHN SMITH
DATE OF CALCULATION :- 3Ø/Ø5/83
****************************************


<1> AREA FROM CO-ORDINATES :-
****************************************
STN.NO.    EASTING         NORTHING
========================================
1          5Ø              75
2          124.76          2Ø1.46
3          246.82          235.47
4          352.91          149.72
5          298.52          51.73
========================================
TOTAL AREA = 362Ø6.414 SQ. UNITS
****************************************


<2> TRAPEZOIDAL RULE :- L = 1Ø
****************************************
OFFSET NO.     DISTANCE
======================
1               16.76
2               19.81
3               2Ø.42
4               18.59
5               16.76
6               17.68
7               17.68
8               17.37
9               16.76
1Ø              17.68
========================================
TOTAL AREA = 1622.9 SQ. UNITS
****************************************


<3> SIMPSON'S RULE :- L = 2Ø
****************************************
OFFSET NO.     DISTANCE
======================
1               Ø
2               5.49
3               9.14
4               8.53
5               1Ø.67
6               12.5
7               9.75
8               4.57
9               1.83
========================================
TOTAL AREA = 1235.4 SQ. UNITS
****************************************
```

4.3 EARTHWORKS

The computation of volumes of earthworks for building or civil engineering projects is a frequent necessity, due to the continual excavation, removal and dumping of material, be it earth, sand or rock. In any construction work, a contractor will be paid on the basis of the volume of material handled and hence it is essential that good estimates are obtained for inclusion in the contract documents at the tender stage.

In road, rail or pipeline contracts, the standard method of volume computation for earthworks is using cross-sections, selected at regular chainages along the centre line of the proposed formation. These cross-sections may be regular or irregular in shape, and be either cut or fill or a combination of both. The first part of the following program gives a choice of six types of cross-sections which should cater for each of the above variations.

Three methods are available for volume computation between cross-sections, the mean area method, the end areas method and the prismoidal rule, the latter being the most accurate, and all are offered in the second part of the following program, commencing at line 2800. Curvature corrections are also included where the centroid of the cross-sectional shape can be readily obtained from the data supplied.

Once volume computations are complete for a stretch of roadworks, a mass-haul diagram is invariably drawn to aid the planning and construction of the roadworks. A mass-haul diagram is really a plotting of the aggregate volumes of earthwork to a horizontal base of distance. Cuttings are taken as positive, and fills as negative with positive volumes plotted above the base line, and negative volumes plotted below the base line. This program, <EARTHWORKS>, can use either end-area volumes or prismoidal volumes to compute aggregate volumes, and this last section commences at line 5000.

All three sections of <EARTHWORKS> can be entered with known cross-sections, or the mass-haul section for aggregate volumes can be entered with known volumes.

4.3.1 Earthworks – subroutine index

	Line numbers	*Function*
(a)	10–90	Initialization and control
(b)	100–180	Screen header display
(c)	300–400	Program selection from menu
(d)	500–680	Entry of earthworks location, etc.
(e)	700–790	Entry of earthworks data
(f)	800–990	Cross-section selection from menu
(g)	1000–1190	<A1> One-level section (cut or fill)
(h)	1200–1440	<A2> Two-level section (cut or fill)
(i)	1450–1690	<A3> Two-level section (cut + + fill)
(j)	1700–1990	<A4> Three-level section (cut or fill)
(k)	2000–2350	<A5> Irregular section (cut or fill)
(l)	2500–2630	<A6> Irregular section (cut ++ fill)

	Line numbers	Function
(m)	2650–2690	Equivalent area (curvature correction) for <A2>
(n)	2700–2780	Equivalent area (curvature correction) for <A3>
(o)	2800–2980	Entry of cross-section data for earthworks
(p)	3000–3080	Volume calculation selection from menu
(q)	3200–3380	<V1> Mean area rule
(r)	3400–3660	<V2> End areas rule
(s)	3700–3890	Extrapolation of cut/fill lengths
(t)	4000–4280	<V3> Prismoidal rule
(u)	5000–5180	Entry of volume data for mass-haul
(v)	5200–5680	Mass-haul aggregate volume computations
(w)	6000–6080	Routine to print earthworks location, etc.
(x)	6100–6390	Data printout for cross-sections <A1–A4>
(y)	6400–6590	Data printout for cross-sections <A5 and A6>
(z)	6600–6720	Printout summary of cross-sectional areas
(aa)	6800–6880	Data printout of volumes for <V1>
(bb)	7000–7150	Data printout of volumes for <V2>
(cc)	7200–7350	Data printout of volumes for <V3>
(dd)	7400–7550	Printout summary of volumes
(ee)	7600–7790	Mass-haul aggregate volume printout
(ff)	9000–9010	Routine to clear screen
(gg)	9100–9130	Routine to display error message
(hh)	9200–9290	Error trace routine
(ii)	9500–9620	Termination of program

(a) Initialization and control

Line numbers 10–90

All numeric variables and required string variables are initialized and dimensioned. Note the maximum array for cross-sections and cross-sectional points is 21. If a larger number is required the arrays should be redimensioned.

(b) Screen header display

Line numbers 100–180

This is the screen header display used for civil engineering students at Strathclyde University. The operator can alter this as required to suit another organization.

(c) Program selection from menu

Line numbers 300–400

A choice of three entry points to the <EARTHWORKS> program is given depending on the data provided:

<1> Areas of cross-sections
<2> Volumes from cross-sections
<3> Mass-haul from volumes.

(d) Entry of earthworks location, etc.

Line numbers 500–680

This routine is provided for record purposes to annotate the printout with a title of location, operator's name and date of calculation.

(e) Entry of earthworks data

Line numbers 700–790

Each of the programs chosen in (c) above uses this routine to enter the number of cross-sections, the chainage of the first cross-section, and the distance between the cross-sections, before branching to the selected program.

(f) Cross-section selection from menu

Line numbers 800–990

A choice of six types of cross-sections is given:

<1> One-level	(cut or fill)
<2> Two-level	(cut or fill)
<3> Two-level	(cut ++ fill)
<4> Three-level	(cut or fill)
<5> Irregular	(cut or fill)
<6> Irregular	(cut ++ fill)

After entry of each cross-section, the program returns to the menu to select the next type of cross-section before offering a printout summary in line 960.

(g) <A1> One-level section (cut or fill)

Line numbers 1000–1190

Where the ground surface is level, <A1> can be used for either cut or fill, where the formation depth/height is H units. If the formation width is B and the side slope $1:M$, as shown in Fig. 4.3, then the width W of excavation/embankment either side of the centre line will be:

$$W = B/2 + MH \tag{4.10}$$

and the total area of the cross-section will be:

$$\begin{aligned} A &= H(B+2W)/2 \\ &= H(B+MH) \end{aligned} \tag{4.11}$$

The program prompts with the cross-section number and chainage and requests whether cut or fill, and asks for B, H and M. The side width W and A are then calculated and displayed with an option for a printout of the results. The final area, either cut or fill, is stored in an array for later volume computations.

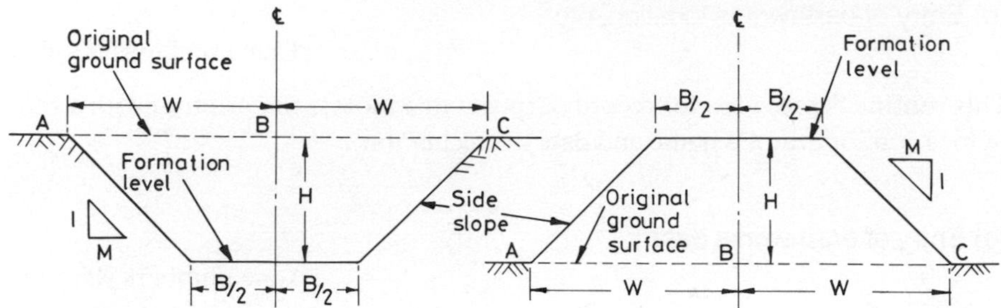

Fig. 4.3. Nomenclature for area calculations from a one-level cross-section (cut or fill).

(h) <A2> Two-level section (cut or fill)

Line numbers 1200–1440

Where the ground surface has a constant slope, transverse gradient or crossfall relative to the centre line, determined by two levels taken at right angles to the centre line, then the side-widths will be unequal, Fig. 4.4. As before, B is the formation width, H the depth/height of cut/fill and M the side slope, where the crossfall of the ground surface is taken as $1{:}K$. To calculate the cross-sectional area it is first necessary to calculate W_1 and W_2 the greater and lesser side-widths as shown in Fig. 4.4, where:

$$W_1 = (B/2 + MH)[K/(K - M)] \tag{4.12}$$

$$W_2 = (B/2 + MH)[K/(K + M)] \tag{4.13}$$

These side-widths are evaluated in lines 1320–1330, and then used to compute the area A in line 1340 from:

$$A = [(B/2 + MH)(W_1 + W_2) - B^2/2]/2M \tag{4.14}$$

After displaying the computed side-widths and area, an option is given to allow curvature corrections to be applied if necessary, as discussed later in (m), followed by a printout of the data and results.

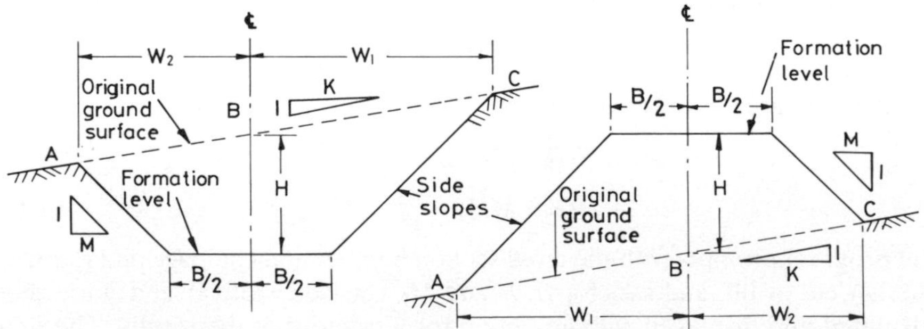

Fig. 4.4. Nomenclature for area calculations from a two-level cross-section (cut or fill).

(i) <A3> Two-level section (cut ++ fill)

Line numbers 1450–1690

The crossfall K is determined in the same way as <A2> above. Depending upon the relationship between the formation level of construction and the original ground surface, there will often be both cut and fill in each cross-section as shown in Fig. 4.5. As the cut and fill side slopes are entered separately they can be varied depending on the ground conditions. The critical information required for this area calculation is whether there is cut or fill at the centre line to ensure

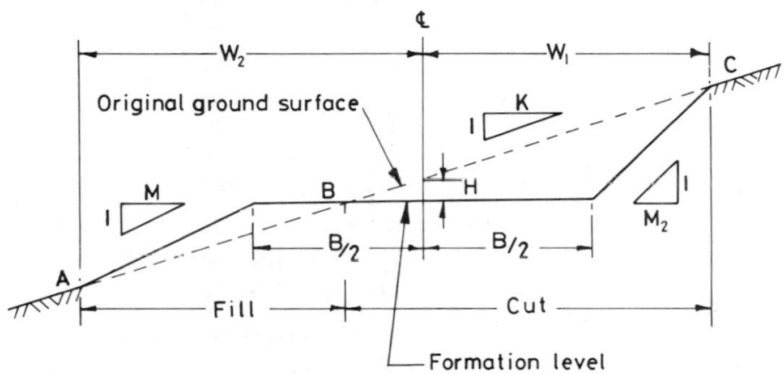

Fig. 4.5. Nomenclature for area calculations from a two-level cross-section consisting of cut and fill.

the correct calculation of the side-widths W_1 and W_2. If there is cut at the centre line then:

$$W_1 = (B/2 + M_2H)[K/(K - M_2)] \tag{4.15}$$

$$W_2 = (B/2 - MH)[K/(K - M)] \tag{4.16}$$

where H is the formation depth (positive), M is fill side slope and M_2 is cut side slope. If there is fill at the centre line and the formation height is H, then the correct values for W_1 and W_2 from Equations 4.15 and 4.16 can be found by changing the sign of H (line 1580). The areas of cut and fill are then stored in arrays for future volume calculations and the results displayed on the screen. As in <A2> a curvature correction is possible together with a printout option for record purposes.

(j) <A4> Three-level section (cut or fill)

Line numbers 1700–1990

This type of section is shown in Fig. 4.6 where there is a change of slope or crossfall at the centre line. The section is given its name from the three levels required to calculate the two slopes and can be used for either cut or fill. Where the ground surface is very irregular, either <A5> or <A6> should be used.

The side-width formulae for W_1 and W_2 can be obtained as for two-level

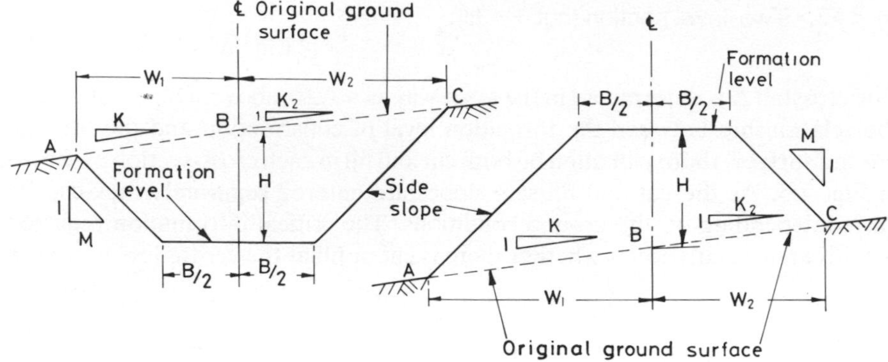

Fig. 4.6. Nomenclature for area calculations from a three-level cross-section (cut or fill).

sections <A2> from Equations 4.12 and 4.13. It should be noted that these two equations assume that W_1 is calculated for a negative slope from the centre line and W_2 calculated for a positive slope from the centre line. The program allows for either positive or negative slopes from the centre line, lines 1820–1900, to ensure the correct evaluation of W_1 and W_2. After displaying the results and storing the cut or fill in an array, an option is given to obtain a printout of the results.

(k) <A5> Irregular section (cut or fill)

Line numbers 2000–2350

Where there is a cut or fill cross-section with a very irregular ground surface, then the area can be computed using an adaptation of the method used for the calculation of a plan area from co-ordinates, as described in Section 4.2 using Equation 4.6. Points are entered in a clockwise direction. The cross-sectional area is calculated by entering the offset from the centre line (x) and the difference in height (y) from the formation level as shown in Fig. 4.7. Points to

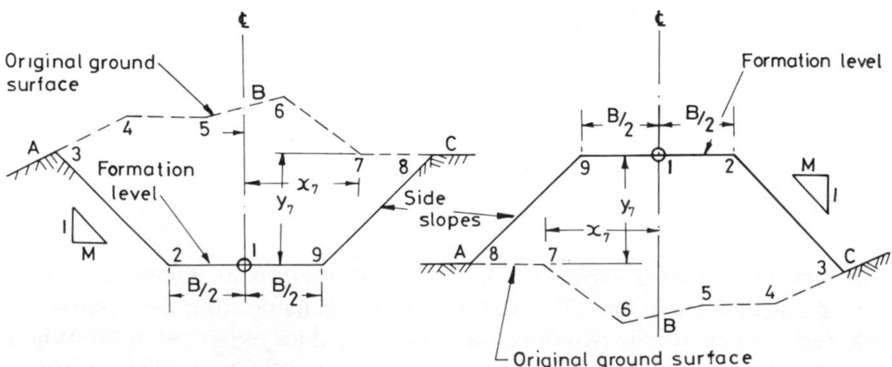

Fig. 4.7. Nomenclature for area calculations from an irregular cross-section (cut or fill).

the left of the centre line have negative offsets, and points below the formation level are negative. After calculation, storage and display, a computer printout of the points and their offsets, etc., can be obtained for record purposes.

(l) <A6> Irregular section (cut ++ fill)

Line numbers 2500–2630

This program follows the lines of <A5> above, but allows both cut and fill areas to be computed in the same irregular cross-section, Fig. 4.8. The points are entered in clockwise order round the cut and fill areas, together with their offsets

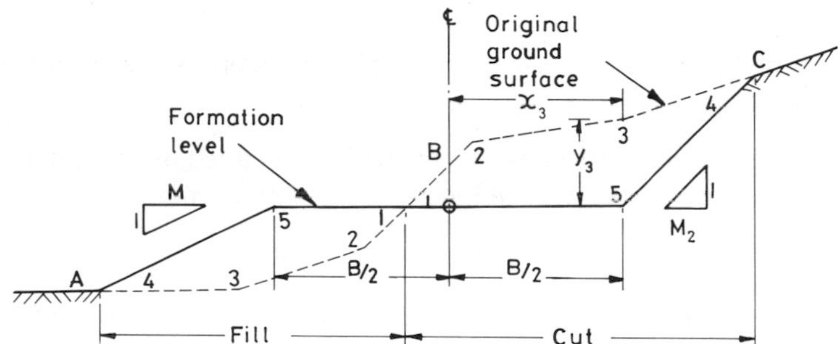

Fig. 4.8. Nomenclature for area calculations from an irregular cross-section consisting of cut and fill.

(x) and heights (y). Aftcr calculation, storage and display, a computer printout of the points and their offsets, etc. can be obtained for record purposes.

(m) Equivalent area (curvature correction) for <A2>

Line numbers 2650–2690

In computing volumes from cross-sections, it is normally assumed that the cross-sections are taken on a straight road. However, where there is a horizontal curve between two straights, the cross-sections will no longer be parallel to one another and errors can result in volume calculations, unless a curvature correction is applied. To ensure that the calculated volumes are as representative of the ground situation as possible, the equivalent area of the cross-section is calculated using *Pappus' theorem.* This theorem states that a volume swept out by a plane constant area revolving about a fixed axis is given by the product of the cross-sectional area and the distance moved by the centre of gravity of the section from the centre line.

In a symmetrical cross-section (one-level), Fig. 4.3, the centroid of the area will lie on the centre line. However for an unsymmetrical cross-section, Fig. 4.4, the centroid of the area will be on the side of the higher ground level, Fig. 4.9.

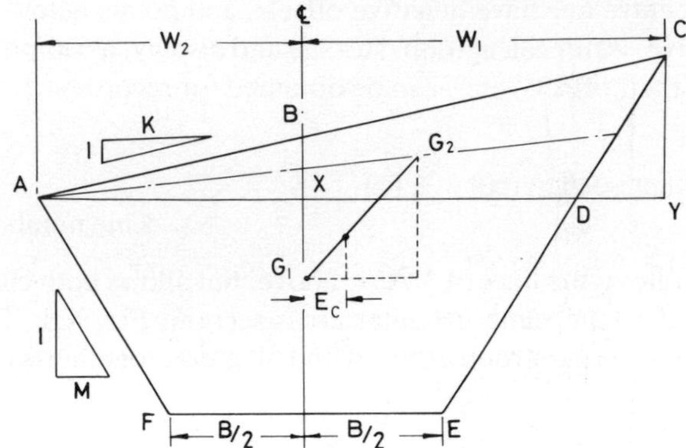

Fig. 4.9. Method of finding the centroid, and the eccentricity of a two-level cross-section.

The distance from the centroid to the centre line is called the eccentricity (E_c) and can be calculated for a two-level section, Fig. 4.9, from:

$$E_c = W_1 W_2 (W_1 + W_2)/3KA \tag{4.17}$$

Now the length of the path of the centroid, from Fig. 4.10, for a road curve of radius R, swinging through angle θ, will be:

$$L = (R + E_c)\theta \,\text{rad} \tag{4.18}$$

From Pappus' theorem, the volume will then be:

$$V = A(R + E_c)\theta \,\text{rad} \tag{4.19}$$

but since

$$\theta \,\text{rad} = L/R$$

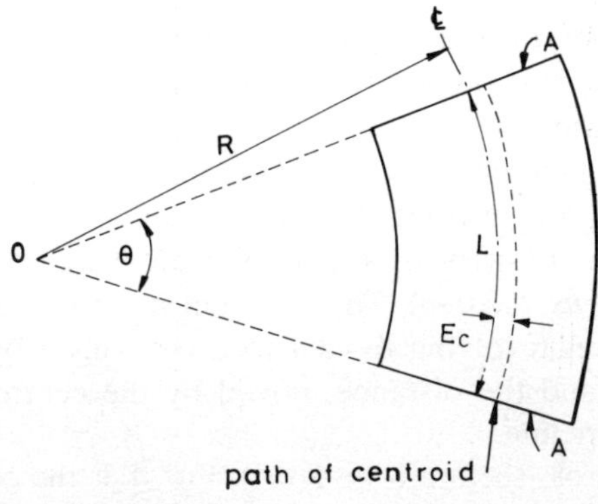

Fig. 4.10. Application of Pappus' theorem to volumes from cross-sections of a circular road curve.

we have

$$V = A(R+E_c)L/R$$
$$= AL(1+E_c/R) \quad (4.20)$$

In Equation 4.20 for V, AL is the volume of the prismoid of length L and the term E_c/R can be regarded as the correction required to be made to the cross-sectional area before using in a volume computation by the end areas or prismoidal method. Hence the equivalent area A_e is given by:

$$A_e = A(1 \pm E_c/R) \quad (4.21)$$

the positive/negative sign depending upon whether E_c lies outside the centre line or inside the centre line with respect to the centre of the curve, as indicated in lines 2670–2690 of the program.

(n) Equivalent area (curvature correction) for <A3>

Line numbers 2700–2780

In this two-level cross-section of the <A3> type, Fig. 4.5, there is both cut and fill. As each of these cut and fill areas takes the form of a triangle, their centroids will lie at the junction of the lines joining the apex and the mid-point on the opposite side. There will now be two centroid calculations, E_1 for the cut area and E_2 for the fill area as shown in Fig. 4.11, given by:

$$E_1 = (W_1 + B/2 - KH)/3 \quad (4.22)$$

$$E_2 = (W_2 + B/2 + KH)/3 \quad (4.23)$$

where it is assumed that the formation level is in cut at the centre line. If in fill, the sign for H is changed as discussed previously. The equivalent areas are then computed using Equation 4.21 where E_c is replaced by E_1 and E_2, the variations in sign depending upon whether the centre of the curve is on the lower (L) or higher (H) side of the cross-section, lines 2730–2780.

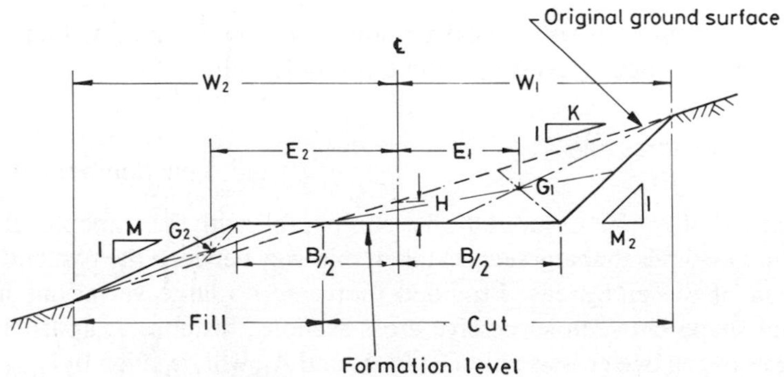

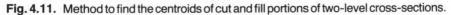

Fig. 4.11. Method to find the centroids of cut and fill portions of two-level cross-sections.

(o) Entry of cross-section data for earthworks

Line numbers 2800–2980

If no cross-sectional areas have been computed, then they can be entered in this section prior to selecting the method of volume computation in the next section (p). If previous cross-sectional areas have been computed, this section is skipped from line 2830.

(p) Volume calculation selection from menu

Line numbers 3000–3080

All volume calculations return to this menu, which allows a selection of volume calculations, or a final mass-haul aggregate volume calculation as shown:

<1> Mean area rule
<2> End areas rule
<3> Prismoidal rule
<4> Mass-haul calculations
<5> Quit

If both end-area volumes and prismoidal rule volumes have been computed, the mass-haul aggregate volumes can be determined for both before quitting the program.

(q) <V1> Mean area rule

Line numbers 3200–3380

This is a simple formula for calculating volume by multiplying the mean of the cross-sectional areas by the distance between the end sections. However it is not very accurate as it assumes that the cross-sections are of similar shape and area. If the areas of the cross-sections are denoted by A_1, A_2, A_3, up to A_N for N cross-sections, and the distance between each cross-section is L, then the total volume V_T will be given by:

$$V_T = (A_1 + A_2 + A_3 + \ldots + A_N)(N-1)L/N \tag{4.24}$$

Since this method of volume calculation only gives the total volume it is not suitable for mass-haul aggregate volume computations.

(r) <V2> End areas rule

Line numbers 3400–3660

This method of volume computation is comparable with the trapezoidal rule for areas, and assumes that any section taken mid-way between the two end areas is the mean of the end areas. Provided there are no large variations in cross-sectional shape between successive cross-sections, distance L apart, then the volume between two cross-sections, say A_1 and A_2, will be given by:

$$V_1 = (A_1 + A_2)L/2 \tag{4.25}$$

For a series of N cross-sections, L apart, the total volume V_T would be given by:

$$V_T = L[(A_1 + A_N)/2 + A_2 + A_3 + \ldots + A_{N-1}] \quad (4.26)$$

This formula is used for the calculation of both cut and fill volumes and the results stored in an array for display and then printout if required, prior to mass-haul aggregate volume calculations.

(s) Extrapolation of cut/fill lengths

Line numbers 3700–3890

In the computation of volumes by the end areas <V2> or prismoidal <V3> methods, should a zero area be encountered in either cut or fill areas, then the program branches to this routine. Either the specific area highlighted, cross-section $CS(I)$, has just reduced to zero, or it could have reached zero somewhere

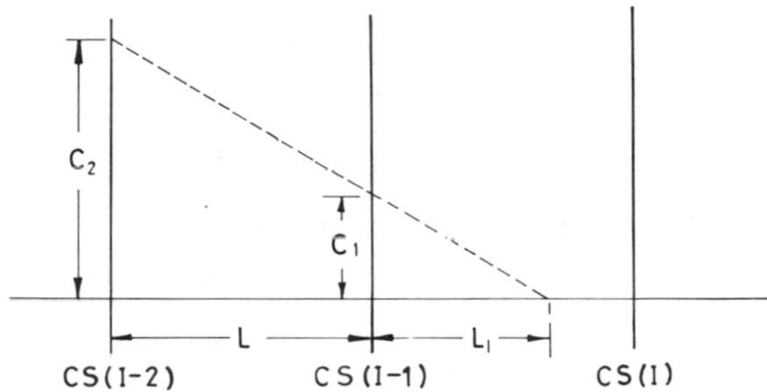

Fig. 4.12. Nomenclature for extrapolation of cut or fill at the end of a volume computation.

between the previous/next cross-section and the present one. If zero is to be used, the program returns to its volume computations.

However if the cut or fill has been steadily increasing/reducing in previous cross-sections, Fig. 4.12, it is possible to extrapolate the new chainage of the zero area, at a distance L_1 from the adjacent cross-section. This extrapolation method will however only work if the cross-sectional area C_2 is more than twice the area C_1 as shown in Fig. 4.12. The length L_1 is evaluated from similar triangles where the ratio of the lengths L_1:L is equal to C_1:$(C_2 - C_1)$ giving:

$$L_1 = LC_1/(C_2 - C_1) \quad (4.27)$$

The volume of the cut or fill in the small segment is then calculated from:

$$V = (C_1 + 0)/2 \quad (4.28)$$

in lines 3780 and 3880 for cut and fill respectively.

(t) <V3> Prismoidal rule

Line numbers 4000–4280

This method of volume computation is more accurate than <V1> and <V2> as the volume of earth between two successive cross-sections is considered as a *prismoid*, which is a solid made up of two end faces parallel to one another. It should be noted that the faces of the prismoid must be formed by straight continuous lines running from one end face to the other. Thus unless the ground profile is regular both transversely and longitudinally, it is likely that errors will be introduced in assuming that the figure is prismoidal over its entire length. However, since these errors are invariably small, the volume obtained using the prismoidal rule can be considered a good approximation of the volume to be excavated or filled.

The prismoidal formula assumes that the cross-sectional area M of the section mid-way between the end faces is also known. If A_1 and A_2 are the end face areas, distance L apart, then the volume, V, is given by:

$$V_{1-2} = (A_1 + 4M + A_2)L/6 \tag{4.29}$$

It is therefore normal practice in earthworks to use the prismoidal rule over three cross-sections A_1, A_2, A_3, again distance L apart, to give:

$$V_{1-3} = (A_1 + 4A_2 + A_3)L/3 \tag{4.30}$$

as used in lines 4070 and 4100 for cut and fill respectively.

Extending Equation 4.28 to the case of N cross-sections, where N must be *odd*, the total volume V_T is given by:

$$\begin{aligned} V_T = [&\text{sum of end areas} + 4 \times \text{sum of even areas} \\ &+ 2 \times \text{sum of odd areas}]L/3 \end{aligned} \tag{4.31}$$

There is therefore a check in line 4050 of the program to see if N entered at the outset for the number of cross-sections is odd. If not, then the program calculates up to the previous cross-section, and the prismoidal cut and/or fill is stored in arrays for subsequent mass-haul aggregate volume computations. All the results are displayed with an opportunity to obtain a printout for record purposes.

(u) Entry of volume data for mass-haul

Line numbers 5000–5180

The program can be entered at this point from the original menu if the volumes are already known. If previous cross-sections or volumes have been entered, this routine is skipped from line 5030.

(v) Mass-haul aggregate volume computations

Line numbers 5200–5680

At the start of this section, there is a choice of using the previously computed end-area or prismoidal volumes. Both can be used to obtain mass-haul aggregate volumes if required. The aggregate volumes are computed on the premise that cut volumes are positive and fill volumes are negative. The computed volumes are tabulated against chainage, to allow a simple drawing of a mass-haul diagram where the positive volumes are plotted above the base line and negative volumes plotted below the base line, Fig. 4.13.

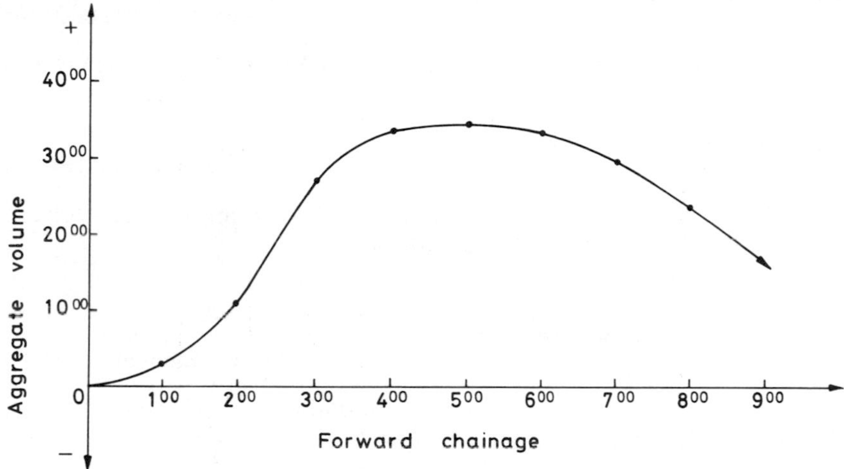

Fig. 4.13. Mass-haul diagram from aggregate volume computations.

(w) Routine to print earthworks location, etc.

Line numbers 6000–6080

This routine is provided for record purposes to annotate the printout, and is based on Applesoft DOS. It may require alteration for different microcomputers and printers.

(x) Data printout for cross-sections <A1–A4>

Line numbers 6100–6390

This standard printout is provided to keep a record of entered data; formation width, depth/height, side-slopes, crossfall, etc., together with the computed areas of cut and/or fill. If a curvature correction has been made, the radius R is also printed.

(y) Data printout for cross-sections <A5–A6>

Line numbers 6400–6590

A separate printout is provided for the irregular cross-sections giving the actual offsets and heights used in the area computations.

(z) Printout summary of cross-sectional areas

Line numbers 6600–6720

This routine is provided at the end of the area calculations prior to the volume computations.

(aa) Data printout of volumes for <V1>

Line numbers 6800–6880

This is a simple printout of the total cut and fill computed for record purposes.

(bb) Data printout of volumes for <V2>

Line numbers 7000–7150

This is the volume printout for the end-area computations between each cross-section. The total cut and fill volumes are also printed.

(cc) Data printout of volumes for <V3>

Line numbers 7200–7350

The prismoidal volumes computed between odd cross-sections are presented, together with the total cut and fill volumes.

(dd) Printout summary of volumes

Line numbers 7400–7550

This routine is provided to record volumes entered directly into the mass-haul aggregate volume computation section.

(ee) Mass-haul aggregate volume printout

Line numbers 7600–7790

This final print routine can be used by either end-area or prismoidal volumes for printing out the computed totals and the aggregate volume for plotting the mass-haul diagram.

(ff) Routine to clear screen

Line numbers 9000–9010

To simplify portability, the Applesoft 'HOME' command to clear the screen has been placed in a subroutine to allow an easy change to another microcomputer.

(gg) Routine to display error message

Line numbers 9100–9130

If a wrong keyboard response has been made, the error is trapped and this error message displayed and re-entry requested.

(hh) Error trace routine

Line numbers 9200–9290

This error trace routine was discussed earlier in Section 2.2.9.

(ii) Termination of program

Line numbers 9500–9620

This program can be terminated either from the initial screen header (line 370) or after the area or volume computations are complete (line 3060). An end-of-program message is displayed to advise the operator that computations are complete.

4.3.2 Earthworks – numeric variables

A	= Area of cross-section
$A0$	= Type of cross-section selection
$A1$	= Area of fill
$A2$	= Area of cut
$A6$	= Flag for program A6
AC	= Area of cut
AF	= Area of fill
B	= Formation breadth of road
C	= Counter in error trace routine
$C1, C2$	= Cut areas used in extrapolation
$C(N-1)$	= Cut area array
DE	= Direct entry volumes for mass-haul
E	= Error number in error trace routine
$F1, F2$	= Fill areas used in extrapolation
$F(N-1)$	= Fill area array
H	= Level difference, formation and point
I	= Integer loop variable
$J, J2$	= Integer loop variables
$K, K2$	= Ground surface slope
L	= Distance between cross-sections
$L0$	= Chainage of first cross-section
$L1$	= Extrapolated chainage length
$M, M2$	= Slope of cut and fill side slopes
MC	= Mass-haul cut volume total
MF	= Mass-haul fill volume total
MH	= Mass-haul aggregate volume total
$MC(N-1)$	= Mass-haul cut volume array
$MF(N-1)$	= Mass-haul fill volume array
$MH(N-1)$	= Mass-haul aggregate volume array
N	= Number of cross-sections
P	= Number of points in cross-section
PC	= Total cut volume, prismoidal rule
PF	= Total fill volume, prismoidal rule
$PC(N-1)$	= Cut volume array, prismoidal rule
$PF(N-1)$	= Fill volume array, prismoidal rule

Table—continued

T	= Number of prismoidal rule cross-sections
$T0$	= Number of cross-sections
$T1$, $T2$	= Storage variables used in area calculations
TC	= Total cut volume, end areas rule
TF	= Total fill volume, end areas rule
U	= Program selection from menu
$V0$	= Type of volume calculation selection
VC	= Total volume cut, end areas rule
VF	= Total volume fill, end areas rule
$VC(N-1)$	= Cut volume array, end areas rule
$VF(N-1)$	= Fill volume array, end areas rule
W–$W2$	= Widths of cut or fill from centre line of cross-section
$X(P-1)$	= Point offset array in cross-section
$Y(P-1)$	= Point height array in cross-section

4.3.3 Earthworks – string variables

A$	Storage for location of earthworks
B$	Storage for operator's name
C$	Storage for date of calculations
D$	= CHR$(4), DOS command, CTRL-D
E0$	= "SORRY, DATA ERROR . . . PLEASE RE-ENTER"
F$	Storage for cut "C", fill "F" or both "B"
G$	= "IS CURVATURE CORRECTION REQUIRED (Y/N)"
H$	Selection of data for mass-haul computations
M$	= "EARTHWORKS PROGRAM – MASS-HAUL"
P$	= "DO YOU WISH PRINTOUT OF ABOVE DATA (Y/N)"
Q$	= Question response (Y/N)
S$	Storage for positive/negative slope (P/N)
T$	Storage for type of cross-section A1–A6
V$	Storage for type of volume computation

4.3.4 Earthworks – BASIC program

```
1Ø  REM  <EARTHWORKS> PROGRAM FOR APPLE II+  USES QUME PRINTER
15  REM  PROGRAM INCLUDES CURVATURE CORRECTIONS
2Ø D$ =  CHR$ (4): ONERR  GOTO 92ØØ
3Ø E = Ø:I = Ø:J = Ø:K = Ø:N = Ø:P1 = Ø:T = Ø:C = Ø
4Ø AØ = Ø:A6 = Ø:B = Ø:H = Ø:K2 = Ø:L = Ø:LØ = Ø:M = Ø:M2 = Ø:P = Ø:T1 =
     Ø:T2 = Ø:W1 = Ø:W2 = Ø
5Ø EØ$ = "SORRY, DATA ERROR ... PLEASE RE-ENTER."
55 P$ = "DO YOU WISH PRINTOUT OF ABOVE DATA (Y/N)"
6Ø  DIM C(2Ø),F(2Ø),X(2Ø),Y(2Ø)
7Ø  DIM VC(2Ø),VF(2Ø),PC(2Ø),PF(2Ø)
8Ø  DIM MC(2Ø),MF(2Ø),MH(2Ø)
9Ø G$ = "IS CURVATURE CORRECTION REQUIRED(Y/N)"
95  REM  ***************************************************
1ØØ  GOSUB 9ØØØ
```

```
110  PRINT "****************************************"
120  PRINT "*                                      *"
130  PRINT "*     UNIVERSITY  OF  STRATHCLYDE      *"
140  PRINT "*  DEPARTMENT  OF  CIVIL  ENGINEERING  *"
150  PRINT "*                                      *"
160  PRINT "*          SURVEYING  SECTION          *"
170  PRINT "*         EARTHWORKS  PROGRAM          *"
180  PRINT "****************************************"
300 :
310  PRINT
320  PRINT "SELECT PROGRAM FROM MENU"
330  PRINT
340  PRINT "    <1> AREAS OF CROSS-SECTIONS"
350  PRINT "    <2> VOLUMES FROM CROSS-SECTIONS"
360  PRINT "    <3> MASS-HAUL FROM VOLUMES"
370  PRINT "    <4> QUIT"
380  INPUT U
390  IF U < 1 OR U > 4 THEN 310
400  IF U = 4 THEN 9500
500  GOSUB 9000
510  PRINT "UNIVERSITY OF STRATHCLYDE"
520  PRINT "EARTHWORKS PROGRAM"
530  PRINT "****************************************"
540  PRINT
550  PRINT "ENTER LOCATION OF WORK, ETC. WHEN       REQUESTED, AND PRESS
     <RETURN>"
560  PRINT "ENTER LOCATION OF EARTHWORKS :-"
570  INPUT A$
580  PRINT "ENTER OPERATOR'S NAME :-"
590  INPUT B$
600  PRINT "ENTER DATE OF CALCULATIONS AS DD/MM/YY"
610  INPUT C$
640  PRINT
650  PRINT "DO YOU WISH PRINTOUT OF LOCATION ETC.,  (Y/N) ";
660  INPUT Q$
670  IF Q$ = "Y" THEN 6000
680  IF Q$ <  > "N" THEN 650
690 :
693  REM  ****************************************
695  REM  ENTER CROSS-SECTION DATA
700  GOSUB 9000
710  PRINT "EARTHWORKS PROGRAM"
720  PRINT "****************************************"
730  PRINT : PRINT "ENTER NUMBER OF CROSS-SECTIONS ";
740  INPUT N
750  PRINT : PRINT "ENTER CHAINAGE OF FIRST CROSS-SECTION"
760  INPUT L0
770  PRINT : PRINT "ENTER DISTANCE BETWEEN CROSS-SECTIONS"
780  INPUT L
790  ON U GOTO 800,2800,5000
793  REM  ****************************************
795  REM  START AREAS OF CROSS-SECTION DATA
800  GOSUB 9000
810  PRINT "EARTHWORKS PROGRAM - AREAS"
820  PRINT "****************************************"
830  FOR I = 0 TO N - 1: REM  START LOOP
840  PRINT : PRINT "SELECT TYPE OF CROSS-SECTION FROM MENU :"
850  PRINT
860  PRINT "    <1>   ONE-LEVEL  (CUT OR FILL)"
```

```
870  PRINT "     <2>   TWO-LEVEL  (CUT OR FILL)"
880  PRINT "     <3>   TWO-LEVEL  (CUT ++ FILL)"
890  PRINT "     <4> THREE-LEVEL  (CUT OR FILL)"
900  PRINT "     <5>   IRREGULAR  (CUT OR FILL)"
910  PRINT "     <6>   IRREGULAR  (CUT ++ FILL)"
920  INPUT A0
930  IF A0 < 1 OR A0 > 6 THEN 840
940 F$ = " ": REM  EMPTY F$
950  ON A0 GOTO 1000,1200,1450,1700,2000,2500
960  PRINT : PRINT "DO YOU WISH PRINTOUT SUMMARY OF CUT/FILL (Y/N)";
970  INPUT Q$: IF Q$ = "Y" THEN 6600
980  IF Q$ < > "N" THEN 960
990  GOTO 2800
993  REM  ****************************************
995  REM  START OF CROSS-SECTION AREA ROUTINES
1000  GOSUB 9000
1010 T$ = "<A1> ONE-LEVEL": PRINT T$
1020  PRINT "****************************************"
1030  PRINT : PRINT "CROSS-SECTION NO. ";I + 1
1040  PRINT "CHAINAGE         = ";L0 + I * L
1050  PRINT : PRINT "ENTER CUT OR FILL (C/F)";: INPUT F$
1060  IF F$ < > "C" AND F$ < > "F" THEN 1050
1070  PRINT : PRINT "ENTER FORMATION WIDTH ";: INPUT B
1080  IF F$ = "C" THEN  PRINT "ENTER FORMATION DEPTH ";: INPUT H
1090  IF F$ = "F" THEN  PRINT "ENTER FORMATION HEIGHT ";: INPUT H
1100  PRINT "ENTER SIDE SLOPE 1:";: INPUT M
1110 A =  INT ((H * (B + M * H)) * 1000 + .5) / 1000:W =  INT ((B / 2 + M
     * H) * 1000 + .5) / 1000
1120  PRINT : PRINT "WIDTH W = ";W;" : AREA = ";A
1130  IF F$ = "C" THEN 1150
1140 F(I) = A:C(I) = 0: GOTO 1160
1150 C(I) = A:F(I) = 0:
1160 W1 = W:W2 = W: PRINT : PRINT P$;: INPUT Q$
1170  IF Q$ = "Y" THEN 6100
1180  IF Q$ < > "N" THEN 1160
1190  NEXT I: GOTO 960
1193  REM  ****************************************
1200  GOSUB 9000:R = 0
1210 T$ = "<A2> TWO-LEVEL": PRINT T$
1220  PRINT "****************************************"
1230  PRINT : PRINT "CROSS-SECTION NO. ";I + 1
1240  PRINT "CHAINAGE         = ";L0 + I * L
1250  PRINT : PRINT "ENTER CUT OR FILL (C/F)";: INPUT F$
1260  IF F$ < > "C" AND F$ < > "F" THEN 1250
1270  PRINT : PRINT "ENTER FORMATION WIDTH ";: INPUT B
1280  IF F$ = "C" THEN  PRINT "ENTER FORMATION DEPTH ";: INPUT H
1290  IF F$ = "F" THEN  PRINT "ENTER FORMATION HEIGHT ";: INPUT H
1300  PRINT "ENTER CROSSFALL  1:";: INPUT K
1310  PRINT "ENTER SIDE SLOPE 1:";: INPUT M
1320 W1 = (B / 2 + M * H) * (K / (K - M))
1330 W2 = (B / 2 + M * H) * (K / (K + M))
1340 A = ((B / 2 + M * H) * (W1 + W2) - B * B / 2) / (2 * M)
1350 A =  INT (A * 1000 + .5) / 1000:W1 =  INT (W1 * 1000 + .5) / 1000:W2
     =  INT (W2 * 1000 + .5) / 1000
1360  PRINT : PRINT "WIDTH W1 = ";W1;" WIDTH W2 = ";W2: PRINT "AREA     =
     ";A
1370  PRINT : PRINT G$;: INPUT Q$: IF Q$ = "Y" THEN 2650
1380  IF Q$ < > "N" THEN 1370
1390  IF F$ = "C" THEN 1410
```

```
1400 F(I) = A:C(I) = 0: GOTO 1420
1410 C(I) = A:F(I) = 0
1420  PRINT : PRINT P$;: INPUT Q$: IF Q$ = "Y" THEN 6100
1430  IF Q$ < > "N" THEN 1420
1440  NEXT I: GOTO 960
1443  REM  ****************************************
1450  GOSUB 9000:R = 0
1460 T$ = "<A3> TWO-LEVEL (CUT ++ FILL)": PRINT T$
1470  PRINT "****************************************"
1480  PRINT : PRINT "CROSS-SECTION NO. ";I + 1
1490  PRINT "CHAINAGE            = ";L0 + I * L
1500  PRINT : PRINT "ENTER WHETHER CUT OR FILL (C/F)": PRINT "AT CENTRE L
     INE ";: INPUT F$
1510  IF F$ < > "C" AND F$ < > "F" THEN 1500
1520  PRINT "ENTER FORMATION WIDTH ";: INPUT B
1530  IF F$ = "C" THEN  PRINT "ENTER FORMATION DEPTH ";: INPUT H
1540  IF F$ = "F" THEN  PRINT "ENTER FORMATION HEIGHT ";: INPUT H
1550  PRINT "ENTER CROSSFALL        1:";: INPUT K
1560  PRINT "ENTER CUT  SIDE SLOPE 1:";: INPUT M2
1570  PRINT "ENTER FILL SIDE SLOPE 1:";: INPUT M
1580  IF F$ = "F" THEN H = ( - 1) * H
1585  REM  CALCULATE AREA OF FILL (A1)
1590 A1 = ((B / 2 - K * H) ^ 2) / (2 * (K - M))
1595 W1 =  INT ((K / (K - M2)) * (B / 2 + M2 * H) * 1000 + .5) / 1000
1600 F(I) =  INT (A1 * 1000 + .5) / 1000
1605  REM  CALCULATE AREA OF CUT (A2)
1610 A2 = ((B / 2 + K * H) ^ 2) / (2 * (K - M2))
1615 W2 =  INT ((K / (K - M)) * (B / 2 - M * H) * 1000 + .5) / 1000
1620 C(I) =  INT (A2 * 1000 + .5) / 1000
1630  PRINT : PRINT "CUT WIDTH = ";W1;" : AREA = ";C(I)
1640  PRINT "FILL WIDTH= ";W2;" : AREA = ";F(I)
1650  PRINT : PRINT G$;: INPUT Q$: IF Q$ = "Y" THEN 2700
1660  IF Q$ < > "N" THEN 1650
1670  PRINT : PRINT P$;: INPUT Q$: IF Q$ = "Y" THEN 6100
1680  IF Q$ < > "N" THEN 1670
1690  NEXT I: GOTO 960
1693  REM  ****************************************
1700  GOSUB 9000
1710 T$ = "<A4> THREE-LEVEL": PRINT T$
1720  PRINT "****************************************"
1730  PRINT : PRINT "CROSS-SECTION NO. ";I + 1
1740  PRINT "CHAINAGE            = ";L0 + I * L
1750  PRINT : PRINT "ENTER WHETHER CUT OR FILL (C/F)": PRINT "AT CENTRE L
     INE ";: INPUT F$
1760  IF F$ < > "C" AND F$ < > "F" THEN 1750
1770  PRINT "ENTER FORMATION WIDTH ";: INPUT B
1780  IF F$ = "C" THEN  PRINT "ENTER FORMATION DEPTH ";: INPUT H
1790  IF F$ = "F" THEN  PRINT "ENTER FORMATION HEIGHT ";: INPUT H
1800  PRINT : PRINT "ENTER SIDE SLOPE     1:";: INPUT M
1810  PRINT : PRINT "ENTER CROSSFALL-LHS 1:";: INPUT K
1820  PRINT "IS SLOPE +VE/-VE FROM C.L. (P/N) ";: INPUT S$
1830  IF S$ < > "P" AND S$ < > "N" THEN 1820
1840  IF S$ = "P" THEN W1 = (B / 2 + M * H) * (K / (K + M))
1850  IF S$ = "N" THEN W1 = (B / 2 + M * H) * (K / (K - M))
1860  PRINT : PRINT "ENTER CROSSFALL-RHS 1:";: INPUT K2
1870  PRINT "IS SLOPE +VE/-VE FROM C.L. (P/N) ";: INPUT S$
1880  IF S$ < > "P" AND S$ < > "N" THEN 1870
1890  IF S$ = "P" THEN W2 = (B / 2 + M * H) * (K2 / (K2 + M))
1900  IF S$ = "N" THEN W2 = (B / 2 + M * H) * (K2 / (K2 - M))
```

```
1910 A = ((W1 + W2) * (M * H + B / 2) - B * B / 2) / (2 * M):A =  INT (A *
     1000 + .5) / 1000
1915 W1 =  INT (W1 * 1000 + .5) / 1000:W2 =  INT (W2 * 1000 + .5) / 1000
1920  PRINT : PRINT "WIDTH W1 = ";W1;" : WIDTH W2 = ";W2: PRINT "AREA
     = ";A
1930  IF F$ = "C" THEN 1950
1940 F(I) = A:C(I) = 0: GOTO 1960
1950 C(I) = A:F(I) = 0
1960  PRINT : PRINT P$;: INPUT Q$
1970  IF Q$ = "Y" THEN 6100
1980  IF Q$ < > "N" THEN 1960
1990  NEXT I: GOTO 960
1993  REM  *****************************************
2000  GOSUB 9000
2010 A6 = 0:T$ = "<A5> IRREGULAR": PRINT T$
2020  PRINT "*****************************************"
2030  PRINT : PRINT "CROSS-SECTION NO. ";I + 1
2040  PRINT "CHAINAGE             = ";L0 + I * L: IF A6 = 2 THEN 2080
2050  PRINT : PRINT "ENTER CUT OR FILL (C/F) ";: INPUT F$
2060  IF F$ < > "C" AND F$ < > "F" THEN 2050
2070  PRINT "ENTER FORMATION WIDTH ";: INPUT B
2080  PRINT : PRINT "ENTER NO. OF POINTS IN CROSS-SECTION";
2090  INPUT P
2100  IF P = > 3 THEN 2140
2110  PRINT  CHR$ (7): GOSUB 9100
2120  PRINT "THREE OR MORE POINTS ARE REQUIRED !"
2130  GOTO 2080
2140  PRINT : PRINT "ENTER OFFSET,HEIGHT (X,Y) OF EACH POINT IN CLOCKWISE
      DIRECTION"
2150  PRINT "LHS OF C.L. AND BELOW FORMATION IS -VE"
2160  PRINT "RHS OF C.L. AND ABOVE FORMATION IS +VE"
2170  FOR J = 0 TO P - 1
2180  PRINT : PRINT "ENTER X,Y OF PT. NO. ";J + 1;: INPUT X(J),Y(J)
2190  NEXT J
2195  REM  CALCULATE AREA FROM VARIABLES T1,T2
2200 T1 = 0:T2 = 0: REM  SET VARIABLES TO ZERO
2210  FOR J2 = 0 TO P - 2
2220 T1 = T1 + (Y(J2)) * (X(J2 + 1))
2230 T2 = T2 + (X(J2)) * (Y(J2 + 1)): NEXT J2
2240 T1 = T1 + (Y(P - 1)) * (X(0))
2250 T2 = T2 + (X(P - 1)) * (Y(0))
2260 A = (T1 - T2) / 2:A =  INT (A * 1000 + .5) / 1000
2270  PRINT : PRINT "AREA = ";A
2280  IF A6 < > 0 THEN 2530
2290  IF F$ = "C" THEN 2310
2300 F(I) = A:C(I) = 0: GOTO 2320
2310 C(I) = A:F(I) = 0
2320  PRINT : PRINT P$;: INPUT Q$
2330  IF Q$ = "Y" THEN 6400
2340  IF Q$ < > "N" THEN 2320
2350  NEXT I: GOTO 960
2493  REM  *****************************************
2500  GOSUB 9000
2510 T$ = "<A6> IRREGULAR (CUT ++ FILL)": PRINT T$
2520 A6 = 1: GOTO 2020
2530  IF F$ = "C" THEN C(I) = A
2540  IF F$ = "F" THEN F(I) = A
2550  PRINT : PRINT P$;: INPUT Q$
```

```
2560  IF Q$ = "Y" THEN 6400
2570  IF Q$ < > "N" THEN 2550
2580  IF A6 = 2 THEN 2630
2590 A6 = 2: REM  COLLECT SECOND AREA
2595  REM  CHECK IF F$="C" OR "F"
2600  IF F$ = "C" THEN 2620
2605  REM  I.E. F$="F" FOR FILL; CHANGE F$
2610 F$ = "C": PRINT "NOW ENTER CUT :-": GOTO 2030
2620 F$ = "F": PRINT "NOW ENTER FILL :- ": GOTO 2030
2625  REM  ARRIVE HERE AT END OF CALCULATIONS
2630  NEXT I: GOTO 960
2640 :
2643  REM  ****************************************
2645  REM  EQUIVALENT AREA (CURVATURE CORRECTION) FOR <A2>
2650 EC = (W1 * W2 * (W1 + W2)) / (3 * K * A)
2660  PRINT : PRINT "ENTER RADIUS OF CURVATURE ";: INPUT R
2670  PRINT : PRINT "IS CENTRE OF CURVE ON LOWER (L) OR      HIGHER (H) S
    IDE OF SLOPE ";: INPUT L$
2675  IF L$ < > "L" AND L$ < > "H" THEN 2670
2680  IF L$ = "L" THEN A =  INT (A * (1 + EC / R) * 1000 + .5) / 1000: GOTO
    1390
2690  IF L$ = "H" THEN A =  INT (A * (1 - EC / R) * 1000 + .5) / 1000: GOTO
    1390
2693  REM  ****************************************
2695  REM  EQUIVALENT AREA (CURVATURE CORRECTION) FOR <A3>
2700 E1 = (W1 + B / 2 - K * H) / 3
2710 E2 = (W2 + B / 2 + K * H) / 3
2720  PRINT : PRINT "ENTER RADIUS OF CURVATURE ";: INPUT R
2730  PRINT : PRINT "IS CENTRE OF CURVE ON LOWER(L) OR      HIGHER(H) SI
    DE OF SLOPE ";: INPUT L$: IF L$ < > "L" AND L$ < > "H" THEN 2730
2740  IF L$ = "H" THEN 2770
2745  REM  I.E. L$="L"
2750 C(I) =  INT ((C(I)) * (1 + E1 / R) * 1000 + .5) / 1000
2760 F(I) =  INT ((F(I)) * (1 - E2 / R) * 1000 + .5) / 1000: GOTO 1670
2770 C(I) =  INT ((C(I)) * (1 - E1 / R) * 1000 + .5) / 1000
2780 F(I) =  INT ((F(I)) * (1 + E2 / R) * 1000 + .5) / 1000: GOTO 1670
2790 :
2793  REM  ****************************************
2795  REM  START VOLUMES ROUTINE
2800  GOSUB 9000
2810  PRINT "EARTHWORKS PROGRAMS - VOLUMES"
2820  PRINT "****************************************"
2830  IF U = 1 THEN 3000
2835  REM  I.E. U=2, ENTER CROSS-SECTIONAL AREAS
2840  FOR I = 0 TO N - 1
2850  PRINT : PRINT "CROSS-SECTION ";I + 1;
2860  PRINT " : CHAINAGE ";L0 + I * L
2870 C(I) = 0:F(I) = 0
2880  PRINT "ENTER CUT/FILL/BOTH (C/F/B) ";
2890  INPUT F$
2900  IF F$ = "C" THEN 2930
2910  IF F$ = "F" THEN 2940
2920  IF F$ < > "B" THEN 2880
2930  PRINT "ENTER CUT  AREA ";: INPUT C(I): IF F$ = "C" THEN 2950
2940  PRINT "ENTER FILL AREA ";: INPUT F(I)
2950  NEXT I
2960  PRINT : PRINT "DO YOU WISH PRINTOUT SUMMARY OF CUT/FILL (Y/N)";: INPUT
    Q$
```

```
2970  IF Q$ = "Y" THEN 6600
2980  IF Q$ < > "N" THEN 2960
2990 :
2993  REM  ****************************************
2995  REM  VOLUME COMPUTATIONS
3000  PRINT : PRINT "SELECT TYPE OF VOLUME CALCULATIONS :-"
3010  PRINT
3020  PRINT "     <1> MEAN AREA RULE "
3030  PRINT "     <2> END AREAS RULE "
3040  PRINT "     <3> PRISMOIDAL RULE"
3050  PRINT "     <4> MASS-HAUL CALC."
3060  PRINT "     <5> QUIT             "
3070  INPUT V0: IF V0 < 1 OR V0 > 5 THEN 3000
3080  ON V0 GOTO 3200,3400,4000,5000,9500
3090 :
3193  REM  ****************************************
3200  GOSUB 9000
3210  PRINT "EARTHWORKS PROGRAM - VOLUMES"
3220 V$ = "<V1> MEAN AREA RULE":AC = 0:AF = 0
3230  PRINT : PRINT V$
3240  PRINT "****************************************"
3250  FOR I = 0 TO N - 1
3255  REM  COLLECT TOTAL CUT AREA IN AC
3260 AC = AC + C(I)
3265  REM  COLLECT TOTAL FILL AREA IN AF
3270 AF = AF + F(I)
3280  NEXT I
3285  REM  TOTAL VOLUME CUT = VC
3290 VC = AC * L * (N - 1) / N
3300 VC =  INT (VC * 1000 + .5) / 1000
3305  REM  TOTAL VOLUME FILL = VF
3310 VF = AF * L * (N - 1) / N
3320 VF =  INT (VF * 1000 + .5) / 1000
3330  PRINT : PRINT "TOTAL VOLUME CUT  = ";VC
3340  PRINT "TOTAL VOLUME FILL = ";VF
3350  PRINT : PRINT P$;: INPUT Q$
3360  IF Q$ = "Y" THEN 6800
3370  IF Q$ < > "N" THEN 3350
3380  GOTO 3000
3390 :
3393  REM  ****************************************
3395  REM  VOLUME ROUTINE V0=2
3400  GOSUB 9000
3410  PRINT "EARTHWORKS PROGRAM - VOLUMES"
3420 V$ = "<V2> END AREAS RULE":TC = 0:TF = 0
3430  PRINT : PRINT V$
3440  PRINT "****************************************"
3450  FOR I = 0 TO N - 2: IF C(I) < > 0 AND C(I + 1) < > 0 THEN 3460
3455  GOTO 3700
3460 VC(I) = (C(I) + C(I + 1)) * L / 2
3465  REM  COLLECT TOTAL VOLUMES IN TC AND TF
3470 TC = TC + VC(I)
3480 VC(I) =  INT (VC(I) * 1000 + .5) / 1000
3490  IF F(I) < > 0 AND F(I + 1) < > 0 THEN 3500
3495  GOTO 3800
3500 VF(I) = (F(I) + F(I + 1)) * L / 2
3510 TF = TF + VF(I):VF(I) =  INT (VF(I) * 1000 + .5) / 1000
3520  NEXT I:TC =  INT (TC * 1000 + .5) / 1000:TF =  INT (TF * 1000 + .5)
      / 1000
```

```
3530  PRINT "CRXS.    CHAINAGE  CUT VOL.  FILL VOL."
3540  PRINT "=============================================="
3550  FOR J = Ø TO N - 2
3560  PRINT J + 1; TAB( 1Ø);LØ + J * L
3570  PRINT  TAB( 2Ø);VC(J); TAB( 3Ø);VF(J)
358Ø  NEXT J
359Ø  PRINT N; TAB( 1Ø);LØ + (N - 1) * L
36ØØ  PRINT "=============================================="
361Ø  PRINT "TOTAL VOLUMES = "; TAB( 2Ø);TC; TAB( 3Ø);TF
362Ø  PRINT "*****************************************"
363Ø  PRINT : PRINT P$;: INPUT Q$
364Ø  IF Q$ = "Y" THEN 7ØØØ
365Ø  IF Q$ < > "N" THEN 363Ø
366Ø  GOTO 3ØØØ
3693  REM  *****************************************
3695  REM  CALCULATE CUT LENGTH
37ØØ  PRINT "CROSS-SECTION ";I + 1;" CUT AREA = ";C(I)
371Ø  PRINT "CROSS-SECTION ";I + 2;" CUT AREA = ";C(I + 1)
372Ø  PRINT : PRINT "ARE THESE VALUES TO BE USED (Y/N)";: INPUT Q$
373Ø  IF Q$ = "Y" THEN 346Ø
3735  REM  I.E. Q$="N" - COMPUTE CORRECT VOLUME
374Ø  PRINT "ENTER PREVIOUS/NEXT CROSS-SECTION AREA": INPUT C2
375Ø  IF C(I) = Ø THEN C1 = C(I + 1)
376Ø  IF C(I + 1) = Ø THEN C1 = C(I)
3765  REM  CALCULATE LENGTH OF CUT
377Ø L1 = L * C1 / (C2 - C1)
378Ø VC(I) = C1 * L1 / 2
379Ø  GOTO 347Ø
3793  REM  *****************************************
3795  REM  CALCULATE FILL DISTANCE L1
38ØØ  PRINT "CROSS-SECTION ";I + 1;" : FILL AREA = ";F(I)
381Ø  PRINT "CROSS-SECTION ";I + 2;" : FILL AREA = ";F(I + 1)
382Ø  PRINT : PRINT "ARE THESE VALUES TO BE USED (Y/N)";: INPUT Q$
383Ø  IF Q$ = "Y" THEN 35ØØ
3835  REM  I.E. Q$="N"-COMPUTE CORRECT VOLUME
384Ø  PRINT "ENTER PREVIOUS/NEXT CROSS-SECTION AREA": INPUT F2
385Ø  IF F(I) = Ø THEN F1 = F(I + 1)
386Ø  IF F(I + 1) = Ø THEN F1 = F(I)
3865  REM  CALCULATE LENGTH OF FILL
387Ø L1 = L * F1 / (F2 - F1)
388Ø VF(I) = F1 * L1 / 2
389Ø  GOTO 351Ø
3993  REM  *****************************************
3995  REM  PRISMOIDAL CALCULATIONS
4ØØØ  GOSUB 9ØØØ
4Ø1Ø  PRINT "EARTHWOKS PROGRAM - VOLUMES"
4Ø2Ø V$ = "<V3> PRISMOIDAL RULE":T = N - 1:PC = Ø:PF = Ø
4Ø3Ø  PRINT : PRINT V$
4Ø4Ø  PRINT "*****************************************"
4Ø45  REM  CHECK THAT N IS ODD
4Ø5Ø  IF (N - 1) / 2 < >  INT ((N - 1) / 2) THEN T = N - 2
4Ø6Ø  FOR I = Ø TO T - 2 STEP 2
4Ø65  REM  CALCULATE CUT VOL BETWEEN I,I+1,I+2
4Ø7Ø PC(I) = (C(I) + 4 * C(I + 1) + C(I + 2)) * L / 3
4Ø8Ø PC = PC + PC(I)
4Ø9Ø PC(I) =  INT (PC(I) * 1ØØØ + .5) / 1ØØØ
4Ø95  REM  CALCULATE FILL VOL. BETWEEN I,I+1,I+2
41ØØ PF(I) = (F(I) + 4 * F(I + 1) + F(I + 2)) * L / 3
411Ø PF = PF + PF(I)
```

```
4120 PF(I) =  INT (PF(I) * 1000 + .5) / 1000
4130  NEXT I
4140 PC =  INT (PC * 1000 + .5) / 1000:PF =  INT (PF * 1000 + .5) / 1000
4150  PRINT "CRXS.     CHAINAGE  CUT VOL.  FILL VOL."
4160  PRINT "=========================================="
4170  FOR J = 0 TO T - 2 STEP 2
4180  PRINT J + 1; TAB( 10);L0 + J * L
4190  PRINT  TAB( 20);PC(J); TAB( 30);PF(J)
4200  NEXT J
4210  PRINT T + 1; TAB( 10);L0 + T * L
4220  PRINT "=========================================="
4230  PRINT "TOTAL VOLUMES"; TAB( 20);PC; TAB( 30);PF
4240  PRINT "****************************************"
4250  PRINT : PRINT P$;: INPUT Q$
4260  IF Q$ = "Y" THEN 7200
4270  IF Q$ < > "N" THEN 4250
4280  GOTO 3000
4290 :
4993  REM  ****************************************
4995  REM  MASS-HAUL COMPUTATIONS
5000  GOSUB 9000:D = 0:MC = 0:MF = 0
5010 M$ = "EARTHWORKS PROGRAM - MASS-HAUL": PRINT M$
5020  PRINT "****************************************"
5030  IF U < > 3 THEN 5200
5035  REM  I.E. U=3, ENTER CRXS AND VOLUME
5040 D = 1: FOR I = 0 TO N - 2
5050  PRINT : PRINT "VOLUME CRXS ";I + 1;" TO ";I + 2
5060  PRINT "CHAINAGE ";L0 + I * L;" TO ";L0 + (I + 1) * L
5070 MC(I) = 0:MF(I) = 0
5080  PRINT : PRINT "ENTER CUT/FILL/BOTH (C/F/B)";
5090  INPUT F$
5100  IF F$ = "C" THEN 5130
5110  IF F$ = "F" THEN 5140
5120  IF F$ < > "B" THEN 5080
5130  PRINT "ENTER CUT  VOL. ";: INPUT MC(I): IF F$ = "C" THEN 5150
5140  PRINT "ENTER FILL VOL. ";: INPUT MF(I)
5150 MC = MC + MC(I):MF = MF + MF(I): NEXT I
5160  PRINT : PRINT "DO YOU WISH A PRINTOUT SUMMARY OF        VOLUMES (Y/N
     ) ";
5170  INPUT Q$: IF Q$ = "Y" THEN 7400
5180  IF Q$ < > "N" THEN 5160
5190 :
5193  REM  ****************************************
5195  REM  MASS-HAUL ROUTINE FOR VOLUMES
5200  GOSUB 9000:H$ = " "
5210  PRINT M$
5220  PRINT "****************************************": IF D = 1 THEN 527
     0
5230  PRINT : PRINT "DO YOU WISH TO USE END AREA (E)         PRISMOIDAL (
     P) VOLUMES, OR QUIT(Q) ";: INPUT H$
5240  IF H$ = "Q" THEN 9500
5250  IF H$ = "P" THEN 5500
5260  IF H$ < > "E" THEN 5230
5265  REM  I.E. END AREA VOLS VC(I) AND VF(I)
5270 MH = 0: FOR I = 0 TO N - 2: IF D = 1 THEN 5290
5280 MC(I) = VC(I):MF(I) = VF(I):MC = TC:MF = TF
5290 MH = MH + MC(I) - MF(I)
5300 MH(I) = MH
```

```
5310  NEXT I: IF D = 1 THEN D = 2
5320  PRINT : PRINT "CHAIN. CUT VOL.  FILL VOL.  AGG.VOL."
5330  PRINT "=========================================="
5340  FOR J = Ø TO N - 2
5350  PRINT LØ + J * L
5360  PRINT  TAB( 8);MC(J); TAB( 18);MF(J); TAB( 29);MH(J)
5370  NEXT J
5380  PRINT LØ + (N - 1) * L
5390  PRINT "=========================================="
5400  PRINT "TOTAL"; TAB( 8);MC; TAB( 18);MF; TAB( 29);MH
5410  PRINT "****************************************"
5420  PRINT : PRINT P$;: INPUT Q$
5430  IF Q$ = "Y" THEN 76ØØ
5440  IF Q$ <  > "N" THEN 542Ø
5450  IF D = Ø THEN 52ØØ
5460  GOTO 95ØØ
5470 :
5493  REM  ****************************************
5495  REM  PRISMOIDAL RULE VOLS PC(I) AND PF(I)
55ØØ MH = Ø: FOR I = Ø TO T - 2 STEP 2
551Ø MC(I) = PC(I):MF(I) = PF(I)
552Ø MH = MH + MC(I) - MF(I)
553Ø MH(I) = MH
554Ø  NEXT I:MC = PC:MF = PF
555Ø  PRINT : PRINT "CHAIN. CUT VOL.  FILL VOL.  AGG.VOL."
556Ø  PRINT "=========================================="
557Ø  FOR J = Ø TO T - 2 STEP 2
558Ø  PRINT LØ + J * L
559Ø  PRINT  TAB( 8);MC(J); TAB( 18);MF(J); TAB( 29);MH(J)
56ØØ  NEXT J
561Ø  PRINT LØ + T * L
562Ø  PRINT "=========================================="
563Ø  PRINT "TOTAL"; TAB( 8);MC; TAB( 18);MF; TAB( 29);MH
564Ø  PRINT "****************************************"
565Ø  PRINT : PRINT P$;: INPUT Q$
566Ø  IF Q$ = "Y" THEN 76ØØ
567Ø  IF Q$ <  > "N" THEN 565Ø
568Ø  GOTO 52ØØ
569Ø :
5985  REM  ****************************************
599Ø  REM  PRINTOUT OF LOCATION, ETC
6ØØØ  PRINT D$;"PR#1": PRINT "LOCATION OF EARTHWORKS :- ";A$
6Ø1Ø  PRINT "************************************************************
     ********************"
6Ø2Ø  PRINT "OPERATOR'S NAME :- ";B$
6Ø3Ø  PRINT "DATE OF CALCULATIONS :- ";C$
6Ø6Ø  PRINT "************************************************************
     ********************"
6Ø7Ø  PRINT : PRINT D$;"PR#Ø"
6Ø8Ø  GOTO 7ØØ
6Ø9Ø :
6Ø93  REM  ****************************************
6Ø95  REM  PRINTOUT FOR CROSS-SECTIONS A1-A4
61ØØ  PRINT D$;"PR#1": POKE 1784 + 1,8Ø
611Ø  PRINT "EARTHWORKS PROGRAM :-"
612Ø  PRINT T$
613Ø  PRINT "****************************************"
614Ø  PRINT "CROSS-SECTION NO. ";I + 1
615Ø  PRINT "CHAINAGE         = ";LØ + I * L
```

```
6160  PRINT "=========================================="
6170  PRINT "FORMATION WIDTH  = ";B
6180  IF F$ = "C" THEN  PRINT "FORMATION DEPTH  = ";H
6190  IF F$ = "F" THEN  PRINT "FORMATION HEIGHT = ";H
6200  IF AØ = 3 THEN 629Ø
6210  PRINT "SIDE SLOPES 1:";M
6220  IF AØ = 1 THEN 634Ø
6230  IF AØ = 4 THEN 626Ø
6240  PRINT "CROSSFALL    1:";K: IF R <  > Ø THEN  PRINT "RADIUS OF CURVAT
     URE = ";R
6250  GOTO 634Ø
6260  PRINT "CROSSFALL-LHS 1:";K
6270  PRINT "CROSSFALL-RHS 1:";K2
6280  GOTO 634Ø
6290  PRINT "CUT  SIDE SLOPE 1:";M2
6300  PRINT "FILL SIDE SLOPE 1:";M
6310  PRINT "CROSSFALL         1:";K: IF R <  > Ø THEN  PRINT "RADIUS OF CU
     RVATURE = ";R
6320  PRINT "CUT  WIDTH = ";W1;" : AREA = ";C(I)
6330  PRINT "FILL WIDTH = ";W2;" : AREA = ";F(I): GOTO 637Ø
6340  PRINT "WIDTH W1 = ";W1;" : WIDTH W2 = ";W2
6350  IF F$ = "C" THEN  PRINT "AREA OF CUT = ";C(I)
6360  IF F$ = "F" THEN  PRINT "AREA OF FILL = ";F(I)
6370  PRINT "****************************************"
6380  PRINT : PRINT D$;"PR#Ø"
6390  NEXT I: GOTO 96Ø
6393  REM  ****************************************
6395  REM  PRINTOUT FOR CROSS-SECTIONS A5-A6
6400  PRINT D$;"PR#1": POKE 1784 + 1,8Ø
6410  PRINT "EARTHWORKS PROGRAM :-": PRINT T$
6420  PRINT "****************************************"
6430  PRINT "CROSS-SECTION NO. ";I + 1
6440  PRINT "CHAINAGE         = ";LØ + I * L
6450  PRINT "=========================================="
6460  PRINT "FORMATION WIDTH = ";B
6470  PRINT "PT. NO.  OFFSET    HEIGHT"
6480  PRINT "=========================================="
6490  FOR J = Ø TO P - 1
6500  PRINT J + 1;: POKE 36,1Ø: PRINT X(J);: POKE 36,2Ø: PRINT Y(J): NEXT
     J
6510  IF AØ = 6 THEN 655Ø
6520  PRINT : IF F$ = "C" THEN  PRINT "AREA OF CUT = ";C(I)
6530  IF F$ = "F" THEN  PRINT "AREA OF FILL = ";F(I)
6540  GOTO 657Ø
6550  PRINT : PRINT "AREA OF CUT  = ";C(I)
6560  PRINT "AREA OF FILL = ";F(I)
6570  PRINT "****************************************"
6580  PRINT : PRINT D$;"PR#Ø": IF A6 = 2 THEN 659Ø
6585  IF AØ = 6 THEN 259Ø
6590  NEXT I: GOTO 96Ø
6593  REM  ****************************************
6595  REM  PRINTOUT SUMMARY OF AREAS
6600  PRINT D$;"PR#1": POKE 1784 + 1,8Ø
6610  PRINT : PRINT "EARTHWORKS PROGRAM :-"
6620  PRINT "****************************************"
6630  PRINT "SUMMARY OF CROSS-SECTIONAL AREAS"
6640  PRINT : PRINT "CRXS.    CHAINAGE  CUT AREA  FILL AREA"
6650  PRINT "=========================================="
6660  FOR I = Ø TO N - 1
```

```
6670  PRINT I + 1;: POKE 36,1Ø: PRINT LØ + I * L;: POKE 36,2Ø: PRINT C(I)
    ;: POKE 36,3Ø: PRINT F(I)
668Ø  NEXT I
669Ø  PRINT "*****************************************"
67ØØ  PRINT : PRINT D$;"PR#Ø"
671Ø  IF U = 1 THEN 28ØØ
672Ø  IF U = 2 THEN 3ØØØ
6793  REM  ****************************************
6795  REM  PRINTOUT OF <V1> RESULTS - TOTAL VOLUME
68ØØ  PRINT D$;"PR#1": POKE 1784 + 1,8Ø
681Ø  PRINT : PRINT "EARTHWORKS PROGRAM - VOLUMES"
682Ø  PRINT "*****************************************"
683Ø  PRINT V$
684Ø  PRINT : PRINT "TOTAL VOLUME CUT  = ";VC
685Ø  PRINT "TOTAL VOLUME FILL = ";VF
686Ø  PRINT "*****************************************"
687Ø  PRINT : PRINT D$;"PR#Ø"
688Ø  GOTO 3ØØØ
689Ø :
6993  REM  ****************************************
6995  REM  PRINTOUT OF <V2> RESULTS - TOTAL VOLUMES
7ØØØ  PRINT D$;"PR#1": POKE 1784 + 1,8Ø
7Ø1Ø  PRINT : PRINT "EARTHWORKS PROGRAM - VOLUMES"
7Ø2Ø  PRINT "*****************************************"
7Ø3Ø  PRINT V$
7Ø4Ø  PRINT : PRINT "CRXS.    CHAINAGE  CUT VOL.  FILL VOL."
7Ø5Ø  PRINT "========================================"
7Ø6Ø  FOR J = Ø TO N - 2
7Ø7Ø  PRINT J + 1;: POKE 36,1Ø: PRINT LØ + J * L
7Ø8Ø  POKE 36,2Ø: PRINT VC(J);: POKE 36,3Ø: PRINT VF(J)
7Ø9Ø  NEXT J
71ØØ  PRINT N;: POKE 36,1Ø: PRINT LØ + (N - 1) * L
711Ø  PRINT "========================================"
712Ø  PRINT "TOTAL VOLUMES = ";: POKE 36,2Ø: PRINT TC;: POKE 36,3Ø: PRINT
    TF
713Ø  PRINT "*****************************************"
714Ø  PRINT : PRINT D$;"PR#Ø"
715Ø  GOTO 3ØØØ
7193  REM  ****************************************
7195  REM  PRINTOUT OF <V3> RESULTS - TOTAL VOLUMES
72ØØ  PRINT D$;"PR#1": POKE 1784 + 1,8Ø
721Ø  PRINT : PRINT "EARTHWORKS PROGRAM - VOLUMES"
722Ø  PRINT "*****************************************"
723Ø  PRINT V$
724Ø  PRINT : PRINT "CRXS.    CHAINAGE  CUT VOL.  FILL VOL."
725Ø  PRINT "========================================"
726Ø  FOR J = Ø TO T - 2 STEP 2
727Ø  PRINT J + 1;: POKE 36,1Ø: PRINT LØ + J * L
728Ø  POKE 36,2Ø: PRINT PC(J);: POKE 36,3Ø: PRINT PF(J)
729Ø  NEXT J
73ØØ  PRINT T + 1;: POKE 36,1Ø: PRINT LØ + T * L
731Ø  PRINT "========================================"
732Ø  PRINT "TOTAL VOLUMES =";: POKE 36,2Ø: PRINT PC;: POKE 36,3Ø: PRINT
    PF
733Ø  PRINT "*****************************************"
734Ø  PRINT : PRINT D$;"PR#Ø"
735Ø  GOTO 3ØØØ
736Ø :
7393  REM  ****************************************
```

```
7395  REM  SUMMARY PRINTOUT OF VOLUMES
7400  PRINT D$;"PR#1": POKE 1784 + 1,80
7410  PRINT : PRINT M$
7420  PRINT "****************************************"
7430  PRINT "SUMMARY OF CUT/FILL VOLUMES"
7440  PRINT : PRINT "CRXS.    CHAINAGE  CUT VOL.  FILL VOL."
7450  PRINT "========================================"
7460  FOR J = 0 TO N - 2
7470  PRINT J + 1;: POKE 36,10: PRINT L0 + J * L
7480  POKE 36,20: PRINT MC(J);: POKE 36,30: PRINT MF(J)
7490  NEXT J
7500  PRINT N;: POKE 36,10: PRINT L0 + (N - 1) * L
7510  PRINT "========================================"
7520  PRINT "TOTAL VOLUMES = ";: POKE 36,20: PRINT MC;: POKE 36,30: PRINT
    MF
7530  PRINT "****************************************"
7540  PRINT : PRINT D$;"PR#0"
7550  GOTO 5200
7560 :
7593  REM  ****************************************
7595  REM  MASS-HAUL PRINTOUT FOR BOTH END AREAS AND PRISMOIDAL
7600  PRINT D$;"PR#1": POKE 1784 + 1,80
7610  PRINT : PRINT M$
7620  PRINT "****************************************"
7630  PRINT : PRINT "CHAIN. CUT VOL.  FILL VOL.  AGG.VOL."
7640  PRINT "========================================"
7650  IF H$ = "P" THEN 7680
7655  REM  I.E. H$="E" OR DIRECT
7660 T0 = N - 1
7670  FOR J = 0 TO N - 2: GOTO 7700
7680 T0 = T
7690  FOR J = 0 TO T - 2 STEP 2
7700  PRINT L0 + J * L
7710  POKE 36,8: PRINT MC(J);: POKE 36,18: PRINT MF(J);: POKE 36,30: PRINT
    MH(J)
7720  NEXT J
7730  PRINT L0 + T0 * L
7740  PRINT "========================================"
7750  PRINT "TOTAL";: POKE 36,8: PRINT MC;: POKE 36,18: PRINT MF;: POKE 3
    6,30: PRINT MH
7760  PRINT "****************************************"
7770  PRINT : PRINT D$;"PR#0"
7780  IF D = 0 THEN 5200
7790  GOTO 9500
8985  REM  ****************************************
8990  REM  GOSUB ROUTINE TO CLEAR SCREEN
9000  HOME
9010  RETURN
9093  REM  ************************************************
9095  REM  GOSUB ROUTINE FOR ERRORS
9100  PRINT
9110  PRINT E0$
9120  PRINT
9130  RETURN
9193  REM  ************************************************
9195  REM  ERROR TRACE ROUTINE
9200  PRINT  CHR$ (7): REM  CTRL-B (BELL)
9210 E =  PEEK (222): REM  GET ERROR NO.
9220  IF C = 1 THEN 9260
```

```
9230  INVERSE
9240  PRINT "ERROR NO. ";E;" FOUND"
9250 C = 1: TRACE : RESUME
9260  PRINT "ERROR ON SECOND LINE NO. "
9270  NORMAL
9280  NOTRACE
9290  STOP
9493  REM  *****************************************************
9495  REM  END OF PROGRAM
9500  GOSUB 9000
9510  PRINT
9520  PRINT "END EARTHWORKS PROGRAM"
9530  PRINT
9540  PRINT "***************************"
9550  REM  PROGRAM PREPARED BY
9560  REM  DR. P.H. MILNE
9570  REM  DEPT. OF CIVIL ENGINEERING
9580  REM  UNIVERSITY OF STRATHCLYDE
9590  REM  GLASGOW G4 ONG
9600  REM  SCOTLAND
9610  REM  ***************************
9620  END
```

4.3.5 Earthworks – computer printout

```
LOCATION OF EARTHWORKS :- STRATHCLYDE UNIVERSITY - STEPPS
*************************************************************************
OPERATOR'S NAME :- JOHN SMITH
DATE OF CALCULATIONS :- 05/06/83
*************************************************************************

EARTHWORKS PROGRAM :-
<A1> ONE-LEVEL
****************************************
CROSS-SECTION NO. 1
CHAINAGE        = 100
========================================
FORMATION WIDTH  = 12
FORMATION DEPTH  = 3
SIDE SLOPES 1:3
WIDTH W1 = 15 : WIDTH W2 = 15
AREA OF CUT = 63
****************************************

EARTHWORKS PROGRAM :-
<A2> TWO-LEVEL
****************************************
CROSS-SECTION NO. 2
CHAINAGE        = 120
========================================
FORMATION WIDTH  = 12
FORMATION DEPTH  = 1.5
SIDE SLOPES 1:3
CROSSFALL   1:20
WIDTH W1 = 12.353 : WIDTH W2 = 9.13
AREA OF CUT = 25.596
****************************************
```

```
EARTHWORKS PROGRAM :-
<A3> TWO-LEVEL (CUT ++ FILL)
****************************************
CROSS-SECTION NO. 3
CHAINAGE          = 140
========================================
FORMATION WIDTH  = 12
FORMATION HEIGHT = -.5
CUT  SIDE SLOPE 1:1
FILL SIDE SLOPE 1:2
CROSSFALL       1:3
CUT  WIDTH = 8.25 : AREA = 5.063
FILL WIDTH = 21 : AREA = 28.125
****************************************

EARTHWORKS PROGRAM :-
<A4> THREE-LEVEL
****************************************
CROSS-SECTION NO. 4
CHAINAGE          = 160
========================================
FORMATION WIDTH  = 12
FORMATION HEIGHT = .25
SIDE SLOPES 1:2
CROSSFALL-LHS 1:12
CROSSFALL-RHS 1:10
WIDTH W1 = 7.8 : WIDTH W2 = 8.125
AREA OF FILL = 7.878
****************************************


EARTHWORKS PROGRAM :-
<A4> THREE-LEVEL
****************************************
CROSS-SECTION NO. 5
CHAINAGE          = 180
========================================
FORMATION WIDTH  = 12
FORMATION HEIGHT = 2
SIDE SLOPES 1:2
CROSSFALL-LHS 1:12
CROSSFALL-RHS 1:10
WIDTH W1 = 12 : WIDTH W2 = 12.5
AREA OF FILL = 43.25
****************************************


EARTHWORKS PROGRAM :-
****************************************
SUMMARY OF CROSS-SECTIONAL AREAS

CRXS.    CHAINAGE  CUT AREA  FILL AREA
========================================
1          100       63        0
2          120       25.596    0
3          140       5.063     28.125
4          160       0         7.878
5          180       0         43.25
****************************************
```

```
EARTHWORKS PROGRAM - VOLUMES
****************************************
<V2> END AREAS RULE

CRXS.    CHAINAGE  CUT VOL.  FILL VOL.
========================================
1        100
                   885.96    0
2        120
                   306.59    281.25
3        140
                   50.63     360.03
4        160
                   0         511.28
5        180
========================================
TOTAL VOLUMES =    1243.18   1152.56
****************************************


EARTHWORKS PROGRAM - MASS-HAUL
****************************************

CHAIN. CUT VOL.  FILL VOL.  AGG.VOL.
========================================
100
       885.96    0          885.96
120
       306.59    281.25     911.3
140
       50.63     360.03     601.900
160
       0         511.28     90.620
180
========================================
TOTAL  1243.18   1152.56    90.620
****************************************


LOCATION OF EARTHWORKS :- GLENMORE PARK
********************************************************************************
OPERATOR'S NAME :- JOHN SMITH
DATE OF CALCULATIONS :- 07/06/83
********************************************************************************


EARTHWORKS PROGRAM :-
<A5> IRREGULAR
****************************************
CROSS-SECTION NO. 1
CHAINAGE        = 30
========================================
FORMATION WIDTH = 12
PT. NO.  OFFSET    HEIGHT
========================================
1        0         0
2        -6        0
3        -12       6
4        -9        7.5
5        -3        7.5
```

```
6          3          8
7          8          6
8          12         6
9          6          Ø

AREA OF CUT = 134.75
****************************************

EARTHWORKS PROGRAM :-
<A6> IRREGULAR (CUT ++ FILL)
****************************************
CROSS-SECTION NO. 2
CHAINAGE          = 6Ø
========================================
FORMATION WIDTH = 12
PT. NO.  OFFSET    HEIGHT
========================================
1          -2         Ø
2          2          3
3          6          4
4          11         5
5          6          Ø

AREA OF CUT  = 3Ø
AREA OF FILL = Ø
****************************************

EARTHWORKS PROGRAM :-
<A6> IRREGULAR (CUT ++ FILL)
****************************************
CROSS-SECTION NO. 2
CHAINAGE          = 6Ø
========================================
FORMATION WIDTH = 12
PT. NO.  OFFSET    HEIGHT
========================================
1          -2         Ø
2          -2         -2
3          -8         -4
4          -14        -4
5          -6         Ø

AREA OF CUT  = 3Ø
AREA OF FILL = 26
****************************************

EARTHWORKS PROGRAM :-
****************************************
SUMMARY OF CROSS-SECTIONAL AREAS

CRXS.    CHAINAGE  CUT AREA  FILL AREA
========================================
1          3Ø         134.75    Ø
2          6Ø         3Ø        26
****************************************
```

```
EARTHWORKS PROGRAM - VOLUMES
****************************************
<V2> END AREAS RULE

CRXS.    CHAINAGE  CUT VOL.  FILL VOL.
----------------------------------------
1          30
                   2471.25   390
2          60
========================================
TOTAL VOLUMES =    2471.25   390
****************************************


LOCATION OF EARTHWORKS :- GLENMORE PARK
************************************************************************
OPERATOR'S NAME :- JOHN SMITH
DATE OF CALCULATIONS :- 08/06/83
************************************************************************


EARTHWORKS PROGRAM :-
****************************************
SUMMARY OF CROSS-SECTIONAL AREAS

CRXS.    CHAINAGE  CUT AREA  FILL AREA
========================================
1          0         0         608.081
2          30        0         1062.626
3          60        0         1618.182
****************************************


EARTHWORKS PROGRAM - VOLUMES
****************************************
<V1> MEAN AREA RULE

TOTAL VOLUME CUT  = 0
TOTAL VOLUME FILL = 65777.78
****************************************


EARTHWORKS PROGRAM - VOLUMES
****************************************
<V2> END AREAS RULE

CRXS.    CHAINAGE  CUT VOL.  FILL VOL.
========================================
1          0
                     0         25060.605
2          30
                     0         40212.12
3          60
========================================
TOTAL VOLUMES =      0         65272.725
****************************************
```

```
EARTHWORKS PROGRAM - VOLUMES
****************************************
<V3> PRISMOIDAL RULE

CRXS.     CHAINAGE  CUT VOL.  FILL VOL.
========================================
1         Ø
                    Ø         64767.67
3         6Ø
========================================
TOTAL VOLUMES =     Ø         64767.67
****************************************


LOCATION OF EARTHWORKS :- GLENMORE PARK
**********************************************************************
OPERATOR'S NAME :- JOHN SMITH
DATE OF CALCULATIONS :- Ø9/Ø6/83
**********************************************************************


EARTHWORKS PROGRAM - MASS-HAUL
****************************************
SUMMARY OF CUT/FILL VOLUMES

CRXS.     CHAINAGE  CUT VOL.  FILL VOL.
========================================
1         Ø
                    29Ø       Ø
2         1ØØ
                    76Ø       Ø
3         2ØØ
                    168Ø      Ø
4         3ØØ
                    62Ø       Ø
5         4ØØ
                    12Ø       2Ø
6         5ØØ
                    Ø         11Ø
7         6ØØ
                    Ø         35Ø
8         7ØØ
                    Ø         6ØØ
9         8ØØ
========================================
TOTAL VOLUMES =     347Ø      1Ø8Ø
****************************************


EARTHWORKS PROGRAM - MASS-HAUL
****************************************

CHAIN. CUT VOL.  FILL VOL.  AGG.VOL.
========================================
Ø
       29Ø       Ø          29Ø
1ØØ
       76Ø       Ø          1Ø5Ø
2ØØ
       168Ø      Ø          273Ø
```

```
300
        620        Ø              3350
400
        120        20             3450
500
        Ø          110            3340
600
        Ø          350            2990
700
        Ø          600            2390
800
========================================
TOTAL   3470       1080           2390
****************************************
```

4.4 VOLUMES

The previous section dealt with volumes for earthworks using cross-sections. Two other useful methods for the computation of earthworks are the determination of volumes from contour lines and spot-levels. The latter method can be adapted to calculate the volume of borrow-pits or spoil dumps, a very useful adjunct to earthwork calculations. The volumes from contour lines program can also be utilized to determine the water storage capacity of reservoirs, etc., for water resources and water management projects.

4.4.1 Volumes – subroutine index

	Line numbers	*Function*
(a)	10–90	Initialization and control
(b)	100–180	Screen header display
(c)	300–400	Program selection from menu
(d)	500–700	Entry of volume location, etc.
(e)	1000–1190	Volumes from contours-data entry
(f)	1200–1280	Volume calculation selection from menu
(g)	1300–1440	<V1> Mean area rule
(h)	1500–1740	<V2> End areas rule
(i)	1750–1990	<V3> Prismoidal rule
(j)	2000–2520	Volumes from spot levels
(k)	3000–3440	Volume of borrow-pit
(l)	6000–6080	Routine to print volume location, etc.
(m)	6100–6190	Routine to print contour areas
(n)	6200–6260	Mean area rule printout
(o)	6300–6440	End areas rule printout
(p)	6500–6640	Prismoidal rule printout
(q)	6700–6890	Volumes from spot-levels printout
(r)	7000–7120	Volume of borrow-pit printout
(s)	9000–9010	Routine to clear screen
(t)	9100–9130	Routine to display error message
(u)	9200–9290	Error trace routine
(v)	9500–9620	Termination of program

(a) Initialization and control

Line numbers 10–90

All numeric variables, arrays and required string variables are initialized and dimensioned. Note the maximum array size is 21. If larger numbers are required, the arrays should be redimensioned.

(b) Screen header display

Line numbers 100–180

This is the screen header display used for students at Strathclyde University. The operator can alter this as required to suit another organization.

(c) Program selection from menu

Line numbers 300–400

A choice of three entry points to the <VOLUMES> program is provided depending on the requirements:

<1> VOLUMES FROM CONTOURS
<2> VOLUMES FROM SPOT-LEVELS
<3> VOLUME OF BORROW-PIT

(d) Entry of volume location, etc.

Line numbers 500–700

This routine is provided for record purposes to annotate the printout with a title of location, operator's name and date of calculation.

(e) Volumes from contours – data entry

Line numbers 1000–1190

This program is similar to the volumes from cross-sections dealt with in the <EARTHWORKS> program, except here the cross-sections are horizontal rather than vertical, and the distance apart of the cross-sections is given by the contour interval, rather than by chainage. The program asks for the number of contours (cross-sections), the level of the lowest contour and the contour interval. The program then uses a loop to enter the areas at each contour line before asking if a printout summary is required for record purposes.

(f) Volume calculation selection from menu

Line numbers 1200–1280

This selection is identical to the <EARTHWORKS> program with a choice of using:

<1> MEAN AREA RULE
<2> END AREAS RULE
<3> PRISMOIDAL RULE

As before, if the prismoidal method is chosen, there must be an *odd* number of contour lines.

(g) <V1> Mean area rule

Line numbers 1300–1440

The total volume computed by the mean area rule uses Equation 4.24, as discussed in Section 4.3.

(h) <V2> End areas rule

Line numbers 1500–1740

The volume computed between adjacent contour level areas, uses Equation 4.25 and the total volume is computed using Equation 4.26 discussed in Section 4.3. After computation, the contour levels, areas and volumes are displayed with the option of a hard copy printout.

(i) <V3> Prismoidal rule

Line numbers 1750–1990

The volumes computed by the prismoidal rule use Equations 4.29–31, discussed in Section 4.3. It should be noted that, as before, the computations will only work for an *odd* number of contour levels. After computation, the alternate contour levels, areas and volumes are displayed with the option of a hard copy printout. All three volume programs (g)–(i) return to the selection menu, line 1200, to allow alternative volume computations on the same data for comparative purposes.

(j) Volumes from spot-levels

Line numbers 2000–2520

During the excavation of foundations for building slabs, basements, bridge abutments, etc., where the sides of the excavation are vertical, this is a very useful technique. The area is first divided up into a regular square, rectangular or triangular grid and spot-levels observed at each corner of the grid. The level difference between the foundation level and the ground level is then calculated for use in the program.

If a square or rectangular grid, Fig. 4.14, is considered, with a base (*x*-direction) dimension B and a height (*y*-direction) dimension H, where the ground level slopes uniformly from one grid point to another, a truncated prism would be formed by the corner points H_1, H_2, H_3 and H_4. The volume of this prism would then be given by the plan area multiplied by the mean height of the prism, using the following equation:

$$V = BH(H_1 + H_2 + H_3 + H_4)/4 \tag{4.32}$$

If this is now extended to a larger grid, Fig. 4.15, it will be seen that Equation

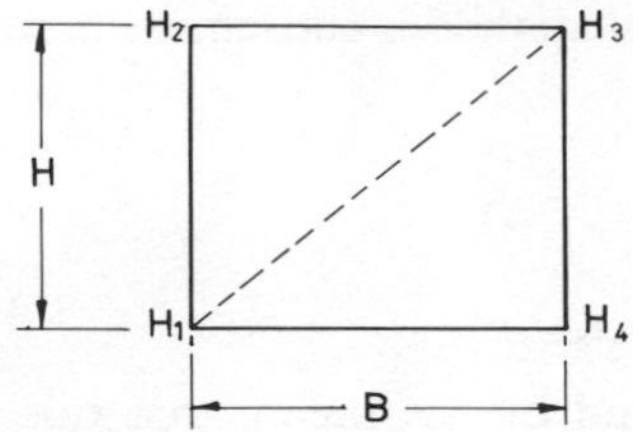

Fig. 4.14. Nomenclature for volumes from spot-levels H_1–H_4 in either a square/rectangular or triangular grid.

4.32 would be used four times. However, several of the corner heights would have to be repeatedly entered on adjacent sides if the prisms were calculated in isolation. The advantage of this computer program is that each level difference is only entered once. Since each point in a square grid can only occur once, twice, thrice or four times, the programs ask for the requisite number of points and then ask for the points in order. The total heights for each number are then stored in the variables T_1, T_2, T_3, T_4 and the final volume computed from:

$$V = BH(T_1 + 2T_2 + 3T_3 + 4T_4)/4 \tag{4.33}$$

as shown in line 2370.

If a triangular grid is used, Fig. 4.14, then there will only be three corner points and the volume will be given by:

$$V = BH(H_1 + H_2 + H_3)/6 \tag{4.34}$$

where the plan area is half the base times the height and the sum of the level differences is divided by three to find the mean.

Extending this as before to a larger grid area, Fig. 4.15, the three corner points

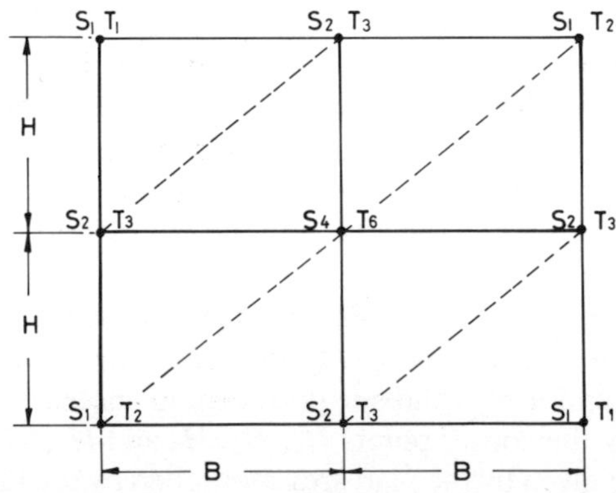

Fig. 4.15. Nomenclature for volumes from spot-levels. In the above figure, S_2 indicates a spot-level used twice in a square/rectangular grid and T_6 indicates a spot-level used six times in a triangular grid.

can now occur in up to six triangles. The entry of the level differences in the triangular routine follows the same procedure as the square/rectangular keyboard entry, where the total volume is computed in line 2350 from:

$$V = BH(T_1 + 2T_2 + 3T_3 + 4T_4 + 5T_5 + 6T_6)/6 \qquad (4.35)$$

On completion, the total volume is displayed with an option to obtain a hard copy printout.

(k) Volume of borrow-pit

Line numbers 3000–3440

The plan area of a borrow-pit does not always split up neatly into either square/rectangular or triangular plan areas with common dimensions. This program is an extension of the volumes from spot-levels, and allows a combination of square, rectangular or triangular areas to be entered separately, each

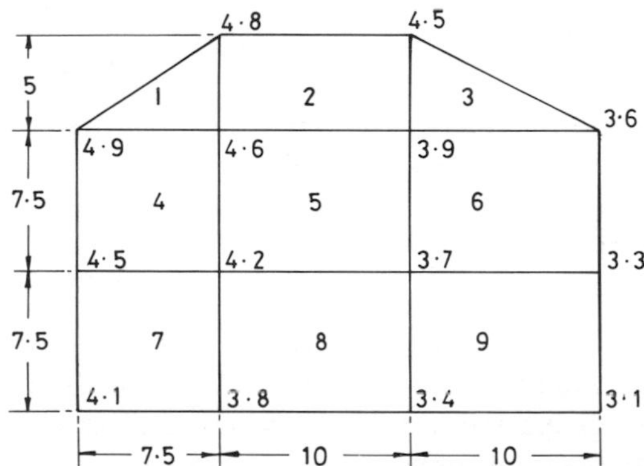

Fig. 4.16. Plan area of a borrow-pit showing unit dimensions and spot-levels used in computer printout.

with their own base and height dimensions, Fig. 4.16. Each volume is computed using either Equation 4.32 or Equation 4.34 depending on its shape. The shape's designation square/triangular, base, height and computed volume are stored in arrays for screen display at line 3300 after the keyboard entries are complete. As before a hard copy printout can be obtained of the results.

(l) Routine to print volume location, etc.

Line numbers 6000–6080

This routine is provided for record purposes to annotate the printout, and is based on Applesoft DOS. It may require alteration for different microcomputers and printers.

(m) Routine to print contour areas

Line numbers 6100–6190

This routine provides a printout of the contour areas entered at the keyboard prior to the volume computations.

(n) Mean area rule printout

Line numbers 6200–6260

This routine is provided for record purposes and gives the volume obtained using the contours in (m) above.

(o) End areas rule printout

Line numbers 6300–6440

In addition to the computed volumes using the end areas rule, this routine also prints out the contour levels and areas, in addition to the total volume.

(p) Prismoidal rule printout

Line numbers 6500–6640

Like (o) above, information on contour levels and areas is included in the printout together with the computed total volume.

(q) Volumes from spot-levels printout

Line numbers 6700–6890

There is a choice of two printout options. The first gives a summary of type of area, base, height, number of points and total volume. The second choice is a complete printout giving the number of points used at each summation with sub-totals, prior to the total volume, etc., as before in the summary.

(r) Volume of borrow-pit printout

Line numbers 7000–7120

This printout gives a complete summary of the number of sections, type with base and height measurements, together with the individual volume and total volume.

(s) Routine to clear screen

Line numbers 9000–9010

To simplify portability, the Applesoft 'HOME' command to clear the screen has been placed in a subroutine to allow an easy change to another microcomputer.

(t) Routine to display error message

Line numbers 9100–9130

If a wrong keyboard entry response has been made, the error is trapped and this error message displayed, and re-entry requested.

(u) Error trace routine

Line numbers 9200–9290

This error trace routine was discussed earlier in Section 2.2.9.

(v) Termination of program

Line numbers 9500–9620

This program can be terminated either from the initial screen header or from the volume computations selection menu. An end-of-program message is displayed to advise the operator that computations are complete.

4.4.2 Volumes – numeric variables

$A(N-1)$	= Area array
B	= Base (x) dimension of grid
H	= Height (y) dimension of grid
$H(N-1)$	= Level difference array
I	= Integer loop variable
J	= Integer loop variable
L	= Contour interval
$L0$	= Lowest contour value
$L1-L4$	= Level differences in grid corners
LT	= Total level differences
N	= Number of contour levels/points
$N1-N6$	= Numbers of grid points
T	= Number of cross-sections for prismoidal calculations
$T1-T6$	= Totals of grid levels
TA	= Total area
U	= Type of volume program
$V0$	= Type of volume calculation
VT	= Volume total
$V(N-1)$	= Volume array

4.4.3 Volumes – string variables

A$	Storage for location of earthwork volumes
B$	Storage for operator's name
C$	Storage for date of calculations
D$	= CHR$(4), DOS command, CTRL-D
E0$	= "SORRY, DATA ERROR . . . PLEASE RE-ENTER"
G$	= Type of spot-level grid
G$(I)	= String array to store type of borrow-pit grid
H$	= "VOLUMES FROM CONTOURS"
P$	= "DO YOU WISH PRINTOUT OF ABOVE DATA (Y/N)"
Q$	= Question response (Y/N)
V$	= Type of volume computation

4.4.4 Volumes – BASIC program

```
10  REM  <VOLUMES> PROGRAM FOR APPLE II+  USES QUME PRINTER
20 D$ =  CHR$ (4): ONERR  GOTO 92ØØ
3Ø E = Ø:I = Ø:J = Ø:K = Ø:N = Ø:P1 = Ø:T = Ø:C = Ø
4Ø AØ = Ø:A6 = Ø:B = Ø:H = Ø:K2 = Ø:L = Ø:LØ = Ø:M = Ø:M2 = Ø:P = Ø:T1 =
     Ø:T2 = Ø:W1 = Ø:W2 = Ø
5Ø EØ$ = "SORRY, DATA ERROR ... PLEASE RE-ENTER."
55 P$ = "DO YOU WISH PRINTOUT OF ABOVE DATA (Y/N)"
6Ø  DIM A(2Ø),H(2Ø),V(2Ø)
7Ø  DIM G$(2Ø)
95  REM ***************************************************
1ØØ  GOSUB 9ØØØ
11Ø  PRINT "****************************************"
12Ø  PRINT "*                                      *"
13Ø  PRINT "*     UNIVERSITY  OF  STRATHCLYDE      *"
14Ø  PRINT "*  DEPARTMENT  OF  CIVIL  ENGINEERING  *"
15Ø  PRINT "*                                      *"
16Ø  PRINT "*           SURVEYING  SECTION         *"
17Ø  PRINT "*           VOLUMES    PROGRAM         *"
18Ø  PRINT "****************************************"
3ØØ :
31Ø  PRINT
32Ø  PRINT "SELECT PROGRAM FROM MENU"
33Ø  PRINT
34Ø  PRINT "     <1> VOLUMES FROM CONTOURS"
35Ø  PRINT "     <2> VOLUMES FROM SPOT-LEVELS"
36Ø  PRINT "     <3> VOLUME OF BORROW-PIT"
37Ø  PRINT "     <4> QUIT"
38Ø  INPUT U
39Ø  IF U < 1 OR U > 4 THEN 31Ø
4ØØ  IF U = 4 THEN 95ØØ
5ØØ  GOSUB 9ØØØ
51Ø  PRINT "UNIVERSITY OF STRATHCLYDE"
52Ø  PRINT "VOLUMES PROGRAM"
53Ø  PRINT "****************************************"
54Ø  PRINT
55Ø  PRINT "ENTER LOCATION OF WORK, ETC. WHEN       REQUESTED, AND PRESS
     <RETURN>"
56Ø  PRINT "ENTER LOCATION OF CALCULATIONS :-"
57Ø  INPUT A$
58Ø  PRINT "ENTER OPERATOR'S NAME :-"
59Ø  INPUT B$
6ØØ  PRINT "ENTER DATE OF CALCULATIONS AS DD/MM/YY"
61Ø  INPUT C$
64Ø  PRINT
65Ø  PRINT "DO YOU WISH PRINTOUT OF LOCATION ETC.,  (Y/N) ";
66Ø  INPUT Q$
67Ø  IF Q$ = "Y" THEN 6ØØØ
68Ø  IF Q$ < > "N" THEN 65Ø
69Ø :
7ØØ  ON U GOTO 1ØØØ,2ØØØ,3ØØØ
993  REM  ****************************************
995  REM  VOLUMES FROM CONTOURS
1ØØØ  GOSUB 9ØØØ
1Ø1Ø  PRINT "VOLUME PROGRAMS"
1Ø2Ø  PRINT "****************************************"
1Ø3Ø H$ = "VOLUMES FROM CONTOURS": PRINT : PRINT H$
1Ø4Ø  PRINT "======================================="
```

```
1050  PRINT : PRINT "ENTER NO. OF CONTOUR LINES ";
1060  INPUT N
1070  PRINT : PRINT "ENTER LEVEL OF LOWEST CONTOUR ";
1080  INPUT LØ
1090  PRINT : PRINT "ENTER CONTOUR INTERVAL ";
1100  INPUT L
1110  PRINT "ENTER AREA ENCLOSED BY EACH CONTOUR :-"
1120  FOR I = Ø TO N - 1
1130  PRINT : PRINT LØ + I * L;" CONTOUR : AREA = ";: INPUT A(I)
1140  NEXT I
1150  PRINT : PRINT "DO YOU WISH PRINTOUT SUMMARY OF AREAS   (Y/N)";
1160  INPUT Q$
1170  IF Q$ = "Y" THEN 6100
1180  IF Q$ <  > "N" THEN 1150
1190 :
1200  GOSUB 9000
1210  PRINT : PRINT "SELECT TYPE OF VOLUME CALCULATION :-"
1220  PRINT
1230  PRINT "    <1> MEAN AREA RULE"
1240  PRINT "    <2> END AREAS RULE"
1250  PRINT "    <3> PRISMOIDAL RULE"
1260  PRINT "    <4> QUIT"
1270  INPUT VØ: IF VØ < 1 OR VØ > 4 THEN 1200
1280  ON VØ GOTO 1300,1500,1750,9500
1290 :
1293  REM  ***************************************
1295  REM  <V1> MEAN AREA CALCULATION
1300  GOSUB 9000
1310  PRINT H$
1320 V$ = "<V1> MEAN AREA RULE":TA = Ø:VT = Ø
1330  PRINT : PRINT V$
1340  PRINT "****************************************"
1350  FOR J = Ø TO N - 1
1355  REM  COLLECT TOTAL AREA IN TA
1360 TA = TA + A(J)
1370  NEXT J
1380 VT = TA * L * (N - 1) / N
1390 VT =  INT (VT * 1000 + .5) / 1000
1400  PRINT : PRINT "TOTAL VOLUME = ";VT
1410  PRINT : PRINT P$;: INPUT Q$
1420  IF Q$ = "Y" THEN 6200
1430  IF Q$ <  > "N" THEN 1410
1440  GOTO 1200
1450 :
1493  REM  ***************************************
1495  REM  <V2> END AREAS CALCULATION
1500  GOSUB 9000
1510  PRINT H$
1520 V$ = "<V2> END AREAS RULE":VT = Ø
1530  PRINT : PRINT V$
1540  PRINT "****************************************"
1550  FOR J = Ø TO N - 2
1560 V(J) = (A(J) + A(J + 1)) * L / 2
1570 VT = VT + V(J)
1580 V(J) =  INT (V(J) * 1000 + .5) / 1000
1590  NEXT J:VT =  INT (VT * 1000 + .5) / 1000
1600  PRINT : PRINT "CONTOUR"; TAB( 10);"AREA"; TAB( 25);"VOLUME"
1610  PRINT "========================================"
1620  FOR I = Ø TO N - 2
```

```
1630  PRINT LØ + I * L; TAB( 1Ø);A(I)
164Ø  PRINT  TAB( 25);V(I)
165Ø  NEXT I
166Ø  PRINT LØ + (N - 1) * L; TAB( 1Ø);A(N - 1)
167Ø  PRINT "========================================"
168Ø  PRINT "TOTAL VOLUME = "; TAB( 25);VT
169Ø  PRINT "****************************************"
17ØØ  PRINT : PRINT P$;: INPUT Q$
171Ø  IF Q$ = "Y" THEN 63ØØ
172Ø  IF Q$ < > "N" THEN 17ØØ
173Ø  GOTO 12ØØ
174Ø :
1743  REM  ****************************************
1745  REM  <V3> PRISMOIDAL RULE CALCULATIONS
175Ø  GOSUB 9ØØØ
176Ø  PRINT H$
177Ø V$ = "<V3> PRISMOIDAL RULE":T = N - 1:VT = Ø
178Ø  PRINT : PRINT V$
179Ø  PRINT "****************************************"
1795  REM  CHECK N IS ODD
18ØØ  IF (N - 1) / 2 < >  INT ((N - 1) / 2) THEN T = N - 2
181Ø  FOR I = Ø TO T - 2 STEP 2
1815  REM  CALCULATE VOL. BETWEEN I,I+1,I+2
182Ø V(I) = (A(I) + 4 * A(I + 1) + A(I + 2)) * L / 3
183Ø VT = VT + V(I)
184Ø V(I) =  INT (V(I) * 1ØØØ + .5) / 1ØØØ
185Ø  NEXT I:VT =  INT (VT * 1ØØØ + .5) / 1ØØØ
186Ø  PRINT : PRINT "CONTOUR"; TAB( 1Ø);"AREA"; TAB( 25);"VOLUME"
187Ø  PRINT "========================================"
188Ø  FOR J = Ø TO T - 2 STEP 2
189Ø  PRINT LØ + J * L; TAB( 1Ø);A(J)
19ØØ  PRINT  TAB( 25);V(J)
191Ø  NEXT J
192Ø  PRINT LØ + T * L; TAB( 1Ø);A(T)
193Ø  PRINT "========================================"
194Ø  PRINT "TOTAL VOLUME = "; TAB( 25);VT
195Ø  PRINT "****************************************"
196Ø  PRINT : PRINT P$;: INPUT Q$
197Ø  IF Q$ = "Y" THEN 65ØØ
198Ø  IF Q$ < > "N" THEN 196Ø
199Ø  GOTO 12ØØ
1993  REM  ****************************************
1995  REM  VOLUMES FROM SPOT-LEVELS
2ØØØ  GOSUB 9ØØØ
2Ø1Ø  PRINT "VOLUME PROGRAMS"
2Ø2Ø V$ = "VOLUMES FROM SPOT-LEVELS":N1 = Ø:N2 = Ø:N3 = Ø:N4 = Ø:N5 =
     6 = Ø:T1 = Ø:T2 = Ø:T3 = Ø:T4 = Ø:T5 = Ø:T6 = Ø
2Ø3Ø  PRINT : PRINT V$
2Ø4Ø  PRINT "****************************************"
2Ø5Ø  PRINT : PRINT "ENTER WHETHER RECTANGULAR/SQUARE(S) OR  TRIANGULAR(T
     ) GRID USED ";
2Ø6Ø  INPUT G$
2Ø7Ø  IF G$ < > "S" AND G$ < > "T" THEN 2Ø5Ø
2Ø8Ø  PRINT : PRINT "ENTER BASE(X) DIMENSION ";: INPUT B
2Ø9Ø  PRINT : PRINT "ENTER HEIGHT(Y) DIMENSION ";: INPUT H
21ØØ  PRINT : PRINT "(1) NO. OF GRID POINTS USED ONCE ";: INPUT N1
21Ø5  IF N1 = Ø THEN 214Ø
211Ø  FOR I = Ø TO N1 - 1
212Ø  PRINT "(1) ENTER LEVEL DIFF. PT. ";I + 1;: INPUT H(I)
```

```
2130 T1 = T1 + H(I): NEXT I
2140  PRINT : PRINT "(2) NO. OF GRID POINTS USED TWICE ";: INPUT N2
2145  IF N2 = Ø THEN 218Ø
215Ø  FOR I = Ø TO N2 - 1
216Ø  PRINT "(2) ENTER LEVEL DIFF. PT. ";I + 1;: INPUT H(I)
217Ø T2 = T2 + II(I): NEXT I
218Ø  PRINT : PRINT "(3) NO. OF GRID POINTS USED * THREE ";: INPUT N3
2185  IF N3 = Ø THEN 222Ø
219Ø  FOR I = Ø TO N3 - 1
22ØØ  PRINT "(3) ENTER LEVEL DIFF. PT. ";I + 1;: INPUT H(I)
221Ø T3 = T3 + H(I): NEXT I
222Ø  PRINT : PRINT "(4) NO. OF GRID POINTS USED * FOUR ";: INPUT N4
2225  IF N4 = Ø THEN 226Ø
223Ø  FOR I = Ø TO N4 - 1
224Ø  PRINT "(4) ENTER LEVEL DIFF. PT. ";I + 1;: INPUT H(I)
225Ø T4 = T4 + H(I): NEXT I
226Ø  IF G$ = "S" THEN 237Ø
2265  REM  I.E. G$="T" COLLECT (5) AND (6)
227Ø  PRINT : PRINT "(5) NO. OF GRID POINTS USED * FIVE ";: INPUT N5
2275  IF N5 = Ø THEN 231Ø
228Ø  FOR I = Ø TO N5 - 1
229Ø  PRINT "(5) ENTER LEVEL DIFF. PT. ";I + 1;: INPUT H(I)
23ØØ T5 = T5 + H(I): NEXT I
231Ø  PRINT : PRINT "(6) NO. OF GRID POINTS USED * SIX ";: INPUT N6
2315  IF N6 = Ø THEN 235Ø
232Ø  FOR I = Ø TO N6 - 1
233Ø  PRINT "(6) ENTER LEVEL DIFF. PT. ";I + 1;: INPUT H(I)
234Ø T6 = T6 + H(I): NEXT I
235Ø V = (B * H / 6) * (T1 + 2 * T2 + 3 * T3 + 4 * T4 + 5 * T5 + 6 * T6)
236Ø  GOTO 238Ø
237Ø V = (B * H / 4) * (T1 + 2 * T2 + 3 * T3 + 4 * T4)
238Ø V =  INT (V * 1ØØØ + .5) / 1ØØØ
239Ø N = N1 + N2 + N3 + N4 + N5 + N6
24ØØ  GOSUB 9ØØØ
241Ø  PRINT V$
242Ø  PRINT "*****************************************"
243Ø  PRINT : IF G$ = "S" THEN  PRINT "SQUARE GRID ";B;" * ";H
244Ø  IF G$ = "T" THEN  PRINT "TRIANGULAR GRID ";B;" * ";H
245Ø  PRINT : PRINT "TOTAL NO. OF GRID POINTS = ";N
246Ø  PRINT : PRINT "TOTAL VOLUME = ";V
247Ø  PRINT : PRINT P$;: INPUT Q$
248Ø  IF Q$ = "N" THEN 3ØØ
249Ø  IF Q$ < > "Y" THEN 247Ø
25ØØ  PRINT : PRINT "DO YOU WISH SUMMARY(S) OR COMPLETE(C)   PRINTOUT ";:
      INPUT T$
251Ø  IF T$ < > "S" AND T$ < > "C" THEN 25ØØ
252Ø  GOTO 67ØØ
253Ø :
2993  REM  *****************************************
2995  REM  BORROW-PIT CALCULATIONS
3ØØØ  GOSUB 9ØØØ
3Ø1Ø  PRINT "VOLUME PROGRAMS"
3Ø2Ø V$ = "VOLUME OF BORROW-PIT":VT = Ø
3Ø3Ø  PRINT : PRINT V$
3Ø4Ø  PRINT "*****************************************"
3Ø5Ø  PRINT : PRINT "ENTER COMBINED NO. OF RECTANGLES/SQUARES AND TRIANGU
     LAR SECTIONS ";
3Ø6Ø  INPUT N
3Ø7Ø  FOR I = Ø TO N - 1
```

```
3Ø8Ø  PRINT : PRINT "SECT.NO. ";I + 1;" : SQUARE(S) OR TRIANG(T)"
3Ø9Ø  INPUT G$(I)
31ØØ  IF G$(I) = "S" THEN 32ØØ
311Ø  IF G$(I) < > "T" THEN 3Ø8Ø
3115  REM  I.E. G$(I)="T", TRIANGULAR
312Ø  PRINT "ENTER BASE,HEIGHT (B,H) ";
313Ø  INPUT B(I),H(I)
314Ø  PRINT "ENTER LEVEL DIFF. AT EACH CORNER         (L1,L2,L3) ";
315Ø  INPUT L1,L2,L3
316Ø LT = L1 + L2 + L3
317Ø V(I) = LT * B(I) * H(I) / 6:VT = VT + V(I)
318Ø V(I) =  INT (V(I) * 1ØØØ + .5) / 1ØØØ
319Ø  NEXT I: GOTO 328Ø
3195  REM  G$(I)="S", SQUARE OR RECTANGULAR
32ØØ  PRINT "ENTER BASE,HEIGHT (B,H) ";
321Ø  INPUT B(I),H(I)
322Ø  PRINT "ENTER LEVEL DIFF. AT EACH CORNER         (L1,L2,L3,L4) ";
323Ø  INPUT L1,L2,L3,L4
324Ø LT = L1 + L2 + L3 + L4
325Ø V(I) = LT * B(I) * H(I) / 4:VT = VT + V(I)
326Ø V(I) =  INT (V(I) * 1ØØØ + .5) / 1ØØØ
327Ø  NEXT I
328Ø VT =  INT (VT * 1ØØØ + .5) / 1ØØØ
329Ø :
33ØØ  GOSUB 9ØØØ
331Ø  PRINT V$
332Ø  PRINT "****************************************"
333Ø  PRINT "SECT.  TYPE  BASE  HEIGHT  VOLUME"
334Ø  PRINT "========================================"
335Ø  FOR J = Ø TO N - 1
336Ø  PRINT J + 1; TAB( 8);G$(J); TAB( 14);B(J); TAB( 2Ø);H(J); TAB( 28);
     V(J)
337Ø  NEXT J
338Ø  PRINT "========================================"
339Ø  PRINT "TOTAL VOLUME = "; TAB( 28);VT
34ØØ  PRINT "****************************************"
341Ø  PRINT : PRINT P$;: INPUT Q$
342Ø  IF Q$ = "Y" THEN 7ØØØ
343Ø  IF Q$ < > "N" THEN 341Ø
344Ø  GOTO 3ØØ
345Ø :
5985  REM  ****************************************
599Ø  REM  PRINTOUT OF LOCATION, ETC
6ØØØ  PRINT D$;"PR#1": PRINT "LOCATION OF EARTHWORKS :- ";A$
6Ø1Ø  PRINT "************************************************************
     ********************"
6Ø2Ø  PRINT "OPERATOR'S NAME :- ";B$
6Ø3Ø  PRINT "DATE OF CALCULATIONS :- ";C$
6Ø5Ø :
6Ø6Ø  PRINT "************************************************************
     ********************"
6Ø7Ø  PRINT : PRINT D$;"PR#Ø"
6Ø8Ø  GOTO 7ØØ
6Ø9Ø :
6Ø93  REM  ****************************************
6Ø95  REM  CONTOUR AREA SUMMARY
61ØØ  PRINT D$;"PR#1": POKE 1784 + 1,8Ø
611Ø  PRINT : PRINT H$
612Ø  PRINT "****************************************"
613Ø  PRINT "CONTOUR  AREA"
```

```
6140  PRINT "========================================"
6150  FOR J = Ø TO N - 1
6160  PRINT LØ + J * L;: POKE 36,1Ø: PRINT A(J)
6170  NEXT J
6180  PRINT "****************************************"
6190  PRINT D$;"PR#Ø": GOTO 12ØØ
6193  REM  ****************************************
6195  REM  <V1> MEAN AREA VOLUME
62ØØ  PRINT D$;"PR#1": POKE 1784 + 1,8Ø
621Ø  PRINT : PRINT H$
622Ø  PRINT "========================================"
623Ø  PRINT V$
624Ø  PRINT "TOTAL VOLUME = ";VT
625Ø  PRINT "****************************************"
626Ø  PRINT D$;"PR#Ø": GOTO 12ØØ
627Ø :
6293  REM  ****************************************
6295  REM  <V2> END AREA RULE VOLUME
63ØØ  PRINT D$;"PR#1": POKE 1784 + 1,8Ø
631Ø  PRINT : PRINT H$
632Ø  PRINT "****************************************"
633Ø  PRINT V$
634Ø  PRINT "CONTOUR";: POKE 36,1Ø: PRINT "AREA";: POKE 36,25: PRINT "VOL
     UME"
635Ø  PRINT "========================================"
636Ø  FOR J = Ø TO N - 2
637Ø  PRINT LØ + J * L;: POKE 36,1Ø: PRINT A(J)
638Ø  POKE 36,25: PRINT V(J)
639Ø  NEXT J
64ØØ  PRINT LØ + (N - 1) * L;: POKE 36,1Ø: PRINT A(N - 1)
641Ø  PRINT "========================================"
642Ø  PRINT "TOTAL VOLUME = ";: POKE 36,25: PRINT VT
643Ø  PRINT "****************************************"
644Ø  PRINT D$;"PR#Ø": GOTO 12ØØ
645Ø :
6493  REM  ****************************************
6495  REM  <V3> PRISMOIDAL RULE VOLUME
65ØØ  PRINT D$;"PR#1": POKE 1784 + 1,8Ø
651Ø  PRINT : PRINT H$
652Ø  PRINT "****************************************"
653Ø  PRINT V$
654Ø  PRINT "CONTOUR";: POKE 36,1Ø: PRINT "AREA";: POKE 36,25: PRINT "VOL
     UME"
655Ø  PRINT "========================================"
656Ø  FOR J = Ø TO T - 2 STEP 2
657Ø  PRINT LØ + J * L;: POKE 36,1Ø: PRINT A(J)
658Ø  POKE 36,25: PRINT V(J)
659Ø  NEXT J
66ØØ  PRINT LØ + T * L;: POKE 36,1Ø: PRINT A(T)
661Ø  PRINT "========================================"
662Ø  PRINT "TOTAL VOLUME = ";: POKE 36,25: PRINT VT
663Ø  PRINT "****************************************"
664Ø  PRINT D$;"PR#Ø": GOTO 12ØØ
665Ø :
6693  REM  ****************************************
6695  REM  VOLUMES FROM SPOT-LEVELS
67ØØ  PRINT D$;"PR#1": POKE 1784 + 1,8Ø
671Ø  PRINT : PRINT V$
672Ø  PRINT "****************************************************"
673Ø  IF G$ = "S" THEN  PRINT "SQUARE GRID ";B;" * ";H
```

```
674Ø  IF G$ = "T" THEN  PRINT "TRIANGULAR GRID ";B;" * ";H
675Ø  IF T$ = "S" THEN 684Ø
6755  REM  I.E. T$="C", COMPLETE PRINTOUT
676Ø  PRINT "(1) NO. OF POINTS   = ";N1;: POKE 36,3Ø: PRINT ": TOTAL = ";
     T1
677Ø  PRINT "(2) NO. OF POINTS   = ";N2;: POKE 36,3Ø: PRINT ": TOTAL = ";
     T2
678Ø  PRINT "(3) NO. OF POINTS   = ";N3;: POKE 36,3Ø: PRINT ": TOTAL = ";
     T3
679Ø  PRINT "(4) NO. OF POINTS   = ";N4;: POKE 36,3Ø: PRINT ": TOTAL = ";
     T4
68ØØ  IF G$ = "S" THEN 683Ø
681Ø  PRINT "(5) NO. OF POINTS   = ";N5;: POKE 36,3Ø: PRINT ": TOTAL = ";
     T5
682Ø  PRINT "(6) NO. OF POINTS   = ";N6;: POKE 36,3Ø: PRINT ": TOTAL = ";
     T6
683Ø  PRINT "====================================================="
684Ø  PRINT "TOTAL NO. OF POINTS = ";N
685Ø  PRINT "TOTAL VOLUME        = ";V
686Ø  PRINT "*****************************************************"
687Ø  PRINT D$;"PR#Ø"
688Ø  GOTO 3ØØ
689Ø :
6993  REM  ****************************************
6995  REM  BORROW-PIT VOLUME PRINTOUT
7ØØØ  PRINT D$;"PR#1": POKE 1784 + 1,8Ø
7Ø1Ø  PRINT : PRINT V$
7Ø2Ø  PRINT "****************************************"
7Ø3Ø  PRINT "SECT.  TYPE  BASE  HEIGHT  VOLUME"
7Ø4Ø  PRINT "========================================"
7Ø5Ø  FOR J = Ø TO N - 1
7Ø6Ø  PRINT J + 1;: POKE 36,8: PRINT G$(J);: POKE 36,14: PRINT B(J);: POKE
     36,2Ø: PRINT H(J);: POKE 36,28: PRINT V(J)
7Ø7Ø  NEXT J
7Ø8Ø  PRINT "========================================"
7Ø9Ø  PRINT "TOTAL VOLUME = ";: POKE 36,28: PRINT VT
71ØØ  PRINT "****************************************"
711Ø  PRINT D$;"PR#Ø"
712Ø  GOTO 3ØØ
8985  REM  *****************************************
899Ø  REM  GOSUB ROUTINE TO CLEAR SCREEN
9ØØØ  HOME
9Ø1Ø  RETURN
9Ø93  REM  ***************************************************
9Ø95  REM  GOSUB ROUTINE FOR ERRORS
91ØØ  PRINT
911Ø  PRINT EØ$
912Ø  PRINT
913Ø  RETURN
9193  REM  ***************************************************
9195  REM  ERROR TRACE ROUTINE
92ØØ  PRINT  CHR$ (7): REM  CTRL-B (BELL)
921Ø E =  PEEK (222): REM  GET ERROR NO.
922Ø  IF C = 1 THEN 926Ø
923Ø  INVERSE
924Ø  PRINT "ERROR NO. ";E;" FOUND"
925Ø C = 1: TRACE : RESUME
926Ø  PRINT "ERROR ON SECOND LINE NO. "
927Ø  NORMAL
928Ø  NOTRACE
```

```
9290  STOP
9493  REM  ***************************************************
9495  REM  END OF PROGRAM
9500  GOSUB 9000
9510  PRINT
9520  PRINT "END VOLUMES PROGRAM"
9530  PRINT
9540  PRINT "***************************"
9550  REM  PROGRAM PREPARED BY
9560  REM  DR. P.H. MILNE
9570  REM  DEPT. OF CIVIL ENGINEERING
9580  REM  UNIVERSITY OF STRATHCLYDE
9590  REM  GLASGOW G4 ONG
9600  REM  SCOTLAND
9610  REM  ***************************
9620  END
```

4.4.5 Volumes – computer printout

```
 LOCATION OF EARTHWORKS :- STRATHCLYDE UNIVERSITY - STEPPS
*************************************************************************
OPERATOR'S NAME :- JOHN SMITH
DATE OF CALCULATIONS :- 11/06/83
*************************************************************************


VOLUMES FROM CONTOURS
****************************************
CONTOUR  AREA
========================================
82         210
84         840
86         1630
88         2460
90         3150
****************************************

VOLUMES FROM CONTOURS
========================================
<V1> MEAN AREA RULE
TOTAL VOLUME = 13264
****************************************

VOLUMES FROM CONTOURS
****************************************
<V2> END AREAS RULE
CONTOUR    AREA           VOLUME
========================================
82         210
                          1050
84         840
                          2470
86         1630
                          4090
88         2460
                          5610
90         3150
========================================
TOTAL VOLUME =            13220
****************************************
```

```
VOLUMES FROM CONTOURS
****************************************
<V3> PRISMOIDAL RULE
CONTOUR   AREA            VOLUME
========================================
82        210
                          3466.667
86        1630
                          9746.667
90        3150
========================================
TOTAL VOLUME =            13213.333
****************************************


LOCATION OF EARTHWORKS :- TOWNHEAD BRIDGE FOUNDATIONS
************************************************************************
OPERATOR'S NAME :- JOHN SMITH
DATE OF CALCULATIONS :- 13/06/83
************************************************************************


VOLUMES FROM SPOT-LEVELS
**************************************************
SQUARE GRID 15 * 12.5
(1) NO. OF POINTS   = 4       : TOTAL = 17.32
(2) NO. OF POINTS   = 4       : TOTAL = 18.71
(3) NO. OF POINTS   = 0       : TOTAL = 0
(4) NO. OF POINTS   = 1       : TOTAL = 4.8
==================================================
TOTAL NO. OF POINTS = 9
TOTAL VOLUME        = 3465.938
**************************************************


VOLUMES FROM SPOT-LEVELS
**************************************************
TRIANGULAR GRID 15 * 12.5
(1) NO. OF POINTS   = 2       : TOTAL = 7.82
(2) NO. OF POINTS   = 2       : TOTAL = 9.5
(3) NO. OF POINTS   = 4       : TOTAL = 18.71
(4) NO. OF POINTS   = 0       : TOTAL = 0
(5) NO. OF POINTS   = 0       : TOTAL = 0
(6) NO. OF POINTS   = 1       : TOTAL = 4.8
==================================================
TOTAL NO. OF POINTS = 9
TOTAL VOLUME        = 3492.188
**************************************************


SPOT-LEVEL GRID :-
============================

3.15        3.7         4.33

3.94        4.8         4.97

5.17        6.1         4.67
```

```
LOCATION OF EARTHWORKS :- TOWNHEAD MOTORWAY
****************************************************************************
OPERATOR'S NAME :- JOHN SMITH
DATE OF CALCULATIONS :- 15/Ø6/83
****************************************************************************


VOLUME OF BORROW-PIT
*****************************************
SECT.  TYPE  BASE  HEIGHT  VOLUME
=========================================
1      T     7.5   5       89.375
2      S     1Ø    5       222.5
3      T     1Ø    5       1ØØ
4      S     7.5   7.5     255.938
5      S     1Ø    7.5     3Ø7.5
6      S     1Ø    7.5     271.875
7      S     7.5   7.5     233.438
8      S     1Ø    7.5     283.125
9      S     1Ø    7.5     253.125
=========================================
TOTAL VOLUME =             2Ø16.875
*****************************************
```

5

Curve ranging

5.1 INTRODUCTION

In the previous chapter, dealing with earthworks, mention was made of the curvature corrections required when a road or rail curve was encountered. When designing a horizontal curve there is a choice of either a circular curve or a transition curve, the latter often with a centre circular portion. The type of curve selected will depend on the design requirements, for example a circular curve would be used for low speed curves in built-up areas, whereas transition curves are essential on high-speed motorways.

After a horizontal curve has been designed, it is necessary to consider the longitudinal profile as mentioned previously when discussing mass-haul diagrams in Section 4.3. At each change of gradient, a vertical curve will be required. If it is found that there is an overlap of horizontal and vertical curve designs, these are normally phased together to ensure compatibility.

The design and setting out of all three types of curves, circular, transition and vertical, are described in this chapter. In addition there is a section on reverse circular curve design. This latter type of curve is often utilized where the tangents are parallel or nearly parallel and a very large radius would be required to join the straights.

5.2 CIRCULAR CURVE DESIGN

In the design of a circular curve, a radius R is used to deflect the road through the angle between the two straights. This angle, known as the *deflection angle* (D_F), is shown in Fig. 5.1 together with the other features essential to the terminology of circular curves. To calculate the distance of the tangent point A from the intersection point (IP) I, it is normally necessary to know the deflection angle (D_F), the radius (R) and the chainage of $I(C_2)$, as illustrated in Fig. 5.1. From triangle AOI, where OA ($=R$) is perpendicular to AI, and A is the first tangent point, the distance AI, commonly termed the *tangent length* (T_L) can be

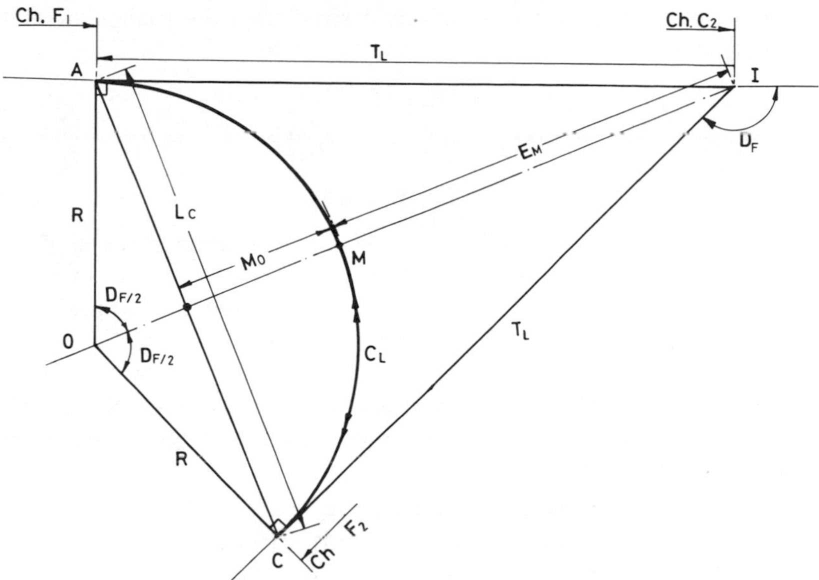

Fig. 5.1. Nomenclature for standard circular curve design.

calculated from:

$$T_L = R\tan(D_F/2) \tag{5.1}$$

If the chainage of I is C_2 then the chainage of $A(F_1)$ will be given by:

$$F_1 = C_2 - T_L \tag{5.2}$$

To carry forward the road chainage of the centre line to the second straight, it is necessary to calculate the *curve length* (C_L) between A and C from:

$$C_L = RD_F/57.3 \tag{5.3}$$

where D_F is in degrees and hence the conversion shown. If high accuracy is required, 57.2957795 should be used. The chainage of $C(F_2)$, the final tangent point of the curve on the second straight, can then be calculated from:

$$F_2 = F_1 + C_L \tag{5.4}$$

Other useful features of the curve can also be calculated knowing R and D_F. The first is the distance between the two tangent points, often called the *long chord* (L_C). The second useful distance is the *mid-ordinate* (M_O) between the long chord and the mid-point on the curve M. The last useful calculation is the external distance (E_M) between I and the mid-point M, and is often used as a check on the setting out of a curve on the ground. These three distances are calculated from the following equations derived from the triangular relationships in Fig. 5.1:

$$L_C = 2R\sin(D_F/2) \tag{5.5}$$

$$M_O = R[1 - \cos(D_F/2)] \tag{5.6}$$

$$E_M = T_L \tan(D_F/4) \tag{5.7}$$

The above Equations 5.1–5.7 give the standard solution as featured in Fig. 5.1. However curve design on site is often complicated, either from the intersection point being inaccessible, or the curve being required to pass through specific points on the ground. The following computer program <CIRCULAR CURVES> allows a curve to be designed in several different ways either from linear measurement only or by theodolite and tape measurements. If both the deflection angle and radius are known, the solution is simple. Other methods included in the program are:

<1> IP INACCESSIBLE OR
LINEAR MEASUREMENT ONLY
<2> TANGENT TO 3 STRAIGHTS
<3> CURVE THROUGH 3 POINTS
<4> CURVE THROUGH 1 POINT

In each case, all the design parameters are computed and listed for reference.

5.2.1 Circular curves – subroutine index

	Line numbers	*Function*
(a)	10–90	Initialization and control
(b)	100–180	Screen header display
(c)	200–240	Selection of angle format
(d)	250–280	Curve information available
(e)	500–680	Entry of curve location, etc.
(f)	700–800	Menu selection for curve design
(g)	1000–1480	<1> IP inaccessible or linear measurement only
(h)	1500–1980	<2> Tangent to 3 straights
(i)	2000–2350	<3> Curve through 3 points
(j)	2500–2990	<4> Curve through 1 point
(k)	3000–3060	Curve design detail check
(l)	3100–3390	IP inaccessible/linear measurement only solution
(m)	3400–3630	Curve through 3 points solution
(n)	4000–4060	Re-run program option
(o)	6000–6080	Routine to print curve location, etc.
(p)	6100–6230	Routine to print data <1>
(q)	6300–6360	Routine to print solution <2>
(r)	6400–6460	Routine to print data <3>
(s)	6500–6590	Routine to print data <4>
(t)	6600–6640	Routine to print solution <4>
(u)	6700–6770	Routine to print solution <1>

	Line numbers	*Function*
(v)	6800–6860	Routine to print solution <3>
(w)	7750–7790	Triangle solution (*A*1, *S*1, *A*3)
(x)	7800–7870	Triangle solution (*S*1, *S*2, *S*3)
(y)	8000–8040	Routine to convert DD.MMSS to decimals
(z)	8050–8140	Routine to convert decimals to DD.MMSS
(aa)	9000–9010	Routine to clear screen
(bb)	9100–9130	Routine to display error message
(cc)	9200–9290	Error trace routine
(dd)	9500–9620	Termination of program

(a) Initialization and control

Line numbers 10–90

All numeric variables and required string variables are initialized and two 'DEF FN' functions used to define arccosine and arcsine as discussed previously in Section 2.2.8.

(b) Screen header display

Line numbers 100–180

This is the screen display used for civil engineering students at Strathclyde University. The operator can alter this as required to suit another organization.

(c) Selection of angle format

Line numbers 200–240

The operator has a choice in line 200 of entering angles in decimal or DD.MMSS format, where the *D*9 flag is set to 1 for decimals and 2 for DD.MMSS in line 210.

(d) Curve information available

Line numbers 250–280

The operator is now asked first if the deflection angle is known and second if either the radius, or a degree curve, or the tangent length is known.

(e) Entry of curve location, etc.

Line numbers 500–680

This routine is provided to keep a record of site location, operator, date, etc., with an opportunity to obtain a printout if required.

(f) Menu selection for curve design

Line numbers 700–800

If both questions in (d) above were affirmative, then the program jumps to line 3000 since it is a simple solution. However if either of the questions in (d) were negative then further information is required and the type of curve problem selected from the following menu:

<1> IP INACCESSIBLE OR
LINEAR MEASUREMENT ONLY
<2> TANGENT TO 3 STRAIGHTS
<3> CURVE THROUGH 3 POINTS
<4> CURVE THROUGH 1 POINT

The menu selection is stored in variable *P*1 and the program branches accordingly from line 800.

(g) IP inaccessible or linear measurement only

Line numbers 1000–1480

On many civil engineering sites the intersection point *I* is inaccessible, either due to construction work or due to it lying in the middle of a water-filled area. If this is the case (line 1030), it is necessary to establish two points R_1 and R_2 on the two straights, where the chainage of the first point is known, Fig. 5.2(a). The distance (H_1) between the points R_1 and R_2 is also required and this can be obtained either from a knowledge of the co-ordinates of the points, or of the measured distance itself. The angles Z_1 and Z_3 at the two reference points as shown in Fig. 5.2(a) are then measured, giving the reciprocal angles U_1 and U_3, whence the deflection angle (D_F) in line 1230 from:

$$D_F = U_1 + U_3 \tag{5.8}$$

In triangle R_1IR_3, three items are now known in the format (*A*1, *S*1, *A*3) as dealt with in Section 3.2 and hence distances H_2 and H_3 can be solved. If the chainage of the first reference point is C_1 then the chainage of the intersection point *I* will be given in line 1260 from:

$$C_2 = C_1 + H_2 \tag{5.9}$$

The previous discussion in this routine has assumed the ability to measure angles Z_1 and Z_2 in the field. On some occasions, if no angle measuring devices are available, it is possible to calculate the deflection angle from linear measurement only. This routine commences at line 1300, and requires access to the intersection point to allow three measurements to take place. Two points *A* and *B* are set out distance H_2 from *I*, with known chainage, as shown in Fig. 5.2(b), and the distance between *A* and *B* measured.

Since triangle *AIB* is isosceles, the angles *IAB* and *IBA* are equal, say U_1. To calculate the deflection angle D_F the line *AB* is bisected in *M* giving a right-

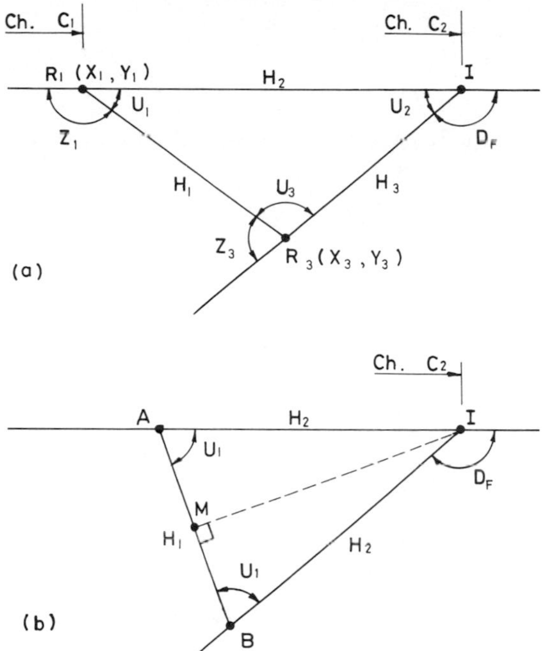

Fig. 5.2. Nomenclature for circular curve design where either (a) the IP is inaccessible, or (b) the deflection angle has to be determined from linear measurements only.

angled triangle *AIM* where *AI* and *AM* are known. The angle U_1 can thus be found using the arccosine definition in line 1360:

$$U_1 = \arccos(H_1/2H_2) \tag{5.10}$$

and hence D_F which is twice U_1.

After displaying the resulting deflection angle, an opportunity is given to obtain a printout of the calculations at line 6100.

(h) <2> Tangent to 3 straights

Line numbers 1500–1980

This special case is shown in Fig. 5.3 where the two main straights *AI* and *IC* intersect at *I* with a deflection angle D_F. The problem posed is one of designing a curve with a constant radius *R* which will also be tangential to a third straight $R_1 R_3$ at the point *B*. The only fixed points at the outset are R_1 and R_2 and it is necessary first to know either their co-ordinates or their distance apart (H_1). The second set of information required is either their distances from *I*, that is H_2 and H_3, or the angles Z_1 and Z_2 as discussed in section (g) above. The chainage of the first reference point R_1 is also required if chainages are to be calculated.

If the distances H_1, H_2, H_3 are known then the triangle R_1IR_3 can be solved using the (S_1, S_2, S_3) solution format described in Section 3.2, giving angles U_1,

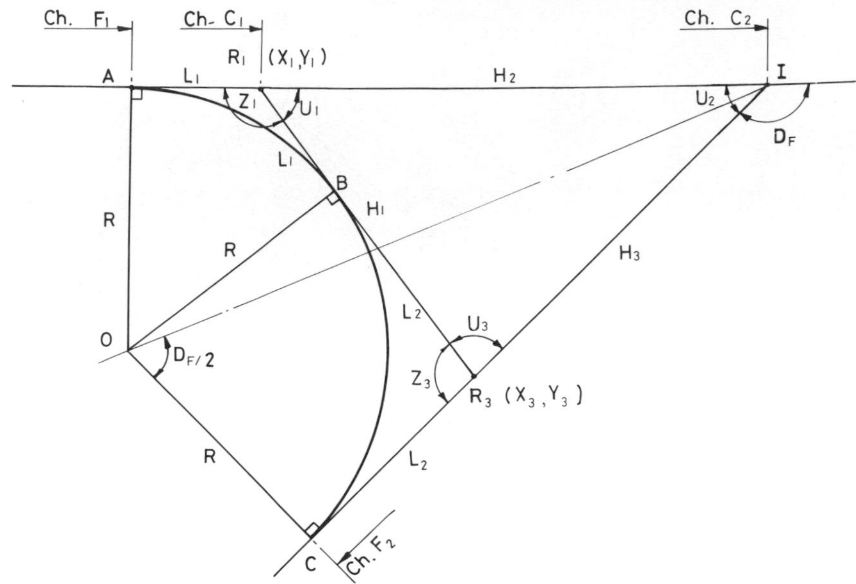

Fig. 5.3. Nomenclature for designing a circular curve tangential to three straights.

U_2 and U_3 as shown in Fig. 5.3 at line 1600. If however the angles Z_1 and Z_2 are known, the earlier routine used for angle entry in (g) is utilized from line numbers 1170–1280, solving for U_1, U_3, H_2 and H_3.

The circular curve ABC is then split into two segments AB and BC and considered as two separate circular curves with intersection points at R_1 and R_2 with tangent lengths L_1 and L_2 respectively as shown in Fig. 5.3. Now from Equation 5.1 we obtain:

$$L_1 = R\tan(U_1/2) \tag{5.11}$$

and

$$L_2 = R\tan(U_3/2) \tag{5.12}$$

But from Fig. 5.3, Equations 5.11 and 12 can be combined to give:

$$\begin{aligned} H_1 &= L_1 + L_2 \\ &= R\tan(U_1/2) + R\tan(U_3/2) \end{aligned}$$

hence

$$R = H_1/[\tan(U_1/2) + \tan(U_3/2)] \tag{5.13}$$

Thus L_1, L_2, R, C_L, F_1, F_2, C_2, T_L, L_C, M_O and E_M can be calculated from Equations 5.1–5.13 to give the complete curve design parameters. These are displayed from line 1800, with an option to obtain a hard copy printout at line 6300.

(i) <3> Curve through 3 points

Line numbers 2000–2350

This special case is shown in Fig. 5.4 where the three points A, B and C are known. Either their co-ordinates or the distances between the points, H_1, H_2 and H_3 are required for the solution. Whichever information is available, it means that in triangle ABC the three side distances are known and therefore a standard

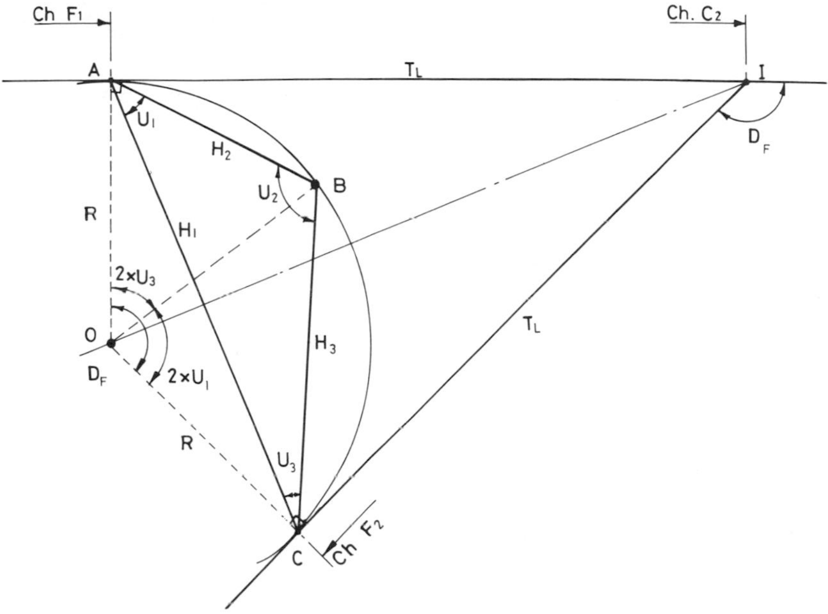

Fig. 5.4. Nomenclature for designing a circular curve through three points A, B and C.

triangle solution with the format ($S1$, $S2$, $S3$) as discussed in Section 3.2. The three angles U_1, U_2, U_3 are then determined and the sine rule used to solve for R where:

$$2R = H_1/\sin(U_2) = H_2/\sin(U_3) = H_3/\sin(U_1) \tag{5.14}$$

(j) <4> Curve through 1 point

Line numbers 2500–2990

This special case is shown in Fig. 5.5 where it is required to pass a circular curve through a point P tangential to two straights where the deflection angle (D_F) is known. The position of the point P is required in relation to the intersection point. Either the x and y distances, X_p and Y_p respectively, or the polar co-ordinates, distance Z_p and the angle (B) from the tangent are required. This is a useful routine where it is necessary to pass a circular curve through a point, say a rock outcrop for a bridge pier in the middle of a river, where the curve is to be tangential to two straights.

Those familiar with co-ordinate geometry will know there are two solutions, the first (internal) where the centre of the curve lies between P and I, and the second (external) where the centre of the curve lies beyond P. Only the second solution is given here. The information requested for triangle with hypotenuse PI, either gives Z_p and B direct or allows them to be solved from X_p and Y_p. Thus

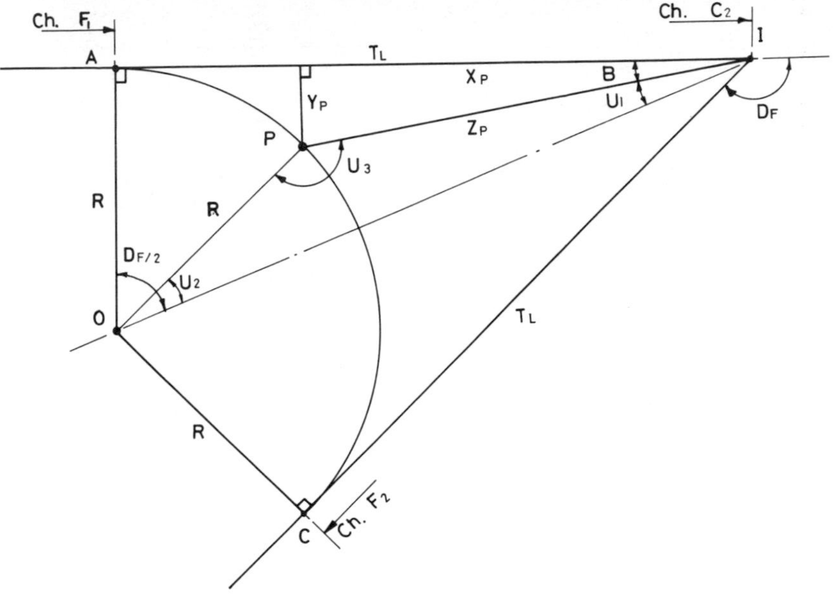

Fig. 5.5. Nomenclature for designing a circular curve tangential to two straights and passing through point P.

from the additional knowledge of the deflection angle (D_F) it is possible to calculate angle U_1 in triangle OPI at line 2670 from:

$$U_1 = 90 - (B + D_F/2) \tag{5.15}$$

If P lies on the line IO, that is $U_1 = 0$, then the point P is co-incident with M the mid-ordinate as shown in Fig. 5.1 and the solution is straightforward. In other circumstances where the point P lies along the curve it is necessary to solve the triangle IOP. Using the sine rule where the design radius is R, and denoting the angles U_1, U_2 and U_3 as shown in Fig. 5.5, gives:

$$R/\sin(U_1) = Z_p/\sin(U_2) = IO/\sin(U_3) \tag{5.16}$$

Utilizing the first and last part of Equation 5.16 to solve for angle U_3 gives:

$$\sin(U_3) = (IO/R)\sin(U_1) \tag{5.17}$$

But IO and R are unknown. However in the right-angled triangle IOA, the angle AIO is known, where:

$$\sin(B+U_1) = R/IO \tag{5.18}$$

Substituting for R/IO from Equation 5.18 in Equation 5.17 thus gives:

$$\sin(U_3) = \sin(U_1)/\sin(B+U_1) \tag{5.19}$$

and hence the value of U_2 and U_3 from lines 2680–2690. R is then found in line 2700 using the sine rule in triangle IOP since all angles are known, as is the side IP:

$$R = Z_p \sin(U_1)/\sin(U_2) \tag{5.20}$$

Once R is known in addition to the deflection angle (D_F) all the other curve design parameters T_L, C_L, L_C, M_O, E_M, plus the chainage of A and C can then be calculated using Equations 5.1–5.7. After screen display, an option is provided for a hard copy printout.

(k) Curve design detail check

Line numbers 3000–3060

All curve design solutions pass through this point. If sufficient information was known at the outset to give a simple circular curve solution ($P1 = 0$) then the program jumps to line 3110 to collect the known data. In special case solutions <2> and <4> all the necessary parameters were calculated and the program jumps to line 4000. If, however, special cases <1> and <3> were selected, more information is required.

(l) IP inaccessible/linear measurement only solution

Line numbers 3100–3390

In both cases for curve design, only the deflection angle (D_F) and the chainage of the intersection point $I(C_2)$ are known, just as in simple curve design. It is therefore necessary to select the type of information available as presented in line 3130.

"ENTER KNOWN DATA <1> RADIUS
<2> DEGREE (100M)
<3> TANGENT LENGTH"

If the radius R is known, then the tangent length T_L is calculated using Equation 5.1 in line 3170 before proceeding to the other curve parameters. If the angle A (in degrees) subtended at the centre of the circle by a 100 m chord is known, then the radius R can be calculated from the relationship:

$$\sin(A/2) = 100/2R$$

i.e.

$$R = 50/\sin(A/2) \tag{5.21}$$

If, however, A is small, then:

$$\sin(A/2) = (A/2)\,\text{rad} \tag{5.22}$$

The substitution from Equation 5.22 is made in Equation 5.21, and the value of R found from Equation 5.23 as shown in line 3230.

$$R = 206265 \times 50/(A/2) \times 3600 \tag{5.23}$$

Finally, if it is the tangent length which is known, it is entered in line 3260 and the radius calculated in line 3270 using Equation 5.1.

The remaining curve design parameters C_L, L_C, M_O, E_M and the chainages of the tangent points are then found from Equations 5.2–5.7 in lines 3280–3320. This is followed by a screen display and hard copy printout if required.

(m) Curve through 3 points solution

Line numbers 3400–3630

This routine can only be used if the points A and B in Fig. 5.4 are tangent points to the straights. If this is the case, then the solution is straightforward. In the earlier calculations for the radius R, the angles of triangle ABC (U_1, U_2 and U_3) were computed, and these are now used to determine the deflection angle D_F. If O is the centre of the circle of radius R passing through A, B and C (Fig. 5.4) then since the chord AB subtends an angle U_3 on the circumference, the angle at O will be twice U_3. Similarly the chord BC which subtends U_1 at A will give twice U_1 at the centre. Since the angle AOC at the centre equals the deflection angle D_F, the solution is given in line 3450 from:

$$D_F = 2(U_1 + U_3) \tag{5.24}$$

The rest of the curve design parameters and chainages are determined as before from Equations 5.1–5.7, prior to screen display and printout if required.

(n) Re-run program option

Line numbers 4000–4060

This program can be run again if required, since there are no dimensioned arrays.

(o) Routine to print curve location, etc.

Line numbers 6000–6080

This routine is provided for record purposes, and is based on Applesoft DOS and may require alteration for different microcomputers and printers.

(p) Routine to print data <1>

Line numbers 6100–6230

This routine is used by both the IP inaccessible and linear measurement only solutions to print out the data for record purposes.

(q) Routine to print solution <2>

Line numbers 6300–6360

This is the complete curve design solution printout and uses the previous screen display from lines 1820–1940 to avoid repetition.

(r) Routine to print data<3>

Line numbers 6400–6460

This routine is used to record the data for the curve through three points problem and uses the previous screen display from lines 2280–2310.

(s) Routine to print data <4>

Line numbers 6500–6590

This routine is used to record the data for the curve through one point problem.

(t) Routine to print solution <4>

Line numbers 6600–6640

This is the complete curve design solution printout for a curve through one point and tangential to two straights. To avoid repetition it uses the previous screen display routine from lines 2870–2960.

(u) Routine to print solution <1>

Line numbers 6700–6770

This routine is used to print out the curve design parameters for each of the IP inaccessible case and the linear measurement only case, and the straightforward case. As before it uses the same screen display as (t) above.

(v) Routine to print solution <3>

Line numbers 6800–6860

This routine prints out the curve design parameters for the curve through three points problem and uses the same screen display as (t) and (u) above.

(w) Triangle solution ($A1$, $S1$, $A3$)

Line numbers 7750–7790

The solution to this triangle configuration, Fig. 3.2, was discussed previously in section <2> of the <TRIANGLE> program in Section 3.2.

(x) Triangle solution ($S1$, $S2$, $S3$)

Line numbers 7800–7870

The solution of this triangle configuration, Fig. 3.2, was discussed previously in section <1> of the <TRIANGLE> program in Section 3.2.

(y) Routine to convert DD.MMSS to decimals

Line numbers 8000–8040

This routine was discussed previously in Section 2.2.6.

(z) Routine to convert decimals to DD.MMSS

Line numbers 8050–8140

This routine was discussed previously in Section 2.2.6.

(aa) Routine to clear screen

Line numbers 9000–9010

To simplify portability, the Applesoft 'HOME' command to clear the screen has been placed in a subroutine to allow an easy change to another microcomputer.

(bb) Routine to display error message

Line numbers 9100–9130

If a wrong keyboard entry has been made, the error is trapped and this error message displayed and re-entry requested.

(cc) Error trace routine

Line numbers 9200–9290

This error trace routine was discussed earlier in Section 2.2.9.

(dd) Termination of program

Line numbers 9500–9620

When the '<QUIT>' selection is made in the program, an end of program message is displayed to advise the operator that computations are complete.

5.2.2 Circular curve design – numeric variables

$A1–A3$	= Angles of triangle
B	= Polar angle for given point
C	= Counter in error trace routine
$C1$, $C2$	= Chainage 1st point and IP
CL	= Curve length

Table—continued

D, *D*0, *D*1, *D*2, *D*5	= Degrees of angles
*D*9	= Angles in decimals <1> or DD.MMSS<2>
DF	= Deflection angle
E	= Error number
EM	= External distance at mid-ordinate
*F*1, *F*2	= Forward chainage, tangent points
*H*1	= Distance *R*1 to *R*3
*L*1, *L*2	= Sub-tangent lengths
LC	= Length of long chord
M, *M*0, *M*1	= Minutes of angles
MO	= Mid-ordinate
P	= 1/2 Perimeter of triangle
*P*1, *P*2, *P*4	= Program selection
PC	= Type of known data
R	= Radius
*R*1, *R*3	= Reference station numbers
*R*9	= 57.2957795
S, *S*0	= Seconds of angles
*S*1–*S*3	= Sides of triangle
*T*1–*T*3	= Temporary angle storage
TL, *TL*1, *TL*2	= Tangent lengths
*U*1–*U*3	= Angles of triangle
*U*8, *U*9	= Angles used in arcsine solutions
*V*1, *V*3	= Computed decimal angles
XP, *YP*, *ZP*	= Distances to point *P*
*Z*1, *Z*3	= Entered horizontal angles

5.2.3 Circular curve design – string variables

A$	Storage for location of curve
B$	Storage for operator's name
C$	Storage for date of calculations
D$	= CHR$(4), CTRL-D
*E*0$	= "SORRY, DATA ERROR . . . PLEASE RE-ENTER"
H$	= "CIRCULAR CURVE DESIGN PROGRAM"
K$	Storage for instrument used
*P*1$	= "I.P. INACCESSIBLE"
*P*2$	= "TANGENT TO 3 STRAIGHTS"
*P*3$	= "CURVE THROUGH 3 POINTS"
*P*4$	= "CURVE THROUGH A GIVEN POINT"
Q$	= Question response (Y/N)
R$	= Question response (Y/N)
S$	= Question response (Y/N)

5.2.4 Circular curve design – BASIC program

```
10  REM  <CIRCULAR-CURVES) PROGRAM FOR APPLE II+  USES QUME PRINTER
20 D$ =  CHR$ (4): ONERR  GOTO 9200
30 A = 0:B = 0:E = 0:G = 0:J = 0:K = 0:N = 0:S = 0:U = 0:V = 0:W = 0
```

```
40 H$ = "CIRCULAR CURVE DESIGN PROGRAM"
50 E0$ = "SORRY, DATA ERROR ... PLEASE RE-ENTER."
55 P$ = "DO YOU WISH PRINTOUT OF ABOVE DATA (Y/N)"
60 R$ = "ENTER REDUCED BEARING ":W$ = "ENTER WHOLE CIRCLE BEARING "
65 S$ = "SELECT <1> WHOLE CIRCLE BEARINGS,        OR     <2> REDUCED BEARI
   NGS."
70 X$ = "ENTER X(E) COORD. OF STN. ":Y$ = "ENTER Y(N) COORD. OF STN. "
80  DEF  FN C9(C9) = (1.5707964 -  ATN (C9 /  SQR (1 - C9 * C9))) * R9: REM
    ARCCOSINE
90  DEF  FN S9(S9) =  ATN (S9 /  SQR (1 - S9 * S9)) * R9: REM  ARCSINE
95  REM  ***************************************************
100  GOSUB 9000
110  PRINT "****************************************"
120  PRINT "*                                      *"
130  PRINT "*      UNIVERSITY  OF  STRATHCLYDE     *"
140  PRINT "*  DEPARTMENT  OF  CIVIL  ENGINEERING  *"
150  PRINT "*                                      *"
160  PRINT "*            SURVEYING  SECTION        *"
170  PRINT "*       CIRCULAR  CURVE  SOLUTIONS     *"
180  PRINT "****************************************"
190  PRINT
200  PRINT "ARE ANGLES IN DECIMALS <1>, DD.MMSS <2> OR DO YOU WISH TO QUI
     T <3>."
210  INPUT D9
220  IF D9 < 1 OR D9 > 3 THEN 200
230  ON D9 GOTO 240,240,9500
240 R9 = 57.2957795:P1 = 0
250  PRINT "IS DEFLECTION ANGLE KNOWN (Y/N)";: INPUT S$
260  IF S$ < > "Y" AND S$ < > "N" THEN 250
270  PRINT : PRINT "ARE EITHER (1) RADIUS, (2) DEGREE(100M),OR (3) TANGEN
     T LENGTH KNOWN (Y/N)";: INPUT R$
280  IF R$ < > "Y" AND R$ < > "N" THEN 270
500  GOSUB 9000
510  PRINT "UNIVERSITY OF STRATHCLYDE"
520  PRINT "CIRCULAR CURVE PROGRAM"
530  PRINT "****************************************"
540  PRINT
550  PRINT "ENTER LOCATION OF WORK, ETC. WHEN       REQUESTED, AND PRESS
     <RETURN>"
560  PRINT "ENTER LOCATION OF CURVE :-"
570  INPUT A$
580  PRINT "ENTER OPERATOR'S NAME :-"
590  INPUT B$
600  PRINT "ENTER DATE OF CALCULATION AS DD/MM/YY :-"
610  INPUT C$
620  PRINT "ENTER INSTRUMENT USED :-"
630  INPUT K$
640  PRINT
650  PRINT "DO YOU WISH PRINTOUT OF LOCATION ETC.,  (Y/N) ";
660  INPUT Q$
670  IF Q$ = "Y" THEN 6000
680  IF Q$ < > "N" THEN 650
690 :
700  IF S$ = "Y" AND R$ = "Y" THEN 3000
710  GOSUB 9000
720  PRINT : PRINT "SELECT CURVE PROGRAM FROM MENU :-"
730  PRINT
740  PRINT "    <1> I.P. INACESSIBLE, OR                      LINEAR MEASUR
     EMENT ONLY"
```

```
750  PRINT "    <2> TANGENT TO 3 STRAIGHTS"
760  PRINT "    <3> CURVE THROUGH 3 POINTS"
770  PRINT "    <4> CURVE THROUGH 1 POINT"
780  PRINT "    <5> QUIT"
790  INPUT P1: IF P1 < 1 OR P1 > 5 THEN 710
800  ON P1 GOTO 1000,1500,2000,2500,9500
993  REM  ************************************************
995  REM  P1=1, I.P. INACESSIBLE OR LINEAR MEASUREMENT ONLY
1000  GOSUB 9000: PRINT H$: PRINT "*************************************
   **"
1010  PRINT : PRINT "SELECT <1> I.P. INACESSIBLE,": PRINT "OR     <2> LIN
   EAR MEASUREMENT ONLY";: INPUT P0: IF P0 < 1 OR P0 > 2 THEN 1010
1020  ON P0 GOTO 1030,1300
1030 P1$ = "<1> I.P. INACESSIBLE": PRINT : PRINT P1$: PRINT "=============
   ======================================"
1040  PRINT : PRINT "ENTER REF. NO. OF 1ST POINT ";: INPUT R1
1050  PRINT "ENTER CHAINAGE OF ";R1;: INPUT C1
1060  PRINT : PRINT "ENTER REF. NO. OF 2ND POINT ";: INPUT R3
1070  PRINT : PRINT "ARE COORDS OF ";R1;" AND ";R3;" KNOWN (Y/N)": INPUT
   Q$
1080  IF Q$ = "N" THEN 1150
1090  IF Q$ < > "Y" THEN 1060
1095  REM  I.E. COORDS ARE KNOWN
1100  PRINT : PRINT X$;R1;: INPUT X1
1110  PRINT Y$;R1;: INPUT Y1
1120  PRINT : PRINT X$;R3;: INPUT X3
1130  PRINT Y$;R3;: INPUT Y3
1140 H1 =  SQR ((X3 - X1) ^ 2 + (Y3 - Y1) ^ 2): GOTO 1160
1150  PRINT : PRINT "ENTER DISTANCE ";R1;"-";R3;: INPUT H1
1160  IF P1 = 2 THEN 1540
1170  PRINT : PRINT "ENTER ANGLE 1ST STR.-";R1;"-";R3;: INPUT Z1
1180  PRINT "ENTER ANGLE 2ND STR.-";R3;"-";R1;: INPUT Z3
1190 V1 = Z1:V3 = Z3: IF D9 = 1 THEN 1220
1195  REM  CHANGE DD.MMSS TO DD.DEC
1200 D5 = Z1: GOSUB 8000:V1 = D5
1210 D5 = Z3: GOSUB 8000:V3 = D5
1220 U1 = 180 - V1:U3 = 180 - V3
1230 DF = U1 + U3
1235  REM  SOLVE TRIANGLE R1-R3-I.P.
1240 A1 = U1:S1 = H1:A3 = U3: GOSUB 7750
1250 H2 = S2:H3 = S3:U2 = A2
1255  REM  CALCULATE CHAINAGE I.P.(CH2)
1260 C2 = C1 + H2
1270 C2 =  INT (C2 * 1000 + .5) / 1000
1280  IF P1 = 2 THEN 1650
1290  GOTO 1370
1295  REM  P0=2, LINEAR MEASUREMENT ONLY
1300 P1$ = "<1> LINEAR MEASUREMENT ONLY": PRINT : PRINT P1$
1310  PRINT "=================================================="
1320  PRINT : PRINT "ENTER CHAINAGE OF I.P.  ";: INPUT C2
1330  PRINT : PRINT "ENTER DIST. ALONG STRAIGHTS ";: INPUT H2
1340  PRINT : PRINT "ENTER DIST. JOINING STRAIGHTS ";: INPUT H1
1350 U9 = H1 / (2 * H2)
1360 U1 =  FN C9(U9):DF = 2 * U1
1365  REM  P0=1 AND 2 MERGE HERE
1370  IF D9 = 1 THEN 1390
1380 D = DF: GOSUB 8050:DF = D: GOTO 1400
1390 DF =  INT (DF * 1000 + .5) / 1000
1400  GOSUB 9000
```

```
1410  PRINT P1$
1420  PRINT "*****************************************"
1430  PRINT "DEFLECTION ANGLE = ";DF
1440  PRINT "CHAINAGE OF I.P. = ";C2
1450  PRINT P$;: INPUT Q$
1460  IF Q$ = "Y" THEN 6100
1470  IF Q$ < > "N" THEN 1450
1480  GOTO 3000
1490 :
1493  REM  **************************************************
1495  REM  P1=2, TANGENT TO 3 LINES
1500  GOSUB 9000:P2$ = "<2> TANGENT TO 3 STRAIGHTS"
1510  PRINT H$: PRINT P2$
1520  PRINT "*****************************************"
1530  GOTO 1040
1535  REM  RETURN WITH R1,R3,C1,H1
1540  PRINT : PRINT "ENTER <1> FOR DISTANCES TO I.P.              <2> FO
     R ANGLES AT ";R1;" AND ";R3
1550  INPUT P2: IF P2 < 1 OR P2 > 2 THEN 1540
1560  IF P2 = 2 THEN 1170
1565  REM  I.E. P2=1, COLLECT DISTANCES
1570  PRINT "ENTER DISTANCE ";R1;"-I.P.";: INPUT H2
1580  PRINT "ENTER DISTANCE ";R3;"-I.P.";: INPUT H3
1585  REM  SOLVE TRIANGLE R1-R2-I.P.
1590 S1 = H1:S2 = H2:S3 = H3: GOSUB 7800
1600 U1 = A1:U2 = A2:U3 = A3
1610 C2 = C1 + H2
1620 C2 =  INT (C2 * 1000 + .5) / 1000
1630 DF = U1 + U3
1640 :
1645  REM  BOTH P2=1 AND 2 REJOIN WITH DF,H1,CH1 ETC.
1650 R = H1 / ( TAN (U1 / (2 * R9)) +  TAN (U3 / (2 * R9)))
1660 L1 =  INT ((R *  TAN (U1 / (2 * R9))) * 1000 + .5) / 1000
1670 L2 =  INT ((R *  TAN (U3 / (2 * R9))) * 1000 + .5) / 1000
1680 CL =  INT ((R * DF / R9) * 1000 + .5) / 1000
1690 F1 = C1 - L1:F2 = F1 + CL
1700 TL =  INT ((R *  TAN (DF / (2 * R9))) * 1000 + .5) / 1000
1710 LC =  INT ((2 * R *  SIN (DF / (2 * R9))) * 1000 + .5) / 1000
1720 MO =  INT ((R * (1 -  COS (DF / (2 * R9)))) * 1000 + .5) / 1000
1730 EM =  INT ((TL *  TAN (DF / (4 * R9))) * 1000 + .5) / 1000
1740 R =  INT (R * 1000 + .5) / 1000
1750  IF D9 = 1 THEN 1770
1760 D = DF: GOSUB 8050:DF = D: GOTO 1780
1770 DF =  INT (DF * 1000 + .5) / 1000
1780 :
1800  GOSUB 9000:P7 = 0
1810  PRINT H$: PRINT P2$: PRINT "****************************************
     *"
1820  PRINT "RADIUS           = ";R
1830  PRINT "DEFLECTION ANGLE = ";DF
1840  PRINT "CHAINAGE I.P.    = ";C2
1850  PRINT "TOTAL TANGENT    = ";TL
1860  PRINT "CHAINAGE TP1     = ";F1
1870  PRINT "CURVE LENGTH     = ";CL
1880  PRINT "CHAINAGE TP2     = ";F2
1890  PRINT "1ST TANGENT L1   = ";L1
1900  PRINT "2ND TANGENT L2   = ";L2
1910  PRINT "DISTANCE R1-R3   = "; INT (H1 * 1000 + .5) / 1000
1920  PRINT "LONG CHORD       = ";LC
```

```
193Ø  PRINT "MID-ORDINATE      = ";MO
194Ø  PRINT "EXTERNAL MO-DIST = ";EM
1945  IF P7 = 1 THEN 634Ø
195Ø  PRINT P$;: INPUT Q$
196Ø  IF Q$ = "Y" THEN 63ØØ
197Ø  IF Q$ < > "N" THEN 195Ø
198Ø  GOTO 3ØØØ
199Ø :
1993  REM  **************************************************
1995  REM  P1=3, CURVE THROUGH 3 POINTS
2ØØØ  GOSUB 9ØØØ:P3$ = "<3> CURVE THROUGH 3 POINTS"
2Ø1Ø  PRINT H$: PRINT P3$
2Ø2Ø  PRINT "****************************************"
2Ø3Ø  FOR I = Ø TO 2
2Ø4Ø  PRINT "ENTER REF. NO. OF POINT ";I + 1;: INPUT R(I)
2Ø5Ø  NEXT I
2Ø6Ø  PRINT "ARE COORDS OF 3 POINTS KNOWN (Y/N)";
2Ø7Ø  INPUT Q$
2Ø8Ø  IF Q$ = "N" THEN 218Ø
2Ø9Ø  IF Q$ < > "Y" THEN 2Ø6Ø
21ØØ  FOR J = Ø TO 2
211Ø  PRINT X$;R(J);: INPUT X(J)
212Ø  PRINT Y$;R(J);: INPUT Y(J)
213Ø  NEXT J
214Ø H1 =  SQR ((X(Ø) - X(2)) ^ 2 + (Y(Ø) - Y(2)) ^ 2)
215Ø H2 =  SQR ((X(1) - X(Ø)) ^ 2 + (Y(1) - Y(Ø)) ^ 2)
216Ø H3 =  SQR ((X(2) - X(1)) ^ 2 + (Y(2) - Y(1)) ^ 2)
217Ø  GOTO 221Ø
218Ø  PRINT "ENTER DISTANCE ";R(Ø);"-";R(2);: INPUT H1
219Ø  PRINT "ENTER DISTANCE ";R(Ø);"-";R(1);: INPUT H2
22ØØ  PRINT "ENTER DISTANCE ";R(1);"-";R(2);: INPUT H3
22Ø5  REM  SOLVE TRIANGLE R(Ø)-R(1)-R(2)
221Ø S1 = H1:S2 = H2:S3 = H3
222Ø  GOSUB 78ØØ
223Ø U1 = A1:U2 = A2:U3 = A3
224Ø R = H1 / (2 *  SIN (U2 / R9))
225Ø R =  INT (R * 1ØØØ + .5) / 1ØØØ
226Ø  PRINT : PRINT P3$:P7 = Ø
227Ø  PRINT "****************************************"
228Ø  PRINT "RADIUS = ";R
229Ø  PRINT "DISTANCE ";R(Ø);"-";R(2);" = "; INT (H1 * 1ØØØ + .5) / 1ØØØ
23ØØ  PRINT "DISTANCE ";R(Ø);"-";R(1);" = "; INT (H2 * 1ØØØ + .5) / 1ØØØ
231Ø  PRINT "DISTANCE ";R(1);"-";R(2);" = "; INT (H3 * 1ØØØ + .5) / 1ØØØ:
      IF P7 = 1 THEN 644Ø
232Ø  PRINT P$;: INPUT Q$
233Ø  IF Q$ = "Y" THEN 64ØØ
234Ø  IF Q$ < > "N" THEN 232Ø
235Ø  GOTO 3ØØØ
2493  REM  **************************************************
2495  REM  P1=4, CURVE THROUGH GIVEN POINT
25ØØ  GOSUB 9ØØØ:P4$ = "<4> CURVE THROUGH A GIVEN POINT"
251Ø  PRINT H$: PRINT P4$
252Ø  PRINT "****************************************"
253Ø  PRINT "ENTER <1> FOR X AND Y DISTANCES          OR <2> FOR POLAR
     COORDS."
254Ø  INPUT P4: IF P4 < 1 OR P4 > 2 THEN 253Ø
255Ø  IF P4 = 2 THEN 259Ø
256Ø  PRINT "ENTER DIST. ALONG TANGENT ";: INPUT XP
257Ø  PRINT "ENTER DIST. PERP. TANGENT ";: INPUT YP
```

```
2580 ZP =  SQR (XP ^ 2 + YP ^ 2):B =  ATN (YP / XP) * R9: GOTO 2630
2590  PRINT "ENTER DIST. FROM I.P. TO POINT ";: INPUT ZP
2600  PRINT "ENTER ANGLE FROM TANGENT ";: INPUT B
2610  IF D9 = 1 THEN 2630
2620 D5 = B: GOSUB 8000:B = D5
2630  PRINT P$;: INPUT Q$: IF Q$ = "Y" THEN 6500
2640  IF Q$ < > "N" THEN 2630
2650  PRINT "ENTER DEFLECTION ANGLE ";: INPUT DF: IF D9 = 1 THEN 2670
2660 D5 = DF: GOSUB 8000:DF = D5
2670 U1 = 90 - (B + DF / 2): IF U1 < .01 THEN 2710
2680 U9 =  SIN (U1 / R9) /  SIN ((B + U1) / R9)
2690 U8 =  FN S9(U9):U2 = U8 - U1:U3 = 180 - U1 - U2
2700 R = ZP *  SIN (U1 / R9) /  SIN (U2 / R9): GOTO 2720
2710 R = ZP / (1 /  COS (DF / (2 * R9)) - 1)
2720 TL =  INT ((R *  TAN (DF / (2 * R9))) * 1000 + .5) / 1000
2730 CL =  INT ((R * DF / R9) * 1000 + .5) / 1000
2740 LC =  INT ((2 * R *  SIN (DF / (2 * R9))) * 1000 + .5) / 1000
2750 MO =  INT ((R * (1 -  COS (DF / (2 * R9)))) * 1000 + .5) / 1000
2760 EM =  INT ((TL *  TAN (DF / (4 * R9))) * 1000 + .5) / 1000
2770  PRINT "IS CHAINAGE OF I.P. KNOWN (Y/N)";: INPUT Q$
2780  IF Q$ = "N" THEN 2830
2790  IF Q$ < > "Y" THEN 2770
2800  PRINT "ENTER CHAINAGE OF I.P. ";: INPUT C2
2810 F1 = C2 - TL
2820 F2 = F1 + CL: GOTO 2840
2830 F1 = 0:C2 = TL:F2 = CL
2840  IF D9 = 1 THEN 2860
2850 D = DF: GOSUB 8050:DF = D
2860  PRINT H$: PRINT P4$:P7 = 0: PRINT "********************************
     ********"
2870  PRINT "RADIUS           = "; INT (R * 1000 + .5) / 1000
2880  PRINT "DEFLECTION ANGLE = ";DF
2890  PRINT "CHAINAGE I.P.    = ";C2
2900  PRINT "TANGENT LENGTH   = ";TL
2910  PRINT "CHAINAGE TP1     = "; INT (F1 * 1000 + .5) / 1000
2920  PRINT "CURVE LENGTH     = ";CL
2930  PRINT "CHAINAGE TP2     = "; INT (F2 * 1000 + .5) / 1000
2940  PRINT "LONG CHORD       = ";LC
2950  PRINT "MID-ORDINATE     = ";MO
2960  PRINT "EXTERNAL MO-DIST = ";EM
2970  IF P7 = 1 THEN  RETURN
2980  PRINT P$;: INPUT Q$: IF Q$ = "Y" THEN 6600
2990  IF Q$ < > "N" THEN 2980
2993  REM  ***********************************************
2995  REM  CURVE DESIGN DETAIL CHECK
3000  GOSUB 9000
3010  PRINT H$
3020  IF P1 = 0 THEN 3040
3030  ON P1 GOTO 3100,4000,3400,4000
3035  REM  BOTH DF PLUS SOME DATA KNOWN, I.E. HAVE COME FROM LINE 290.
3040  PRINT : PRINT "ENTER DEFLECTION ANGLE ";: INPUT DF
3050  PRINT : PRINT "ENTER CHAINAGE OF I.P. ";: INPUT C2
3055  REM  COLLECT REST OF DATA
3060  GOTO 3110
3070 :
3093  REM  ***********************************************
3095  REM  P1=1, I.P. INACESSIBLE
3100  PRINT P1$: PRINT "*******************************************"
3105  REM  KNOW DF AND C2 ONLY
```

```
3110  IF D9 = 1 THEN 3130
3120 D5 = DF: GOSUB 8000:DF = D5
3130  PRINT : PRINT "ENTER KNOWN DATA <1> RADIUS,           <2> DEGREE(1
    00M), <3> TANGENT LENGTH";
3140  INPUT PC: IF PC < 1 OR PC > 3 THEN 3130
3150  ON PC GOTO 3160,3200,3260
3155  REM  PC=1, RADIUS KNOWN
3160  PRINT : PRINT "ENTER RADIUS ";: INPUT R
3170 TL =  INT (R *  TAN (DF / (2 * R9)) * 1000 + .5) / 1000
3180  GOTO 3280
3190 :
3195  REM  PC=2, DEGREE(100M) CHORD
3200  PRINT : PRINT "ENTER ANGLE SUBTENDED BY 100M CHORD";: INPUT A
3210  IF D9 = 1 THEN 3230
3220 D5 = A: GOSUB 8000:A = D5
3230 R =  INT ((R9 * 100 / A) * 1000 + .5) / 1000
3240 TL =  INT (R *  TAN (DF / (2 * R9)) * 1000 + .5) / 1000
3250  GOTO 3280
3255  REM  PC=3, TANGENT LENGTH
3260  PRINT : PRINT "ENTER TANGENT LENGTH ";: INPUT TL
3270 R =  INT ((TL /  TAN (DF / (2 * R9))) * 1000 + .5) / 1000
3275  REM  ARRIVE WITH DF,C2,R,TL
3280 CL =  INT ((R * DF / R9) * 1000 + .5) / 1000
3290 F1 = C2 - TL:F2 = F1 + CL
3300 LC =  INT ((2 * R *  SIN (DF / (2 * R9))) * 1000 + .5) / 1000
3310 MO =  INT ((R * (1 -  COS (DF / (2 * R9)))) * 1000 + .5) / 1000
3320 EM =  INT ((TL *  TAN (DF / (4 * R9))) * 1000 + .5) / 1000
3330  IF D9 = 1 THEN 3340
3335 D = DF: GOSUB 8050:DF = D
3340  GOSUB 9000: PRINT H$:P7 = 1: PRINT "*******************************
    *********"
3350  IF PC = 2 THEN  PRINT "100M CHORD ANGLE = ";A
3360  GOSUB 2870: PRINT "****************************************"
3370  PRINT P$;: INPUT Q$: IF Q$ = "Y" THEN 6700
3380  IF Q$ < > "N" THEN 3370
3390  GOTO 4000
3393  REM  ************************************************
3395  REM  P1=3, CURVE THROUGH 3 POINTS
3400  PRINT P3$
3410  PRINT "****************************************"
3420  PRINT : PRINT "ARE ";R(0);" AND ";R(2);" TANGENT POINTS": PRINT "(Y
    /N)";
3430  INPUT Q$: IF Q$ = "N" THEN 4000
3440  IF Q$ < > "Y" THEN 3420
3450 DF = 2 * (U1 + U3)
3460 TL =  INT (R *  TAN (DF / (2 * R9)) * 1000 + .5) / 1000
3470 CL =  INT ((R * DF / R9) * 1000 + .5) / 1000
3480  PRINT : PRINT "ENTER CHAINAGE OF ";R(0);: INPUT F1
3490 C2 = F1 + TL:F2 = F1 + CL
3500 LC =  INT ((2 * R *  SIN (DF / (2 * R9))) * 1000 + .5) / 1000
3510 MO =  INT ((R * (1 -  COS (DF / (2 * R9)))) * 1000 + .5) / 1000
3520 EM =  INT ((TL *  TAN (DF / (4 * R9))) * 1000 + .5) / 1000
3530  IF D9 = 1 THEN 3550
3540 D = DF: GOSUB 8050:DF = D: GOTO 3560
3550 DF =  INT (DF * 1000 + .5) / 1000
3560  GOSUB 9000
3570  PRINT H$: PRINT P3$
3580  PRINT "****************************************"
3590 P7 = 1: GOSUB 2870
```

```
3600  PRINT "****************************************"
3610  PRINT P$;: INPUT Q$: IF Q$ = "Y" THEN 6800
3620  IF Q$ < > "N" THEN 3610
3630  GOTO 4000
3640 :
4000  GOSUB 9000
4010  PRINT H$
4020  PRINT "****************************************"
4030  PRINT : PRINT "DO YOU WISH TO RUN PROGRAM AGAIN (Y/N)"
4040  INPUT Q$: IF Q$ = "Y" THEN 100
4050  IF Q$ < > "N" THEN 4000
4060  GOTO 9500
4070 :
5985  REM  **************************************************************
    **
5990  REM  PRINTOUT OF LOCATION, ETC
6000  PRINT D$;"PR#1": PRINT "LOCATION OF CURVE :- ";A$
6010  PRINT "***********************************************************
    ********************"
6020  PRINT "OPERATOR'S NAME :- ";B$
6030  PRINT "DATE OF CALCS.  :-";C$
6040  PRINT "INSTRUMENT USED :- ";K$
6060  PRINT "***********************************************************
    ********************"
6070  PRINT : PRINT D$;"PR#0"
6080  GOTO 700
6093  REM  *************************************************
6095  REM  PRINTOUT FOR P1=1
6100  PRINT D$;"PR#1": POKE 1784 + 1,80
6110  PRINT H$: PRINT P1$
6120  PRINT "****************************************": IF P0 = 2 THEN 62
    00
6130  PRINT "ANGLE 1ST STR-";R1;"-";R3;" = ";Z1
6140  PRINT "ANGLE 2ND STR-";R3;"-";R1;" = ";Z3
6150  PRINT "DEFLECTION ANGLE = ";DF
6160  PRINT "DISTANCE ";R1;"-";R3;" = "; INT (H1 * 1000 + .5) / 1000
6170  PRINT "DISTANCE ";R1;"-IP = "; INT (H2 * 1000 + .5) / 1000
6180  PRINT "DISTANCE ";R3;"-IP = "; INT (H3 * 1000 + .5) / 1000
6190  PRINT "CHAINAGE OF ";R1;" = ";C1
6200  PRINT "CHAINAGE OF IP = ";C2: IF P0 = 1 THEN 6240
6210  PRINT "DISTANCE ALONG STRAIGHTS   = ";H2
6220  PRINT "DISTANCE JOINING STRAIGHTS = ";H1
6230  PRINT "DEFLECTION ANGLE           = ";DF
6240  PRINT "========================================"
6293  REM  *************************************************
6295  REM  PRINTOUT P1=2
6300  PRINT D$;"PR#1": POKE 1784 + 1,80
6310  PRINT H$: PRINT P2$
6320  PRINT "****************************************"
6330 P7 = 1: GOTO 1820
6340  PRINT "****************************************"
6350  PRINT : PRINT D$;"PR#0"
6360  GOTO 3000
6393  REM  ************************************************
6395  REM  PRINTOUT P1=3
6400  PRINT D$;"PR#1": POKE 1784 + 1,80
6410  PRINT H$: PRINT P3$
6420  PRINT "****************************************"
6430 P7 = 1: GOTO 2280
```

```
6440  PRINT "****************************************"
6450  PRINT : PRINT D$;"PR#0"
6460  GOTO 3000
6470 :
6493  REM  **************************************************
6495  REM  PRINTOUT P1=4 DATA INPUT
6500  PRINT D$;"PR#1": POKE 1784 + 1,80
6510  PRINT H$: PRINT P4$
6520  PRINT "****************************************"
6530  IF P4 = 2 THEN 6560
6540  PRINT "DIST. ALONG TANGENT = ";XP
6550  PRINT "DIST. PERP. TANGENT = ";YP
6560  PRINT "DIST. FROM  I.P.    = ";ZP
6570  PRINT "ANGLE FROM TANGENT = "; INT (B * 1000 + .5) / 1000;" (DD.DEC
    )"
6580  PRINT "========================================"
6590  PRINT : PRINT D$;"PR#0": GOTO 2650
6593  REM  **************************************************
6595  REM  PRINTOUT P1=4 RESULTS
6600  PRINT D$;"PR#1": POKE 1784 + 1,80
6610 P7 = 1: GOSUB 2870
6620  PRINT "****************************************"
6630  PRINT : PRINT D$;"PR#0"
6640  GOTO 3000
6693  REM  **************************************************
6695  REM  PRINTOUT CURVE DESIGN P1=1
6700  PRINT D$;"PR#1": POKE 1784 + 1,80
6710  PRINT H$:P7 = 1
6720  PRINT "****************************************"
6730  IF PC = 2 THEN  PRINT "100M CHORD ANGLE = ";A
6740  GOSUB 2870
6750  PRINT "****************************************"
6760  PRINT : PRINT D$;"PR#0"
6770  GOTO 4000
6780 :
6793  REM  **************************************************
6795  REM  PRINTOUT CURVE DESIGN P1=3
6800  PRINT D$;"PR#1": POKE 1784 + 1,80
6810  PRINT H$: PRINT P3$:P7 = 1
6820  PRINT "****************************************"
6830  GOSUB 2870
6840  PRINT "****************************************"
6850  PRINT : PRINT D$;"PR#0"
6860  GOTO 4000
6870 :
7743  REM  **************************************************
7745  REM  <TRIANGLE> SOLUTION (A1,S1,A3)
7750 T2 =  -  COS ((A3 + A1) / R9)
7760 A2 =  FN C9(T2)
7770 S2 = S1 *  SIN (A3 / R9) /  SIN (A2 / R9)
7780 S3 = S1 *  COS (A3 / R9) + S2 *  COS (A2 / R9)
7790  RETURN
7793  REM  **************************************************
7795  REM  <TRIANGLE> SOLUTION (S1,S2,S3)
7800 P = (S1 + S2 + S3) / 2
7810 T3 =  SQR (P * (P - S2) / (S1 * S3))
7820 A3 = 2 *  FN C9(T3)
7830 T2 =  SQR (P * (P - S1) / (S2 * S3))
7840 A2 = 2 *  FN C9(T2)
```

```
7850 T1 =  - COS ((A3 + A2) / R9)
7860 A1 =  FN C9(T1)
7870  RETURN
7993  REM  **************************************************
7995  REM  ROUTINE DD.MMSS TO DECIMALS
8ØØØ DØ =  INT (D5)
8ØØ5  REM  FIND NO. OF MINUTES,MØ
8Ø1Ø M1 = D5 * 1ØØ - DØ * 1ØØ:MØ =  INT (M1 + .5)
8Ø15  REM  FIND NO. OF SECONDS,SØ
8Ø2Ø SØ =  INT (M1 * 1ØØ - MØ * 1ØØ + .5)
8Ø25  REM  COLLECT DEGS,MINS,SECS
8Ø3Ø D5 = DØ + MØ / 6Ø + SØ / 36ØØ
8Ø4Ø  RETURN
8Ø43  REM  **************************************************
8Ø45  REM  ROUTINE DECIMAL TO DD.MMSS
8Ø5Ø D1 =  INT (D)
8Ø55  REM  FIND TOTAL NO. OF SECONDS
8Ø6Ø D2 = (D - D1) * 36ØØ
8Ø65  REM  FIND NO. OF MINUTES
8Ø7Ø M =  INT (D2 / 6Ø)
8Ø75  REM  FIND NO. OF SECONDS
8Ø8Ø S = D2 - 6Ø * M + .5
8Ø9Ø  IF S < 6Ø THEN 813Ø
81ØØ M = M + 1:S = Ø
811Ø  IF M < 6Ø THEN 813Ø
812Ø D1 = D1 + 1:M = Ø
813Ø D = D1 * 1ØØØØ + M * 1ØØ + S
814Ø D =  INT (D) / 1ØØØØ: RETURN
8985  REM  *****************************************
899Ø  REM  GOSUB ROUTINE TO CLEAR SCREEN
9ØØØ  HOME
9Ø1Ø  RETURN
9Ø93  REM  **************************************************
9Ø95  REM  GOSUB ROUTINE FOR ERRORS
91ØØ  PRINT
911Ø  PRINT EØ$
912Ø  PRINT
913Ø  RETURN
9193  REM  **************************************************
9195  REM  ERROR TRACE ROUTINE
92ØØ  PRINT  CHR$ (7): REM  CTRL-B (BELL)
921Ø E =  PEEK (222): REM  GET ERROR NO.
922Ø  IF C = 1 THEN 926Ø
923Ø  INVERSE
924Ø  PRINT "ERROR NO. ";E;" FOUND"
925Ø C = 1: TRACE : RESUME
926Ø  PRINT "ERROR ON SECOND LINE NO. "
927Ø  NORMAL
928Ø  NOTRACE
929Ø  STOP
9493  REM  **************************************************
9495  REM  END OF PROGRAM
95ØØ  GOSUB 9ØØØ
951Ø  PRINT
952Ø  PRINT "END CURVE DESIGN PROGRAM"
953Ø  PRINT
954Ø  PRINT "************************"
955Ø  REM  PROGRAM PREPARED BY
956Ø  REM  DR. P.H. MILNE
```

```
957Ø  REM  DEPT. OF CIVIL ENGINEERING
958Ø  REM  UNIVERSITY OF STRATHCLYDE
959Ø  REM  GLASGOW G4 ONG
96ØØ  REM  SCOTLAND
961Ø  REM  ***************************
962Ø  END
```

5.2.5 Circular curve design – computer printout

```
 LOCATION OF CURVE :- STRATHCLYDE MOTORWAY
*************************************************************************
OPERATOR'S NAME :- JOHN SMITH
DATE OF CALCS.  :-15/Ø6/83
INSTRUMENT USED :- ZEISS ZENA T2ØA
*************************************************************************

CIRCULAR CURVE DESIGN PROGRAM
<1> I.P. INACESSIBLE
****************************************
ANGLE 1ST STR-1-2 = 119.4337
ANGLE 2ND STR-2-1 = 1Ø3.Ø9Ø7
DEFLECTION ANGLE = 137.Ø716
DISTANCE 1-2 = 728.5Ø7
DISTANCE 1-IP = 1Ø42.541
DISTANCE 2-IP = 929.728
CHAINAGE OF 1 = 1ØØØ
CHAINAGE OF IP = 2Ø42.541
========================================

CIRCULAR CURVE DESIGN PROGRAM
****************************************
RADIUS           = 5ØØ
DEFLECTION ANGLE = 137.Ø716
CHAINAGE I.P.    = 2Ø42.541
TANGENT LENGTH   = 1273.269
CHAINAGE TP1     = 769.272
CURVE LENGTH     = 1196.6Ø7
CHAINAGE TP2     = 1965.879
LONG CHORD       = 93Ø.8Ø4
MID-ORDINATE     = 317.241
EXTERNAL MO-DIST = 867.923
****************************************


CIRCULAR CURVE DESIGN PROGRAM
<1> LINEAR MEASUREMENT ONLY
****************************************
CHAINAGE OF IP = 3ØØØ
DISTANCE ALONG STRAIGHTS   = 2ØØ
DISTANCE JOINING STRAIGHTS = 175.693
DEFLECTION ANGLE           = 127.5325
========================================

CIRCULAR CURVE DESIGN PROGRAM
****************************************
RADIUS           = 488.918
DEFLECTION ANGLE = 127.5325
CHAINAGE I.P.    = 3ØØØ
TANGENT LENGTH   = 1ØØØ
```

```
CHAINAGE TP1      = 2000
CURVE LENGTH      = 1091.317
CHAINAGE TP2      = 3091.317
LONG CHORD        = 878.462
MID-ORDINATE      = 274.17
EXTERNAL MO-DIST = 624.204
****************************************

CIRCULAR CURVE DESIGN PROGRAM
<2> TANGENT TO 3 STRAIGHTS
****************************************
RADIUS            = 530.284
DEFLECTION ANGLE = 137.0716
CHAINAGE I.P.     = 2042.541
TOTAL TANGENT     = 1350.388
CHAINAGE TP1      = 692.153
CURVE LENGTH      = 1269.084
CHAINAGE TP2      = 1961.237
1ST TANGENT L1    = 307.847
2ND TANGENT L2    = 420.66
DISTANCE R1-R3    = 728.507
LONG CHORD        = 987.182
MID-ORDINATE      = 336.456
EXTERNAL MO-DIST = 920.491
****************************************

CIRCULAR CURVE DESIGN PROGRAM
<3> CURVE THROUGH 3 POINTS
****************************************
RADIUS = 530.285
DISTANCE 1-3 = 987.182
DISTANCE 1-2 = 597.356
DISTANCE 2-3 = 597.356
****************************************

CIRCULAR CURVE DESIGN PROGRAM
<3> CURVE THROUGH 3 POINTS
****************************************
RADIUS            = 530.285
DEFLECTION ANGLE = 137.0715
CHAINAGE I.P.     = 2042.534
TANGENT LENGTH    = 1350.381
CHAINAGE TP1      = 692.153
CURVE LENGTH      = 1269.083
CHAINAGE TP2      = 1961.236
LONG CHORD        = 987.182
MID-ORDINATE      = 336.455
EXTERNAL MO-DIST = 920.484
****************************************

CIRCULAR CURVE DESIGN PROGRAM
<4> CURVE THROUGH A GIVEN POINT
****************************************
DIST. FROM  I.P.     = 920.49
ANGLE FROM TANGENT  = 21.439 (DD.DEC)
========================================

RADIUS            = 530.288
DEFLECTION ANGLE = 137.0715
CHAINAGE I.P.     = 2042.539
```

```
TANGENT LENGTH     = 1350.39
CHAINAGE TP1       = 692.149
CURVE LENGTH       = 1269.092
CHAINAGE TP2       = 1961.241
LONG CHORD         = 987.189
MID-ORDINATE       = 336.457
EXTERNAL MO-DIST = 920.49
****************************************
```

5.3 CIRCULAR CURVE – SETTING OUT

Once all the circular curve design parameters have been determined from Section 5.2, it is necessary to choose the method of setting out the curve. The ultimate choice will depend on the type of terrain encountered on site, but basically it resolves into either linear measurements only or combined angular and linear measurements. For very long curves, large setting out errors can accumulate in the linear measurements only method, whereas only short chord distances need to be measured in the combined angular/linear method. If only a small radius curve is being set out in an urban location, the linear measurement only technique is often satisfactory, with measurements either from the tangent or the long chord, where the curve has to be set out in two halves. The combined angular/linear method, often termed the tangential deflection angle case, normally only requires to be set out from one point, either the first or last tangent point.

This program <CIRCULAR CURVES – SETTING OUT> presents a menu at line 900 to allow the operator to select the desired setting out method:

<1> Offsets from tangent
<2> Offsets from long chord
<3> Tangential deflection angles.

5.3.1 Circular curve – setting out – subroutine index

	Line numbers	*Function*
(a)	10–90	Initialization and control
(b)	100–180	Screen header display
(c)	200–240	Selection of angle format
(d)	500–680	Entry of curve location, etc.
(e)	700–880	Entry of curve design data
(f)	900–990	Menu selection for setting out
(g)	1000–1340	<1> Offsets from tangent
(h)	2000–2380	<2> Offsets from long chord
(i)	3000–3590	<3> Tangential deflection angles
(j)	6000–6080	Routine to print curve location, etc.
(k)	6100–6390	Printout for <1>
(l)	6400–6580	Printout for <2>
(m)	6600–6940	Printout for <3> – right-hand curve
(n)	7000–7340	Printout for <3> – left-hand curve

(o)	8000–8040	Routine to convert DD.MMSS to decimals
(p)	8050–8140	Routine to convert decimals to DD.MMSS
(q)	9000–9010	Routine to clear screen
(r)	9100–9130	Routine to display error message
(s)	9200–9290	Error trace routine
(t)	9500–9620	Termination of program

(a) Initialization and control

Line numbers 10–90

The numeric variables and required string variables are initialized and two 'DEF FN' functions used to define arccosine and arcsine as discussed previously in Section 2.2.8. Several arrays for chainage, etc. are also dimensioned. Note the maximum array in the program is 21. If a larger number is required, the arrays should be redimensioned.

(b) Screen header display

Line numbers 100–180

This is the screen display used for civil engineering students at Strathclyde University. The operator can alter this as required to suit another organization.

(c) Selection of angle format

Line numbers 200–240

The operator has a choice in line 200 of entering angles in decimal or DD.MMSS format, where the *D*9 flag is set to 1 for decimals and 2 for DD.MMSS in line 210.

(d) Entry of curve location, etc.

Line numbers 500–680

This routine is provided to keep a record of site location, operator, date, etc., with an opportunity to obtain a printout if required.

(e) Entry of curve design data

Line numbers 700–880

Only three data entries, the deflection angle (D_F), the radius (R), and the chainage of the intersection point (F_3), are required to calculate the curve design parameters from Equations 5.1–5.7.

(f) Menu selection for setting out

Line numbers 900–990

The required method for setting out the curve is now chosen from the menu:

<1> OFFSETS FROM TANGENT
<2> OFFSETS FROM LONG CHORD
<3> TANGENTIAL DEFLECTION ANGLES.

The choice is stored in the variable $P1$ and the program branches accordingly from line 990.

(g) <1> Offsets from tangent

Line numbers 1000–1340

This is the simpler of the two linear setting-out methods, since all measurements are either along or perpendicular to the straight joining the tangent point A to the intersection point I (Fig. 5.6).

Consider a point P lying on the circular curve with tangential distance U from the tangent point A and with a tangential offset of V. The required tangential offset ($BP = V$) can be determined from triangle OPQ where PQ is parallel to the tangent and perpendicular to the radius ($OA = R$), so that in triangle OPQ:

$$OP^2 = OQ^2 + PQ^2 \tag{5.25}$$

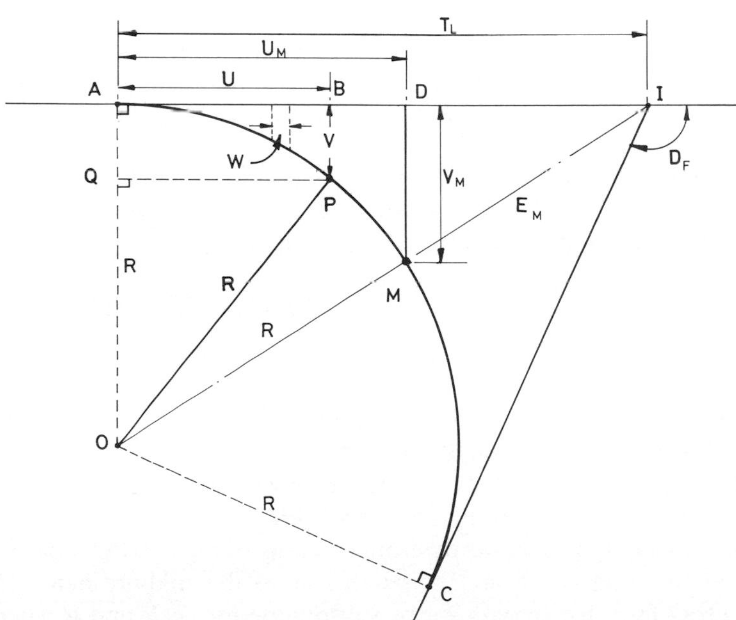

Fig. 5.6. Nomenclature for setting out a circular curve by linear measurements – offsets from the tangent.

Substituting for R, U and V in Equation 5.25 gives:

$$R^2 = (R-V)^2 + U^2 \tag{5.26}$$

which when re-arranged gives V in terms of R and U:

$$V = R - (R^2 - U^2)^{1/2} \tag{5.27}$$

Hence for a chosen value of U, knowing R, the tangential offset V can be calculated from Equation 5.27.

It is therefore necessary first of all to determine the maximum tangential distance (U_M) to the mid-ordinate M in Fig. 5.6. This can be determined by considering the two similar triangles AOI and DMI where:

$$V_M/R = E_M/(E_M + R) = (T_L - U_M)/T_L \tag{5.28}$$

From the above relationships, U_M and V_M can be calculated, where:

$$V_M = E_M R/(E_M + R) \tag{5.29}$$

$$U_M = T_L R/(E_M + R) \tag{5.30}$$

The operator is then advised of the maximum tangential distance U_M, and asked to determine the interval spacing (W) of the tangential offsets V_M. The total number (N) of tangential offsets is then determined in line 1070 and arrays $U(I)$ and $V(I)$ used to store the tangential distances and tangential offsets respectively.

To calculate the chainage of each of the points, the chord length is determined from $U(I)$ and $V(I)$ and the angle subtended at the centre calculated, $D_F(I)$. From this, the corresponding curve length $C_L(I)$ is computed using Equation 5.3 (lines 1190–1210).

In the screen display, listing the setting-out points, the mid-ordinate M (Fig. 5.6) is also included together with the first ($TP1$) and last ($TP2$) tangent points. A computer printout of the results is also available.

(h) <2> Offsets from long chord

Line numbers 2000–2380

In this case, the long chord (L_C) between the tangent points A and C must be capable of being marked out on the ground.

Consider a point P lying on the circular curve with chord distance U from the centre (D) of the long chord, Fig. 5.7, and with chord offset V. The required chord offset ($BP = V$) can be determined from triangle OPQ where PQ is parallel to the long chord and perpendicular to the mid-ordinate. Thus in triangle OPQ from Equation 5.25, by substituting for U, V and R where OD equals the radius (R) minus the mid-ordinate (M_O), gives:

$$R^2 = U^2 + [(R - M_O) + V]^2 \tag{5.31}$$

Rearranging the terms to give V results in:

$$V = (R^2 - U^2)^{1/2} - (R - M_O) \tag{5.32}$$

Hence for a chosen value of U, knowing R and M_O, the chord offset V can be calculated from Equation 5.32.

It is therefore necessary first of all to determine the maximum chord distance which is equal to half the long chord ($L_C/2$). The operator is then advised of this distance and asked to select the interval spacing (W) of the chord offsets. The total number (N) of chord offsets is then determined in line 2050 and arrays $U(I)$

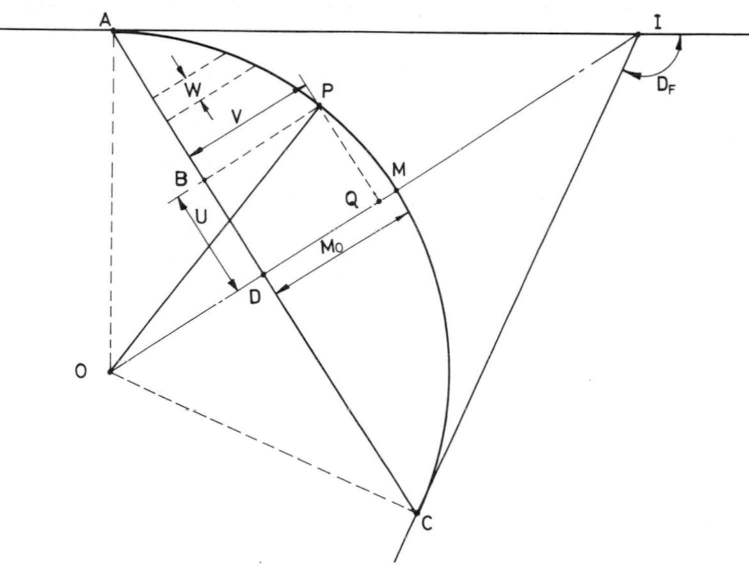

Fig. 5.7. Nomenclature for setting out a circular curve by linear measurements – offsets from long chord.

and $V(I)$ used to store the chord distances and chord offsets respectively. To compute the curve length $C_L(I)$ to each point, Equation 5.3 was used with the angle subtended by the point P at the centre O, as shown in lines 2100–2120.

In the screen display of the results, listing all the points, the mid-ordinate M (Fig. 5.7) is also included together with the first ($TP1$) and last ($TP2$) tangent points. A computer printout of the results is also available.

(i) <3> Tangential deflection angles

Line numbers 3000–3590

The tangential deflection angle method of setting out, although requiring the use of a theodolite at a tangent point for angle measurement, is more accurate than the linear measurement only techniques described in (g) and (h) above. This method assumes that the distance along the arc of a circle is equal to the chord.

This assumption will only be true for short chords, and depends on the accuracy required in the setting out. The operator is advised at the outset of the length of the curve, and asked to select a suitable chord length. Advice is presented for three regularly used orders of accuracy where the equivalent chord lengths are:

$$1/2500 \text{ Accuracy}, C < R/10$$
$$1/5000 \text{ Accuracy}, C < R/14$$
$$1/10000 \text{ Accuracy}, C < R/20,$$

and the operator asked to choose a chord length (C_H).

As it is seldom that the first tangent point is at zero chainage, it means that if pegs are to be set out round the curve, a short chord will be required at the outset in order to establish the pegs at standard chainages, which are multiple intervals of the chord length. The chainage of the first point on the curve is therefore calculated as a multiple of the chord length C_H, in line 3090, and the length of the first chord C_1, computed in line 3100. The chainage of the last point on the curve is then calculated using a similar process in line 3110, and a short chord at the end of the curve (C_2) used to bring the curve to the final tangent point. The number (N) of chords C_H long required to set out the curve can now be computed in line 3130.

The deflection angles from the tangent can then be calculated for each of the chord lengths. At the outset, chord C_1 is used with angle A_1 as shown in Fig. 5.8.

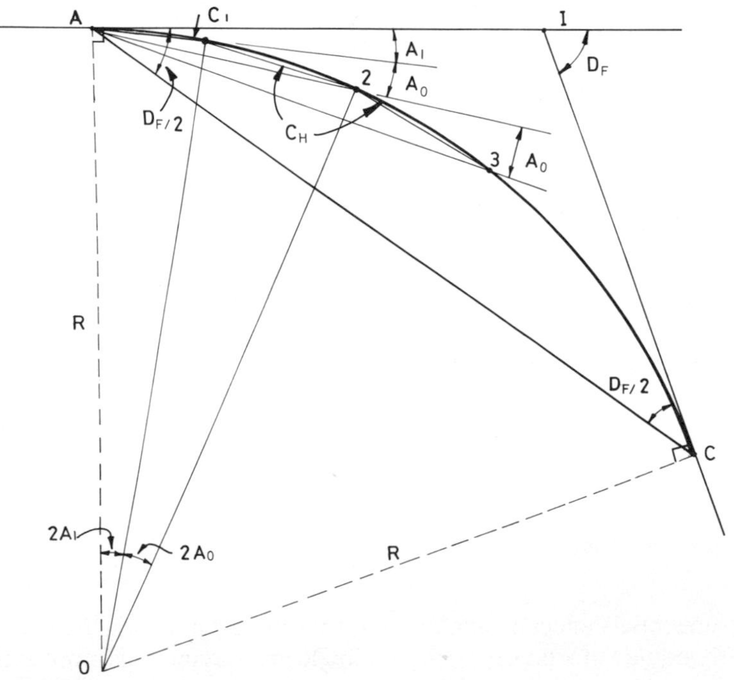

Fig. 5.8. Nomenclature for setting out a circular curve by tangential deflection angles.

Now the chord C_1 will subtend an angle $2A_1$ at the centre, thus using the relationship determined in Equations 5.21–5.23, knowing the chord length and the radius, angle A_1 can be determined from Equation 5.23. This equation is the equivalent of that used in the field where the angle A_1 in minutes would be given by:

$$A_1 = 1718.8\, C_1/R \tag{5.33}$$

Similarly the angles A_0 and A_2 are found from Equation 5.33 using the standard chord C_H and the final short chord C_2.

Before computing the chainages and calculating the relevant setting-out angles, the operator is asked to specify whether it is a left-hand or right-hand curve. The latter is set out clockwise from zero, whilst the former is set out anti-clockwise from 360°. The screen display then gives all the relevant details and setting-out angles with the option of a hard copy printout if required. A check is also made that the final angle set out is equal to half the deflection angle, as shown in Fig. 5.8.

The above screen display and subsequent printout assume that the complete curve has been set out from the first tangent point A (Fig. 5.8). Should there be an obstruction, preventing the whole curve being set out from A, then the same setting-out table can still be used. If, for example, the line of sight to point 3 in Fig. 5.8 was obstructed, then point 2 would be set out in the normal way. The theodolite would be moved to point 2, and the angle reading on A from point 2 would be set to 180° plus the angle used to set out point 2. The theodolite should then be transitted, that is turned through 180°, and point 3 set out using the same table as before. This procedure can be extremely useful when setting out a long curve.

(j) Routine to print curve location, etc.

Line numbers 6000–6080

This routine is provided for record purposes, and is based on Applesoft DOS and may require alteration for different microcomputers and printers.

(k) Printout for <1>

Line numbers 6100–6390

The first part of this printout, lines 6120–6220, are accessed by each of the printout routines, to list the basic curve design parameters and relevant chainages. The latter part of the printout details the point numbers, chainages, tangential distances and tangential offsets for setting out the complete curve.

(l) Printout for <2>

Line numbers 6400–6580

After listing the curve design parameters in (k) above, the printout gives the point numbers, chainages, chord distances and chord offsets for setting out the complete curve.

(m) Printout for <3> right hand curve

Line numbers 6600–6940

After listing the curve design parameters in (k) above, the point numbers, chainages, chords and cumulative setting-out angles from zero, for a right-hand curve, are printed.

(n) Printout for <3> left hand curve

Line numbers 7000–7340

This routine is similar to (m) above, except the cumulative setting-out angles are calculated from 360° for a left-hand curve.

(o) Routine to convert DD.MMSS to decimals

Line numbers 8000–8040

This routine was discussed previously in Section 2.2.6.

(p) Routine to convert decimals to DD.MMSS

Line numbers 8050–8140

This routine was discussed previously in Section 2.2.6.

(q) Routine to clear screen

Line numbers 9000–9010

To simplify portability, the Applesoft 'HOME' command to clear the screen has been placed in a subroutine to allow an easy change to another microcomputer.

(r) Routine to display error message

Line numbers 9100–9130

If a wrong keyboard entry has been made, the error is trapped and this error message displayed and re-entry requested.

(s) Error trace routine

Line numbers 9200–9290

This error trace routine was discussed earlier in Section 2.2.9.

(t) Termination of program

Line numbers 9500–9620

When the '<QUIT>' selection is made in the program, an end-of-program message is displayed to advise the operator that computations are complete.

5.3.2 Circular curve – setting out – numeric variables

A0–A2 = Chord setting-out angles
A9 = Angle solution for arcsine
A(N+1) = Array containing setting-out angles
C = Counter in error trace routine
C1 = First chord on curve
C2 = Last chord on curve
C9 = Chainage of last point on curve
CH = Chord length
CH(N) = Array for chainage of points on curve
CL = Curve length
CL(N) = Array for distance along curve
D, D0, D1, D2, D5 = Degrees of angles
D9 = Angles in decimals <1> or DD.MMSS <2>
DA(N) = Array of angles used in offsets from long chord solution
DF = Deflection angle
DX = Half deflection angle
E = Error number
EM = External distance at mid-ordinate
F1, F2 = Chainages of tangent points
F3 = Chainage of IP
F4 = Chainage of *M*, mid-ordinate
I = Loop integer
J = Loop integer
K = Loop integer
LC = Long chord
M, M0, M1 = Minutes of angle
MO = Mid-ordinate
N = Number of chords or offsets
P1 = Program selection number
P7 = Counter in printout routine
R = Radius of curve
R9 = 57.2957795
S, S0 = Seconds of angle
TL = Tangent length
UM = Maximum tangent distance at mid-ordinate
U(N) = Array of chord or tangent distances
VM = Tangent offset at mid-ordinate
V(N) = Array of chord or tangent offset distances
W = Offset interval

5.3.3 Circular curve – setting out – string variables

A$ Storage for location of curve
B$ Storage for operator's name
C$ Storage for date of calculations
*D$* = CHR$(4), CTRL-D
E0$ = "SORRY, DATA ERROR . . . PLEASE RE-ENTER"
H$ = "CIRCULAR CURVE SETTING OUT PROGRAM"

Table—continued

K$	Storage for instrument used
L$	= Either "L" left handed or "R" for right handed curve
P$	= "DO YOU WISH PRINTOUT OF ABOVE DATA (Y/N)"
*P*1$	= "<1> OFFSETS FROM TANGENT"
*P*2$	= "<2> OFFSETS FROM LONG CHORD"
*P*3$	= "<3> TANGENTIAL DEFLECTION ANGLES"
Q$	= Answer response to question (Y/N)

5.3.4 Circular curve – setting out – BASIC program

```
1Ø  REM  <CIRCULAR CURVES - SETTING OUT> PROGRAM FOR APPLE II+
15  REM  USES QUME PRINTER
2Ø D$ =  CHR$ (4): ONERR  GOTO 92ØØ
3Ø A = Ø:B = Ø:E = Ø:G = Ø:J = Ø:K = Ø:N = Ø:S = Ø:U = Ø:V = Ø:W = Ø
35  DIM A(2Ø),CH(2Ø),CL(2Ø),DA(2Ø),DF(2Ø),U(2Ø),V(2Ø)
4Ø H$ = "CIRCULAR CURVE SETTING OUT PROGRAM"
5Ø EØ$ = "SORRY, DATA ERROR ... PLEASE RE-ENTER."
55 P$ = "DO YOU WISH PRINTOUT OF ABOVE DATA (Y/N)"
8Ø  DEF  FN C9(C9) = (1.57Ø7964 -  ATN (C9 /  SQR (1 - C9 * C9))) * R9: REM
    ARCCOSINE
9Ø  DEF  FN S9(S9) =  ATN (S9 /  SQR (1 - S9 * S9)) * R9: REM  ARCSINE
95  REM  ***************************************************
1ØØ  GOSUB 9ØØØ
11Ø  PRINT "***************************************"
12Ø  PRINT "*                                     *"
13Ø  PRINT "*     UNIVERSITY  OF  STRATHCLYDE     *"
14Ø  PRINT "*  DEPARTMENT  OF  CIVIL  ENGINEERING *"
15Ø  PRINT "*                                     *"
16Ø  PRINT "*           SURVEYING  SECTION        *"
17Ø  PRINT "*    CIRCULAR  CURVE  SETTING  OUT    *"
18Ø  PRINT "***************************************"
19Ø  PRINT
2ØØ  PRINT "ARE ANGLES IN DECIMALS <1>, DD.MMSS <2> OR DO YOU WISH TO QUI
     T <3>."
21Ø  INPUT D9
22Ø  IF D9 < 1 OR D9 > 3 THEN 2ØØ
23Ø  ON D9 GOTO 24Ø,24Ø,95ØØ
24Ø R9 = 57.2957795:P1 = Ø
5ØØ  GOSUB 9ØØØ
51Ø  PRINT "UNIVERSITY OF STRATHCLYDE"
52Ø  PRINT "CIRCULAR CURVE PROGRAM"
53Ø  PRINT "***************************************"
54Ø  PRINT
55Ø  PRINT "ENTER LOCATION OF WORK, ETC. WHEN       REQUESTED, AND PRESS
     <RETURN>"
56Ø  PRINT "ENTER LOCATION OF CURVE :-"
57Ø  INPUT A$
58Ø  PRINT "ENTER OPERATOR'S NAME :-"
59Ø  INPUT B$
6ØØ  PRINT "ENTER DATE OF CALCULATION AS DD/MM/YY :-"
61Ø  INPUT C$
62Ø  PRINT "ENTER INSTRUMENT USED :-"
63Ø  INPUT K$
64Ø  PRINT
65Ø  PRINT "DO YOU WISH PRINTOUT OF LOCATION ETC.,  (Y/N) ";
66Ø  INPUT Q$
```

```
670  IF Q$ = "Y" THEN 6000
680  IF Q$ < > "N" THEN 650
690 :
700  GOSUB 9000
710  PRINT H$
720  PRINT : PRINT "****************************************"
730  PRINT : PRINT "ENTER DEFLECTION ANGLE ";: INPUT DF
740  IF DF < 0 OR DF > 180 THEN 730
750  PRINT : PRINT "ENTER RADIUS OF CURVE ";: INPUT R
760  IF R < 0 THEN 750
770  PRINT : PRINT "ENTER CHAINAGE OF 1ST TANGENT POINT"
780  INPUT F1
790  PRINT : PRINT "CALCULATING DESIGN CONSTANTS ...."
800  IF D9 = 1 THEN 820
805  REM  CONVERT DF IN DD.MMSS TO DECIMALS
810 D5 = DF: GOSUB 8000:DF = D5
820 TL =  INT ((R *  TAN (DF / (2 * R9))) * 1000 + .5) / 1000
830 LC =  INT ((2 * R *  SIN (DF / (2 * R9))) * 1000 + .5) / 1000
840 CL =  INT ((R * DF / R9) * 1000 + .5) / 1000
850 MO =  INT ((R * (1 -  COS (DF / (2 * R9)))) * 1000 + .5) / 1000
860 EM =  INT ((TL *  TAN (DF / (4 * R9))) * 1000 + .5) / 1000
870 F2 = F1 + CL
880 F3 = F1 + TL
890 :
900  GOSUB 9000: PRINT H$
910  PRINT : PRINT "SELECT SETTING OUT METHOD FROM MENU :-"
920  PRINT
930  PRINT "    <1> OFFSETS FROM TANGENT"
940  PRINT "    <2> OFFSETS FROM LONG CHORD"
950  PRINT "    <3> TANGENTIAL DEFLECTION ANGLES"
960  PRINT "    <4> QUIT"
970  INPUT P1
980  IF P1 < 1 OR P1 > 4 THEN 900
990  ON P1 GOTO 1000,2000,3000,9500
993  REM  ************************************************
995  REM  <1> LINEAR MEASUREMENTS FROM TANGENT
1000  GOSUB 9000:P1$ = "<1> OFFSETS FROM TANGENT"
1010  PRINT H$: PRINT P1$
1020  PRINT "****************************************"
1025  REM  KNOW DF,R,F1,F2,TL,LC,CL,MO,EM
1030 UM =  INT ((TL * R / (EM + R)) * 1000 + .5) / 1000
1040 VM =  INT ((EM * R / (EM + R)) * 1000 + .5) / 1000
1050  PRINT : PRINT "MAX. OFFSET AT MID-ORDINATE = ";UM
1060  PRINT : PRINT "SELECT OFFSET INTERVAL ";: INPUT W
1070 N =  INT (UM / W)
1080  IF D9 = 1 THEN 1100
1090 D = DF: GOSUB 8050:DF = D
1100  GOSUB 9000
1110  PRINT H$: PRINT P1$
1120  PRINT "****************************************"
1130  PRINT "PT.  CHAINAGE      TAN.DIST  TAN.OFFSET"
1140  PRINT "========================================"
1150  PRINT "TP1"; TAB( 6);F1; TAB( 20);"0"; TAB( 30);"0"
1160  FOR I = 0 TO N - 1
1170 U(I) = (I + 1) * W
1180 V(I) =  INT ((R -  SQR (R * R - U(I) * U(I))) * 1000 + .5) / 1000
1190 A9 =  SQR (U(I) * U(I) + V(I) * V(I)) / (2 * R)
1200 DF(I) = 2 *  FN S9(A9)
1210 CL(I) =  INT ((R * DF(I) / R9) * 1000 + .5) / 1000
```

```
1220  PRINT (I + 1); TAB( 6);(F1 + CL(I)); TAB( 20);U(I); TAB( 30);V(I)
1230  NEXT I
1240 :
1250  PRINT "MO"; TAB( 6); INT ((F1 + CL / 2) * 1000 + .5) / 1000; TAB( 2
     0);UM; TAB( 30);VM
1260  FOR J = N - 1 TO 0 STEP  - 1
1270  PRINT (2 * N - J); TAB( 6);F2 - CL(J); TAB( 20);U(J); TAB( 30);V(J)

1280  NEXT J
1290  PRINT "TP2"; TAB( 6);F2; TAB( 20);"0"; TAB( 30);"0"
1300  PRINT "****************************************"
1310  PRINT P$;: INPUT A$
1320  IF A$ = "Y" THEN 6100
1330  IF A$ <  > "N" THEN 1310
1340  GOTO 100
1993  REM  **************************************************
1995  REM  <2> LINEAR MEASUREMENT FROM LONG CHORD
2000  GOSUB 9000:P2$ = "<2> OFFSETS FOM LONG CHORD"
2010  PRINT H$: PRINT P2$
2020  PRINT "****************************************"
2025  REM  KNOW DF,R,F1,F2,F3,TL,LC,CL,MO,EM
2030  PRINT : PRINT "MAX. OFFSET AT MID-ORDINATE = ";LC / 2
2040  PRINT : PRINT "SELECT OFFSET INTERVAL ";: INPUT W
2050 N =  INT (LC / (2 * W))
2055  REM  CALCULATE CHAINAGE OF M = F4
2060 F4 =  INT ((F1 + CL / 2) * 1000 + .5) / 1000
2070  FOR I = 0 TO N - 1
2080 U(I) = (I + 1) * W
2090 V(I) =  INT (( SQR (R * R - U(I) * U(I)) - (R - MO)) * 1000 + .5) /
     1000
2100 A9 = U(I) / R
2110 DA(I) =  FN S9(A9)
2120 CL(I) =  INT ((R * DA(I) / R9) * 1000 + .5) / 1000
2130  NEXT I
2140 :
2180  IF D9 = 1 THEN 2200
2190 D = DF: GOSUB 8050:DF = D
2200  GOSUB 9000
2210  PRINT H$: PRINT P2$
2220  PRINT "****************************************"
2230  PRINT "PT.  CHAINAGE      CH.DIST   CH.OFFSET"
2240  PRINT "========================================"
2250  PRINT "TP1"; TAB( 6);F1; TAB( 20);LC / 2; TAB( 30);"0"
2260  FOR J = N - 1 TO 0 STEP  - 1
2270  PRINT (N - J); TAB( 6);(F4 - CL(J)); TAB( 20);U(J); TAB( 30);V(J)
2280  NEXT J
2290  PRINT "MO"; TAB( 6);F4; TAB( 20);"0"; TAB( 30);MO
2300  FOR K = 0 TO N - 1
2310  PRINT (N + 1 + K); TAB( 6);(F4 + CL(K)); TAB( 20);U(K); TAB( 30);V(
     K)
2320  NEXT K
2330  PRINT "TP2"; TAB( 6);F2; TAB( 20);LC / 2; TAB( 30);"0"
2340  PRINT "****************************************"
2350  PRINT : PRINT P$;: INPUT A$
2360  IF A$ = "Y" THEN 6400
2370  IF A$ <  > "N" THEN 2350
2380  GOTO 100
2993  REM  **************************************************
2995  REM  <3> ANGULAR MEASUREMENTS FROM TANGENT POINT
```

```
3000  GOSUB 9000:P3$ = "<3> TANGENTIAL DEFLECTION ANGLES"
3010  PRINT H$: PRINT P3$
3020  PRINT "****************************************"
3025  REM  KNOW DF,R,F1,F2,F3,TL,LC,CL,MO,EM
3030  PRINT : PRINT "TOTAL CURVE LENGTH - ";CL
3040  PRINT : PRINT "SELECT CHORD LENGTH C FROM :-"
3050  PRINT "  1/2500  ACCURACY, C < "; INT (R / 10)
3060  PRINT "  1/5000  ACCURACY, C < "; INT (R / 14)
3070  PRINT "  1/10000 ACCURACY, C < "; INT (R / 20)
3080  INPUT CH
3085  REM  CALCULATE CHAINAGE OF 1ST POINT ON CURVE
3090 CH(0) = ( INT (F1 / CH + .01) + 1) * CH
3095  REM  CALCULATE LENGTH OF 1ST CHORD
3100 C1 =  INT ((CH(0) - F1) * 1000 + .5) / 1000
3105  REM  CALCULATE CHAINAGE OF LAST POINT ON CURVE
3110 C9 = ( INT (F2 / CH)) * CH
3115  REM  CALCULATE LENGTH OF LAST CHORD
3120 C2 =  INT ((F2 - C9) * 1000 + .5) / 1000
3125  REM  CALCULATE NO.(N) OF CHORDS (CH)
3130 N =  INT ((CL - C1 - C2) / CH + .5)
3135  REM  CALCULATE TANGENTIAL ANGLES FOR CHORDS IN DECIMAL DEGREES
3140 A0 = 1718.8 * CH / (R * 60)
3150 A1 = 1718.8 * C1 / (R * 60):A(0) = A1
3160 A2 = 1718.8 * C2 / (R * 60)
3170  PRINT : PRINT "IS CURVE LEFT OR RIGHT HANDED (L/R)";: INPUT L$
3180  IF L$ = "L" THEN 3400
3190  IF L$ <  > "R" THEN 3170
3195  REM  SET OUT CURVE RIGHT HANDED FROM ZERO DEGREES
3200  GOSUB 9000
3210  PRINT H$: PRINT P3$
3220  PRINT "****************************************"
3230  PRINT "PT.  CHAINAGE      CHORD     ANGLE(DEC)"
3240  PRINT "========================================"
3250  PRINT "TP1"; TAB( 6);F1; TAB( 30);"0.000"
3260  PRINT "1"; TAB( 6);CH(0); TAB( 20);C1; TAB( 30); INT (A1 * 1000 + .
     5) / 1000
3270  FOR I = 1 TO N
3280 CH(I) = CH(0) + I * CH
3290 A(I) = A(0) + I * A0
3300  PRINT (I + 1); TAB( 6);CH(I); TAB( 20);CH; TAB( 30); INT (A(I) * 10
     00 + .5) / 1000
3310  NEXT I
3320  PRINT "TP2"; TAB( 6);F2; TAB( 20);C2; TAB( 30); INT ((A(N) + A2) *
     1000 + .5) / 1000
3330  PRINT "========================================"
3340  PRINT "FINAL ANGLE AT TP2 SHOULD BE "; INT ((DF / 2) * 1000 + .5) /
     1000
3350  PRINT "****************************************"
3360  PRINT : PRINT P$;: INPUT A$
3370  IF A$ = "Y" THEN 6600
3380  IF A$ <  > "N" THEN 3360
3390  GOTO 100
3395  REM  SET OUT CURVE LEFT HANDED FROM 360 DEGREES
3400  GOSUB 9000
3410  PRINT H$: PRINT P3$
3420  PRINT "****************************************"
3430  PRINT "PT.  CHAINAGE      CHORD     ANGLE(DEC)"
3440  PRINT "========================================"
3450  PRINT "TP1"; TAB( 6);F1; TAB( 30);"360.000"
```

```
3460  PRINT "1"; TAB( 6);CH(Ø); TAB( 2Ø);C1; TAB( 3Ø); INT ((36Ø - A1) *
     1ØØØ + .5) / 1ØØØ
3470  FOR I = 1 TO N
3480 CH(I) = CH(Ø) + I * CH
3490 A(I) = (36Ø - A(Ø)) - (I * AØ)
3500  PRINT (I + 1); TAB( 6);CH(I); TAB( 2Ø);CH; TAB( 3Ø); INT (A(I) * 1Ø
     ØØ + .5) / 1ØØØ
3510  NEXT I
3520  PRINT "TP2"; TAB( 6);F2; TAB( 2Ø);C2; TAB( 3Ø); INT ((A(N) - A2) *
     1ØØØ + .5) / 1ØØØ
3530  PRINT "========================================"
3540  PRINT "FINAL ANGLE AT TP2 SHOULD BE "; INT ((36Ø - DF / 2) * 1ØØØ +
     .5) / 1ØØØ
3550  PRINT "****************************************"
3560  PRINT : PRINT P$;: INPUT A$
3570  IF A$ = "Y" THEN 7ØØØ
3580  IF A$ < > "N" THEN 356Ø
3590  GOTO 1ØØ
5985  REM  ************************************************
5990  REM  PRINTOUT OF LOCATION, ETC
6000  PRINT D$;"PR#1": PRINT "LOCATION OF CURVE :- ";A$
6010  PRINT "***********************************************"
6020  PRINT "OPERATOR'S NAME :- ";B$
6030  PRINT "DATE OF CALCS.  :- ";C$
6040  PRINT "INSTRUMENT USED :- ";K$
6060  PRINT "***********************************************"
6070  PRINT : PRINT D$;"PR#Ø"
6080  GOTO 7ØØ
6093  REM  ************************************************
6095  REM  PRINTOUT FOR P1=1
6100  PRINT D$;"PR#1": POKE 1784 + 1,8Ø
6110  PRINT H$: PRINT P1$:P7 = Ø
6120  PRINT "****************************************"
6130  PRINT "RADIUS           = ";R
6140  PRINT "DEFLECTION ANGLE = ";DF
6150  PRINT "CHAINAGE I.P.    = ";F3
6160  PRINT "TANGENT LENGTH   = ";TL
6170  PRINT "CHAINAGE TP1     = ";F1
6180  PRINT "CURVE LENGTH     = ";CL
6190  PRINT "CHAINAGE TP2     = ";F2
6200  PRINT "LONG CHORD       = ";LC
6210  PRINT "MID-ORDINATE     = ";MO
6220  PRINT "EXTERNAL MO-DIST = ";EM
6230  PRINT : IF P7 = 1 THEN  RETURN
6240  PRINT "========================================"
6250  PRINT "PT.  CHAINAGE      TAN.DIST  TAN.OFFSET"
6260  PRINT "========================================"
6270  PRINT "TP1";: POKE 36,6: PRINT F1;: POKE 36,2Ø: PRINT "Ø";: POKE 36
     ,3Ø: PRINT "Ø"
6280  FOR I = Ø TO N - 1
6290  PRINT (I + 1);: POKE 36,6: PRINT (F1 + CL(I));: POKE 36,2Ø: PRINT U
     (I);: POKE 36,3Ø: PRINT V(I)
6300  NEXT I
6310  PRINT "MO";: POKE 36,6: PRINT  INT ((F1 + CL / 2) * 1ØØØ + .5) / 1Ø
     ØØ;: POKE 36,2Ø: PRINT UM;: POKE 36,3Ø: PRINT VM
6320  FOR J = N - 1 TO Ø STEP  - 1
6330  PRINT (2 * N - J);: POKE 36,6: PRINT (F2 - CL(J));: POKE 36,2Ø: PRINT
     U(J);: POKE 36,3Ø: PRINT V(J)
6340  NEXT J
```

```
6350  PRINT "TP2";: POKE 36,6: PRINT F2;: POKE 36,20: PRINT "Ø";: POKE 36
     3Ø: PRINT "Ø"
636Ø  PRINT "****************************************"
637Ø  PRINT
638Ø  PRINT D$;"PR#Ø"
639Ø  GOTO 1ØØ
6393  REM  **************************************************
6395  REM  PRINTOUT FOR P1=2
64ØØ  PRINT D$;"PR#1": POKE 1784 + 1,8Ø
641Ø  PRINT H$: PRINT P2$
642Ø P7 = 1: GOSUB 612Ø
643Ø  PRINT "=========================================="
644Ø  PRINT "PT.  CHAINAGE      CH.DIST   CH.OFFSET"
645Ø  PRINT "=========================================="
646Ø  PRINT "TP1";: POKE 36,6: PRINT F1;: POKE 36,2Ø: PRINT LC / 2;: POKE
     36,3Ø: PRINT "Ø"
647Ø  FOR J = N - 1 TO Ø STEP  - 1
648Ø  PRINT (N - J);: POKE 36,6: PRINT (F4 - CL(J));: POKE 36,2Ø: PRINT U
     (J);: POKE 36,3Ø: PRINT V(J)
649Ø  NEXT J
65ØØ  PRINT "MO";: POKE 36,6: PRINT F4;: POKE 36,2Ø: PRINT "Ø";: POKE 36,
     3Ø: PRINT MO
651Ø  FOR K = Ø TO N - 1
652Ø  PRINT (N + 1 + K);: POKE 36,6: PRINT (F4 + CL(K));: POKE 36,2Ø: PRINT
     U(K);: POKE 36,3Ø: PRINT V(K)
653Ø  NEXT K
654Ø  PRINT "TP2";: POKE 36,6: PRINT F2;: POKE 36,2Ø: PRINT LC / 2;: POKE
     36,3Ø: PRINT "Ø"
655Ø  PRINT "****************************************"
656Ø  PRINT
657Ø  PRINT D$;"PR#Ø"
658Ø  GOTO 1ØØ
6593  REM  **************************************************
6595  REM  <3> RIGHT HANDED CURVE PRINTOUT
66ØØ  PRINT D$;"PR#1": POKE 1784 + 1,8Ø
661Ø  PRINT H$: PRINT P3$:P7 = 1: IF D9 = 1 THEN 664Ø
662Ø D = DF / 2: GOSUB 8Ø5Ø:DX = D
663Ø D = DF: GOSUB 8Ø5Ø:DF = D
664Ø  GOSUB 612Ø
665Ø  PRINT "=========================================="
666Ø  PRINT "PT.  CHAINAGE       CHORD     ANGLE"
667Ø  PRINT "=========================================="
668Ø  PRINT "TP1";: POKE 36,6: PRINT F1;: POKE 36,3Ø: PRINT "Ø.ØØØ"
669Ø  IF D9 = 2 THEN 68ØØ
6695  REM  D9=1, ANGLES IN DECIMALS
67ØØ  PRINT "1";: POKE 36,6: PRINT CH(Ø);: POKE 36,2Ø: PRINT C1;: POKE 36
     ,3Ø: PRINT  INT (A1 * 1ØØØ + .5) / 1ØØØ
671Ø  FOR J = 1 TO N
672Ø  PRINT (J + 1);: POKE 36,6: PRINT CH(J);: POKE 36,2Ø: PRINT CH;: POKE
     36,3Ø: PRINT  INT (A(J) * 1ØØØ + .5) / 1ØØØ
673Ø  NEXT J
674Ø  PRINT "TP2";: POKE 36,6: PRINT F2;: POKE 36,2Ø: PRINT C2;: POKE 36,
     3Ø: PRINT  INT ((A(N) + A2) * 1ØØØ + .5) / 1ØØØ
675Ø  PRINT "=========================================="
676Ø  PRINT "FINAL ANGLE AT TP2 SHOULD BE  "; INT ((DF / 2) * 1ØØØ + .5) /
     1ØØØ
677Ø  PRINT "****************************************"
678Ø  GOTO 692Ø
679Ø :
```

```
6795  REM  D9=2, ANGLES REQUIRED IN DD.MMSS
6800 D = (A(N) + A2): GOSUB 8050:A(N + 1) = D
6810  FOR J = 0 TO N
6820 D = A(J): GOSUB 8050:A(J) = D
6830  NEXT J
6840  PRINT "1";: POKE 36,6: PRINT CH(0);: POKE 36,20: PRINT C1;: POKE 36
     ,30: PRINT A(0)
6850  FOR K = 1 TO N
6860  PRINT (K + 1);: POKE 36,6: PRINT CH(K);: POKE 36,20: PRINT CH;: POKE
     36,30: PRINT A(K)
6870  NEXT K
6880  PRINT "TP2";: POKE 36,6: PRINT F2;: POKE 36,20: PRINT C2;: POKE 36,
     30: PRINT A(N + 1)
6890  PRINT "=========================================="
6900  PRINT "FINAL ANGLE AT TP2 SHOULD BE  ";DX
6910  PRINT "******************************************"
6920  PRINT
6930  PRINT D$;"PR#0"
6940  GOTO 100
6993  REM  ***************************************************
6995  REM  <3> LEFT HANDED CURVE PRINTOUT
7000  PRINT D$;"PR#1": POKE 1784 + 1,80
7010  PRINT H$: PRINT P3$:P7 = 1: IF D9 = 1 THEN 7040
7020 D = (360 - DF / 2): GOSUB 8050:DX = D
7030 D = DF: GOSUB 8050:DF = D
7040  GOSUB 6120
7050  PRINT "=========================================="
7060  PRINT "PT.  CHAINAGE      CHORD     ANGLE"
7070  PRINT "=========================================="
7080  PRINT "TP1";: POKE 36,6: PRINT F1;: POKE 36,30: PRINT "360.0000"
7090  IF D9 = 2 THEN 7200
7095  REM  D9=1, ANGLES IN DECIMALS
7100  PRINT "1";: POKE 36,6: PRINT CH(0);: POKE 36,20: PRINT C1;: POKE 36
     ,30: PRINT  INT ((360 - A1) * 1000 + .5) / 1000
7110  FOR J = 1 TO N
7120  PRINT (J + 1);: POKE 36,6: PRINT CH(J);: POKE 36,20: PRINT CH;: POKE
     36,30: PRINT  INT (A(J) * 1000 + .5) / 1000
7130  NEXT J
7140  PRINT "TP2";: POKE 36,6: PRINT F2;: POKE 36,20: PRINT C2;: POKE 36,
     30: PRINT  INT ((A(N) - A2) * 1000 + .5) / 1000
7150  PRINT "=========================================="
7160  PRINT "FINAL ANGLE AT TP2 SHOULD BE  "; INT ((360 - DF / 2) * 1000 +
     .5) / 1000
7170  PRINT "******************************************"
7180  GOTO 7320
7190 :
7195  REM  D9=2, ANGLES REQUIRED IN DD.MSS
7200 D = (A(N) - A2): GOSUB 8050:A(N + 1) = D
7205 A(0) = 360 - A1
7210  FOR J = 0 TO N
7220 D = A(J): GOSUB 8050:A(J) = D
7230  NEXT J
7240  PRINT "1";: POKE 36,6: PRINT CH(0);: POKE 36,20: PRINT C1;: POKE 36
     ,30: PRINT A(0)
7250  FOR K = 1 TO N
7260  PRINT (K + 1);: POKE 36,6: PRINT CH(K);: POKE 36,20: PRINT CH;: POKE
     36,30: PRINT A(K)
7270  NEXT K
```

```
7280  PRINT "TP2";: POKE 36,6: PRINT F2;: POKE 36,20: PRINT C2;: POKE 36,
     30: PRINT A(N + 1)
7290  PRINT "========================================"
7300  PRINT "FINAL ANGLE AT TP2 SHOULD BE  ";DX
7310  PRINT "****************************************"
7320  PRINT
7330  PRINT D$;"PR#Ø"
7340  GOTO 1ØØ
7993  REM  **************************************************
7995  REM  ROUTINE DD.MMSS TO DECIMALS
8ØØØ DØ =  INT (D5)
8ØØ5  REM  FIND NO. OF MINUTES,MØ
8Ø1Ø M1 = (D5 - DØ) * 1ØØ:MØ =  INT (M1 + .1)
8Ø15  REM  FIND NO. OF SECONDS,SØ
8Ø2Ø SØ =  INT (M1 * 1ØØ - MØ * 1ØØ + .5)
8Ø25  REM  COLLECT DEGS,MINS,SECS
8Ø3Ø D5 = DØ + MØ / 6Ø + SØ / 36ØØ
8Ø4Ø  RETURN
8Ø43  REM  **************************************************
8Ø45  REM  ROUTINE DECIMAL TO DD.MMSS
8Ø5Ø D1 =  INT (D)
8Ø55  REM  FIND TOTAL NO. OF SECONDS
8Ø6Ø D2 = (D - D1) * 36ØØ
8Ø65  REM  FIND NO. OF MINUTES
8Ø7Ø M =  INT (D2 / 6Ø)
8Ø75  REM  FIND NO. OF SECONDS
8Ø8Ø S = D2 - 6Ø * M + .5
8Ø9Ø  IF S < 6Ø THEN 813Ø
81ØØ M = M + 1:S = Ø
811Ø  IF M < 6Ø THEN 813Ø
812Ø D1 = D1 + 1:M = Ø
813Ø D = D1 * 1ØØØØ + M * 1ØØ + S
814Ø D =  INT (D) / 1ØØØØ: RETURN
8985  REM  ******************************************
8990  REM  GOSUB ROUTINE TO CLEAR SCREEN
9ØØØ  HOME
9Ø1Ø  RETURN
9Ø93  REM  **************************************************
9Ø95  REM  GOSUB ROUTINE FOR ERRORS
91ØØ  PRINT
911Ø  PRINT EØ$
912Ø  PRINT
913Ø  RETURN
9193  REM  **************************************************
9195  REM  ERROR TRACE ROUTINE
92ØØ  PRINT  CHR$ (7): REM  CTRL-B (BELL)
921Ø E =  PEEK (222): REM  GET ERROR NO.
922Ø  IF C = 1 THEN 926Ø
923Ø  INVERSE
924Ø  PRINT "ERROR NO. ";E;" FOUND"
925Ø C = 1: TRACE : RESUME
926Ø  PRINT "ERROR ON SECOND LINE NO. "
927Ø  NORMAL
928Ø  NOTRACE
929Ø  STOP
9493  REM  **************************************************
9495  REM  END OF PROGRAM
95ØØ  GOSUB 9ØØØ
```

```
9510  PRINT
9520  PRINT "END CURVE SETTING OUT PROGRAM"
9530  PRINT
9540  PRINT "******************************"
9550  REM  PROGRAM PREPARED BY
9560  REM  DR. P.H. MILNE
9570  REM  DEPT. OF CIVIL ENGINEERING
9580  REM  UNIVERSITY OF STRATHCLYDE
9590  REM  GLASGOW G4 0NG
9600  REM  SCOTLAND
9610  REM  ***************************
9620  END
```

5.3.5 Circular curve – setting out – computer printout

```
LOCATION OF CURVE :- STRATHCLYDE UNIVERSITY - STEPPS
**************************************************
OPERATOR'S NAME :- JOHN SMITH
DATE OF CALCS.  :- 18/06/83
INSTRUMENT USED :- ZEISS ZENA T20A
**************************************************

CIRCULAR CURVE SETTING OUT PROGRAM
<1> OFFSETS FROM TANGENT
***************************************
RADIUS            = 300
DEFLECTION ANGLE = 140.302
CHAINAGE I.P.     = 1835.697
TANGENT LENGTH    = 835.697
CHAINAGE TP1      = 1000
CURVE LENGTH      = 735.685
CHAINAGE TP2      = 1735.685
LONG CHORD        = 564.715
MID-ORDINATE      = 198.639
EXTERNAL MO-DIST = 587.913
```

PT.	CHAINAGE	TAN.DIST	TAN.OFFSET
TP1	1000	0	0
1	1030.05	30	1.504
2	1060.407	60	6.061
3	1091.408	90	13.818
4	1123.455	120	25.045
5	1157.08	150	40.192
6	1193.05	180	60
7	1232.619	210	85.757
8	1278.189	240	120
9	1335.931	270	169.233
MO	1367.843	282.358	198.639
10	1399.754	270	169.233
11	1457.496	240	120
12	1503.066	210	85.757
13	1542.635	180	60
14	1578.605	150	40.192
15	1612.23	120	25.045
16	1644.277	90	13.818

17	1675.278	60	6.061
18	1705.635	30	1.504
TP2	1735.685	0	0

****

LOCATION OF CURVE :- STRATHCLYDE UNIVERSITY - STEPPS
****
OPERATOR'S NAME :- JOHN SMITH
DATE OF CALCS. :- 19/06/83
INSTRUMENT USED :- ZEISS ZENA T20A
****

CIRCULAR CURVE SETTING OUT PROGRAM
<2> OFFSETS FOM LONG CHORD
****
RADIUS = 300
DEFLECTION ANGLE = 140.302
CHAINAGE I.P. = 1835.697
TANGENT LENGTH = 835.697
CHAINAGE TP1 = 1000
CURVE LENGTH = 735.685
CHAINAGE TP2 = 1735.685
LONG CHORD = 564.715
MID-ORDINATE = 198.639
EXTERNAL MO-DIST = 587.913

PT.	CHAINAGE	CH.DIST	CH.OFFSET
TP1	1000	282.3575	0
1	1031.912	270	29.406
2	1089.654	240	78.639
3	1135.224	210	112.882
4	1174.793	180	138.639
5	1210.763	150	158.447
6	1244.388	120	173.594
7	1276.435	90	184.821
8	1307.436	60	192.578
9	1337.793	30	197.135
MO	1367.843	0	198.639
10	1397.893	30	197.135
11	1428.25	60	192.578
12	1459.251	90	184.821
13	1491.298	120	173.594
14	1524.923	150	158.447
15	1560.893	180	138.639
16	1600.462	210	112.882
17	1646.032	240	78.639
18	1703.774	270	29.406
TP2	1735.685	282.3575	0

****

LOCATION OF CURVE :- STRATHCLYDE UNIVERSITY - STEPPS
****
OPERATOR'S NAME :- JOHN SMITH
DATE OF CALCS. :- 20/06/83
INSTRUMENT USED :- ZEISS ZENA T20A
****

```
CIRCULAR CURVE SETTING OUT PROGRAM
<3> TANGENTIAL DEFLECTION ANGLES
****************************************
RADIUS            = 3ØØ
DEFLECTION ANGLE = 21.452
CHAINAGE I.P.     = 1Ø57.65
TANGENT LENGTH    = 57.65
CHAINAGE TP1      = 1ØØØ
CURVE LENGTH      = 113.912
CHAINAGE TP2      = 1113.912
LONG CHORD        = 113.229
MID-ORDINATE      = 5.39
EXTERNAL MO-DIST = 5.489
========================================
PT.  CHAINAGE        CHORD     ANGLE
========================================
TP1   1ØØØ                      Ø.ØØØ
1     1ØØ5           5          .2839
2     1Ø2Ø           15         1.5435
3     1Ø35           15         3.2Ø32
4     1Ø5Ø           15         4.4628
5     1Ø65           15         6.1224
6     1Ø8Ø           15         7.3821
7     1Ø95           15         9.Ø417
8     111Ø           15         1Ø.3Ø14
TP2   1113.912       3.912      1Ø.5238
========================================
FINAL ANGLE AT TP2 SHOULD BE  1Ø.524
****************************************

CIRCULAR CURVE SETTING OUT PROGRAM
<3> TANGENTIAL DEFLECTION ANGLES
****************************************
RADIUS            = 3ØØ
DEFLECTION ANGLE = 21.452
CHAINAGE I.P.     = 1Ø57.65
TANGENT LENGTH    = 57.65
CHAINAGE TP1      = 1ØØØ
CURVE LENGTH      = 113.912
CHAINAGE TP2      = 1113.912
LONG CHORD        = 113.229
MID-ORDINATE      = 5.39
EXTERNAL MO-DIST = 5.489
========================================
PT.  CHAINAGE       CHORD     ANGLE
========================================
TP1   1ØØØ                      36Ø.ØØØØ
1     1ØØ5           5          359.3121
2     1Ø2Ø           15         358.Ø525
3     1Ø35           15         356.3928
4     1Ø5Ø           15         355.1332
5     1Ø65           15         353.4736
6     1Ø8Ø           15         352.2139
7     1Ø95           15         35Ø.5543
8     111Ø           15         349.2946
TP2   1113.912       3.912      349.Ø722
========================================
FINAL ANGLE AT TP2 SHOULD BE  349.Ø72
****************************************
```

5.4 REVERSE CIRCULAR CURVE DESIGN

In the previous circular curve design cases dealt with so far in this chapter, there has been a definite intersection point with a known or calculable chainage and deflection angle. On occasions, however, where it is desired to join two parallel or nearly parallel straights it is more convenient to use reverse circular curves, rather than in the latter case a very long circular curve. These curves consist of two consecutive curves of the same or different radii without any intervening straight section where their centres of curvature lie on opposite sides of the common tangent.

Due to the complexity of the design calculations, trial and error sketching methods were often used in the past to select suitable radii. With the advent of the microcomputer, several calculations can be carried out very quickly to ascertain the best radii to choose for a specific site terrain, especially where obstacles have to be avoided. Four separate sub-programs are included in the program <REVERSE CIRCULAR CURVES> with a choice from:

<1> Tangents parallel – radii equal
<2> Tangents parallel – radii unequal
<3> Tangents not parallel – radii equal
<4> Tangents not parallel – radii unequal.

Once the reverse circular curve radii, deflection angles and curve lengths have been calculated, one of the methods of setting out should be selected from Section 5.3 described earlier.

5.4.1 Reverse circular curves – subroutine index

	Line numbers	*Function*
(a)	10–90	Initialization and control
(b)	100–180	Screen header display
(c)	200–240	Selection of angle format
(d)	500–680	Entry of curve location, etc.
(e)	700–840	Menu selection for curve conditions
(f)	1000–1290	<1> Tangents parallel, radii equal solution
(g)	1500–1820	<2> Tangents parallel, radii unequal solution
(h)	2000–2170	Routine to calculate curve lengths for <1> and <2>
(i)	3000–3970	<3> Tangents not parallel, radii equal solution
(j)	4000–4270	<4> Tangents not parallel, radii unequal solution
(k)	5000–5450	Routine to calculate curve lengths for <3> and <4>
(l)	5500–5540	Program breakdown routine
(m)	6000–6080	Routine to print curve location, etc.
(n)	6100–6180	Printout of data and results for <1> and <2>
(o)	6200–6260	Printout of data for <3> and <4>
(p)	6300–6390	Printout of results for <3> and <4>
(q)	8000–8040	Routine to convert DD.MMSS to decimals
(r)	8050–8140	Routine to convert decimals to DD.MMSS
(s)	9000–9010	Routine to clear screen

	Line numbers	*Function*
(t)	9100–9130	Routine to display error message
(u)	9200–9290	Error trace routine
(v)	9500–9620	Termination of program

(a) Initialization and control

Line numbers 10–90

All numeric variables and required string variables are initialized and two 'DEF FN' functions used to define arccosine and arcsine as discussed previously in Section 2.2.8.

(b) Screen header display

Line numbers 100–180

This is the screen display used for civil engineering students at Strathclyde University. The operator can alter this as required to suit another organization.

(c) Selection of angle format

Line numbers 200–240

The operator has a choice in line 200 of entering angles in decimal or DD.MMSS format where the *D*9 flag is set to 1 for decimals and 2 for DD.MMSS in line 210.

(d) Entry of curve location, etc.

Line numbers 500–680

This routine is provided to keep a record of site location, operator, date, etc., with an opportunity to obtain a printout if required.

(e) Menu selection for curve conditions

Line numbers 700–840

The operator has a choice of four different design conditions:

Tangents	*Radii*
<1> PARALLEL	EQUAL
<2> PARALLEL	UNEQUAL
<3> NOT PARALLEL	EQUAL
<4> NOT PARALLEL	UNEQUAL

The menu selection is stored in the variable $P1$ and the program branches accordingly from line 840.

(f) <1> Tangents parallel radii equal solution

Line numbers 1000–1290

This is the simplest reverse circular curve design problem as shown in Fig. 5.9. The required data for the curve solution are the distance (H_1) between the tangent points A and C and the distance between the centre lines (L_1). If the two equal radii are R_1 and R_2 with centres of curvature O_1 and O_2 where AC and O_1O_2 meet at D, then D will be the mid-point of AC. Now from triangles O_1AP

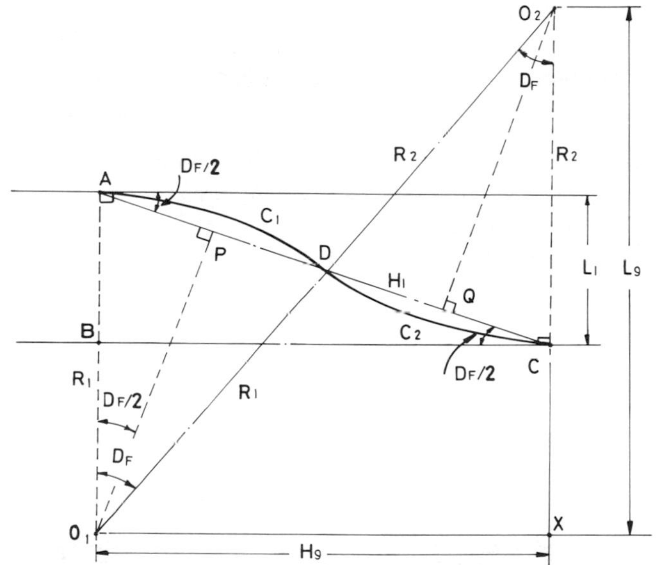

Fig. 5.9. Nomenclature for designing a reverse circular curve with parallel tangents and equal radii.

and O_2CQ (Fig. 5.9), if the angle AO_1P equals $D_F/2$, that is the same angle between the line joining the two tangents AC and the first straight, then:

$$AD = 2AP = 2R_1\sin(D_F/2) \tag{5.34}$$

and

$$DC = 2CQ = 2R_2\sin(D_F/2) \tag{5.35}$$

also

$$\begin{aligned} DC &= \mathrm{AC} - \mathrm{AD} \\ &= H_1 - 2R_1\sin(D_F/2) \end{aligned} \tag{5.36}$$

Equating 5.35 and 5.36 therefore gives:

$$R_1 + R_2 = H_1/[2\sin(D_F/2)] \tag{5.37}$$

where since the radii are equal ($R_1 = R_2$), the solution is:

$$R_1 = H_1/[4\sin(D_F/2)] \tag{5.38}$$

where in triangle ABC,

$$\sin(D_F/2) = L_1/H_1 \tag{5.39}$$

Once the radii and deflection angles (D_F) for each curve are known, a GOSUB routine is used at line 2000 to work out the exact curve lengths between the tangent points, before the screen display of results.

(g) <2> Tangents parallel, radii unequal solution

Line numbers 1500–1820

In this case, either the initial radius R_1 or the final radius R_2 must be known (Fig. 5.10). The solution is similar to (f) above, and uses the same Equations 5.34–5.37 with H_2 substituted for H_1. If R_1 is known, then from Equation 5.37, the value of R_2 will be given by:

$$R_2 = H_2/[2\sin(D_F/2)] - R_1 \tag{5.40}$$

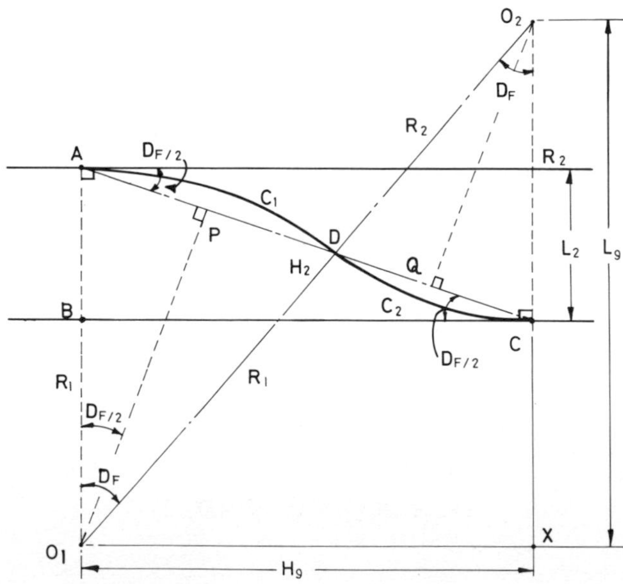

Fig. 5.10. Nomenclature for designing a reverse circular curve with parallel tangents and unequal radii.

and vice versa for R_1, where the deflection angle is determined as before from Equation 5.39. Once the radii and deflection angles (D_F) for each curve are known, a GOSUB routine is used at line 2000 to work out the exact curve lengths between the tangent points before the screen display of results.

(h) Routine to calculate curve lengths for <1> and <2>

Line numbers 2000–2170

The deflection angles for both cases <1> and <2> with parallel tangents are identical. The deflection angle (D_F) can be calculated from triangles AO_1P (Equation 5.39) or O_1XO_2 where either ($D_F/2$) or (D_F) can be solved directly. In this solution, triangle O_1XO_2 is used so that the distance O_1O_2 calculated using H_9 and L_9 in Figs 5.9 and 5.10 can be compared with the combined radii distance and any error stored in the variable E_9. The curve lengths C_1 and C_2 relating to the initial and final curve lengths are then computed using Equation 5.3.

(i) <3> Tangents not parallel, radii equal solution

Line numbers 3000–3970

If the tangents are not parallel, then their difference in direction can be specified in two ways from a knowledge of the location of the tangent points. Either the whole circle bearings of the straights, or the position of another point on the straight must be known. If the whole circle bearings are known, these are entered as W_1 and W_2, Fig. 5.11. This is followed by the co-ordinates of the tangent points $A(E_1, N_1)$ and $C(E_3, N_3)$, thus allowing the distance (H_1) between the tangent points AC to be computed. If other points on the straights are known, these are then entered where $B(E_2, N_2)$ lies on the first straight, and $D(E_4, N_4)$ lies on the second straight.

If the radii are equal then $R_1 = R_2 = R$, Fig. 5.11. To solve for R draw O_1S parallel to AC where O_2QS and O_1P are perpendicular to AC. Now consider triangle AO_1P where angle AO_1P equals V_1 which is the angle of inclination of the line joining the tangent points and the first straight. The value of V_1 is given by the difference in bearings of the line $AC(T_T)$ and the first straight (T_1), thus:

$$AP = R\sin V_1 \tag{5.41}$$

$$PO_1 = R\cos V_1 \tag{5.42}$$

Also in triangle CO_2Q where V_2 is determined in the same way as V_1:

$$CQ = R\sin V_2 \tag{5.43}$$

$$O_2Q = R\cos V_2 \tag{5.44}$$

In Fig. 5.11, consider line O_2S:

$$\begin{aligned} O_2S &= O_2Q + QS \\ &= O_2Q + PO_1 \end{aligned}$$

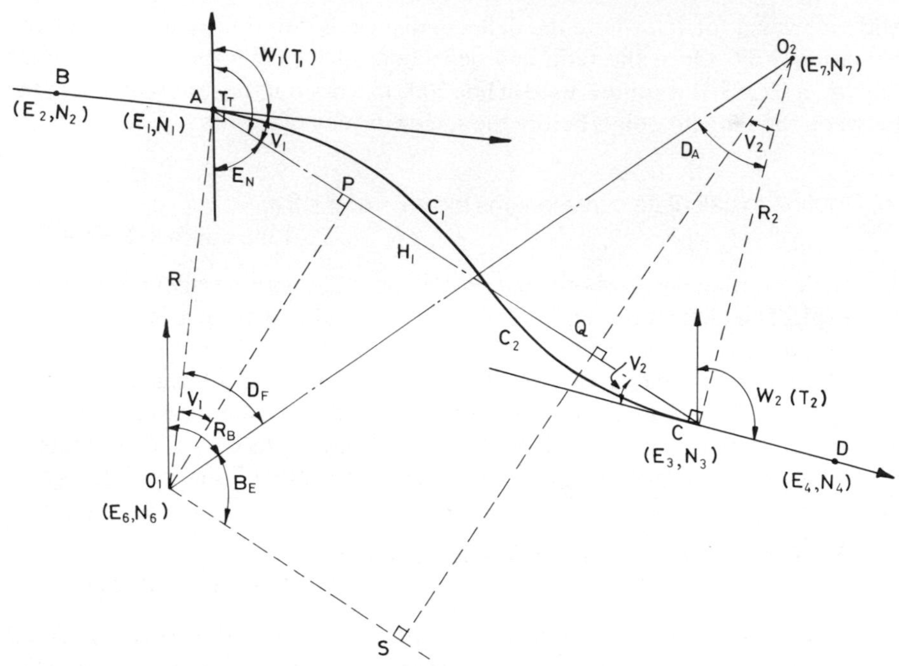

Fig. 5.11. Nomenclature for designing a reverse circular curve where the tangents are not parallel and the radii are equal.

where by substitution from Equations 5.42 and 5.44:

$$O_2S = R(\cos V_1 + \cos V_2) \tag{5.45}$$

Now in triangle O_1SO_2:

$$\begin{aligned}\sin B_E &= O_2S/O_1O_2\\ &= R(\cos V_1 + \cos V_2)/2R\\ &= (\cos V_1 + \cos V_2)/2 \end{aligned} \tag{5.46}$$

and

$$O_1S = O_1O_2 \cos B_E$$

$$= 2R\cos B_E \tag{5.47}$$

$$= AC - (AP + QC)$$

$$= H_1 - R(\sin V_1 + \sin V_2) \tag{5.48}$$

where by combining Equations 5.47 and 5.48, and solving the angle B_E from Equation 5.46 gives:

$$R = H_1/(2\cos B_E + \sin V_1 + \sin V_2) \tag{5.49}$$

These calculations are carried out in lines 3390–3540. On completion, the

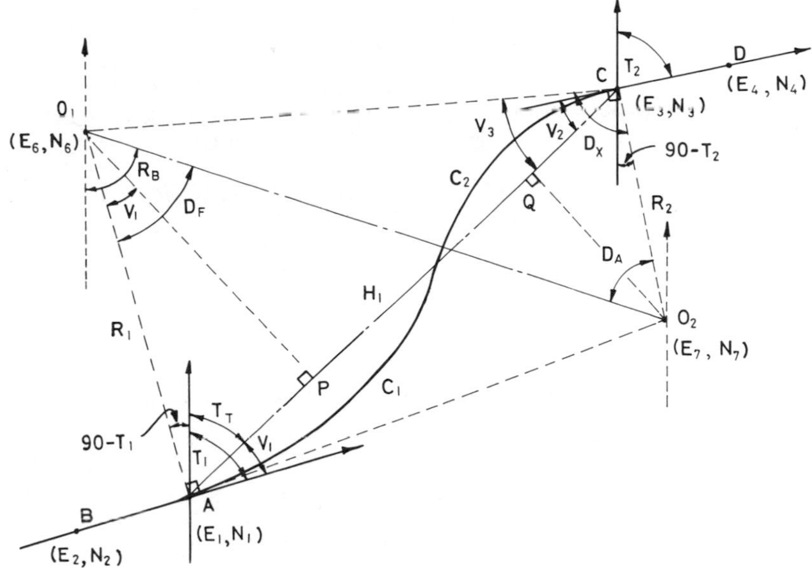

Fig. 5.12. Nomenclature for designing a reverse circular curve where the tangents are not parallel and the radii are unequal.

GOSUB routine at line 5000 is used to compute the deflection angles and curve lengths before the screen display of data and results with the option of printouts of both if desired.

(j) <4> Tangents not parallel, radii unequal solution

Line numbers 4000–4270

The directions of the two straights are found by the same method as described in the first paragraph of (i) above for <3> and uses the same subroutine from lines 3030–3510. On completion, the previous subroutine for <2> is used in lines 1570–1640 to determine if it is the initial or final radius which is known.

If the initial radius R_1 is known, then drop a perpendicular from O_1 to AC at P in Fig. 5.12. The angles V_1 and V_2 between the straights and AC are determined as in (i) above. Now join O_1C where in the triangle PO_1C the angle PCO_1 is denoted by V_3. Using Equations 5.41 and 5.42 where R is replaced by R_1, consider triangle PO_1C where:

$$PC = AC - AP$$

$$= H_1 - R_1 \sin V_1 \qquad (5.50)$$

and

$$\tan V_3 = O_1P/PC$$

$$= [R_1 \cos V_1/(H_1 - R_1 \sin V_1)] \qquad (5.51)$$

with

$$O_1C = O_1P/\sin V_3$$
$$= R_1\cos V_1/\sin V_3 \quad (5.52)$$

Now consider triangle O_1CO_2, where using the cosine rule:

$$O_1O_2^{\,2} = O_2C^2 + O_1A^2 - 2O_2C \times O_1C\cos D_X \quad (5.53)$$

The values of O_1O_2, O_2C, O_1C and D_X (Fig. 5.12) are now substituted in terms of R_1, R_2, V_1, V_2, V_3 from the previous equations into Equation 5.53 to solve for R_2.

$$(R_1 + R_2)^2 = R_2^{\,2} + [R_1^{\,2}\cos^2 V_1/\sin^2 V_3]$$
$$- [2R_2R_1\cos V_1\cos\{90° - (V_2 - V_3)\}/\sin V_3]$$
$$R_1^{\,2} + 2R_1R_2 + R_2^{\,2} = R_2^{\,2} + [\{R_1^{\,2}\cos^2 V_1$$
$$- 2R_1R_2\cos V_1\sin(V_2 - V_3)\sin V_3\}/\sin^2 V_3] \quad (5.54)$$

The R_2 terms are now collected together from Equation 5.54 to give:

$$R_2 = [R_1(\cos^2 V_1 - \sin^2 V_3)]/[2\sin V_3\{\sin V_3 + \cos V_1\sin(V_2 - V_3)\}] \quad (5.55)$$

In the computer program, having solved for angle V_3 from Equation 5.51 in line 4120, the numerator and denominator of Equation 5.55 are evaluated separately in variables RX and RY before being combined to solve for R_2 in line 4160. If, however, it is R_2 which is known, then a perpendicular is dropped from O_2 to AC at Q with a similar procedure to the above for R_2 to find R_1. A simpler computer procedure is to invert the angles V_1 and V_2 and radii R_1 and R_2 and consider the curve backwards as shown in line 4180 which allows the same Equation 5.55 to be used again. Once the other radius has been found the GOSUB routine at line 5000 is used to determine the curve length and deflection angles before returning to the screen display of data and results, with a printout if desired.

(k) Routine to calculate curve lengths for <3> and <4>

Line numbers 5000–5450

This is a much more complicated process than that used for the parallel tangents case. Therefore from a knowledge of the co-ordinates of the tangent points and the whole circle bearings of the tangents, together with the radii, it is possible to determine the centres of the two circles $O_1(E_6, N_6)$ and $O_2(E_7, N_7)$. Using this data, the reduced bearing (R_B) of the line joining the centres of the circles O_1O_2 can be found. In addition, the co-ordinates of O_1 and O_2 can be used to determine the distance O_1O_2 as a check against the computed total of the radii R_1 and R_2. Any error is stored in variable E_9 in line 5180.

The two deflection angles D_F for the initial curve and D_A for the final curve are then calculated knowing the bearings of the straights (T_1 and T_2) and of the line joining the centres of the circles (R_B), as shown in lines 5260–5270. The final calculation for the curve lengths uses Equation 5.3. If faulty data has been entered, the program will stop after two attempts at line 5210 and jump to line 5500.

(l) Program breakdown routine

Line numbers 5500–5540

The program will only jump to this routine if there is no solution to the entered data. If this occurs then either faulty bearings, co-ordinates or radii have been entered. An option is given to re-run the program with different data.

(m) Routine to print curve location etc.

Line numbers 6000–6080

This routine is provided for record purposes, and is based on Applesoft DOS and may require alteration for different microcomputers and printers.

(n) Printout of data and results for <1> and <2>

Line numbers 6100–6180

This printout is used for both cases of radii equal or unequal where the tangents are parallel. If it is an unequal radii solution, a note is added to say which radii were specified at the outset. The print routine uses the previous screen display from lines 1130–1240 to avoid repetition.

(o) Printout of data for <3> and <4>

Line numbers 6200–6260

This routine is available for record purposes to print the data entered at the keyboard during the tangents not parallel solutions, and uses the previous screen display from lines 3600–3750.

(p) Printout of results for <3> and <4>

Line numbers 6300–6390

The complete printout of the results for both radii equal and unequal cases for the tangents not parallel solutions uses the previous screen display from lines 3830–3920.

(q) Routine to convert DD.MMSS to decimals

Line numbers 8000–8040

This routine was discussed previously in Section 2.2.6.

(r) Routine to convert decimals to DD.MMSS

Line numbers 8050–8140

This routine was discussed previously in Section 2.2.6.

(s) Routine to clear screen

Line numbers 9000–9010

To simplify portability, the Applesoft 'HOME' command to clear the screen has been placed in a subroutine to allow an easy change to another microcomputer.

(t) Routine to display error message

Line numbers 9100–9130

If a wrong keyboard entry has been made in entering data, the error is trapped and this error message displayed and re-entry requested.

(u) Error trace routine

Line numbers 9200–9290

This error trace routine was discussed earlier in Section 2.2.9.

(v) Termination of program

Line numbers 9500–9620

When the '<QUIT>' selection is made in the program, an end-of-program message is displayed to advise the operator that computations are complete.

5.4.2 Reverse circular curves – numeric variables

BE	= Angle between circle centres and bearing of line joining tangent points
C	= Counter in error trace routine
*C*1, *C*2	= Curve lengths
*C*9	= Arccosine definition
CC	= Distance between curve centres
D–*D*5	= Degrees of angle
*D*9	= Angles in decimals <1>, DD.MMSS <2>
DA	= Deflection angle of second curve
DF	= Deflection angle of first curve
DT	= Difference between tangent bearings
DX	= Angle used in cosine rule
E	= Error number
*E*1–*E*8	= Eastings of points
*E*9	= Error in radius calculation
EN	= Gradient of line joining tangent points

Table—continued

*H*1–*H*3 = Distance between tangent points
*H*9 = Distance between curve centres along tangents
*L*1–*L*3 = Distance between parallel tangents
*L*9 = Distance between curve centres perpendicular to tangents
M–*M*1 = Minutes of angle
N = Number of times routine run
*N*1–*N*8 = Northings of points
*P*1 = Program selection (tangents/radii)
*P*2 = Program selection (bearing/points)
*P*7, *P*8 = Gosub routine flags
*R*1–*R*3 = Curve radii
*R*9 = 57.2957795
RB = Bearing of line joining curve centres
RX, *RY* = Precalculation variable storage
S, *S*0 = Seconds of angle
*S*9 = Arcsine definition
*T*1–*T*4 = Bearings of tangent lines
*T*8, *T*9 = Precalculation variable storage
TT = Bearing of line joining tangent points
*V*1, *V*3 = Angles between tangents
*V*9 = Precalculation variable storage
*W*1, *W*2 = Whole circle bearings of tangents
*Z*1–*Z*5 = Multiplying constant (±1) for curve length routine

5.4.3 Reverse circular curves – string variables

A$ Storage for location of curves
B$ Storage for operator's name
C$ Storage for date of calculations
*D$* = CHR$(4), CTRL-D
E0$ = "SORRY, DATA ERROR . . . PLEASE RE-ENTER"
F$ = "I" (Initial) or "F" (Final) radius
FF$*, *FI$ = "INITIAL" or "FINAL" depending on *F$*
H$ = "REVERSE CIRCULAR CURVE DESIGN"
K$ Storage for instrument used
P$ = "DO YOU WISH PRINTOUT OF ABOVE DATA (Y/N)"
PP$ = "<1> TANGENTS PARALLEL, RADII EQUAL", or "<2> TANGENTS PARALLEL, RADII UNEQUAL", depending on whether *P*1 = 1 or 2 respectively.
PN$ = "<3> TANGENTS NOT PARALLEL, RADII EQUAL", or "<4> TANGENTS NOT PARALLEL, RADII UNEQUAL", depending on whether *P*1 = 3 or 4 respectively.
Q$ = Question response (Y/N)
W$ = "ENTER WHOLE CIRCLE BEARING"
X$ = "ENTER X(E) COORD. OF "
Y$ = "ENTER Y(N) COORD. OF "

5.4.4 Reverse circular curves – BASIC program

```
10  REM  <REVERSE CIRCULAR CURVES> PROGRAM FOR APPLE II+
20 D$ =  CHR$ (4): ONERR  GOTO 92ØØ
3Ø C = Ø:E = Ø:N = Ø:S = Ø:P1 = Ø:P2 = Ø:P7 = Ø:P8 = Ø
4Ø H$ = "REVERSE CIRCULAR CURVE DESIGN"
5Ø EØ$ = "SORRY, DATA ERROR ... PLEASE RE-ENTER."
55 P$ = "DO YOU WISH PRINTOUT OF ABOVE DATA (Y/N)"
6Ø W$ = "ENTER WHOLE CIRCLE BEARING "
7Ø X$ = "ENTER X(E) COORD. OF ":Y$ = "ENTER Y(N) COORD. OF "
8Ø  DEF  FN C9(C9) = (1.57Ø7964 -  ATN (C9 /  SQR (1 - C9 * C9))) * R9: REM
     ARCCOSINE
9Ø  DEF  FN S9(S9) =  ATN (S9 /  SQR (1 - S9 * S9)) * R9: REM  ARCSINE
95  REM  ****************************************************
1ØØ  GOSUB 9ØØØ
11Ø  PRINT "****************************************"
12Ø  PRINT "*                                      *"
13Ø  PRINT "*      UNIVERSITY  OF  STRATHCLYDE      *"
14Ø  PRINT "*  DEPARTMENT  OF  CIVIL  ENGINEERING  *"
15Ø  PRINT "*                                      *"
16Ø  PRINT "*           SURVEYING  SECTION          *"
17Ø  PRINT "*   REVERSE  CIRCULAR  CURVE  DESIGN   *"
18Ø  PRINT "****************************************"
19Ø  PRINT
2ØØ  PRINT "ARE ANGLES IN DECIMALS <1>, DD.MMSS <2> OR DO YOU WISH TO QUI
     T <3>."
21Ø  INPUT D9
22Ø  IF D9 < 1 OR D9 > 3 THEN 2ØØ
23Ø  ON D9 GOTO 24Ø,24Ø,95ØØ
24Ø R9 = 57.2957795:P1 = Ø
5ØØ  GOSUB 9ØØØ
51Ø  PRINT "UNIVERSITY OF STRATHCLYDE"
52Ø  PRINT H$
53Ø  PRINT "****************************************"
54Ø  PRINT
55Ø  PRINT "ENTER LOCATION OF WORK, ETC. WHEN        REQUESTED, AND PRESS
     <RETURN>"
56Ø  PRINT "ENTER LOCATION OF CURVE :-"
57Ø  INPUT A$
58Ø  PRINT "ENTER OPERATOR'S NAME :-"
59Ø  INPUT B$
6ØØ  PRINT "ENTER DATE OF CALCULATION AS DD/MM/YY :"
61Ø  INPUT C$
62Ø  PRINT "ENTER INSTRUMENT USED :-"
63Ø  INPUT K$
64Ø  PRINT
65Ø  PRINT "DO YOU WISH PRINTOUT OF LOCATION ETC.,  (Y/N) ";
66Ø  INPUT Q$
67Ø  IF Q$ = "Y" THEN 6ØØØ
68Ø  IF Q$ < > "N" THEN 65Ø
69Ø :
7ØØ  GOSUB 9ØØØ
71Ø  PRINT H$
72Ø  PRINT : PRINT "****************************************"
73Ø  PRINT : PRINT "SELECT DESIGN CONDITIONS :-"
74Ø  PRINT
75Ø  PRINT "          TANGENTS      RADII   "
76Ø  PRINT "          ============  ======= "
77Ø  PRINT "     <1> PARALLEL      EQUAL    "
```

```
780  PRINT "    <2> PARALLEL      UNEQUAL "
790  PRINT "    <3> NOT PARALLEL  EQUAL   "
800  PRINT "    <4> NOT PARALLEL  UNEQUAL "
810  PRINT "    <5> QUIT                  "
820  INPUT P1
830  IF P1 < 1 OR P1 > 5 THEN 700
840  ON P1 GOTO 1000,1500,3000,4000,9500
993  REM  **************************************************
995  REM  <1> TANGENTS PARALLEL, RADII EQUAL
1000  GOSUB 9000:PP$ = "<1> TANGENTS PARALLEL, RADII EQUAL"
1010  PRINT H$: PRINT PP$
1020  PRINT "****************************************"
1030  PRINT : PRINT "ENTER DISTANCE BETWEEN TANGENT POINTS"
1040  INPUT H1
1050  PRINT : PRINT "ENTER DISTANCE BETWEEN CENTRE LINES"
1060  INPUT L1
1070 R1 = (H1 / ((4 * L1) / H1))
1080 R2 = R1
1090 H3 = H1:L3 = L1
1095  REM  CALCULATE CURVE LENGTHS
1100  GOSUB 2000
1105  REM  SCREEN DISPLAY OF RESULTS
1110  GOSUB 9000
1120  PRINT H$: PRINT PP$:P7 = 0
1130  PRINT "****************************************"
1140  PRINT "DISTANCE TANGENT POINTS = ";H3
1150  PRINT "DISTANCE CENTRE  LINES  = ";L3
1160  PRINT
1170  PRINT "INITIAL  CURVE   RADIUS = ";R1
1180  PRINT "FINAL    CURVE   RADIUS = ";R2
1190  PRINT "DEFLECTION       ANGLES = ";DF
1200  PRINT "INITIAL  CURVE   LENGTH = ";C1
1210  PRINT "FINAL    CURVE   LENGTH = ";C2
1220  IF E9 < .001 THEN 1240
1230  PRINT "RADIUS ERROR            = "; INT (E9 * 1000 + .5) / 1000
1240  PRINT "****************************************"
1250  IF P7 = 1 THEN  RETURN
1260  PRINT : PRINT P$;: INPUT Q$
1270  IF Q$ = "Y" THEN 6100
1280  IF Q$ < > "N" THEN 1260
1290  GOTO 100
1493  REM  **************************************************
1495  REM  <2> TANGENTS PARALLEL, RADII UNEQUAL
1500  GOSUB 9000:PP$ = "<2> TANGENTS PARALLEL, RADII UNEQUAL"
1510  PRINT H$: PRINT PP$
1520  PRINT "****************************************"
1530  PRINT : PRINT "ENTER DISTANCE BETWEEN TANGENT POINTS"
1540  INPUT H2
1550  PRINT : PRINT "ENTER DISTANCE BETWEEN CENTRE LINES"
1560  INPUT L2
1570  PRINT : PRINT "ENTER WHETHER INITIAL(I) OR FINAL(F)    RADIUS KNOWN
      (I/F) ";
1580  INPUT F$
1590  IF F$ < > "I" AND F$ < > "F" THEN 1570
1600  IF F$ = "I" THEN FI$ = "INITIAL"
1610  IF F$ = "I" THEN FF$ = "FINAL"
1620  IF F$ = "F" THEN FI$ = "FINAL"
1630  IF F$ = "F" THEN FF$ = "INITIAL"
1640  IF P1 = 4 THEN  RETURN
```

```
1650  IF F$ = "F" THEN 1690
1655  REM  F$="I", INITIAL RADIUS KNOWN
1660  PRINT : PRINT "ENTER THE ";FI$;" RADIUS ";: INPUT R1
1670 R2 = ABS ((H2 / ((2 * L2) / H2)) - R1)
1680  GOTO 1720
1685  REM  F$="F", FINAL RADIUS KNOWN
1690  PRINT : PRINT "ENTER THE ";FI$;" RADIUS ";
1700  INPUT R2
1710 R1 = ABS (R2 - (H2 / ((2 * L2) / H2)))
1720 H3 = H2:L3 = L2
1725  REM  CALCULATE CURVE LENGTHS
1730  GOSUB 2000
1735  REM  SCREEN DISPLAY OF RESULTS
1740  GOSUB 9000
1750  PRINT H$: PRINT PP$:P7 = 1
1760  GOSUB 1130
1770  PRINT "RADII DETERMINED BY ";FI$;" RADIUS"
1780  PRINT "****************************************"
1790  PRINT : PRINT P$;: INPUT Q$
1800  IF Q$ = "Y" THEN 6100
1810  IF Q$ < > "N" THEN 1790
1820  GOTO 100
1993  REM  **************************************************
1995  REM  ROUTINE TO CALCULATE CURVE LENGTHS FOR PARALLEL TANGENTS
2000 H9 = SQR (H3 * H3 - L3 * L3)
2010 L9 = (R1 + R2 - L3)
2020 DF = ATN (H9 / L9) * R9
2025  REM  ANY ERROR IN CALCULATION IS E9
2030 E9 = ABS ((R1 + R2) - SQR (H9 * H9 + L9 * L9))
2040  IF P1 = 2 THEN 2100
2045  REM  I.E. P1=1, RADII EQUAL
2050 C1 = R1 * DF / R9
2060 C1 = INT (C1 * 1000 + .5) / 1000
2070 C2 = C1
2080  GOTO 2120
2090  :
2100 C1 = INT ((R1 * DF / R9) * 1000 + .5) / 1000
2110 C2 = INT ((R2 * DF / R9) * 1000 + .5) / 1000
2120 R1 = INT (R1 * 1000 + .5) / 1000
2130 R2 = INT (R2 * 1000 + .5) / 1000
2140  IF D9 = 1 THEN 2160
2150 D = DF: GOSUB 8050:DF = D: GOTO 2170
2160 DF = INT (DF * 1000 + .5) / 1000
2170  RETURN
2993  REM  **************************************************
2995  REM  <3> TANGENTS NOT PARALLEL, RADII EQUAL
3000  GOSUB 9000:PN$ = "<3> TANGENTS NOT PARALLEL, RADII EQUAL"
3010  PRINT H$: PRINT PN$
3020  PRINT "****************************************"
3030  PRINT : PRINT "SELECT DATA INPUT :-"
3040  PRINT "   <1> TANGENT POINTS + LINE BEARINGS"
3050  PRINT "   <2> TANGENT POINTS + 2 OTHER POINTS"
3060  INPUT P2
3070  IF P2 < 1 OR P2 > 2 THEN 3030
3080  ON P2 GOTO 3100,3220
3090  :
3095  REM  TANGENT POINTS + LINE BEARINGS
3100  PRINT : PRINT W$;" 1ST TANGENT"
3110  INPUT W1
```

```
3120  PRINT : PRINT W$;" 2ND TANGENT"
3130  INPUT W2
3140 T1 = W1:T2 = W2: IF D9 = 1 THEN 3170
3145  REM  D9=2, CONVERT DD.MMSS TO DECIMALS
3150 D5 = T1: GOSUB 8000:T1 = D5
3160 D5 = T2: GOSUB 8000:T2 = D5
3170 DT =  ABS (T2 - T1)
3180  IF DT = 0 THEN 700
3190  IF DT = 180 THEN 700
3200  IF T1 > 180 THEN T1 = T1 - 180
3210  IF T2 > 180 THEN T2 = T2 - 180
3220  GOSUB 9000
3230  PRINT "ENTRY OF POINT CO-ORDINATES :-"
3240  PRINT "******************************************"
3250  PRINT : PRINT "1ST TANGENT LINE"
3260  PRINT "================="
3270  PRINT X$;"TANG. PT. ";: INPUT E1
3280  PRINT Y$;"TANG. PT. ";: INPUT N1
3290  IF P2 = 1 THEN 3320
3295  REM  P2=2, COLLECT OTHER POINT
3300  PRINT : PRINT X$;"OTHER PT. ";: INPUT E2
3310  PRINT Y$;"OTHER PT. ";: INPUT N2
3320  PRINT : PRINT "2ND TANGENT LINE"
3330  PRINT "================="
3340  PRINT X$;"TANG. PT. ";: INPUT E3
3350  PRINT Y$;"TANG. PT. ";: INPUT N3
3360  IF P2 = 1 THEN 3450
3365  REM  P2=2, COLLECT OTHER POINT
3370  PRINT : PRINT X$;"OTHER PT. ";: INPUT E4
3380  PRINT Y$;"OTHER PT. ";: INPUT N4
3390 T9 =  ABS ((E2 - E1) / (N2 - N1))
3400 T1 =  ATN (T9) * R9
3410  IF N2 > N1 THEN T1 = 180 - T1
3420 T8 =  ABS ((E4 - E3) / (N4 - N3))
3430 T2 =  ATN (T8) * R9
3440  IF N4 < N3 THEN T2 = 180 - T2
3445  REM  P2=1 AND 2 REJOIN HERE
3450 E5 = E3 - E1:N5 = N3 - N1
3460 H1 =  SQR (E5 * E5 + N5 * N5)
3470 EN =  ABS (E5 / N5)
3480 TT = 180 -  ATN (EN) * R9
3490  IF N3 > N1 THEN TT =  ATN (EN) * R9
3500 V1 =  ABS (TT - T1):V2 =  ABS (TT - T2)
3510  IF P1 = 4 THEN  RETURN
3520 V9 = ( COS (V1 / R9) +  COS (V2 / R9)) / 2
3530 BE =  FN S9(V9)
3540 R3 = H1 / (2 *  COS (BE / R9) +  SIN (V1 / R9) +  SIN (V2 / R9))
3550 R1 = R3:R2 = R3
3555  REM  GOTO ROUTINE TO CALCULATE CURVE LENGTHS
3560  GOSUB 5000
3565  REM  SCREEN DISPLAY OF RESULTS
3570  GOSUB 9000
3580  PRINT H$: PRINT PN$:P7 = 0:P8 = 0
3590  PRINT "******************************************"
3600  PRINT "1ST TANGENT LINE": PRINT "================="
3610  PRINT "X(E) COORD. TANG. PT. = ";E1
3620  PRINT "Y(N) COORD. TANG. PT. = ";N1
3630  IF P2 = 2 THEN 3650
3640  PRINT "LINE BEARING          = ";W1: GOTO 3670
```

```
3650  PRINT "X(E) COORD. OTHER PT. = ";E2
3660  PRINT "Y(N) COORD. OTHER PT. = ";N2
3670  PRINT "2ND TANGENT LINE"
3680  PRINT "================"
3690  PRINT "X(E) COORD. TANG. PT. = ";E3
3700  PRINT "Y(N) COORD. TANG. PT. = ";N3
3710  IF P2 = 2 THEN 3730
3720  PRINT "LINE BEARING           = ";W2: GOTO 3750
3730  PRINT "X(E) COORD. OTHER PT. = ";E4
3740  PRINT "Y(N) COORD. OTHER PT. = ";N4
3750  PRINT "========================================"
3760  IF P7 = 1 THEN  RETURN
3770  PRINT P$;: INPUT Q$
3780  IF Q$ = "Y" THEN 6200
3790  IF Q$ <  > "N" THEN 3770
3800  GOSUB 9000
3810  PRINT H$: PRINT PN$:P7 = 0
3820  PRINT "****************************************"
3830  PRINT "DISTANCE TANGENT POINTS = ";H1
3840  PRINT "INITIAL  CURVE   RADIUS = ";R1
3850  PRINT "FINAL    CURVE   RADIUS = ";R2
3860  PRINT "1ST DEFLECTION   ANGLE  = ";DF
3870  PRINT "2ND DEFLECTION   ANGLE  = ";DA
3880  PRINT "INITIAL  CURVE   LENGTH = ";C1
3890  PRINT "FINAL    CURVE   LENGTH = ";C2
3900  IF E9 < .001 THEN 3920
3910  PRINT "RADIUS ERROR            = "; INT (E9 * 1000 + .5) / 1000
3920  PRINT "****************************************"
3930  IF P7 = 1 THEN  RETURN
3940  PRINT : PRINT P$;: INPUT Q$
3950  IF Q$ = "Y" THEN 6300
3960  IF Q$ <  > "N" THEN 3940
3970  GOTO 100
3990 :
3993  REM  ************************************************
3995  REM  <4> TANGENTS NOT PARALLEL, RADII UNEQUAL
4000  GOSUB 9000:PN$ = "<4> TANGENTS NOT PARALLEL, RADII UNEQUAL"
4010  PRINT H$: PRINT PN$
4020  PRINT "****************************************"
4025  REM  USE ROUTINE IN <3> TO COLLECT DATA FROM LINES 3030-3510
4030  GOSUB 3030
4035  REM  USE ROUTINE IN <2> TO SPECIFY KNOWN RADIUS FROM LINES 1570-164
     0
4040  GOSUB 1570
4045  REM  NOW ENTER SPECIFIED RADIUS
4050  IF F$ = "F" THEN 4080
4055  REM  F$="I", INITIAL RADIUS KNOWN
4060  PRINT : PRINT "ENTER THE ";FI$;" RADIUS ";: INPUT R1
4070  GOTO 4100
4075  REM  F$="F", FINAL RADIUS KNOWN
4080  PRINT : PRINT "ENTER THE ";FI$;" RADIUS ";: INPUT R2
4090 :
4100  IF F$ = "F" THEN 4180
4105  REM  F$="I", INITIAL RADIUS KNOWN
4110 R3 = R1
4120 V3 =  ATN ((R3 *  COS (V1 / R9)) / (H1 - (R3 *  SIN (V1 / R9)))) * R
     9
4130 RX = ( COS (V1 / R9) ^ 2 -  SIN (V3 / R9) ^ 2)
```

```
4140 RY = 2 *  SIN (V3 / R9) * ( SIN (V3 / R9) +  COS (V1 / R9) *  SIN ((
     V2 - V3) / R9))
4150  IF F$ = "F" THEN 4190
4155  REM  F$="I", INITIAL RADIUS KNOWN
4160 R2 = R3 * (RX / RY)
4170  GOTO 4200
4175  REM  F$="F", FINAL RADIUS KNOWN
4180 R3 = R2:V9 = V1:V1 = V2:V2 = V9: GOTO 4120
4190 R1 = R3 * (RX / RY)
4195  REM  CALCULATE CURVE LENGTHS
4200  GOSUB 5000: GOSUB 9000
4205  REM  SCREEN DISPLAY OF RESULTS
4210  PRINT H$: PRINT PN$:P7 = 1:P8 = Ø
4220  PRINT "****************************************"
4225  REM  USE SCREEN DISPLAY FOR <3> IN LINES 3600-3760
4230  GOSUB 3600
4240  PRINT : PRINT P$;: INPUT Q$
4250  IF Q$ = "Y" THEN 6200
4260  IF Q$ < > "N" THEN 4240
4265  REM  USE DISPLAY OF RESULTS FOR <3> IN LINES 3800-3920
4270  GOTO 3800
4993  REM  ************************************************
4995  REM  ROUTINE TO CALCULATE CURVE LENGTHS FOR TANGENTS NOT PARALLEL
5000 Z1 = 1:Z2 = 1:Z3 = 1:Z4 = 1:Z5 = 1
5010 T3 = T1:T4 = T2:N = Ø
5020  IF T1 < 90 THEN 5060
5025  REM  I.E. T1>90
5030 Z1 =  - 1:Z2 =  - 1
5040 T1 = 180 - T1:T2 = 180 - T2
5050  GOTO 5100
5055  REM  I.E. T1<90
5060 Z2 =  - 1:Z3 =  - 1: GOTO 5100
5070 Z1 = 1:Z2 = 1:Z3 =  - 1:Z4 =  - 1
5080 T1 = 180 - T3:T2 = 180 - T4: GOTO 5100
5090 Z1 =  - 1:Z2 = 1:Z3 = 1:Z4 =  - 1:T1 = T3:T2 = T4
5095  REM  CALCULATE COORDS OF CIRCLE CENTRES
5100 E6 = E1 + Z1 * R1 *  COS (T1 / R9)
5110 N6 = N1 + Z2 * R1 *  SIN (T1 / R9)
5120 E7 = E3 + Z3 * R2 *  COS (T2 / R9)
5130 N7 = N3 + Z4 * R2 *  SIN (T2 / R9)
5140 E8 =  ABS (E7 - E6)
5150 N8 =  ABS (N7 - N6)
5160 RB =  ATN (E8 / N8) * R9
5170 CC =  SQR (E8 * E8 + N8 * N8)
5180 E9 =  ABS (R1 + R2 - CC)
5190  IF E9 < .ØØ1 THEN 5240
5200 N = N + 1
5210  IF N > 1 THEN 5500
5220  IF T3 < 90 THEN 5090
5230  IF T3 > 90 THEN 5070
5240  IF N = Ø THEN 5260
5250 Z5 =  - 1
5260 DF = RB + Z5 * (90 - T3)
5270 DA = RB + Z5 * (90 - T4)
5280 C1 =  INT ((R1 * DF / R9) * 1000 + .5) / 1000
5290 C2 =  INT ((R2 * DA / R9) * 1000 + .5) / 1000
5300 R1 =  INT (R1 * 1000 + .5) / 1000
5310 R2 =  INT (R2 * 1000 + .5) / 1000
```

```
5320  IF D9 = 1 THEN 5400
5325  REM  D9=2, ANGLES REQUIRED IN DD.MMSS
5330 D = DF: GOSUB 8050:DF = D
5340 D = DA: GOSUB 8050:DA = D
5350 D = T1: GOSUB 8050:T1 = D
5360 D = T2: GOSUB 8050:T2 = D
5370 H1 =  INT (H1 * 1000 + .5) / 1000
5380  RETURN
5390 :
5395  REM  D9=1, ANGLES IN DECIMALS
5400 DF =  INT (DF * 1000 + .5) / 1000
5410 DA =  INT (DA * 1000 + .5) / 1000
5420 T1 =  INT (T1 * 1000 + .5) / 1000
5430 T2 =  INT (T2 * 1000 + .5) / 1000
5440 H1 =  INT (H1 * 1000 + .5) / 1000
5450  RETURN
5493  REM  **************************************************
5495  REM  PROGRAM BREAKDOWN-NO SOLUTION DUE TO ERRONEOUS INPUT OF DATA
5500  PRINT : PRINT E0$
5510  PRINT : PRINT "DO YOU WISH TO RE-RUN (Y/N) ";: INPUT Q$
5520  IF Q$ = "Y" THEN 700
5530  IF Q$ < > "N" THEN 5510
5540  GOTO 9500
5985  REM  **************************************************
5990  REM  PRINTOUT OF LOCATION, ETC
6000  PRINT D$;"PR#1": PRINT "LOCATION OF CURVE :- ";A$
6010  PRINT "****************************************************"
6020  PRINT "OPERATOR'S NAME :- ";B$
6030  PRINT "DATE OF CALCS.  :- ";C$
6040  PRINT "INSTRUMENT USED :- ";K$
6060  PRINT "****************************************************"
6070  PRINT : PRINT D$;"PR#0"
6080  GOTO 700
6093  REM  **************************************************
6095  REM  PRINTOUT FOR <1> AND <2>
6100  PRINT D$;"PR#1"
6110  PRINT H$: PRINT PP$:P7 = 1
6120  GOSUB 1130
6130  IF P1 = 1 THEN 6160
6140  PRINT "RADII DETERMINED BY ";FI$;" RADIUS"
6150  PRINT "****************************************"
6160  PRINT
6170  PRINT D$;"PR#0"
6180  GOTO 100
6190 :
6193  REM  **************************************************
6195  REM  PRINTOUT OF DATA FOR <3> AND <4>
6200  PRINT D$;"PR#1"
6210  PRINT H$: PRINT PN$:P7 = 1:P8 = 1
6220  PRINT "****************************************"
6225  REM  REUSE SCREEN DISPLAY LINES 3600-3760
6230  GOSUB 3600
6240  PRINT
6250  PRINT D$;"PR#0"
6260  GOTO 3800
6293  REM  **************************************************
6295  REM  PRINTOUT OF RESULTS FOR <3> AND <4>
6300  PRINT D$;"PR#1":P7 = 1
6310  IF P8 = 1 THEN 6340
```

```
6320  PRINT H$: PRINT PN$
6330  PRINT "****************************************"
6335  REM  REUSE SCREEN DISPLAY LINES 3830-3930
6340  GOSUB 3830
6350  IF P1 = 3 THEN 6380
6360  PRINT "RADII DETERMINED BY ";FI$;" RADIUS"
6370  PRINT "****************************************"
6380  PRINT
6390  PRINT D$;"PR#Ø": GOTO 1ØØ
7993  REM  **************************************************
7995  REM  ROUTINE DD.MMSS TO DECIMALS
8ØØØ DØ =  INT (D5)
8ØØ5  REM  FIND NO. OF MINUTES,MØ
8Ø1Ø M1 = D5 * 1ØØ - DØ * 1ØØ:MØ =  INT (M1 + .1)
8Ø15  REM  FIND NO. OF SECONDS,SØ
8Ø2Ø SØ =  INT (M1 * 1ØØ - MØ * 1ØØ + .5)
8Ø25  REM  COLLECT DEGS,MINS,SECS
8Ø3Ø D5 = DØ + MØ / 6Ø + SØ / 36ØØ
8Ø4Ø  RETURN
8Ø43  REM  **************************************************
8Ø45  REM  ROUTINE DECIMAL TO DD.MMSS
8Ø5Ø D1 =  INT (D)
8Ø55  REM  FIND TOTAL NO. OF SECONDS
8Ø6Ø D2 = (D - D1) * 36ØØ
8Ø65  REM  FIND NO. OF MINUTES
8Ø7Ø M =  INT (D2 / 6Ø)
8Ø75  REM  FIND NO. OF SECONDS
8Ø8Ø S = D2 - 6Ø * M + .5
8Ø9Ø  IF S < 6Ø THEN 813Ø
81ØØ M = M + 1:S = Ø
811Ø  IF M < 6Ø THEN 813Ø
812Ø D1 = D1 + 1:M = Ø
813Ø D = D1 * 1ØØØØ + M * 1ØØ + S
814Ø D =  INT (D) / 1ØØØØ: RETURN
8985  REM  ****************************************
899Ø  REM  GOSUB ROUTINE TO CLEAR SCREEN
9ØØØ  HOME
9Ø1Ø  RETURN
9Ø93  REM  **************************************************
9Ø95  REM  GOSUB ROUTINE FOR ERRORS
91ØØ  PRINT
911Ø  PRINT EØ$
912Ø  PRINT
913Ø  RETURN
9193  REM  **************************************************
9195  REM  ERROR TRACE ROUTINE
92ØØ  PRINT  CHR$ (7): REM  CTRL-B (BELL)
921Ø E =  PEEK (222): REM  GET ERROR NO.
922Ø  IF C = 1 THEN 926Ø
923Ø  INVERSE
924Ø  PRINT "ERROR NO. ";E;" FOUND"
925Ø C = 1: TRACE : RESUME
926Ø  PRINT "ERROR ON SECOND LINE NO. "
927Ø  NORMAL
928Ø  NOTRACE
929Ø  STOP
9493  REM  **************************************************
9495  REM  END OF PROGRAM
95ØØ  GOSUB 9ØØØ
```

```
951Ø  PRINT
952Ø  PRINT "END REVERSE CIRCULAR CURVE DESIGN"
953Ø  PRINT
954Ø  PRINT "**********************************"
955Ø  REM  PROGRAM PREPARED BY
956Ø  REM  DR. P.H. MILNE
957Ø  REM  DEPT. OF CIVIL ENGINEERING
958Ø  REM  UNIVERSITY OF STRATHCLYDE
959Ø  REM  GLASGOW G4 ONG
96ØØ  REM  SCOTLAND
961Ø  REM  ***************************
962Ø  END
```

5.4.5 Reverse circular curves – computer printout

```
LOCATION OF CURVE :- GLENMORE
**************************************************
OPERATOR'S NAME :- JOHN MACMILLAN
DATE OF CALCS.  :- 26/Ø7/83
INSTRUMENT USED :- APPLE II+
**************************************************

REVERSE CIRCULAR CURVE DESIGN
<1> TANGENTS PARALLEL, RADII EQUAL
****************************************
DISTANCE TANGENT POINTS = 12Ø
DISTANCE CENTRE  LINES  = 3Ø

INITIAL  CURVE   RADIUS = 12Ø
FINAL    CURVE   RADIUS = 12Ø
DEFLECTION       ANGLES = 28.5718
INITIAL  CURVE   LENGTH = 6Ø.643
FINAL    CURVE   LENGTH = 6Ø.643
****************************************

REVERSE CIRCULAR CURVE DESIGN
<2> TANGENTS PARALLEL, RADII UNEQUAL
****************************************
DISTANCE TANGENT POINTS = 12Ø
DISTANCE CENTRE  LINES  = 3Ø

INITIAL  CURVE   RADIUS = 1ØØ
FINAL    CURVE   RADIUS = 14Ø
DEFLECTION       ANGLES = 28.5718
INITIAL  CURVE   LENGTH = 5Ø.536
FINAL    CURVE   LENGTH = 7Ø.75
****************************************
RADII DETERMINED BY INITIAL RADIUS
****************************************

REVERSE CIRCULAR CURVE DESIGN
<2> TANGENTS PARALLEL, RADII UNEQUAL
****************************************
DISTANCE TANGENT POINTS = 12Ø
DISTANCE CENTRE  LINES  = 3Ø

INITIAL  CURVE   RADIUS = 1ØØ
FINAL    CURVE   RADIUS = 14Ø
```

```
DEFLECTION        ANGLES = 28.5718
INITIAL  CURVE    LENGTH = 50.536
FINAL    CURVE    LENGTH = 70.75
****************************************
RADII DETERMINED BY FINAL RADIUS
****************************************


REVERSE CIRCULAR CURVE DESIGN
<3> TANGENTS NOT PARALLEL, RADII EQUAL
****************************************
1ST TANGENT LINE
================
X(E) COORD. TANG. PT. = 1125.66
Y(N) COORD. TANG. PT. = 1491.28
LINE BEARING          = 96.45
2ND TANGENT LINE
================
X(E) COORD. TANG. PT. = 2401.37
Y(N) COORD. TANG. PT. = 650.84
LINE BEARING          = 105.3
========================================

DISTANCE TANGENT POINTS = 1527.67
INITIAL  CURVE    RADIUS = 1001.379
FINAL    CURVE    RADIUS = 1001.379
1ST DEFLECTION    ANGLE  = 49.1701
2ND DEFLECTION    ANGLE  = 40.3201
INITIAL  CURVE    LENGTH = 861.346
FINAL    CURVE    LENGTH = 708.419
****************************************


REVERSE CIRCULAR CURVE DESIGN
<4> TANGENTS NOT PARALLEL, RADII UNEQUAL
****************************************
1ST TANGENT LINE
================
X(E) COORD. TANG. PT. = 248.86
Y(N) COORD. TANG. PT. = 422.62
X(E) COORD. OTHER PT. = 103.61
Y(N) COORD. OTHER PT. = 204.82
2ND TANGENT LINE
================
X(E) COORD. TANG. PT. = 866.34
Y(N) COORD. TANG. PT. = 406.61
X(E) COORD. OTHER PT. = 900
Y(N) COORD. OTHER PT. = 544.31
========================================

DISTANCE TANGENT POINTS = 617.688
INITIAL  CURVE    RADIUS = 200
FINAL    CURVE    RADIUS = 140.096
1ST DEFLECTION    ANGLE  = 124.0906
2ND DEFLECTION    ANGLE  = 144.0652
INITIAL  CURVE    LENGTH = 433.371
FINAL    CURVE    LENGTH = 352.379
****************************************
RADII DETERMINED BY INITIAL RADIUS
****************************************
```

```
REVERSE CIRCULAR CURVE DESIGN
<4> TANGENTS NOT PARALLEL, RADII UNEQUAL
****************************************
DISTANCE TANGENT POINTS = 617.688
INITIAL  CURVE   RADIUS = 200
FINAL    CURVE   RADIUS = 140.096
1ST DEFLECTION   ANGLE  = 124.0906
2ND DEFLECTION   ANGLE  = 144.0652
INITIAL  CURVE   LENGTH = 433.371
FINAL    CURVE   LENGTH = 352.38
****************************************
RADII DETERMINED BY FINAL RADIUS
****************************************
```

5.5 TRANSITION CURVE DESIGN

Mention was made in Section 5.1 that circular curves were normally selected for built-up areas whereas transition curves were essential for high-speed motorways. The reason why circular curves cannot be used at high speed is that a vehicle, of say mass M(kg), travelling at a speed v(m s^{-1}), round a circular arc of radius R(m) is subjected to a radial acceleration (v^2/R), which creates a centrifugal force P acting on the vehicle. If g is the acceleration due to gravity, then the centrifugal force P will be given by:

$$P = Mv^2/gR \tag{5.56}$$

Now since there is no centrifugal force along a straight road, a vehicle encountering a circular curve of radius R would experience an instantaneous sideways force or lateral shock as the tangent point was passed. As v increases in Equation 5.56 so does the centrifugal force, which for a high-speed road would cause skidding or sideslipping. To prevent such an occurrence, a transition curve, either a clothoid or cubic parabola, is inserted between the straight (radius ∞) and the circular curve of radius R. This circular curve will have a minimum radius depending upon the design speed and the width of the road. Most countries publish design criteria for road design, for example, in the United Kingdom the information is available from the Department of the Environment.

By introducing a transition curve at either end of a circular curve, the centrifugal force will gradually increase up to the maximum when the circular curve is reached, and then decrease again until back on the straight. This considerably reduces any tendency of the vehicle to skid and also minimizes any discomfort experienced by the passengers. Sometimes, there is insufficient space to insert a circular curve to meet the specified design conditions, and the two transition curves therefore meet at the centre of the new curve.

To avoid any shock forces on the vehicle in negotiating the curve, the rate of turning the steering wheel should be constant. Thus the angle turned through (θ) with respect to the distance (S) from the tangent point will depend on the curvature at that point as shown in Fig. 5.13, therefore:

$$1/R = \mathrm{d}\theta/\mathrm{d}S \tag{5.57}$$

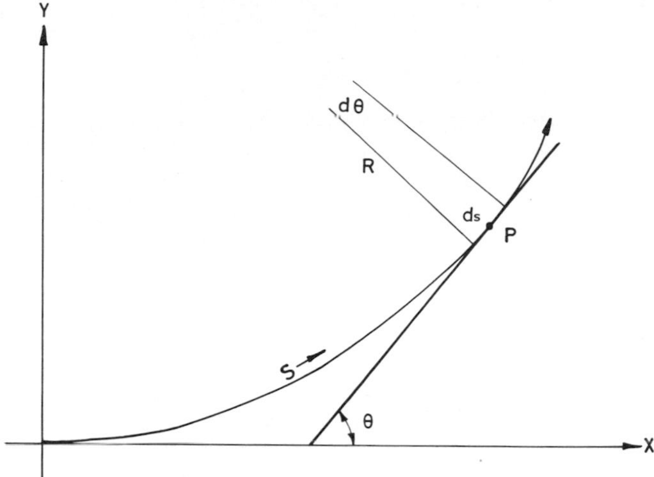

Fig. 5.13. Nomenclature for designing a transition curve to reduce centrifugal forces.

If the curvature $(1/R)$ is proportional to time, and the velocity is constant, then the distance along the curve (S) will also be proportional to time, thus:

$$1/R \propto S \tag{5.58}$$

and

$$RS = K^2 \tag{5.59}$$

where K is a multiplying constant.

From Equation 5.57, θ can be found:

$$\begin{aligned} d\theta/dS &= 1/R = S/K^2 \\ \theta &= \textstyle\int_0^S (S/K^2)dS \\ &= S^2/2K^2 \end{aligned} \tag{5.60}$$

Now consider a small element (ds) at point P where the tangent makes angle θ with the x-axis, and the line joining the tangent point to P makes angle ϕ with the x-axis, Fig. 5.14. The small element ds is expanded as dx and dy parallel to the axes giving:

$$dx = ds \cos\theta \tag{5.61}$$

and

$$dy = ds \sin\theta \tag{5.62}$$

If the point P lies at distance S from the tangent then θ can be replaced by $(S^2/2K^2)$ from Equation 5.60, and Equations 5.61 and 5.62 can then be integrated

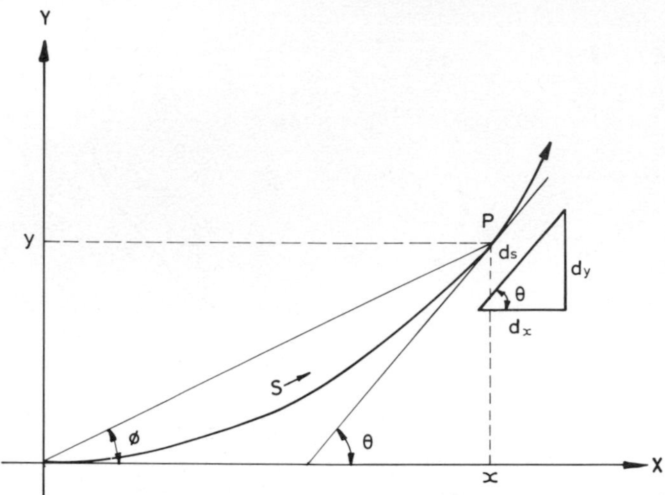

Fig. 5.14. Nomenclature for designing a transition curve using polar co-ordinates.

with respect to S, where the sin and cos expressions are written in terms of S:

$$dx = ds\cos\theta$$

$$x = \int_0^S\{1-(\theta)^2/2! + (\theta)^4/4! + \ldots\}ds$$

$$x = \int_0^S\{1-(S^2/2K^2)^2/2 + (S^2/2K^2)^4/24 + \ldots\}ds$$

$$x = [S - S^5/40K^4 + S^9/3456K^8 + \ldots] \qquad (5.63)$$

and

$$dy = ds\sin\theta$$

$$y = \int_0^S\{\theta-(\theta)^3/3! + (\theta)^5! + \ldots\}ds$$

$$y = \int_0^S\{(S^2/2K^2)-(S^2/2K^2)^3/3! + (S^2/2K^2)^5/5! + \ldots\}ds$$

$$y = [S^3/6K^2 - S^7/336K^6 + S^{11}/42240K^{10} + \ldots] \qquad (5.64)$$

Where polar co-ordinates are required to specify the point P rather than rectangular co-ordinates as defined above in Equations 5.63 and 5.64, then the angle ϕ can be found from:

$$\tan\phi = y/x \qquad (5.65)$$

Expressing the angle ϕ in terms of s and by rewriting ϕ in terms of $\tan\phi$ gives:

$$\phi = \tan\phi - (\tan\phi)^3/3 + (\tan\phi)^5/5$$

$$\phi = s^2/6 - s^6/2835 \text{ rad}$$

$$\phi = 573s^2 - 1.213s^6 \text{min} \qquad (5.66)$$

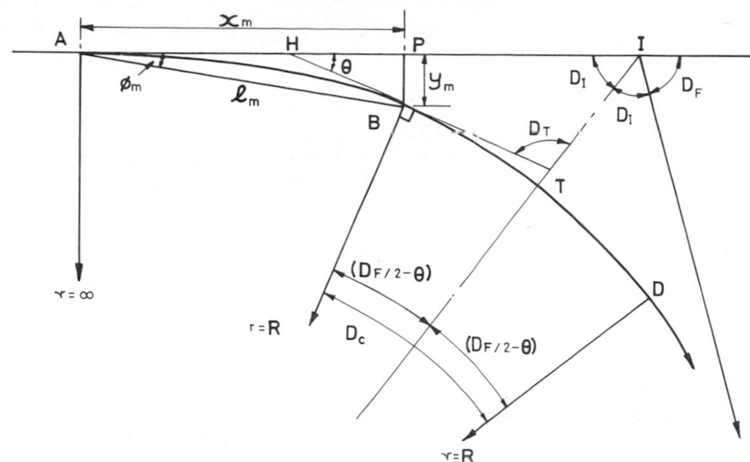

Fig. 5.15. Nomenclature for designing a transition curve where $\theta < D_F/2$.

and if l is the distance from the tangent point to the point P then,

$$l = (x^2 + y^2)^{1/2} \tag{5.67}$$

Thus the setting-out points of a transition curve can be determined either in rectangular co-ordinates from Equations 5.63 and 5.64 or by polar co-ordinates from Equations 5.66 and 5.67, as shown in Fig. 5.15.

If the deviation angle ϕ between the tangent and the curve is less than 12° there is no difference between the clothoid and the cubic parabola. Thus for manual calculations, the cubic parabola is often used up to 12° and the clothoid thereafter.

To set out the curve on the ground, the value of K, the multiplying constant has to be evaluated from the design criteria. Ideally transition curves are defined for:

(i) A suitable speed (V),
(ii) A permissible centripetal ratio (F), and
(iii) A permissible rate of change of radial acceleration (C).

If each of these is specified then K can be determined. If V is constant, and $1/R$ is proportional to time, then V^2/R will be proportional to time. Therefore the rate of change of radial acceleration C, which is a constant, will be given by:

$$C = \mathrm{d}/\mathrm{d}T(V^2/R)$$

$$= \mathrm{d}/\mathrm{d}R(V^2/R)\mathrm{d}R/\mathrm{d}S\,\mathrm{d}S/\mathrm{d}T$$

where

$$R = K^2/S \text{ and } \mathrm{d}S/\mathrm{d}T = V$$

thus

$$C = (-V^2/R^2)(-K^2/S^2)V$$

$$= V^3/K^2 \tag{5.68}$$

Thus if C and V are given, K can be found. Similarly if F is the centripetal ratio in Equation 5.56, then,

$$F = V^2/Rg = V^2S/gK^2 = CS/Vg \tag{5.69}$$

Therefore if F, C and V are known then K can be found together with the maximum values of θ and S and the minimum value of R. Sometimes a minimum value of R is specified rather than F.

When θ is evaluated it should be compared with the deflection angle (D_F) and if θ is less than $D_F/2$ then there is a circular curve portion between two transition curves. If θ is greater than $D_F/2$ then the curve cannot be set out to the design constants, and an alternative solution must be found.

One such alternative is to join two transition curves at the centre using a cubic parabola with no circular curve portion, and to design a model curve with length s and multiplying constant K where the full scale length $S = Ks$. Substituting in Equation 5.60 gives:

$$\theta = s^2/2$$

but

$$\theta = D_F/2 \tag{5.70}$$

therefore

$$s = (D/57.3)^{1/2} \tag{5.71}$$

where

$$x = s - s^5/40 \tag{5.72}$$

and

$$y = s^3/6 - s^7/336 \tag{5.73}$$

Now in Fig. 5.16 the tangent length $AI(=T_D)$ is made up of AP and PI where AP is the maximum X value and PI can be found from triangle BPI where BP is the maximum Y value and angle PBI equals $\theta(=D_F/2)$, hence:

$$T_D = x_m + y_m \tan(D_F/2) \tag{5.74}$$

Thus if a suitable tangent length can be measured on site, say T_L, then the multiplying constant K can be found from:

$$K = T_L/T_D \tag{5.75}$$

The values of C and F already entered are used again, and the design speed of the curve, plus the other parameters, then computed.

An alternative approach, especially if no site measurements are available for T_D is to specify a minimum radius R, after which it reverts to the earlier procedure for the 'F, C, R' input.

Each of the above alternative solutions is presented in the following program <TRANSITION CURVES> together with the setting-out information using

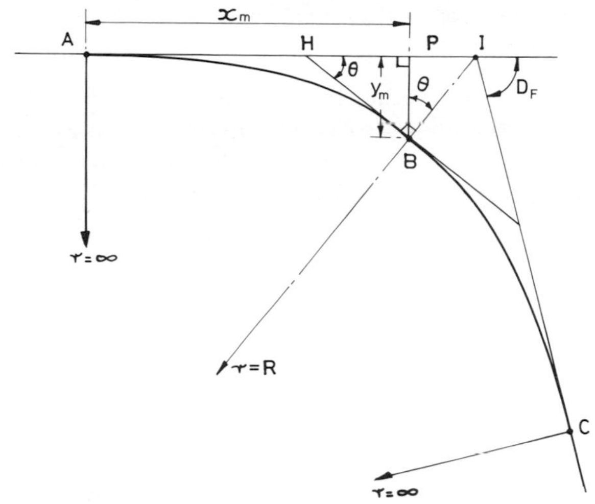

Fig. 5.16. Nomenclature for designing a transition curve where $\theta = D_F/2$.

either tangential offsets (Equations 5.63 and 5.64) or polar co-ordinates (Equations 5.66 and 5.67).

5.5.1 Transition curve design – subroutine index

	Line numbers	*Function*
(a)	10–90	Initialization and control
(b)	100–180	Screen header display
(c)	200–240	Selection of angle format
(d)	500–680	Entry of curve location, etc.
(e)	1000–1090	Selection of curve design criteria
(f)	1100–1190	Entry of curve design criteria
(g)	1200–1290	Calculation of curve parameters
(h)	1300–1490	Alternative solutions for $\theta > DF/2$
(i)	1500–1830	Screen display of curve design
(j)	1900–1990	Curve setting out selection
(k)	2000–2740	First transition calculations
(l)	2750–3330	Second transition calculations
(m)	6000–6080	Printout of curve location, etc.
(n)	6100–6150	Printout of curve design criteria
(o)	6200–6390	Printout of first transition curve
(p)	6500–6630	Printout of second transition curve
(q)	8000–8040	Routine to convert DD.MMSS to decimals
(r)	8050–8140	Routine to convert decimals to DD.MMSS
(s)	9000–9010	Routine to clear screen
(t)	9100–9130	Routine to display error message
(u)	9200–9290	Error trace routine
(v)	9500–9620	Termination of program

(a) Initialization and control

Line numbers 10–90

All numeric variables, arrays and required string variables are initialized and two 'DEF FN' functions used to define arccosine and arcsine as discussed previously in Section 2.2.8. If more than twenty setting-out points on the transition curve are required, the arrays should be altered accordingly.

(b) Screen header display

Line numbers 100–180

This is the screen display for civil engineering students at Strathclyde University. The operator can alter this as required to suit another organization.

(c) Selection of angle format

Line numbers 200–240

The operator has a choice in line 200 of entering angles in decimal or DD.MMSS format where the $D9$ flag is set to 1 for decimals and 2 for DD.MMSS in line 210.

(d) Entry of curve location, etc.

Line numbers 500–680

This routine is provided to keep a record of site location, operator, date, etc., with an opportunity to obtain a printout if required.

(e) Selection of curve design criteria

Line numbers 1000–1090

The operator is given a choice of design criteria where three out of four parameters are required from velocity (V), permissible centripetal ratio (F), permissible rate of change of radial acceleration (C) and minimum radius (R). The deflection angle (D_F) between the straights is also entered here.

(f) Entry of curve design criteria

Line numbers 1100–1190

The three specified design parameters are entered from the keyboard and depending on which selection ($P1$) was made in (e) above, the program branches accordingly from line 1190.

(g) Calculation of curve parameters

Line numbers 1200–1290

The missing parameters are computed and then the multiplying constant K evaluated together with the total length of the transition curve (S_F) and the angle $\theta(T_H)$. If the angle $\theta > D_F/2$ then the curve cannot be set out to the design parameters given, and the program branches to line 1300. If $\theta < D_F/2$ then there is a circular curve portion with deflection angle D_C and a circular curve length C_L. The program continues at line 1500.

(h) Alternative solution for $\theta > D_F/2$

Line numbers 1300–1490

The program only branches to this routine if the curve cannot be set out to given design constants. A choice is given in line 1310 to either return to the beginning and enter new design criteria, or to force a solution either with $\theta = D_F/2$ or a new specified minimum radius. If the former alternative is chosen, Equations 5.70–5.75 are used in conjunction with a physical measurement for the tangent length from the site. If the minimum radius alternative is chosen then the radius is entered and the program goes back to the previous computations at line 1240 for the $P1 = 3$ solution since F and C are assumed constant.

(i) Screen display of curve design

Line numbers 1500–1830

All the curve design parameters are then displayed on the screen as listed in lines 1620–1780, with the option of a printout for record purposes.

(j) Curve setting out selection

Line numbers 1900–1990

Having designed the curve, an option is given to prepare a setting-out table for the transition curve portions using either rectangular or polar co-ordinates. If it is desired to set out the curve, the chainage of the IP, the length of the setting-out chords and whether the curve is left or right handed are requested.

(k) First transition calculations

Line numbers 2000–2740

If an alternative solution was obtained in section (h) above (that is, $\theta = D_F/2$) then the tangent length is known and the program jumps to line 2100. In the standard solution, the tangent length AI in Fig. 5.15 is computed by splitting AI into three sections AP, HP and HI where AP is the maximum tangential distance (x_m) and BP is the maximum tangential offset (y_m). Therefore HP and HB can be found in terms of angle θ from triangle HPB:

$$HP = y_m/\tan\theta \tag{5.76}$$

$$HB = y_m/\sin\theta \tag{5.77}$$

Also in triangle OBT, BT can be found knowing the radius R and the angle at the centre of the circular arc:

$$BT = R\tan(D_F/2 - \theta) \tag{5.78}$$

Now in triangle HIT all the angles can be found since,

$$D_I = (180 - D_F)/2 \tag{5.79}$$

and

$$D_T = 180 - \theta - D_I \tag{5.80}$$

thus from the sine rule HI can be found:

$$HI = HT\sin(D_T)/\sin(D_I) \tag{5.81}$$

Hence the total tangent length AI ($=T_L$) from:

$$T_L = HI + x_m - HP \tag{5.82}$$

Knowing the chainage of the IP (C_I) the chainage of the first tangent point (F_1) can be found and thus all the chainages of the transition curve points round to the second straight. As mentioned previously, either rectangular or polar co-ordinates can be displayed. Both are in fact calculated and printed, but only one presented on the 40-column screen. If the operator is using an 80-column screen, both could be displayed.

The technique of finding the first and last multiple chord interval points was described earlier in Section 5.3.1(i) for tangential deflection angles. Once the end points are established, the number of chords N is calculated (line 2170) and then an array used to calculate each of the rectangular co-ordinates (x_I, y_I) and polar co-ordinates (l_I, ϕ_I) from Fig. 5.15. At the end of each calculation, the results are displayed on the screen. On completion, the chainages of the ends of the second transition curve are also displayed for information.

(l) Second transition calculations

Line numbers 2750–3330

The procedure for the second transition is the same as for (k) above except the deflection angles will be measured in the opposite direction, since a transition curve has to be set out from both ends, whereas a circular curve can be set out from one point. If there is a centre circular portion, the design parameters should be fed into the program <CIRCULAR CURVES – SETTING OUT> discussed previously in Section 5.3.

(m) Printout of curve location, etc.

Line numbers 6000–6080

This routine is provided for record purposes, and is based on Applesoft DOS and may require alteration for different microcomputers and printers.

(n) Printout of curve design criteria

Line numbers 6100–6150

This printout uses the previous screen display from line numbers 1620–1790.

(o) Printout of first transition curve

Line numbers 6200–6390

This printout combines the rectangular and polar co-ordinate computations and also lists the chainage of the IP and the tangent length for reference.

(p) Printout of second transition curve

Line numbers 6500–6630

This is similar to (o) above.

(q) Routine to convert DD.MMSS to decimals

Line numbers 8000–8040

This routine was discussed previously in Section 2.2.6.

(r) Routine to convert DD.MMSS to decimals

Line numbers 8050–8140

This routine was discussed previously in Section 2.2.6.

(s) Routine to clear screen

Line numbers 9000–9010

To simplify portability, the Applesoft 'HOME' command to clear the screen has been placed in a subroutine to allow an easy change to another microcomputer.

(t) Routine to display error message

Line numbers 9100–9130

If a wrong keyboard entry has been made in entering data, the error is trapped and this error message displayed and re-entry requested.

(u) Error trace routine

Line numbers 9200–9290

This error trace routine was discussed earlier in Section 2.2.9.

(v) Termination of program

Line numbers 9500–9620

When the '<QUIT>' selection is made in the program, an end-of-program message is displayed to advise the operator that computations are complete.

5.5.2 Transition curve design – numeric variables

A	= 0 or 360 degrees
BT	= Distance *BT* in Fig. 5.15
*C*0	= Counter in error trace routine
C	= Rate of change of radial acceleration
*C*1	= First chord in transition
CI	= Chainage of IP
CH	= Standard chord length
CL	= Circular curve length
D–*D*5	= Degrees of angle
*D*9	= Flag for decimals <1> or DD.MMSS <2>
DC	= Angular arc for circular curve
DF	= Deflection angle at IP
DI	= Internal angle at IP
DT	= Angle in triangle *HIT*
E	= Error number
F	= Centripetal force ratio
*F*1–*F*4	= Chainages of transition curve points
G	= Acceleration due to gravity
HB, *HI*, *HT*, *HP*	= Distances in Fig. 5.15
I, *IN*	= Integer loop variables
J, *JN*	= Integer loop variables
K	= Multiplying constant
LF	= Polar distance to end point
LF(*I*)	= Polar distance array
M–*M*1	= Minutes of angle
N, *N*1	= Number of chords in first and second transitions
*P*1, *P*2	= Program selection numbers
*P*7	= Counter in GOSUB routine
PH	= Total angle ϕ at end point
PT	= Total angle ϕ set out
PH(*I*)	= Polar angle array
R	= Minimum radius
*R*9	= 57.2957795
S, *S*0	= Seconds of angle
SF	= Full scale transition curve length
SM	= Model scale value of *SF*
TC	= Total transition curve length
TD	= Tangent distance in model
TH	= Angle θ
TL	= Tangent distance at full scale
V, *V*0	= Design velocity of curve
XF, *YF*	= Full scale values of *X* and *Y* at end of transition
XM, *YM*	= Model scale values of *X* and *Y* at end of transition
XF(*I*)	= Array of *X* measurements
YF(*I*)	= Array of *Y* measurements

5.5.3 Transition curve design – string variables

A$	Storage for location of curve
B$	Storage for operator's name
C$	Storage for date of calculations
D$	= CHR$(4), CTRL-D
*E*0$	= "SORRY, DATA ERROR . . . PLEASE RE-ENTER"
H$	= "TRANSITION CURVE DESIGN"
K$	Storage for instrument used
L$	= "L" for left and "R" for right handed curve
P$	= "DO YOU WISH PRINTOUT OF ABOVE DATA (Y/N)"
Q$	= Answer response to question (Y/N)
R$	= "R" for rectangular and "P" for polar co-ordinates

5.5.4 Transition curve design – BASIC program

```
10  REM  <TRANSITION CURVES> PROGRAM FOR APPLE II+.
20 D$ =  CHR$ (4): ONERR  GOTO 9200
30 A = 0:C0 = 0:CL = 0:G = 9.81:P1 = 0:P2 = 0:TL = 0:V0 = 0
35  DIM CH(20),CL(20),LF(20),PH(20),SM(20),XF(20),XM(20),YF(20),YM(20)
40 H$ = "TRANSITION CURVE DESIGN"
50 E0$ = "SORRY, DATA ERROR ... PLEASE RE-ENTER."
55 P$ = "DO YOU WISH PRINTOUT OF ABOVE DATA (Y/N)"
80  DEF  FN C9(C9) = (1.5707964 -  ATN (C9 /  SQR (1 - C9 * C9))) * R9: REM
     ARCCOSINE
90  DEF  FN S9(S9) =  ATN (S9 /  SQR (1 - S9 * S9)) * R9: REM  ARCSINE
95  REM  *****************************************************
100  GOSUB 9000
110  PRINT "****************************************"
120  PRINT "*                                      *"
130  PRINT "*     UNIVERSITY  OF  STRATHCLYDE      *"
140  PRINT "*  DEPARTMENT  OF  CIVIL  ENGINEERING  *"
150  PRINT "*                                      *"
160  PRINT "*            SURVEYING  SECTION        *"
170  PRINT "*       TRANSITION  CURVE  DESIGN      *"
180  PRINT "****************************************"
190  PRINT
200  PRINT "ARE ANGLES IN DECIMALS <1>, DD.MMSS <2> OR DO YOU WISH TO QUI
     T <3>."
210  INPUT D9
220  IF D9 < 1 OR D9 > 3 THEN 200
230  ON D9 GOTO 240,240,9500
240 R9 = 57.2957795:P1 = 0:TL = 0
500  GOSUB 9000
510  PRINT "UNIVERSITY OF STRATHCLYDE"
520  PRINT H$
530  PRINT "****************************************"
540  PRINT
550  PRINT "ENTER LOCATION OF WORK, ETC. WHEN       REQUESTED, AND PRESS
     <RETURN>"
560  PRINT "ENTER LOCATION OF CURVE :-"
570  INPUT A$
580  PRINT "ENTER OPERATOR'S NAME :-"
590  INPUT B$
600  PRINT "ENTER DATE OF CALCULATION AS DD/MM/YY -"
610  INPUT C$
```

```
620  PRINT "ENTER INSTRUMENT USED :-"
630  INPUT K$
640  PRINT
650  PRINT "DO YOU WISH PRINTOUT OF LOCATION ETC.,  (Y/N) ";
660  INPUT Q$
670  IF Q$ = "Y" THEN 6000
680  IF Q$ < > "N" THEN 650
690 :
993  REM  ************************************************
995  REM  KEYBOARD ENTRY OF DESIGN CRITERIA
1000  GOSUB 9000: PRINT H$: PRINT "*************************************
     **"
1010  PRINT : PRINT "SELECT CURVE DESIGN CRITERIA :-"
1020  PRINT "VELOCITY (V),"
1030  PRINT "RATE OF CHANGE OF RADIAL ACC. (C),"
1040  PRINT "CENTRIPETAL RATIO (F), MIN. RADIUS (R)"
1050  PRINT : PRINT "<1> V,C,F : <2> V,C,R : <3> F,C,R ";
1060  INPUT P1: IF P1 < 1 OR P1 > 3 THEN 1000
1070  PRINT : PRINT "ENTER DEFLECTION ANGLE ";: INPUT DF
1080  IF D9 = 1 THEN 1100
1090 D5 = DF: GOSUB 8000:DF = D5
1100  GOSUB 9000: PRINT H$
1110  PRINT "****************************************"
1120  IF P1 = 3 THEN 1140
1130  PRINT : PRINT "ENTER DESIGN VELOCITY (KM/HR) ";: INPUT V0
1140  PRINT : PRINT "ENTER RATE OF CHANGE OF RADIAL          ACCELERATION
     (M/S3) ";: INPUT C
1150  IF P1 = 2 THEN 1180
1160  PRINT : PRINT "ENTER CENTRIPETAL RATIO ";: INPUT F
1170  IF P1 = 1 THEN 1190
1180  PRINT : PRINT "ENTER MINIMUM RADIUS (M) ";: INPUT R
1190  ON P1 GOTO 1200,1220,1240
1200 V = V0 * 10 / 36:R = V * V / (F * G)
1210  GOTO 1250
1220 V = V0 * 10 / 36:F = V * V / (R * G)
1230  GOTO 1250
1240 V =  SQR (F * R * G)
1250 K =  SQR (V ^ 3 / C):SF = F * V * G / C
1260 TH = (SF * SF / (2 * K * K)) * R9
1270  IF TH > (DF / 2) THEN 1300
1280 DC = DF - 2 * TH:CL = R * DC / R9
1290 TC = CL + 2 * SF: GOTO 1500
1295  REM  FAULTY DESIGN CRITERIA
1300  GOSUB 9100
1310  PRINT "DO YOU WISH TO ENTER NEW CURVE DESIGN   CRITERIA (Y/N): IF (
     N) TRANSITIONS WILL MEET AT CENTRE ";: INPUT Q$
1320  IF Q$ = "Y" THEN 1000
1330  IF Q$ < > "N" THEN 1310
1340 TH = DF / 2:SM =  SQR (DF / R9)
1350 XM = SM - (SM ^ 5 / 40):YM = SM ^ 3 / 6 - (SM ^ 7 / 336)
1360 TD = XM + YM *  TAN (TH / R9)
1370  PRINT : PRINT "IS TANGENT DISTANCE FROM I.P. KNOWN      (Y/N) ";: INPUT
     Q$
1380  IF Q$ = "N" THEN 1450
1390  IF Q$ < > "Y" THEN 1370
1400  PRINT : PRINT "ENTER TANGENT DISTANCE (M) ";: INPUT TL
1410 P2 = 1:K = TL / TD
1415  REM  ASSUME C AND F AS BEFORE
1420 V = (K * K * C) ^ (1 / 3):R = V * V / (F * G)
```

```
1430 SF = SM * K:XF = K * XM:YF = K * YM
1440 TC = 2 * SF: GOTO 1500
1450  PRINT : PRINT "DO YOU WISH TO SPECIFY MIN. RADIUS (Y/N) IF (N) PROG
     RAM WILL REVERT TO START";
1460  INPUT Q$: IF Q$ = "N" THEN 100
1470  IF Q$ < > "Y" THEN 1450
1480  PRINT : PRINT "ENTER MINIMUM RADIUS (M) ";: INPUT R
1485  REM  PROGRAM REVERTS TO P1=3 CRITERIA
1490 P1 = 3: GOTO 1240
1495  REM  PREPARE ANGLES FOR DISPLAY
1500  IF D9 = 1 THEN 1550
1505  REM  CONVERT DECIMAL TO DD.MMSS
1510 D = DF: GOSUB 8050:DF = D
1520 D = TH: GOSUB 8050:TH = D
1530 D = DC: GOSUB 8050:DC = D
1540  GOTO 1600
1550 TH =  INT (TH * 1000 + .5) / 1000
1560 DC =  INT (DC * 1000 + .5) / 1000
1595  REM  SCREEN DISPLAY OF DESIGN DATA
1600  GOSUB 9000: PRINT H$:P7 = 0
1610  PRINT "****************************************"
1620  PRINT "DESIGN CRITERIA FOR CURVE :-"
1630  IF V0 = 0 THEN 1650
1640  PRINT "VELOCITY (KPH)                = ";V0
1650  PRINT "VELOCITY (MPS)                = "; INT (V * 1000 + .5) / 1000
1660  PRINT "RADIAL ACCELERATION (MPS3) = ";C
1670  PRINT "CENTRIPETAL RATIO             = "; INT (F * 1000 + .5) / 1000
1680  PRINT "MINIMUM RADIUS (M)            = "; INT (R * 1000 + .5) / 1000
1690  PRINT
1700  PRINT "MULTIPLYING CONSTANT          = "; INT (K * 1000 + .5) / 1000
1710  PRINT "DEFLECTION ANGLE AT I.P.      = ";DF
1720  PRINT "TRANSITION CURVE ANGLE        = ";TH
1730  PRINT "CIRCULAR CURVE ANGLE          = ";DC
1740  PRINT "SPIRAL TRANSITION LENGTH      = "; INT (SF * 1000 + .5) / 1000
1750  PRINT "CIRCULAR CURVE LENGTH         = "; INT (CL * 1000 + .5) / 1000
1760  PRINT "TOTAL CURVE LENGTH            = "; INT (TC * 1000 + .5) / 1000
1770  IF TL = 0 THEN 1790
1780  PRINT "MEASURED TANGENT LENGTH       = ";TL
1790  PRINT "****************************************"
1800  IF P7 = 1 THEN  RETURN
1810  PRINT : PRINT P$;: INPUT Q$
1820  IF Q$ = "Y" THEN 6100
1830  IF Q$ < > "N" THEN 1810
1840 :
1895  REM  SETTING OUT ?
1900  GOSUB 9000: PRINT H$
1910  PRINT "****************************************"
1920  PRINT : PRINT "DO YOU WISH TO SET OUT CURVE (Y/N) ";: INPUT Q$
1930  IF Q$ = "N" THEN 100
1940  IF Q$ < > "Y" THEN 1900
1950  PRINT : PRINT "SELECT POLAR (P) OR RECTANGULAR (R)     CO-ORDINATES
      (P/R) ";: INPUT R$
1960  IF R$ < > "P" AND R$ < > "R" THEN 1950
1970  PRINT : PRINT "ENTER CHAINAGE OF I.P. ";: INPUT CI
1980  PRINT : PRINT "ENTER MULTIPLE CHORD INTERVALS (M) ";: INPUT CH
1990  PRINT : PRINT "IS CURVE LEFT OR RIGHT HANDED (L/R) ";: INPUT L$: IF
     L$ < > "L" AND L$ < > "R" THEN 1990
1995  REM  FIND CHAINAGE OF 1ST TANGENT POINT
2000  IF P2 = 1 THEN 2100
```

```
2010 XF = SF - SF ^ 5 / (40 * K ^ 4):YF = SF ^ 3 / (6 * K ^ 2) - SF ^ 7 /
     (336 * K ^ 6)
2020  IF D9 = 1 THEN 2050
2025  REM  D9=2, CONVERT ANGLES DECIMALS
2030 D5 = DF: GOSUB 8000:DF = D5
2040 D5 = TH: GOSUB 8000:TH = D5
2045  REM  SOLVE TRIANGLE HPB
2050 HP = YF / ( TAN (TH / R9)):HB = YF / ( SIN (TH / R9))
2055  REM  SOLVE TRIANGLE OBT
2060 BT = R *  TAN ((DF / 2 - TH) / R9):HT = HB + BT
2065  REM  SOLVE TRIANGLE HIT USING SINE RULE
2070 DI = (180 - DF) / 2:DT = (180 - TH - DI)
2080 HI = (HT *  SIN (DT / R9)) /  SIN (DI / R9)
2090 TL = HI + XF - HP
2095  REM  FIND CHAINAGE OF A, START TRANSITION
2100 F1 = CI - TL
2105  REM  FIND CHAINAGE AT END 1ST TRANSITION
2110 F2 = F1 + SF
2115  REM  FIND CHAINAGE AT START 2ND TRANSITION
2120 F3 = F2 + CL
2125  REM  FIND CHAINAGE AT NEW STRAIGHT
2130 F4 = F1 + TC
2135  REM  FIND 1ST POINT ON CURVE
2140 CH(0) = ( INT (F1 / CH) + 1) * CH
2145  REM  FIND LENGTH FIRST CHORD
2150 C1 =  INT ((CH(0) - F1) * 1000 + .5) / 1000
2155  REM  FIND LAST POINT ON CURVE
2160 C9 = ( INT (F2 / CH)) * CH
2165  REM  FIND NUMBER OF CHORDS
2170 N =  INT ((SF - C1) / CH)
2180 CL(0) = C1
2190 F1 =  INT (F1 * 1000 + .5) / 1000:F2 =  INT (F2 * 1000 + .5) / 1000:
     F3 =  INT (F3 * 1000 + .5) / 1000:F4 =  INT (F4 * 1000 + .5) / 1000
2195  REM  SCREEN DISPLAY OF SETTING OUT DETAILS
2200 A = 0:P7 = 0
2210  GOSUB 9000: PRINT H$: PRINT "**************************************
     **"
2220  IF R$ = "P" THEN 2240
2230  PRINT "CHAINAGE  LENGTH   TANG.DIST  TANG.OFF": GOTO 2250
2240  PRINT "CHAINAGE  LENGTH   POLAR DIST  ANGLE"
2250  PRINT "========================================": IF P7 = 1 THEN  RETURN

2260  IF R$ = "P" THEN 2280
2270  PRINT F1; TAB( 11);"0"; TAB( 20);"0"; TAB( 31);"0": GOTO 2300
2280  IF L$ = "L" THEN A = 360
2290  PRINT F1; TAB( 11);"0"; TAB( 20);"0"; TAB( 31);A
2295  REM  START LOOP FOR CURVE SETTING OUT POINTS
2300  FOR I = 0 TO N
2310 CH(I) = CH(0) + CH * I
2320 CL(I) = CL(0) + CH * I
2330 :
2340 XF(I) = CL(I) - (CL(I)) ^ 5 / (40 * K ^ 4)
2350 YF(I) = (CL(I)) ^ 3 / (6 * K ^ 2) - (CL(I)) ^ 7 / (336 * K ^ 6)
2355  REM  FIND POLAR CO-ORDINATES
2360 LF(I) =  INT (( SQR ((XF(I)) ^ 2 + (YF(I)) ^ 2)) * 1000 + .5) / 1000

2370 PH(I) =  ATN (YF(I) / XF(I)) * R9
2380  IF L$ = "L" THEN 2430
2390  IF D9 = 1 THEN 2410
```

```
24ØØ D = PH(I): GOSUB 8Ø5Ø:PH(I) = D: GOTO 247Ø
241Ø PH(I) =  INT ((PH(I)) * 1ØØØ + .5) / 1ØØØ
242Ø  GOTO 247Ø
243Ø PH(I) = 36Ø - PH(I)
244Ø  IF D9 = 1 THEN 246Ø
245Ø D = PH(I): GOSUB 8Ø5Ø:PH(I) = D: GOTO 247Ø
246Ø PH(I) =  INT ((PH(I)) * 1ØØØ + .5) / 1ØØØ
247Ø XF(I) =  INT ((XF(I)) * 1ØØØ + .5) / 1ØØØ
248Ø YF(I) =  INT ((YF(I)) * 1ØØØ + .5) / 1ØØØ
2485  REM  ARRIVE HERE WITH BOTH "R" AND "P" RESULTS
249Ø  IF R$ = "P" THEN 252Ø
2495  REM  RECTANGULAR CO-ORDINATES DISPLAY
25ØØ  PRINT CH(I); TAB( 11);CL(I); TAB( 2Ø);XF(I); TAB( 31);YF(I)
251Ø  GOTO 253Ø
2515  REM  POLAR CO-ORDINATES DISPLAY
252Ø  PRINT CH(I); TAB( 11);CL(I); TAB( 2Ø);LF(I); TAB( 31);PH(I)
253Ø  NEXT I
254Ø SF =  INT (SF * 1ØØØ + .5) / 1ØØØ
255Ø XF =  INT (XF * 1ØØØ + .5) / 1ØØØ:YF =  INT (YF * 1ØØØ + .5) / 1ØØØ
256Ø  IF R$ = "P" THEN 258Ø
257Ø  PRINT F2; TAB( 11);SF; TAB( 2Ø);XF; TAB( 31);YF
2575  REM  FIND POLAR ANGLE
258Ø LF =  INT (( SQR (XF ^ 2 + YF ^ 2)) * 1ØØØ + .5) / 1ØØØ
259Ø PH = (R9 * 1Ø * (SF / K) ^ 2 - 1.213 * (SF / K) ^ 6) / 6Ø
26ØØ  IF L$ = "L" THEN 264Ø
261Ø  IF D9 = 1 THEN 263Ø
262Ø D = PH: GOSUB 8Ø5Ø:PT = D: GOTO 268Ø
263Ø PT =  INT (PH * 1ØØØ + .5) / 1ØØØ: GOTO 268Ø
264Ø PT = 36Ø - PH
265Ø  IF D9 = 1 THEN 267Ø
266Ø D = PT: GOSUB 8Ø5Ø:PT = D: GOTO 268Ø
267Ø PT =  INT (PT * 1ØØØ + .5) / 1ØØØ
268Ø  IF R$ = "R" THEN 27ØØ
269Ø  PRINT F2; TAB( 11);SF; TAB( 2Ø);LF; TAB( 31);PT
27ØØ  PRINT F3; TAB( 11);"START OF 2ND SPIRAL": PRINT F4; TAB( 11);"FINAL
      TANGENT POINT"
271Ø  PRINT "CHAINAGE OF I.P."; TAB( 2Ø);CI: PRINT "LENGTH OF TANGENT "; TAB(
     2Ø); INT (TL * 1ØØØ + .5) / 1ØØØ
272Ø  PRINT "========================================"
273Ø  PRINT : PRINT P$;: INPUT Q$: IF Q$ = "Y" THEN 62ØØ
274Ø  IF Q$ < > "N" THEN 273Ø
275Ø  GOSUB 9ØØØ: PRINT H$
276Ø  PRINT "****************************************"
277Ø  PRINT : PRINT "DO YOU WISH DATA FOR 2ND SPIRAL (Y/N)";: INPUT Q$
278Ø  IF Q$ = "N" THEN 1ØØ
279Ø  IF Q$ < > "Y" THEN 277Ø
2795  REM  DATA REQUIRED FOR 2ND SPIRAL
28ØØ A = Ø:P7 = 1
281Ø  GOSUB 221Ø
282Ø  IF R$ = "P" THEN 284Ø
283Ø  PRINT F3; TAB( 11);SF; TAB( 2Ø);XF; TAB( 31);YF
284Ø  IF L$ = "R" THEN 289Ø
285Ø  IF D9 = 1 THEN 287Ø
286Ø D = PH: GOSUB 8Ø5Ø:PT = D: GOTO 293Ø
287Ø PT =  INT (PH * 1ØØØ + .5) / 1ØØØ
288Ø  GOTO 293Ø
289Ø A = 36Ø:PT = A - PH
29ØØ  IF D9 = 1 THEN 292Ø
291Ø D = PT: GOSUB 8Ø5Ø:PT = D: GOTO 293Ø
```

```
2920 PT =  INT (PT * 1000 + .5) / 1000
2930  IF R$ = "R" THEN 2950
2940  PRINT F3; TAB( 11);SF; TAB( 20);LF; TAB( 31);PT
2945  REM  FIND CHAINAGES OF POINTS ON CURVE
2950 C8 = ( INT (F3 / CH) + 1) * CH
2960 C7 = ( INT (F4 / CH)) * CH
2965  REM  FIND NUMBER OF CHORDS
2970 N1 =  INT ((C7 - C8) / CH + .5)
2990 :
3000  FOR IN = 0 TO N1
3010 CH(IN) = C8 + IN * CH
3020 CL(IN) = F4 - CH(IN)
3030 SM(IN) = CL(IN) / K:CL(IN) =  INT ((CL(IN)) * 1000 + .5) / 1000
3040 XM(IN) = SM(IN) - ((SM(IN)) ^ 5 / 40):XF(IN) = K * XM(IN)
3050 YM(IN) = (SM(IN)) ^ 3 / 6 - (SM(IN)) ^ 7 / 336:YF(IN) = K * YM(IN)
3060 LF(IN) =  INT (( SQR ((XF(IN)) ^ 2 + (YF(IN)) ^ 2)) * 1000 + .5) / 1
     000
3070 XF(IN) =  INT ((XF(IN)) * 1000 + .5) / 1000:YF(IN) =  INT ((YF(IN)) *
     1000 + .5) / 1000
3080 PH(IN) = (R9 * 10 * (SM(IN)) ^ 2 - 1.213 * (SM(IN)) ^ 6) / 60
3090  IF L$ = "R" THEN 3150
3100  IF D9 = 1 THEN 3130
3110 D = PH(IN): GOSUB 8050:PH(IN) = D:
3120  GOTO 3190
3130 PH(IN) =  INT ((PH(IN)) * 1000 + .5) / 1000
3140  GOTO 3190
3150 PH(IN) = 360 - PH(IN)
3160  IF D9 = 1 THEN 3180
3170 D = PH(IN): GOSUB 8050:PH(IN) = D: GOTO 3190
3180 PH(IN) =  INT ((PH(IN)) * 1000 + .5) / 1000
3190  IF R$ = "P" THEN 3220
3195  REM  RECTANGULAR CO-ORDINATE DISPLAY
3200  PRINT CH(IN); TAB( 11);CL(IN); TAB( 20);XF(IN); TAB( 31);YF(IN)
3210  GOTO 3230
3215  REM  POLAR CO-ORDINATE DISPLAY
3220  PRINT CH(IN); TAB( 11);CL(IN); TAB( 20);LF(IN); TAB( 31);PH(IN)
3230  NEXT IN
3240  IF R$ = "P" THEN 3270
3250  PRINT F4; TAB( 11);"0"; TAB( 20);"0"; TAB( 31);"0"
3260  GOTO 3280
3270  PRINT F4; TAB( 11);"0"; TAB( 20);"0"; TAB( 31);A
3280  PRINT "***************************************"
3290  PRINT
3300  PRINT P$;: INPUT Q$
3310  IF Q$ = "Y" THEN 6500
3320  IF Q$ < > "N" THEN 3300
3330  GOTO 100
5985  REM  ************************************************
5990  REM  PRINTOUT OF LOCATION, ETC
6000  PRINT D$;"PR#1": PRINT "LOCATION OF CURVE :- ";A$
6010  PRINT "************************************************"
6020  PRINT "OPERATOR'S NAME :- ";B$
6030  PRINT "DATE OF CALCS.  :- ";C$
6040  PRINT "INSTRUMENT USED :- ";K$
6060  PRINT "************************************************"
6070  PRINT : PRINT D$;"PR#0"
6080  GOTO 1000
6093  REM  ************************************************
6095  REM  PRINTOUT OF DESIGN CRITERIA ETC.
```

```
6100  PRINT D$;"PR#1": POKE 1784 + 1,80
6110  PRINT H$:P7 = 1
6120  PRINT "****************************************"
6125  REM  USE SCREEN DISPLAY LINES 1620-1790
6130  GOSUB 1620
6140  PRINT : PRINT D$;"PR#0"
6150  GOTO 1900
6193  REM  *************************************************
6195  REM  PRINTOUT FOR FIRST TRANSITION
6200  PRINT D$;"PR#1": POKE 1784 + 1,80
6210  PRINT H$:A = 0:P7 = 0
6220  PRINT "***********************************************************
      ****************"
6230  PRINT "CHAINAGE OF I.P.  = ";: POKE 36,24: PRINT CI
6240  PRINT "LENGTH OF TANGENT = ";: POKE 36,24: PRINT  INT (TL * 1000 +
      .5) / 1000
6250  PRINT "***********************************************************
      ****************"
6260  PRINT "CHAINAGE";: POKE 36,12: PRINT "LENGTH";: POKE 36,24: PRINT "
      TANG.DIST";: POKE 36,36: PRINT "TANG.OFF";: POKE 36,48: PRINT "POLAR
       DIST";
6270  POKE 36,60: PRINT "ANGLE": IF P7 = 1 THEN  RETURN
6280  PRINT "===========================================================
      ================": IF L$ = "L" THEN A = 360
6290  PRINT F1;: POKE 36,12: PRINT "0";: POKE 36,24: PRINT "0";: POKE 36,
      36: PRINT "0";: POKE 36,48: PRINT "0";: POKE 36,60: PRINT A
6300  FOR J = 0 TO N
6310  PRINT CH(J);: POKE 36,12: PRINT CL(J);: POKE 36,24: PRINT XF(J);: POKE
      36,36: PRINT YF(J);: POKE 36,48: PRINT LF(J);: POKE 36,60: PRINT PH(
      J)
6320  NEXT J
6330  PRINT F2;: POKE 36,12: PRINT SF;: POKE 36,24: PRINT XF;: POKE 36,36
      : PRINT YF;: POKE 36,48: PRINT LF;: POKE 36,60: PRINT PT
6340  PRINT F3;: POKE 36,12: PRINT "START OF 2ND SPIRAL"
6350  PRINT F4;: POKE 36,12: PRINT "FINAL TANGENT POINT"
6360  PRINT "===========================================================
      ================"
6370  PRINT
6380  PRINT D$;"PR#0"
6390  GOTO 2750
6493  REM  *************************************************
6495  REM  PRINTOUT FOR 2ND SPIRAL
6500  PRINT D$;"PR#1": POKE 1784 + 1,80
6510  PRINT H$:A = 0:P7 = 1
6520  GOSUB 6250
6530  PRINT "===========================================================
      ================"
6540  IF L$ = "R" THEN A = 360
6550  PRINT F3;: POKE 36,12: PRINT SF;: POKE 36,24: PRINT XF;: POKE 36,36
      : PRINT YF;: POKE 36,48: PRINT LF;: POKE 36,60: PRINT PT
6560  FOR JN = 0 TO N1
6570  PRINT CH(JN);: POKE 36,12: PRINT CL(JN);: POKE 36,24: PRINT XF(JN);
      : POKE 36,36: PRINT YF(JN);: POKE 36,48: PRINT LF(JN);: POKE 36,60: PRINT
      PH(JN)
6580  NEXT JN
6590  PRINT F4;: POKE 36,12: PRINT "0";: POKE 36,24: PRINT "0";: POKE 36,
      36: PRINT "0";: POKE 36,48: PRINT "0";: POKE 36,60: PRINT A
6600  PRINT "***********************************************************
      ****************"
```

```
6610  PRINT
6620  PRINT D$;"PR#0"
6630  GOTO 100
7993  REM  ****************************************************
7995  REM  ROUTINE DD.MMSS TO DECIMALS
8000 D0 =  INT (D5)
8005  REM  FIND NO. OF MINUTES,M0
8010 M1 = (D5 - D0) * 100:M0 =  INT (M1 + .1)
8015  REM  FIND NO. OF SECONDS,S0
8020 S0 =  INT (M1 * 100 - M0 * 100 + .5)
8025  REM  COLLECT DEGS,MINS,SECS
8030 D5 = D0 + M0 / 60 + S0 / 3600
8040  RETURN
8043  REM  ****************************************************
8045  REM  ROUTINE DECIMAL TO DD.MMSS
8050 D1 =  INT (D)
8055  REM  FIND TOTAL NO. OF SECONDS
8060 D2 = (D - D1) * 3600
8065  REM  FIND NO. OF MINUTES
8070 M =  INT (D2 / 60)
8075  REM  FIND NO. OF SECONDS
8080 S = D2 - 60 * M + .5
8090  IF S < 60 THEN 8130
8100 M = M + 1:S = 0
8110  IF M < 60 THEN 8130
8120 D1 = D1 + 1:M = 0
8130 D = D1 * 10000 + M * 100 + S
8140 D =  INT (D) / 10000: RETURN
8985  REM  *******************************************
8990  REM  GOSUB ROUTINE TO CLEAR SCREEN
9000  HOME
9010  RETURN
9093  REM  ****************************************************
9095  REM  GOSUB ROUTINE FOR ERRORS
9100  PRINT
9110  PRINT E0$
9120  PRINT
9130  RETURN
9193  REM  ****************************************************
9195  REM  ERROR TRACE ROUTINE
9200  PRINT  CHR$ (7): REM  CTRL-B (BELL)
9210 E =  PEEK (222): REM  GET ERROR NO.
9220  IF C0 = 1 THEN 9260
9230  INVERSE
9240  PRINT "ERROR NO. ";E;" FOUND"
9250 C0 = 1: TRACE : RESUME
9260  PRINT "ERROR ON SECOND LINE NO. "
9270  NORMAL
9280  NOTRACE
9290  STOP
9493  REM  ****************************************************
9495  REM  END OF PROGRAM
9500  GOSUB 9000
9510  PRINT
9520  PRINT "END TRANSITION CURVE PROGRAM"
9530  PRINT
9540  PRINT "****************************"
9550  REM  PROGRAM PREPARED BY
9560  REM  DR. P.H. MILNE
```

```
9570  REM  DEPT. OF CIVIL ENGINEERING
9580  REM  UNIVERSITY OF STRATHCLYDE
9590  REM  GLASGOW G4 ONG
9600  REM  SCOTLAND
9610  REM  ***************************
9620  END
```

5.5.5 Transition curve design – computer printout

```
LOCATION OF CURVE :- GLENMORE
***************************************************
OPERATOR'S NAME :- JOHN MACMILLAN
DATE OF CALCS.  :- 09/08/83
INSTRUMENT USED :- APPLE II+
***************************************************


TRANSITION CURVE DESIGN
*****************************************
DESIGN CRITERIA FOR CURVE :-
VELOCITY (KPH)                = 160
VELOCITY (MPS)                = 44.444
RADIAL ACCELERATION (MPS3) = .46
CENTRIPETAL RATIO             = .25
MINIMUM RADIUS (M)            = 805.427

MULTIPLYING CONSTANT          = 436.865
DEFLECTION ANGLE AT I.P.      = 75
TRANSITION CURVE ANGLE        = 8.2542
CIRCULAR CURVE ANGLE          = 58.0837
SPIRAL TRANSITION LENGTH      = 236.957
CIRCULAR CURVE LENGTH         = 817.344
TOTAL CURVE LENGTH            = 1291.257
*****************************************


TRANSITION CURVE DESIGN
***************************************************************************
CHAINAGE OF I.P.  =      1862.3
LENGTH OF TANGENT =      738.643
***************************************************************************
```

CHAINAGE	LENGTH	TANG.DIST	TANG.OFF	POLAR DIST	ANGLE
1123.657	0	0	0	0	0
1140	16.343	16.343	4E-03	16.343	4.8E-03
1170	46.343	46.343	.087	46.343	.0627
1200	76.343	76.341	.389	76.342	.173
1230	106.343	106.334	1.05	106.339	.3357
1260	136.343	136.311	2.213	136.329	.5548
1290	166.343	166.256	4.018	166.304	1.2304
1320	196.343	196.143	6.605	196.254	1.5543
1350	226.343	225.935	10.113	226.161	2.3347
1360.613	236.957	236.444	11.601	236.728	2.4832
2177.958	START OF 2ND SPIRAL				
2414.914	FINAL TANGENT POINT				

```
TRANSITION CURVE DESIGN
***********************************************************************
CHAINAGE    LENGTH      TANG.DIST   TANG.OFF    POLAR DIST  ANGLE
=======================================================================
2177.958    236.957     236.444     11.601      236.728     357.1128
2190        224.914     224.519     9.923       224.738     357.2809
2220        194.914     194.721     6.462       194.828     358.0557
2250        164.914     164.83      3.915       164.877     358.3821
2280        134.914     134.883     2.144       134.9       359.0521
2310        104.914     104.905     1.008       104.91      359.2657
2340        74.914      74.912      .367        74.913      359.4309
2370        44.914      44.914      .079        44.914      359.5357
2400        14.914      14.914      3E-03       14.914      359.592
2414.914    0           0           0           0           360
***********************************************************************

LOCATION OF CURVE :- GLENMORE
************************************************
OPERATOR'S NAME :- JOHN MACMILLAN
DATE OF CALCS.  :- 11/08/83
INSTRUMENT USED :- APPLE II+
************************************************

TRANSITION CURVE DESIGN
***************************************
DESIGN CRITERIA FOR CURVE :-
VELOCITY (KPH)              = 160
VELOCITY (MPS)              = 46.052
RADIAL ACCELERATION (MPS3)  = .46
CENTRIPETAL RATIO           = .25
MINIMUM RADIUS (M)          = 864.73

MULTIPLYING CONSTANT        = 460.774
DEFLECTION ANGLE AT I.P.    = 15
TRANSITION CURVE ANGLE      = 7.3
CIRCULAR CURVE ANGLE        = 0
SPIRAL TRANSITION LENGTH    = 235.761
CIRCULAR CURVE LENGTH       = 0
TOTAL CURVE LENGTH          = 471.523
MEASURED TANGENT LENGTH     = 236.71
***************************************

TRANSITION CURVE DESIGN
***********************************************************************
CHAINAGE OF I.P.  =     1500
LENGTH OF TANGENT =     236.71
***********************************************************************
CHAINAGE    LENGTH      TANG.DIST   TANG.OFF    POLAR DIST  ANGLE
=======================================================================
1263.29     0           0           0           0           0
1290        26.71       26.71       .015        26.71       .0156
1320        56.71       56.71       .143        56.71       .0841
1350        86.71       86.707      .512        86.709      .2017
1380        116.71      116.698     1.248       116.705     .3646
1410        146.71      146.672     2.478       146.693     .5805
1440        176.71      176.614     4.33        176.668     1.2416
1470        206.71      206.501     6.929       206.617     1.5518
1499.051    235.761     235.357     10.274      235.581     2.2959
1499.051    START OF 2ND SPIRAL
1734.813    FINAL TANGENT POINT
=======================================================================
```

TRANSITION CURVE DESIGN

CHAINAGE	LENGTH	TANG.DIST	TANG.OFF	POLAR DIST	ANGLE
1499.Ø51	235.761	235.357	1Ø.274	235.581	357.3ØØ1
15ØØ	234.813	234.417	1Ø.151	234.637	357.3114
153Ø	2Ø4.813	2Ø4.613	6.74	2Ø4.724	358.Ø648
156Ø	174.813	174.722	4.192	174.773	358.3732
159Ø	144.813	144.778	2.384	144.797	359.Ø325
162Ø	114.813	114.8Ø2	1.188	114.8Ø8	359.2426
165Ø	84.813	84.811	.479	84.812	359.4Ø35
168Ø	54.813	54.813	.129	54.813	359.5154
171Ø	24.813	24.813	.Ø12	24.813	359.582
1734.813	Ø	Ø	Ø	Ø	36Ø

5.6 VERTICAL CURVE DESIGN

All the previous sections in this chapter have dealt with linking straights using curves in the horizontal plane. However, due to level changes in the topography there are also changes in gradient along the straight. These changes are dictated by studying the mass-haul diagram, which was discussed in Section 4.3 under Earthworks. As with horizontal curves, vertical curves are designed for a particular design speed, and a parabolic arc is fitted at the change of gradient. In the United Kingdom the Department of the Environment specifies the maximum permissible gradients allowable on roads, which are normally 3% for motorways and 4% for rural roads. The percentage method is to be preferred for expressing a gradient, that is 4% rather than 1 in 25. Accordingly, a road which rises 'm' metres in 100 m in the direction of chainage would have a gradient of 'm'%. If the road rises in the direction of chainage it has a positive gradient (i.e. +m%), and if the road falls in the direction of chainage it has a negative gradient (e.g. −n%).

Where there is a change of sign from plus to minus between two gradients there will be a summit or *crest* curve, and if from minus to plus there will be a valley or *sag* curve. In addition to these two specific curves, there are also four other combinations of gradient, making a total of six different types of vertical curve as shown in Fig. 5.17(a–f). In calculating the difference in gradients it should be noted that it is the *algebraic* difference (A) of the gradients which is required.

In designing a vertical curve, several criteria require to be met. The first is the provision of adequate visibility for vehicles at the design speed to be able to stop or overtake safely. This requirement was met in the past by the use of *sight distances*, but now mostly replaced by tables of *K-values* discussed later. The second and third criteria to be met are that of passenger comfort and vehicle safety. A sharp crest such as a hump-back bridge can cause a vehicle to either leave the road or to straddle the crest. Similarly a sharp sag curve could cause problems with a long low-loader vehicle. Hence in curve design, it is essential to restrict the gradients and introduce the vertical curve gradually.

For vertical curves, the equation of a parabola is used, where C is a constant:

$$x = Cy^2 \tag{5.83}$$

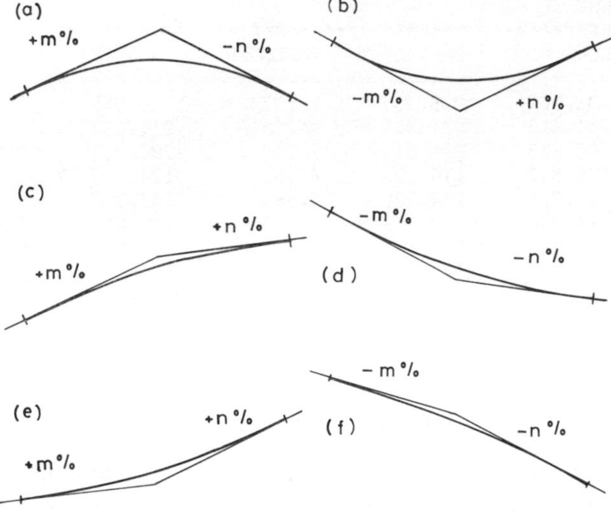

Fig. 5.17. Six possible configurations (a–f) of vertical curves depending on gradients.

i.e.

$$dx/dy = 2Cy \tag{5.84}$$

and

$$d^2x/dy^2 = 2C = \text{constant} \tag{5.85}$$

Therefore if a parabolic curve is used between the two straights there will be a uniform change of gradient and a gradual introduction of the vertical radial force. If the length of the parabolic curve is denoted by L, and the algebraic difference in the gradient by A, then for the design speed there will be a specific relationship between L and A, given by a constant K, where:

$$L = KA \tag{5.86}$$

Once the algebraic difference in gradients is known, the minimum length L of the vertical curve can be found from Equation 5.86, using tables of K-values published by the Department of the Environment.

These tables for K-values specify minimum values of K firstly for overtaking on two-way roads and secondly for stopping and comfort which apply to dual carriageways and motorways. In addition, a minimum value is also specified for the length of the vertical curve depending on the design speed.

As mentioned previously, in the section on horizontal curves, if a change of gradient occurs during a horizontal curve, then the vertical curve is phased in at the same tangent points for compatibility, as shown in Fig. 5.18. In such a case, the minimum lengths specified by K may be only of academic interest due to the much greater horizontal curve lengths required.

In the following program <VERTICAL CURVES>, after finding the

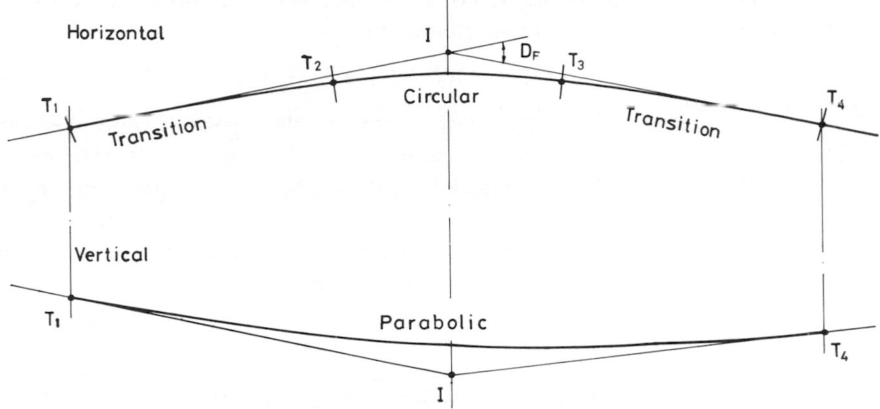

Fig. 5.18. Phasing in of tangent points for both horizontal and vertical curves.

minimum length of the vertical curve for a specific design speed, the reduced levels of the multiple chord interval points are tabulated for setting out on the ground. This microcomputer solution assumes equal tangent lengths ($=L/2$) on either side of the gradient intersection point, and the distances and levels are computed from the first tangent point A as shown in Fig. 5.19. Since the gradients are very slight, it is assumed that the distance along the vertical curve is the same as that along the horizontal.

The differences in level between I the intersection point, A the first tangent and C the second tangent point are found from:

$$H_{AI} = (m/100)(L/2) = mL/200 \tag{5.87}$$

and

$$H_{CI} = (n/100)(L/2) = nL/200 \tag{5.88}$$

To set out the curve the line AI is considered extended past I and the parabolic

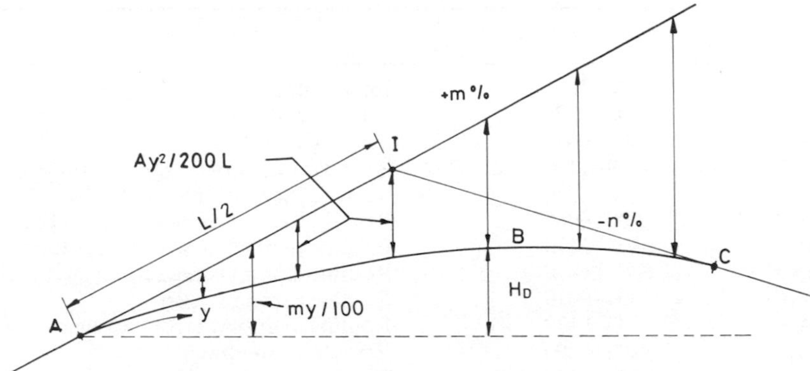

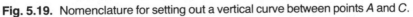

Fig. 5.19. Nomenclature for setting out a vertical curve between points A and C.

shape computed, and then the height difference (H_D) at the multiple chord points along the vertical curve computed from:

$$H_D = [(m)y/100 - (A)y^2/200L] \qquad (5.89)$$

where H_D can be positive or negative depending on the algebraic signs of m and A. It should be noted that the above assumes that the tangential offset from the first straight is equal to the vertical height interval; a negligible error for gradients below 3–4%.

In addition to setting out the multiple chord points, it is normal to calculate the highest point of a crest curve or the lowest point of a sag curve. It is therefore necessary to find the maximum or minimum value of H_D which will occur when $d/dy(H_D)$ equals zero:

$$d/dy(H_D) = m/100 - Ay/100L = 0 \qquad (5.90)$$

i.e.

$$m/100 = Ay/100L$$

and

$$y = Lm/A \qquad (5.91)$$

Therefore the maximum or minimum value for H_D above or below A will be given by substituting for y from Equation 5.91 in Equation 5.89:

$$(H_D)_{max/min} = [(m)(Lm/A)/100 - (A)(Lm/A)^2/200L]$$

i.e.

$$(H_D)_{max/min} = Lm^2/200A \qquad (5.92)$$

5.6.1 Vertical curve design – subroutine index

	Line numbers	*Function*
(a)	10–90	Initialization and control
(b)	100–230	Screen header display
(c)	500–680	Entry of curve location, etc.
(d)	1000–1190	Keyboard entry of design criteria
(e)	1200–1520	Vertical curve setting out details
(f)	6000–6080	Routine to print curve location, etc.
(g)	6100–6270	Routine to print curve design criteria
(h)	6300–6470	Routine to print curve setting out details
(i)	9000–9010	Routine to clear screen
(j)	9100–9130	Routine to display error message
(k)	9200–9290	Error trace routine
(l)	9500–9620	Termination of program

(a) Initialization and control

Line numbers 10–90

All numeric variables and required string variables are initialized, and four arrays dimensioned. If more than 20 points are required on each curve, the dimensions should be changed accordingly.

(b) Screen header display

Line numbers 100–230

This is the screen display used for civil engineering students at Strathclyde University. The operator can alter this as required to suit another organization. An option to jump the curve location subroutine is included in this display menu.

(c) Entry of curve location, etc.

Line numbers 500–680

This routine is provided to keep a record of site location, operator, date, etc., with an opportunity to obtain a printout if required.

(d) Keyboard entry of design criteria

Line numbers 1000–1190

All the necessary data required to compute a vertical curve design solution are requested, namely, gradients of both straights, *K*-value, acceptable minimum length of curve, chainage and level of the intersection point. The chainage of the two tangent points is also calculated. A printout option is provided for data record purposes.

(e) Vertical curve setting-out details

Line numbers 1200–1520

The only keyboard entry requested is that of the chord interval to be used to set out the curve. The chainage of the first point on the curve, the first chord length and the number of chords are then calculated in a similar manner to that used in Sections 5.3 and 5.5.

The levels of the two tangent points A and C are calculated using Equations 5.87 and 5.88, and the level and chainage of B found from Equations 5.91 and 5.92. An array is then set up using the distance along the curve y to compute the chainage, height difference and reduced level of each setting-out point, where the height difference is found from Equation 5.89. Each of these parameters is then displayed on the screen with an option to obtain a printout.

(f) Routine to print curve location, etc.

Line numbers 6000–6080

This routine is provided for record purposes, and is based on Applesoft DOS and may require alteration for different microcomputers and printers.

(g) Routine to print curve design criteria

Line numbers 6100–6270

This printout provides a record of data entered at the keyboard.

(h) Routine to print curve setting-out details

Line numbers 6300–6470

The complete setting-out details for each point, chainage, distance, height difference and reduced level are printed for record purposes.

(i) Routine to clear screen

Line numbers 9000–9010

To simplify portability, the Applesoft 'HOME' command to clear the screen has been placed in a subroutine to allow an easy change to another microcomputer.

(j) Routine to display error message

Line numbers 9100–9130

If a wrong keyboard entry has been made in entering data, the error is trapped and this error message displayed and re-entry requested.

(k) Error trace routine

Line numbers 9200–9290

This error trace routine was discussed earlier in Section 2.2.9.

(l) Termination of program

Line numbers 9500–9620

When the '<QUIT>' selection is made in the program, an end-of-program message is displayed to advise the operator that computations are complete.

5.6.2 Vertical curve design – numeric variables

A	= Algebraic difference in gradients
*C*0	= Counter in error trace routine
*C*1	= First chord in vertical circle
*C*9	= Chainage of last point on curve
CB	= Chainage to *B*, highest/lowest point
CH	= Multiple chord interval
CI	= Chainage of IP
CH(*I*)	= Chainage array along curve
E	= Error number

Table—continued

*F*1–*F*3 = Chainages of points on vertical curve

HB = Vertical height difference at *B*

HD(*I*) = Vertical height difference array along curve

I = Integer loop variable

J = Integer loop variable

K = *K*-value

L, *L*0 = Length of vertical curve

LM = Minimum length of vertical curve

M = Gradient of first straight

N = Gradient of second straight

NC = Number of setting out chords

*P*5, *P*7 = Counters in printout routines

RA, *RB*, *RC*, *RI* = Reduced levels of *A*, *B*, *C* and *I*

RL(*I*) = Reduced level array

U = Menu selection at start

Y(*I*) = Distance along curve array

5.6.3 Vertical curve design – string variables

A$ Storage for location of curve

B$ Storage for operator's name

C$ Storage for date of calculation

D$ = CHR$(4), CTRL-D

*E*0$ – "SORRY, DATA ERROR . . . PLEASE RE-ENTER"

H$ = "VERTICAL CURVE DESIGN"

K$ Storage for instrument used

P$ = "DO YOU WISH PRINTOUT OF ABOVE DATA (Y/N)"

Q$ = Answer response to question (Y/N)

5.6.4 Vertical curve design – BASIC program

```
10  REM  <VERTICAL CURVES> PROGRAM FOR APPLE II+  USES QUME PRINTER
20 D$ =  CHR$ (4): ONERR  GOTO 92ØØ
3Ø A = Ø:C = Ø:C1 = Ø:CH = Ø:K = Ø:L = Ø:LØ = Ø:LM = Ø:M = Ø:N = Ø:NC = Ø
     :P5 = Ø:P7 = Ø:U = Ø
35  DIM CH(2Ø),HD(2Ø),RL(2Ø),Y(2Ø)
4Ø H$ = "VERTICAL CURVE DESIGN"
5Ø EØ$ = "SORRY, DATA ERROR ... PLEASE RE-ENTER."
55 P$ = "DO YOU WISH PRINTOUT OF ABOVE DATA (Y/N)"
95  REM  ******************************************************
1ØØ  GOSUB 9ØØØ
11Ø  PRINT "****************************************"
12Ø  PRINT "*                                      *"
13Ø  PRINT "*     UNIVERSITY  OF  STRATHCLYDE      *"
14Ø  PRINT "*  DEPARTMENT  OF  CIVIL  ENGINEERING  *"
15Ø  PRINT "*                                      *"
16Ø  PRINT "*           SURVEYING  SECTION         *"
17Ø  PRINT "*         VERTICAL  CURVE  DESIGN      *"
18Ø  PRINT "****************************************"
19Ø  PRINT
```

```
200  PRINT "DO YOU WISH TO ENTER LOCATION <1>,      NEW DATA <2> OR QUIT
     <3>"
210  INPUT U
220  IF U < 1 OR U > 3 THEN 200
230  ON U GOTO 500,1000,9500
500  GOSUB 9000
510  PRINT "UNIVERSITY OF STRATHCLYDE"
520  PRINT H$
530  PRINT "****************************************"
540  PRINT
550  PRINT "ENTER LOCATION OF WORK, ETC. WHEN       REQUESTED, AND PRESS
     <RETURN>"
560  PRINT "ENTER LOCATION OF CURVE :-"
570  INPUT A$
580  PRINT "ENTER OPERATOR'S NAME :-"
590  INPUT B$
600  PRINT "ENTER DATE OF CALCULATION AS DD/MM/YY -"
610  INPUT C$
620  PRINT "ENTER INSTRUMENT USED :-"
630  INPUT K$
640  PRINT
650  PRINT "DO YOU WISH PRINTOUT OF LOCATION ETC.,  (Y/N) ";
660  INPUT Q$
670  IF Q$ = "Y" THEN 6000
680  IF Q$ < > "N" THEN 650
690 :
993  REM  ************************************************
995  REM  KEYBOARD ENTRY OF DESIGN CRITERIA
1000  GOSUB 9000: PRINT H$
1010  PRINT "****************************************"
1020  PRINT : PRINT "ENTER GRADIENT OF 1ST STRAIGHT ";: INPUT M
1030  PRINT : PRINT "ENTER GRADIENT OF 2ND STRAIGHT ";: INPUT N
1035  REM  ALGEBRAIC DIFF.= A
1040 A = M - N
1050  PRINT : PRINT "ENTER K-VALUE ";: INPUT K
1055  REM  MINIMUM CURVE LENGTH = L
1060 LM = K *  ABS (A)
1070  PRINT : PRINT "MINIMUM LENGTH OF CURVE IS = "; INT (LM * 1000 + .5)
      / 1000
1080  PRINT : PRINT "IS THIS ACCEPTABLE (Y/N) ";: INPUT Q$: IF Q$ = "Y" THEN
      1120
1090  IF Q$ < > "N" THEN 1080
1100  PRINT : PRINT "ENTER ACCEPTABLE LENGTH ";: INPUT L0: IF L0 < LM THEN
      1100
1110 L = L0
1120  GOSUB 9000: PRINT H$: IF L0 < LM THEN L = LM
1130  PRINT : PRINT "ENTER CHAINAGE OF I.P. ";: INPUT CI
1135  REM  F1 IS CHAINAGE OF 1ST TANGENT PT.
1140 F1 =  INT ((CI - L / 2) * 1000 + .5) / 1000
1145  REM  F2 IS CHAINAGE OF 2ND TANGENT PT.
1150 F3 =  INT ((CI + L / 2) * 1000 + .5) / 1000
1160  PRINT : PRINT "ENTER LEVEL OF I.P. ";: INPUT RI
1170  PRINT : PRINT P$;: INPUT Q$
1180  IF Q$ = "Y" THEN 6100
1190  IF Q$ < > "N" THEN 1170
1195  REM  START CALCULATIONS
1200 RA =  INT ((RI - (M * L / 200)) * 1000 + .5) / 1000
1205  REM  REDUCED LEVELS OF A AND C ARE RA AND RC
1210 RC =  INT ((RI + (N * L / 200)) * 1000 + .5) / 1000
```

```
1220  PRINT : PRINT "ENTER MULTIPLE CHORD INTERVALS ";: INPUT CH
1225  REM  FIND CHAINAGE 1ST POINT ON CURVE
1230 CH(Ø) = ( INT (F1 / CH) + 1) * CH
1235  REM  FIND LENGTH 1ST CHORD
1240 C1 =  INT ((CH(Ø) - F1) * 1ØØØ + .5) / 1ØØØ:Y(Ø) = C1
1245  REM  FIND LAST POINT ON CURVE
1250 C9 = ( INT (F3 / CH)) * CH
1255  REM  FIND NUMBER OF CHORDS
1260 NC =  INT ((L - C1) / CH)
1265  REM  CALCULATE HIGHEST/LOWEST POINT ON CURVE AT CB, REDUCED LEVEL R
    B
1270 CB =  INT ((L *  ABS (M) /  ABS (A)) * 1ØØØ + .5) / 1ØØØ:F2 = F1 + C
    B
1280 HB = L * M *  ABS (M) / (2ØØ *  ABS (A)):RB = RA + HB
1290 HB =  INT (HB * 1ØØØ + .5) / 1ØØØ:RB =  INT (RB * 1ØØØ + .5) / 1ØØØ
1295  REM  SCREEN DISPLAY
1300  GOSUB 9ØØØ: PRINT H$:P5 = Ø
1310  PRINT "****************************************"
1320  PRINT "CHAINAGE  LENGTH   HT.DIFF.   RED.LEVEL"
1330  PRINT "========================================"
1340  PRINT F1; TAB( 12);"Ø"; TAB( 21);"Ø"; TAB( 31);RA
1345  REM  COLLECT DATA IN ARRAY
1350  FOR I = Ø TO NC
1360 Y(I) = Y(Ø) + I * CH
1370 CH(I) = CH(Ø) + I * CH
1380 HD(I) = (M * Y(I) / 1ØØ) - (A * (Y(I)) ^ 2 / (2ØØ * L))
1390 RL(I) =  INT ((RA + HD(I)) * 1ØØØ + .5) / 1ØØØ
1400 HD(I) =  INT ((HD(I)) * 1ØØØ + .5) / 1ØØØ
1410  IF P5 = 1 THEN 145Ø
1420  IF F2 > CH(I) THEN 145Ø
1430  PRINT F2; TAB( 12);CB; TAB( 21);HB; TAB( 31);RB
1440 P5 = 1
1450  PRINT CH(I); TAB( 12);Y(I); TAB( 21);HD(I); TAB( 31);RL(I)
1460  NEXT I
1470  PRINT F3; TAB( 12);L; TAB( 21);RC - RA; TAB( 31);RC
1480  PRINT "****************************************"
1490  PRINT : PRINT P$;: INPUT Q$
1500  IF Q$ = "Y" THEN 63ØØ
1510  IF Q$ < > "N" THEN 149Ø
1520  GOTO 1ØØ
5985  REM  ************************************************
5990  REM  PRINTOUT OF LOCATION, ETC
6000  PRINT D$;"PR#1": PRINT "LOCATION OF CURVE :- ";A$
6010  PRINT "************************************************"
6020  PRINT "OPERATOR'S NAME :- ";B$
6030  PRINT "DATE OF CALCS.  :- ";C$
6040  PRINT "INSTRUMENT USED :- ";K$
6060  PRINT "************************************************"
6070  PRINT : PRINT D$;"PR#Ø"
6080  GOTO 1ØØØ
6093  REM  ************************************************
6095  REM  PRINTOUT OF DESIGN CRITERIA ETC.
6100  PRINT D$;"PR#1": POKE 1784 + 1,8Ø
6110  PRINT H$
6120  PRINT "****************************************"
6130  PRINT "GRADIENT OF 1ST STRAIGHT = ";M
6140  PRINT "GRADIENT OF 2ND STRAIGHT = ";N
6150  PRINT "ALGEBRAIC DIFFERENCE M-N = ";A
6160  PRINT "K-VALUE                  = ";K
```

```
6170  PRINT "MINIMUM LENGTH OF CURVE  = ";LM
6180  PRINT "CURVE LENGTH CHOSEN      = ";L
6190  PRINT
6200  PRINT "CHAINAGE OF INTERSECTION = ";CI
6210  PRINT "REDUCED LEVEL OF I.P.    = ";RI
6220  PRINT "CHAINAGE OF 1ST TANG.PT. = ";F1
6230  PRINT "CHAINAGE OF 2ND TANG.PT. = ";F3
6240  PRINT "****************************************"
6250 P7 = 1
6260  PRINT : PRINT D$;"PR#0"
6270  GOTO 1200
6293  REM  ************************************************
6295  REM  PRINTOUT OF RESULTS
6300  PRINT D$;"PR#1": POKE 1784 + 1,80:P5 = 0
6310  IF P7 = 1 THEN 6340
6320  PRINT H$
6330  PRINT "****************************************"
6340  PRINT "CHAINAGE  LENGTH   HT.DIFF.   RED.LEVEL"
6350  PRINT "========================================"
6360  PRINT F1;: POKE 36,11: PRINT "0";: POKE 36,20: PRINT "0";: POKE 36,
     30: PRINT RA
6370  FOR J = 0 TO NC
6380  IF P5 = 1 THEN 6420
6390  IF F2 > CH(J) THEN 6420
6400  PRINT F2;: POKE 36,11: PRINT CB;: POKE 36,20: PRINT HB;: POKE 36,30
     : PRINT RB
6410 P5 = 1
6420  PRINT CH(J);: POKE 36,11: PRINT Y(J);: POKE 36,20: PRINT HD(J);: POKE
     36,30: PRINT RL(J)
6430  NEXT J
6440  PRINT F3;: POKE 36,11: PRINT L;: POKE 36,20: PRINT RC - RA;: POKE 3
     6,30: PRINT RC
6450  PRINT "****************************************"
6460  PRINT : PRINT D$;"PR#0"
6470  GOTO 100
8985  REM  ****************************************
8990  REM  GOSUB ROUTINE TO CLEAR SCREEN
9000  HOME
9010  RETURN
9093  REM  ************************************************
9095  REM  GOSUB ROUTINE FOR ERRORS
9100  PRINT
9110  PRINT E0$
9120  PRINT
9130  RETURN
9193  REM  ************************************************
9195  REM  ERROR TRACE ROUTINE
9200  PRINT  CHR$ (7): REM  CTRL-B (BELL)
9210 E =  PEEK (222): REM  GET ERROR NO.
9220  IF C0 = 1 THEN 9260
9230  INVERSE
9240  PRINT "ERROR NO. ";E;" FOUND"
9250 C0 = 1: TRACE : RESUME
9260  PRINT "ERROR ON SECOND LINE NO. "
9270  NORMAL
9280  NOTRACE
9290  STOP
9493  REM  ************************************************
9495  REM  END OF PROGRAM
```

```
9500  GOSUB 9000
9510  PRINT
9520  PRINT "END VERTICAL CURVE PROGRAM"
9530  PRINT
9540  PRINT "**************************"
9550  REM  PROGRAM PREPARED BY
9560  REM  DR. P.H. MILNE
9570  REM  DEPT. OF CIVIL ENGINEERING
9580  REM  UNIVERSITY OF STRATHCLYDE
9590  REM  GLASGOW G4 ONG
9600  REM  SCOTLAND
9610  REM  ***************************
9620  END
```

5.6.5 Vertical curve design – computer printout

```
 LOCATION OF CURVE :- GLENMORE
*************************************************
OPERATOR'S NAME :- JOHN MACMILLAN
DATE OF CALCS.  :- 13/08/83
INSTRUMENT USED :- APPLE II+
*************************************************

VERTICAL CURVE DESIGN
***************************************
GRADIENT OF 1ST STRAIGHT = 1
GRADIENT OF 2ND STRAIGHT = -.5
ALGEBRAIC DIFFERENCE M-N = 1.5
K-VALUE                  = 90
MINIMUM LENGTH OF CURVE  = 135
CURVE LENGTH CHOSEN      = 135

CHAINAGE OF INTERSECTION = 671.34
REDUCED LEVEL OF I.P.    = 93.6
CHAINAGE OF 1ST TANG.PT. = 603.84
CHAINAGE OF 2ND TANG.PT. = 738.84
***************************************

CHAINAGE  LENGTH   HT.DIFF.   RED.LEVEL
=======================================
603.84     0        0          92.925
620        16.16    .147       93.072
640        36.16    .289       93.214
660        56.16    .386       93.311
680        76.16    .439       93.364
693.84     90       .45        93.375
700        96.16    .448       93.373
720        116.16   .412       93.337
738.84     135      .338       93.263
***************************************

LOCATION OF CURVE :- GLENMORE
*************************************************
OPERATOR'S NAME :- JOHN MACMILLAN
DATE OF CALCS.  :- 15/08/83
INSTRUMENT USED :- APPLE II+
*************************************************
```

```
VERTICAL CURVE DESIGN
****************************************
GRADIENT OF 1ST STRAIGHT = -6
GRADIENT OF 2ND STRAIGHT = 1
ALGEBRAIC DIFFERENCE M-N = -7
K-VALUE                  = 9Ø
MINIMUM LENGTH OF CURVE  = 63Ø
CURVE LENGTH CHOSEN      = 65Ø

CHAINAGE OF INTERSECTION = 2Ø1Ø
REDUCED LEVEL OF I.P.    = 58.62
CHAINAGE OF 1ST TANG.PT. = 1685
CHAINAGE OF 2ND TANG.PT. = 2335
****************************************

CHAINAGE  LENGTH   HT.DIFF.   RED.LEVEL
========================================
1685      Ø        Ø          78.12
171Ø      25       -1.466     76.654
174Ø      55       -3.137     74.983
177Ø      85       -4.711     73.4Ø9
18ØØ      115      -6.188     71.932
183Ø      145      -7.568     7Ø.552
186Ø      175      -8.851     69.269
189Ø      2Ø5      -1Ø.Ø37    68.Ø83
192Ø      235      -11.126    66.994
195Ø      265      -12.119    66.ØØ1
198Ø      295      -13.Ø14    65.1Ø6
2Ø1Ø      325      -13.812    64.3Ø8
2Ø4Ø      355      -14.514    63.6Ø6
2Ø7Ø      385      -15.119    63.ØØ1
21ØØ      415      -15.626    62.494
213Ø      445      -16.Ø37    62.Ø83
216Ø      475      -16.351    61.769
219Ø      5Ø5      -16.568    61.552
222Ø      535      -16.688    61.432
2242.143  557.143  -16.714    61.4Ø6
225Ø      565      -16.711    61.4Ø9
228Ø      595      -16.637    61.483
231Ø      625      -16.466    61.654
2335      65Ø      -16.25     61.87
****************************************
```

6

Observational errors and their adjustment

6.1 INTRODUCTION

In each of the previous microcomputer programs, just sufficient observations were obtained to give a unique solution. If, however, more measurements are taken than are actually required, it will invariably be found that there are several solutions due to observational errors. Since land surveying is almost wholly concerned with site measurements and setting out to specifications, the question, 'Which value or measurement of an angle/distance is the correct one?', is of real importance. In many instances it is almost as important to know the accuracy of the result as to know the result itself.

6.1.1 Observational errors

Observational errors can be classified under five headings:

(1) *Mistakes* are avoidable errors which do not conform to any law or pattern, for example, misreading or misbooking. Systematic procedures and careful checking must eliminate these errors.
(2) *Constant errors* are of constant magnitude and sign, for example, tape standardization or vertical index error. The effect of constant errors must be minimized by systematic computation or observation.
(3) *Systematic errors* are of varying magnitude and may be either of constant sign, for example, the misalignment of a tape or trunnion axis error, or of varying sign in the form of periodic errors, for example, graduation errors. Both types can be minimized by systematic observations and reduction techniques.
(4) *Periodic errors* are of varying magnitude and sign but they obey systematic laws, for example, tape and circle graduation errors. Their effects can be minimized by repeated measurements on different parts of the graduations.
(5) *Random (or accidental) errors* are compensating and are generally avoidable. They represent the residual errors (r) left after all the others have been

eliminated. Unfortunately they are present in all observations and are caused by imperfections in the equipment and the observer, together with varying conditions. They usually conform to the laws of probability, and their effect can be minimized by taking the mean of repeated observations.

6.1.2 Arithmetic mean

The *arithmetic mean* ($\bar{x}$) of a set of observations $x_1, x_2 \ldots x_n$ is therefore given by:

$$\bar{x} = \Sigma x/n \tag{6.1}$$

where n is the number of observations. It should be noted that some text-books refer to $\bar{x}$ as the *most probable value* (*mpv*). It follows from the earlier section on random errors that the residual error or deviation in any individual observation (r) can now be found from:

$$r = (x - \bar{x}) \tag{6.2}$$

For readers unfamiliar with the concept of theory of errors applied to surveying, it is suggested they consult one of the text-books listed in the Bibliography, for the derivation of the various formulae presented in this chapter.

6.1.3 Principle of least squares

The arithmetic mean or most probable value of $\bar{x}$ found from Equation 6.1 is the same value as that found using the principle of least squares which states:

> 'the most probable value of any observed quantity is such that the sum of the squares of the deviations (residuals) of the observations from this value is least.'

6.1.4 Standard deviation and standard error

Previous mention was made of observational errors conforming to the laws of probability. Whereas the arithmetic mean discussed earlier allows residuals to be calculated, it does not give a measure of the spread of the observations, with an indication of the accuracy of the measurements. This spread or deviation of the observations for n measurements is given by the sample *standard deviation* (s), where:

$$s = [\Sigma(x - \bar{x})^2/(n-1)]^{1/2} \tag{6.3}$$

The reliability of a set of random observations can then be determined from the relationship of the reliability of the arithmetic mean ($\bar{x}$) to the sample size. The standard deviation of the mean is usually referred to as the *standard error of the mean* ($SE_{\bar{x}}$), where:

$$SE_{\bar{x}} = [\Sigma(x - \bar{x})^2/n(n-1)]^{1/2} \tag{6.4}$$

Comparing Equations 6.3 and 6.4, it will be seen that the latter can be rewritten as:

$$SE_{\bar{x}} = s/(n)^{1/2} \tag{6.5}$$

6.1.5 Probable error

In discussing laws of probability it is usual to assume a Gaussian distribution. In many textbooks the term *probable error* represents 50% of the area under the distribution curve. In modern times this term is synonymous with *50% uncertainty* and *50% confidence interval*, and its value (E) in a single observation can be found from:

$$E_x = 0.6745s \tag{6.6}$$

and the *probable error of the mean* from:

$$E_{\bar{x}} = 0.6745SE_{\bar{x}} \tag{6.7}$$

The above of course only refers to a single measurement; if for example a distance is measured in several lengths, the probable error of the sum of n measurements which are statistically independent will be given by:

$$E_x = [E_1^{\,2} + E_2^{\,2} + E_3^{\,2} \ldots E_n^{\,2}]^{1/2} \tag{6.8}$$

6.1.6 Weighted observations

Until now it has been assumed that all observations of an angle or distance have been to the same precision, that is by the same operator using the same equipment under similar conditions. If, however, the above does not apply, that is the observed values are not of equal precision, then a *weight* may be attributed to the values. This weight is a measure of the reliability of one value compared with the others, where the better the precision, the higher the weight (w). If in observations $x_1, x_2 \ldots x_n$ the corresponding weights are $w_1, w_2 \ldots w_n$, then the most probable value of x, known as the *weighted mean* ($\bar{x}_w$), will be given by:

$$\bar{x}_w = (w_1x_1 + w_2x_2 + \ldots w_nx_n)/(w_1 + w_2 + \ldots w_n) \tag{6.9}$$

Substituting for integer i with values 1 to n gives:

$$\bar{x}_w = \Sigma w_i x_i / \Sigma w_i \tag{6.10}$$

The *standard error of the weighted mean* ($SE_{\bar{x}w}$) and the *probable error of the weighted mean* ($E_{\bar{x}w}$) are given by:

$$SE_{\bar{x}w} = [\Sigma w_i(x_i - \bar{x}_w)^2/\Sigma w_i(n-1)]^{1/2} \tag{6.11}$$

and

$$E_{\bar{x}w} = 0.6745SE_{\bar{x}w} \tag{6.12}$$

6.1.7 Combination of errors

The formulae presented so far in this chapter all refer to individual measurements, either angles or distances but not a combination of both. However, on site the surveying data is dependent on a combination of angles and distances, for example a distance and bearing to compute the co-ordinates of a point, or the multiplication of several lengths to give an area. As each measurement contains an error, it is essential to consider the combined effect of these errors on the derived quantity.

The general procedure is to differentiate with respect to each of the observed (statistically independent) quantities in turn. Thus if:

$$x = f(a, b, c, \ldots)$$

where $a, b, c, \ldots$ each contain an error δa, δb, δc then the total error in x will be:

$$\delta x = \mathrm{d}f/\mathrm{d}a\,\delta a + \mathrm{d}f/\mathrm{d}b\,\delta b + \mathrm{d}f/\mathrm{d}c\,\delta c + \ldots \qquad (6.13)$$

If it is required to find the standard error in x due to standard or probable errors in a, b, c, etc., then it would be given by:

$$SE_x(f) = [(\mathrm{d}f/\mathrm{d}a SE_a)^2 + (\mathrm{d}f/\mathrm{d}b SE_b)^2 + (\mathrm{d}f/\mathrm{d}c SE_c)^2 + \ldots]^{1/2} \qquad (6.14)$$

6.2 OBSERVATIONAL ERROR ANALYSIS

The Equations 6.1–6.14 presented in the previous section, to enable an error analysis of numerous observations to be made, can beneficially be adopted for the modern microcomputer, with a considerable saving in time. The following program <ERROR ANALYSIS>, gives the operator a choice from the initial menu of:

<1> UNWEIGHTED OBSERVATIONS
<2> WEIGHTED OBSERVATIONS
<3> MULTIPLE OBSERVATIONS
<4> AREA CALCULATIONS

where it is possible to enter data in one section before transferring to another if required.

6.2.1 Error analysis – subroutine index

	Line numbers	*Function*
(a)	10–90	Initialization and control
(b)	100–230	Screen header display
(c)	500–680	Entry of data location, etc.
(d)	700–880	Program selection from menu
(e)	1000–1450	<1> Unweighted observations
(f)	2000–2530	<2> Weighted observations
(g)	3000–3440	<3> Multiple observations

	Line numbers	*Function*
(h)	4000–4290	<4> Area calculations
(i)	6000–6080	Routine to print data location, etc.
(j)	6100–6190	Printout for <1>
(k)	6200–6290	Printout for <2>
(l)	6300–6420	Printout for <3>
(m)	6500–6580	Printout for <4>
(n)	8000–8040	Routine to convert DD.MMSS to decimals
(o)	8050–8140	Routine to convert decimals to DD.MMSS
(p)	9000–9010	Routine to clear screen
(q)	9100–9130	Routine to display error message
(r)	9200–9290	Error trace routine
(s)	9500–9620	Termination of program

(a) Initialization and control

Line numbers 10–90

All numeric variables, arrays and required string variables are initialized. It should be noted that if any sample of observations is expected to contain more than twenty-one readings, the arrays should be redimensioned.

(b) Screen header display

Line numbers 100–230

This is the screen display used by civil engineering students at Strathclyde University. The operator can alter this as required to suit another organization. Options are given to enter location, new data or to quit.

(c) Entry of data location, etc.

Line numbers 500–680

This routine is only accessed if requested from (b) above, and is provided to keep a record of the data location, operator's name, date, etc., with an opportunity to obtain a printout if required.

(d) Program selection from menu

Line numbers 700–880

If the 'new data' key was selected from the screen header display, the program jumps to this routine to select the type of error analysis required:

<1> UNWEIGHTED OBSERVATIONS
<2> WEIGHTED OBSERVATIONS
<3> MULTIPLE OBSERVATIONS
<4> AREA CALCULATIONS

If <4> is selected, the program jumps directly to line 4000. If, however, one of the other options is selected, information is requested as to whether the measurements are of angle, level or distance. If angles have been observed, there is a choice of using either a decimal or DD.MMSS format before branching to the program selection at line 890.

(e) <1> Unweighted observations

Line numbers 1000–1450

After entering the number of measurements (*N*), a loop is set up to enter each observation into the array *M*(*I*), which is then transferred to array *A*(*I*) before computation. The arithmetic mean (Equation 6.1) is found in line 1140 and then the residuals (Equation 6.2) found in line 1160. The standard deviation (Equation 6.3) is calculated in line 1200, together with the standard error of the mean (Equations 6.4 and 6.5) at line 1210. Other parameters also computed are the probable error (Equation 6.6) at line 1220, and the probable error of the mean (Equation 6.7) at line 1230.

The results are presented from line 1300 with an option to obtain a printout for record purposes. The printout also includes a list of the observations.

(f) <2> Weighted observations

Line numbers 2000–2530

As in (e) above, *N* is the number of measurements, which are first entered in array *M*(*I*) and then transferred to array *A*(*I*). When weighted measurements are analysed it is essential to differentiate between level observations and angle/distance observations. The latter use a direct weight, *W*(*I*), entered at line 2190. In the case of the former (level observations) the weight of each reading is inversely proportional to the number of set-ups of the instrument, entered as *S*(*I*) in line 2160 and converted to *W*(*I*) in line 2170.

The weighted mean (Equations 6.9 and 6.10) is found in line 2230 and the residuals determined in line 2250. The standard error of the weighted mean (Equation 6.11) is calculated in line 2290, and the probable error of the weighted mean (Equation 6.12) at line 2300.

The results are presented from line 2400 with an option to obtain a printout for record purposes. The printout also includes a list of observations and weights.

(g) <3> Multiple observations

Line numbers 3000–3440

As in (e) and (f) above, *N* is the number of measurements which are first entered in array *M*(*I*) and then transferred to array *A*(*I*). The probable error for each measurement (discussed in Section 6.1.5) is then entered as *P*(*I*). If the observations were of angles, the probable error is entered in degrees for the decimal format and seconds for the DD.MMSS format so that either imperial or metric units can be entered. However, the units for *M*(*I*) and *P*(*I*) must be of the same order.

The probable error of the multiple observations (Equation 6.8) is computed in line 3220 and each individual reading, plus the total, is presented from line 3300. The printout option provided records all keyboard entries for the measurements and probable errors.

(h) <4> Area calculations

Line numbers 4000–4290

This program provides error calculations for either rectangles or triangles where a base (B) and height (H) measurement have been made, together with their probable errors. The concept of combined error analysis computations was discussed in Section 6.1.7, and the standard error in the area calculation is found from Equation 6.14. The area and probable error for a rectangle ($x = BH$) is calculated in lines 4100–4120 and for a triangle ($x = BH/2$) in lines 4140–4160, with a screen display of the results from line 4200. A printout option is also provided for record purposes.

(i) Routine to print data location, etc.

Line numbers 6000–6080

This routine is provided for record purposes, and is based on Applesoft DOS and may require alteration for different microcomputers and printers.

(j) Printout for <1>

Line numbers 6100–6190

This printout includes all the observations and re-uses the previous screen display discussed in (e) above.

(k) Printout for <2>

Line numbers 6200–6290

This printout includes all the observations and weights, and re-uses the previous screen display discussed in (f) above.

(l) Printout for <3>

Line numbers 6300–6420

This printout includes all the observations and their probable errors, together with the total angle/distance and probable error from (g) above.

(m) Printout for <4>

Line numbers 6500–6580

This printout records the type of area calculation, the measurements and probable errors for the computed area, as discussed in (h) above.

(n) Routine to convert DD.MMSS to decimals

Line numbers 8000–8040

This routine was discussed previously in Section 2.2.6.

(o) Routine to convert decimals to DD.MMSS

Line numbers 8050–8140

This routine was discussed previously in Section 2.2.6.

(p) Routine to clear screen

Line numbers 9000–9010

To simplify portability, the Applesoft 'HOME' command to clear the screen has been placed in a subroutine to allow an easy change to another microcomputer.

(q) Routine to display error message

Line numbers 9100–9130

If a wrong keyboard entry has been made, the error is trapped and this error message displayed and re-entry requested.

(r) Error trace routine

Line numbers 9200–9290

This error trace routine was discussed earlier in Section 2.2.9.

(s) Termination of program

Line numbers 9500–9620

When the '<QUIT>' selection is made in the program, an end-of-program message is displayed to advise the operator that computations are complete.

6.2.2 Error analysis – numeric variables

A(*I*)	= Array storage for measurements
C	= Counter in error trace routine
D, *D*0, *D*1, *D*2, *D*5	= Degrees of angles
*D*9	= <1> for decimal degrees or <2> for DD.MMSS
E	= Error number
I	= Integer loop variable
J	= Integer loop variable
K	= Integer loop variable
LD	= <1> for level or <2> for distance
M, *M*0, *M*1	= Minutes of angles

Table—continued

M(I)	= Array storage for measurements
N	= Number of measurements
*P*1	= Program selection number
*P*7	= Printer code number
PE	= Probable error
PM	= Probable error of mean
PT	= Probable error total
P(I)	= Array storage for probable errors
*R*9	= 57.2957795
R(I)	= Array storage for residuals
*R*2*(I)*	= Array storage for *R(I)* squared
S, *S*0, *S*1	= Seconds of angles
SD	= Standard deviation
SE	= Standard error of weighted mean
SM	= Standard error of mean
S(I)	= Array for set-ups in levelling
TM	= Total sum of measurements
TR	= Total sum of weighted residuals
U	= Initial menu selection
WB	= Weighted mean
WT	= Total sum of weights
WX	= Total sum of weighted observations
W(I)	= Array storage for weights of observations
XB	= Arithmetic mean

6.2.3 Error analysis – string variables

A$	Storage for location of data
B$	Storage for operator's name
C$	Storage for date of calculations
D$	= CHR$(4), CTRL-D
*E*0$	= "SORRY DATA ERROR . . . PLEASE RE-ENTER"
H$	= "ERROR ANALYSIS OF OBSERVATIONS"
K$	Storage for instrument used
M$	= "A" for angle or "D" for distance
N$	= "ANGLE" or "DISTANCE" in <1>; also "LEVEL" in <2>; "(SECS)" or "(DEGS)" in <3>; "RECTANGLE" or "TRIANGLE" in <4>
P$	= "DO YOU WISH PRINTOUT OF ABOVE DATA (Y/N)"
*P*1$	= "<1> UNWEIGHTED OBSERVATIONS"
*P*2$	= "<2> WEIGHTED OBSERVATIONS"
*P*3$	= "<3> MULTIPLE OBSERVATIONS"
*P*4$	= "<4> AREA CALCULATIONS"
Q$	= Question response (Y/N)

6.2.4 Error analysis – BASIC program

```
10  REM  <ERROR ANALYSIS> PROGRAM FOR APPLE II+  USES QUME PRINTER
20 D$ =  CHR$ (4): ONERR  GOTO 9200
30 C = 0:E = 0:P1 = 0:P7 = 0
35  DIM A(20),M(20),P(20),R(20),R2(20),S(20),W(20)
40 H$ = "ERROR ANALYSIS OF OBSERVATIONS"
```

```
50 EØ$ = "SORRY, DATA ERROR ... PLEASE RE-ENTER."
55 P$ = "DO YOU WISH PRINTOUT OF ABOVE DATA (Y/N)"
60 L$ = " ":M$ = " ":N$ = " "
95  REM  **************************************************
100  GOSUB 9000
110  PRINT "****************************************"
120  PRINT "*                                      *"
130  PRINT "*     UNIVERSITY  OF  STRATHCLYDE      *"
140  PRINT "*  DEPARTMENT  OF  CIVIL  ENGINEERING  *"
150  PRINT "*                                      *"
160  PRINT "*          SURVEYING  SECTION          *"
170  PRINT "*    ERROR ANALYSIS OF OBSERVATIONS    *"
180  PRINT "****************************************"
190  PRINT
200  PRINT "DO YOU WISH TO ENTER LOCATION <1>,      NEW DATA <2> OR QUIT
     <3>"
210  INPUT U
220  IF U < 1 OR U > 3 THEN 200
230  ON U GOTO 500,700,9500
500  GOSUB 9000
510  PRINT "UNIVERSITY OF STRATHCLYDE"
520  PRINT H$
530  PRINT "****************************************"
540  PRINT
550  PRINT "ENTER LOCATION OF WORK, ETC. WHEN       REQUESTED, AND PRESS
     <RETURN>"
560  PRINT "ENTER LOCATION OF DATA :-"
570  INPUT A$
580  PRINT "ENTER OPERATOR'S NAME :-"
590  INPUT B$
600  PRINT "ENTER DATE OF CALCULATION AS DD/MM/YY -"
610  INPUT C$
620  PRINT "ENTER INSTRUMENT USED :-"
630  INPUT K$
640  PRINT
650  PRINT "DO YOU WISH PRINTOUT OF LOCATION ETC.,  (Y/N) ";
660  INPUT Q$
670  IF Q$ = "Y" THEN 6000
680  IF Q$ < > "N" THEN 650
690 :
695  REM  PROGRAM MENU SELECTION
700  GOSUB 9000: PRINT H$
710  PRINT "****************************************"
720  PRINT "SELECT PROGRAM FROM MENU :-"
730  PRINT
740  PRINT "   <1> UNWEIGHTED OBSERVATIONS"
750  PRINT "   <2> WEIGHTED   OBSERVATIONS"
760  PRINT "   <3> MULTIPLE   OBSERVATIONS"
770  PRINT "   <4> AREA       CALCULATIONS"
780  INPUT P1
790  IF P1 < 1 OR P1 > 4 THEN 700
800  IF P1 = 4 THEN 4000
810  PRINT : PRINT "ARE MEASUREMENTS OF ANGLE (A) OR LEVEL/ DISTANCE (D);
820  INPUT M$
830  IF M$ = "D" THEN 880
840  IF M$ < > "A" THEN 810
850  PRINT : PRINT "ARE ANGLES IN DECIMALS<1> OR DD.MMSS<2>"
860  INPUT D9
870  IF D9 < 1 OR D9 > 2 THEN 850
880  ON P1 GOTO 1000,2000,3000
```

```
995 REM ************************************************
1ØØØ  GOSUB 9ØØØ:P1$ = "<1> UNWEIGHTED OBSERVATIONS"
1Ø1Ø  PRINT H$: PRINT P1$:TM = Ø:TR = Ø
1Ø2Ø  PRINT "****************************************"
1Ø3Ø  PRINT : PRINT "ENTER NUMBER OF MEASUREMENTS ";
1Ø4Ø  INPUT N
1Ø5Ø  FOR I = Ø TO N - 1
1Ø6Ø  PRINT : PRINT "ENTER MEASUREMENT NO. ";(I + 1);
1Ø7Ø  INPUT M(I)
1Ø8Ø  IF M$ = "D" THEN 111Ø
1Ø9Ø  IF D9 = 1 THEN 111Ø
11ØØ D5 = M(I): GOSUB 8ØØØ:A(I) = D5: GOTO 112Ø
111Ø A(I) = M(I)
1115  REM  TM IS TOTAL SUM OF MEASUREMENTS
112Ø TM = TM + A(I)
113Ø  NEXT I
1135  REM  XB IS ARITHMETIC MEAN
114Ø XB = TM / N
1145  REM  FIND RESIDUALS
115Ø  FOR J = Ø TO N - 1
116Ø R(J) = A(J) - XB
1165  REM  R2(J) IS SQUARE OF RESIDUAL
117Ø R2(J) = (R(J)) ^ 2
1175  REM  TR IS TOTAL OF SQUARES OF RESIDUALS
118Ø TR = TR + R2(J)
119Ø  NEXT J
1195  REM  SD IS STANDARD DEVIATION
12ØØ SD =  SQR (TR / (N - 1))
12Ø5  REM  SM IS STANDARD ERROR OF MEAN
121Ø SM = SD /  SQR (N)
1215  REM  PE IS PROBABLE ERROR
122Ø PE = .6745 * SD
1225  REM  PM IS PROBABLE ERROR OF MEAN
123Ø PM = .6745 * SM: IF M$ = "D" THEN 128Ø
1235  REM  I.E. ANGLE MEASUREMENT
124Ø  IF D9 = 1 THEN 128Ø
1245  REM  ANGLE IN DD.MMSS, FIND ERROR IN SECS
125Ø SD =  INT ((SD * 36ØØ) * 1ØØ + .5) / 1ØØ:SM =  INT ((SM * 36ØØ) * 1Ø
     Ø + .5) / 1ØØ
126Ø PE =  INT ((PE * 36ØØ) * 1ØØ + .5) / 1ØØ:PM =  INT ((PM * 36ØØ) * 1Ø
     Ø + .5) / 1ØØ
127Ø D = XB: GOSUB 8Ø5Ø:XB = D: GOTO 13ØØ
128Ø SD =  INT (SD * 1ØØØØ + .5) / 1ØØØØ:SM =  INT (SM * 1ØØØØ + .5) / 1Ø
     ØØØ
129Ø PE =  INT (PE * 1ØØØØ + .5) / 1ØØØØ:PM =  INT (PM * 1ØØØØ + .5) / 1Ø
     ØØØ:XB =  INT (XB * 1ØØØØ + .5) / 1ØØØØ
13ØØ  GOSUB 9ØØØ:P7 = Ø
131Ø  PRINT H$: PRINT P1$: PRINT "***************************************
     *"
132Ø  IF M$ = "A" THEN N$ = " ANGLE "
133Ø  IF M$ = "D" THEN N$ = " DISTANCE "
134Ø  PRINT : PRINT "RESULTS OF ";N;N$;"MEASUREMENTS :-"
135Ø  PRINT "ARITHMETIC MEAN         =     ";XB
136Ø  PRINT "STANDARD DEVIATION      =     ";SD
137Ø  PRINT "STANDARD ERROR OF MEAN = +/-";SM
138Ø  PRINT "PROBABLE ERROR          = +/-";PE
139Ø  PRINT "PROBABLE ERROR OF MEAN = +/-";PM
14ØØ  PRINT "****************************************"
141Ø  IF P7 = 1 THEN  RETURN
142Ø  PRINT : PRINT P$;: INPUT Q$
```

```
1430  IF Q$ = "Y" THEN 6100
1440  IF Q$ < > "N" THEN 1420
1450  GOTO 100
1995  REM  **************************************************
2000  GOSUB 9000:P2$ = "<2> WEIGHTED OBSERVATIONS"
2010  PRINT H$: PRINT P2$:LD = 3:TR = 0:WT = 0:WX = 0
2020  PRINT "****************************************"
2030  PRINT : PRINT "ENTER NUMBER OF MEASUREMENTS ";
2040  INPUT N
2050  FOR I = 0 TO N - 1
2060  PRINT : PRINT "ENTER MEASUREMENT NO. ";(I + 1);
2070  INPUT M(I)
2080  IF M$ = "D" THEN 2110
2090  IF D9 = 1 THEN 2110
2100 D5 = M(I): GOSUB 8000:A(I) = D5: GOTO 2120
2110 A(I) = M(I)
2120  IF M$ = "A" THEN 2180
2130  ON LD GOTO 2160,2180,2140
2140  PRINT : PRINT "ENTER (1) FOR LEVEL OR (2) FOR DISTANCE": INPUT LD: IF
     LD < 1 OR LD > 2 THEN 2140
2150  GOTO 2130
2160  PRINT : PRINT "ENTER NUMBER OF SET-UPS TAKEN DURING     LEVEL MEASUR
     EMENT NO. ";(I + 1);: INPUT S(I)
2170 W(I) =  INT ((1 / S(I)) * 1000 + .5) / 1000: GOTO 2200
2180  PRINT : PRINT "ENTER WEIGHT OF ABOVE MEASUREMENT ";
2190  INPUT W(I)
2195  REM  WX IS TOTAL SUM OF W(I)*A(I)
2200 WX = WX + W(I) * A(I)
2205  REM  WT IS TOTAL SUM OF W(I)
2210 WT = WT + W(I)
2220  NEXT I
2225  REM  WB IS WEIGHTED MEAN
2230 WB = WX / WT
2235  REM  IND RESIDUALS
2240  FOR J = 0 TO N - 1
2250 R(J) = A(J) - WB
2255  REM  R2(J) IS R(J) SQUARED * W(J)
2260 R2(J) = W(J) * (R(J)) ^ 2
2265  REM  TR IS TOTAL OF R2(J)'S
2270 TR = TR + R2(J)
2280  NEXT J
2285  REM  SE IS STANDARD ERROR OF WEIGHTED MEAN
2290 SE =  SQR (TR / (WT * (N - 1)))
2295  REM  PE IS PROBABLE ERROR OF WEIGHTED MEAN
2300 PE = .6745 * SE
2310  IF M$ = "D" THEN 2360
2315  REM  ANGLE MEASUREMENT
2320  IF D9 = 1 THEN 2360
2325  REM  ANGLES IN DD.MMSS, FIND ERROR IN SECS
2330 SE =  INT ((SE * 3600) * 100 + .5) / 100:PE =  INT ((PE * 3600) * 10
     0 + .5) / 100
2340 D = WB: GOSUB 8050:WB = D
2350  GOTO 2400
2360 SE =  INT (SE * 10000 + .5) / 10000
2370 PE =  INT (PE * 10000 + .5) / 10000
2380 WB =  INT (WB * 10000 + .5) / 10000
2390 :
2400  GOSUB 9000:P7 = 0
2410  PRINT H$: PRINT P2$: PRINT "***************************************
     *"
```

```
242Ø  IF M$ = "A" THEN N$ = " ANGLE "
243Ø  IF LD = 1 THEN N$ = " LEVEL "
244Ø  IF LD = 2 THEN N$ = " DISTANCE "
245Ø  PRINT : PRINT "RESULTS OF ";N;N$;"MEASUREMENTS :-"
246Ø  PRINT "WEIGHTED MEAN              -    ",WB
247Ø  PRINT "STANDARD ERROR WT.MEAN = +/-";SE
248Ø  PRINT "PROBABLE ERROR WT.MEAN = +/-";PE
249Ø  PRINT "****************************************": IF P7 = 1 THEN  RETURN
25ØØ  PRINT P$;: INPUT Q$
251Ø  IF Q$ = "Y" THEN 62ØØ
252Ø  IF Q$ < > "N" THEN 25ØØ
253Ø  GOTO 1ØØ
2995  REM  **************************************************
3ØØØ  GOSUB 9ØØØ:P3$ = "<3> MULTIPLE OBSERVATIONS"
3Ø1Ø  PRINT H$: PRINT P3$:PE = Ø:PT = Ø:TM = Ø
3Ø2Ø  PRINT "****************************************"
3Ø3Ø  PRINT : PRINT "ENTER NUMBER OF MEASUREMENTS ";
3Ø4Ø  INPUT N
3Ø5Ø  FOR I = Ø TO N - 1
3Ø6Ø  PRINT : PRINT "ENTER MEASUREMENT NO. ";(I + 1);
3Ø7Ø  INPUT M(I)
3Ø8Ø  IF M$ = "D" THEN 311Ø
3Ø9Ø  IF D9 = 1 THEN 311Ø
31ØØ D5 = M(I): GOSUB 8ØØØ:A(I) = D5: GOTO 312Ø
311Ø A(I) = M(I)
3115  REM  TM IS TOTAL SUM OF MEASUREMENTS
312Ø TM = TM + A(I)
313Ø  IF M$ = "D" THEN 319Ø
314Ø  IF D9 = 1 THEN 317Ø
315Ø  PRINT : PRINT "ENTER PROBABLE ERROR IN SECONDS ";: INPUT P(I)
316Ø  GOTO 32ØØ
317Ø  PRINT : PRINT "ENTER PROBABLE ERROR IN DEGREES ";: INPUT P(I)
318Ø  GOTO 32ØØ
319Ø  PRINT : PRINT "ENTER PROBABLE ERROR ";: INPUT P(I)
3195  REM  PT IS TOTAL OF SQUARE OF PROBABLE ERRORS
32ØØ PT = PT + (P(I) ^ 2)
321Ø  NEXT I
3215  REM  PE IS PROBABLE ERROR
322Ø PE =  SQR (PT)
323Ø  IF M$ = "D" THEN 329Ø
324Ø  IF D9 = 1 THEN 328Ø
325Ø D = TM: GOSUB 8Ø5Ø:TM = D
326Ø PE =  INT (PE * 1ØØ + .5) / 1ØØ
327Ø N$ = "(SECS)": GOTO 33ØØ
328Ø N$ = "(DEGS)"
329Ø PE =  INT (PE * 1ØØØØ + .5) / 1ØØØØ
33ØØ  GOSUB 9ØØØ: PRINT H$: PRINT P3$
331Ø  PRINT "****************************************"
332Ø  IF M$ = "A" THEN 334Ø
333Ø  PRINT "MEASUREMENT   DISTANCE  PROB.ERR.": GOTO 335Ø
334Ø  PRINT "MEASUREMENT   ANGLE     PROB.ERR.";N$
335Ø  PRINT "========================================"
336Ø  FOR J = Ø TO N - 1
337Ø  PRINT (J + 1); TAB( 15);M(J); TAB( 25);"+/-";P(J)
338Ø  NEXT J
339Ø  PRINT : PRINT "TOTAL"; TAB( 15);TM; TAB( 25);"+/-";PE
34ØØ  PRINT "****************************************"
341Ø  PRINT : PRINT P$;: INPUT Q$
342Ø  IF Q$ = "Y" THEN 63ØØ
```

```
343Ø  IF Q$ <  > "N" THEN 342Ø
344Ø  GOTO 1ØØ
3995  REM  ************************************************
4ØØØ  GOSUB 9ØØØ:P4$ = "<4> AREA CALCULATIONS"
4Ø1Ø  PRINT H$: PRINT P4$
4Ø2Ø  PRINT "****************************************"
4Ø3Ø  PRINT : PRINT "ARE MEASUREMENTS FOR RECTANGLE (R) OR   TRIANGLE (T)
     ";: INPUT M$
4Ø4Ø  IF M$ <  > "R" AND M$ <  > "T" THEN 4Ø3Ø
4Ø5Ø  PRINT : PRINT "ENTER BASE LENGTH      ";: INPUT B
4Ø6Ø  PRINT : PRINT "ENTER PROB.ERROR BASE ";: INPUT PB
4Ø7Ø  PRINT : PRINT "ENTER HEIGHT          ";: INPUT H
4Ø8Ø  PRINT : PRINT "ENTER PROB.ERROR HT.  ";: INPUT PH
4Ø9Ø  IF M$ = "T" THEN 414Ø
4Ø95  REM  TA IS TOTAL AREA RECTANGLE
41ØØ TA = B * H
41Ø5  REM  PE IS PROBABLE ERROR
411Ø PE =  SQR ((H * PB) ^ 2 + (B * PH) ^ 2)
412Ø N$ = "RECTANGLE"
413Ø  GOTO 417Ø
4135  REM  TA IS TOTAL AREA TRIANGLE
414Ø TA = B * H / 2
4145  REM  PE IS PROBABLE ERROR
415Ø PE =  SQR ((H * PB / 2) ^ 2 + (B * PH / 2) ^ 2)
416Ø N$ = "TRIANGLE"
417Ø TA =  INT (TA * 1ØØØ + .5) / 1ØØØ
418Ø PE =  INT (PE * 1ØØØ + .5) / 1ØØØ
419Ø :
42ØØ  GOSUB 9ØØØ: PRINT H$: PRINT P4$
421Ø  PRINT "****************************************"
422Ø  PRINT N$; TAB( 12);"MEASUREMENTS AND PROB.ERROR"
423Ø  PRINT : PRINT "BASE"; TAB( 12);B; TAB( 29);"+/-";PB
424Ø  PRINT "HEIGHT"; TAB( 12);H; TAB( 29);"+/-";PH
425Ø  PRINT : PRINT "AREA"; TAB( 12);TA; TAB( 29);"+/-";PE
426Ø  PRINT "****************************************"
427Ø  PRINT : PRINT P$;: INPUT Q$: IF Q$ = "Y" THEN 65ØØ
428Ø  IF Q$ <  > "N" THEN 427Ø
429Ø  GOTO 1ØØ
5985  REM  *********************************************************
     **
599Ø  REM  PRINTOUT OF LOCATION, ETC
6ØØØ  PRINT D$;"PR#1": PRINT "LOCATION OF DATA :-";A$
6Ø1Ø  PRINT "**********************************************************
     ********************"
6Ø2Ø  PRINT "OPERATOR'S NAME :- ";B$
6Ø3Ø  PRINT "DATE OF CALCS.  :-";C$
6Ø4Ø  PRINT "INSTRUMENT USED :- ";K$
6Ø6Ø  PRINT "**********************************************************
     ********************"
6Ø7Ø  PRINT : PRINT D$;"PR#Ø"
6Ø8Ø  GOTO 7ØØ
6Ø93  REM  ************************************************
6Ø95  REM  PRINTOUT FOR <1> UNWEIGHTED OBSERVATIONS
61ØØ  PRINT D$;"PR#1": POKE 1784 + 1,8Ø:P7 = 1
611Ø  PRINT H$: PRINT P1$
612Ø  PRINT "**********************************************************
     ********************"
613Ø  PRINT "FIELD MEASUREMENTS :-"
614Ø  FOR K = Ø TO N - 1: IF K = 7 THEN 616Ø
615Ø  PRINT M(K);", ";: GOTO 617Ø
616Ø  PRINT M(K);","
```

```
6165  REM  USE SCREEN DISPLAY AT LINES 134Ø-141Ø
617Ø  NEXT K: GOSUB 134Ø
618Ø  PRINT : PRINT D$;"PR#Ø"
619Ø  GOTO 1ØØ
6193  REM  **************************************************
6195  REM  PRINTOUT FOR <2> WEIGHTED OBSERVATIONS
62ØØ  PRINT D$;"PR#1": POKE 1784 + 1,8Ø:P7 = 1
621Ø  PRINT H$: PRINT P2$
622Ø  PRINT "***********************************************************
     ********************"
623Ø  PRINT "FIELD MEASUREMENTS :-"
624Ø  FOR K = Ø TO N - 1: IF K = 4 THEN 626Ø
625Ø  PRINT M(K);",(";W(K);"), ";: GOTO 627Ø
626Ø  PRINT M(K);",(";W(K);"), "
6265  REM  USE SCREEN DISPLAY AT LINES 245Ø-249Ø
627Ø  NEXT K: GOSUB 245Ø
628Ø  PRINT : PRINT D$;"PR#Ø"
629Ø  GOTO 1ØØ
6293  REM  **************************************************
6295  REM  PRINTOUT FOR <3> MULTIPLE OBSERVATIONS
63ØØ  PRINT D$;"PR#1": POKE 1784 + 1,8Ø
631Ø  PRINT H$: PRINT P3$: PRINT "****************************************
     *"
632Ø  IF M$ = "A" THEN 634Ø
633Ø  PRINT "MEASUREMENT   DISTANCE  PROB.ERR.": GOTO 635Ø
634Ø  PRINT "MEASUREMENT   ANGLE     PROB.ERR.";N$
635Ø  PRINT "========================================="
636Ø  FOR K = Ø TO N - 1
637Ø  PRINT (K + 1);: POKE 36,15: PRINT M(K);: POKE 36,25: PRINT "+/-";P(
     K)
638Ø  NEXT K
639Ø  PRINT : PRINT "TOTAL";: POKE 36,15: PRINT TM;: POKE 36,25: PRINT "+
     /-";PE
64ØØ  PRINT "****************************************"
641Ø  PRINT : PRINT D$;"PR#Ø"
642Ø  GOTO 1ØØ
6493  REM  **************************************************
6495  REM  PRINTOUT FOR <4> AREA CALCULATIONS
65ØØ  PRINT D$;"PR#1": POKE 1784 + 1,8Ø
651Ø  PRINT H$: PRINT P4$: PRINT "****************************************
     *"
652Ø  PRINT N$;: POKE 36,12: PRINT "MEASUREMENTS AND PROB.ERROR"
653Ø  PRINT : PRINT "BASE";: POKE 36,12: PRINT B;: POKE 36,29: PRINT "+/-
     ";PB
654Ø  PRINT "HEIGHT";: POKE 36,12: PRINT H;: POKE 36,29: PRINT "+/-";PH
655Ø  PRINT : PRINT "AREA";: POKE 36,12: PRINT TA;: POKE 36,29: PRINT "+/
     -";PE
656Ø  PRINT "****************************************"
657Ø  PRINT : PRINT D$;"PR#Ø"
658Ø  GOTO 1ØØ
7993  REM  **************************************************
7995  REM  ROUTINE DD.MMSS TO DECIMALS
8ØØØ DØ =  INT (D5)
8ØØ5  REM  FIND NO. OF MINUTES,MØ
8Ø1Ø M1 = (D5 - DØ) * 1ØØ:MØ =  INT (M1 + .1)
8Ø15  REM  FIND NO. OF SECONDS,SØ
8Ø2Ø SØ =  INT (M1 * 1ØØ - MØ * 1ØØ + .5)
8Ø25  REM  COLLECT DEGS,MINS,SECS
8Ø3Ø D5 = DØ + MØ / 6Ø + SØ / 36ØØ
8Ø4Ø  RETURN
8Ø43  REM  **************************************************
```

```
8Ø45  REM  ROUTINE DECIMAL TO DD.MMSS
8Ø5Ø D1 =  INT (D)
8Ø55  REM  FIND TOTAL NO. OF SECONDS
8Ø6Ø D2 = (D - D1) * 36ØØ
8Ø65  REM  FIND NO. OF MINUTES
8Ø7Ø M =  INT (D2 / 6Ø)
8Ø75  REM  FIND NO. OF SECONDS
8Ø8Ø S = D2 - 6Ø * M + .5
8Ø9Ø  IF S < 6Ø THEN 813Ø
81ØØ M = M + 1:S = Ø
811Ø  IF M < 6Ø THEN 813Ø
812Ø D1 = D1 + 1:M = Ø
813Ø D = D1 * 1ØØØØ + M * 1ØØ + S
814Ø D =  INT (D) / 1ØØØØ: RETURN
8985  REM  *****************************************
899Ø  REM  GOSUB ROUTINE TO CLEAR SCREEN
9ØØØ  HOME
9Ø1Ø  RETURN
9Ø93  REM  **************************************************
9Ø95  REM  GOSUB ROUTINE FOR ERRORS
91ØØ  PRINT
911Ø  PRINT EØ$
912Ø  PRINT
913Ø  RETURN
9193  REM  **************************************************
9195  REM  ERROR TRACE ROUTINE
92ØØ  PRINT  CHR$ (7): REM  CTRL-B (BELL)
921Ø E =  PEEK (222): REM  GET ERROR NO.
922Ø  IF C = 1 THEN 926Ø
923Ø  INVERSE
924Ø  PRINT "ERROR NO. ";E;" FOUND"
925Ø C = 1: TRACE : RESUME
926Ø  PRINT "ERROR ON SECOND LINE NO. "
927Ø  NORMAL
928Ø  NOTRACE
929Ø  STOP
9493  REM  **************************************************
9495  REM  END OF PROGRAM
95ØØ  GOSUB 9ØØØ
951Ø  PRINT
952Ø  PRINT "END ERROR ANALYSIS PROGRAM"
953Ø  PRINT
954Ø  PRINT "**************************"
955Ø  REM  PROGRAM PREPARED BY
956Ø  REM  DR. P.H. MILNE
957Ø  REM  DEPT. OF CIVIL ENGINEERING
958Ø  REM  UNIVERSITY OF STRATHCLYDE
959Ø  REM  GLASGOW G4 ONG
96ØØ  REM  SCOTLAND
961Ø  REM  **************************
962Ø  END
```

6.2.5 Error analysis – computer printout

```
 LOCATION OF DATA :-STRATHCLYDE UNIVERSITY - STEPPS
*************************************************************************
OPERATOR'S NAME :- GORDON BLACK
DATE OF CALCS.  :-25/Ø8/83
INSTRUMENT USED :- APPLE II+
*************************************************************************
```

```
ERROR ANALYSIS OF OBSERVATIONS
<1> UNWEIGHTED OBSERVATIONS
****************************************************************************
FIELD MEASUREMENTS :-
123.1415, 123.1417, 123.1413, 123.1421, 123.1414, 123.1421, 123.1414, 123.1417,

RESULTS OF 8 ANGLE MEASUREMENTS :-
ARITHMETIC MEAN        =    123.1417
STANDARD DEVIATION     =    3.12
STANDARD ERROR OF MEAN = +/-1.1
PROBABLE ERROR         = +/-2.1
PROBABLE ERROR OF MEAN = +/-.74
***************************************

ERROR ANALYSIS OF OBSERVATIONS
<1> UNWEIGHTED OBSERVATIONS
****************************************************************************
FIELD MEASUREMENTS :-
432.33, 432.36, 432.31,
RESULTS OF 3 DISTANCE MEASUREMENTS :-
ARITHMETIC MEAN        =    432.3333
STANDARD DEVIATION     =    .Ø252
STANDARD ERROR OF MEAN = +/-.Ø145
PROBABLE ERROR         = +/-.Ø17
PROBABLE ERROR OF MEAN = +/-9.8E-Ø3
***************************************

ERROR ANALYSIS OF OBSERVATIONS
<2> WEIGHTED OBSERVATIONS
****************************************************************************
FIELD MEASUREMENTS :-
123.1225,(2), 123.123,(5), 123.1235,(4), 123.124,(3),
RESULTS OF 4 ANGLE MEASUREMENTS :-
WEIGHTED MEAN          =    123.1233
STANDARD ERROR WT.MEAN = +/-2.83
PROBABLE ERROR WT.MEAN = +/-1.91
***************************************

ERROR ANALYSIS OF OBSERVATIONS
<2> WEIGHTED OBSERVATIONS
****************************************************************************
FIELD MEASUREMENTS :-
29.492,(.125), 29.44,(.1), 29.485,(.Ø77),
RESULTS OF 3 LEVEL MEASUREMENTS :-
WEIGHTED MEAN          =    29.473
STANDARD ERROR WT.MEAN = +/-.Ø165
PROBABLE ERROR WT.MEAN = +/-.Ø112
***************************************

ERROR ANALYSIS OF OBSERVATIONS
<2> WEIGHTED OBSERVATIONS
*********************************************************************
FIELD MEASUREMENTS :-
432.33,(2), 432.36,(4), 432.31,(2),
RESULTS OF 3 DISTANCE MEASUREMENTS :-
WEIGHTED MEAN          =    432.34
STANDARD ERROR WT.MEAN = +/-.Ø15
PROBABLE ERROR WT.MEAN = +/-.Ø1Ø1
***************************************
```

```
ERROR ANALYSIS OF OBSERVATIONS
<3> MULTIPLE OBSERVATIONS
****************************************
MEASUREMENT   ANGLE     PROB.ERR.(SECS)
========================================
1              34.4522  +/-2
2              45.2345  +/-3
3              52.3212  +/-2

TOTAL          132.4119 +/-4.12
****************************************

ERROR ANALYSIS OF OBSERVATIONS
<3> MULTIPLE OBSERVATIONS
****************************************
MEASUREMENT   DISTANCE  PROB.ERR.
========================================
1              961.22   +/-.Ø44
2              433.12   +/-.Ø31
3              1545.9   +/-.Ø6
4              355.4    +/-.Ø21
5              1252.54  +/-.1
6              32Ø.4    +/-.Ø75

TOTAL          4868.58  +/-.15Ø2
****************************************

ERROR ANALYSIS OF OBSERVATIONS
<4> AREA CALCULATIONS
****************************************
RECTANGLE    MEASUREMENTS AND PROB.ERROR

BASE         85.72          +/-.Ø5
HEIGHT       55.28          +/-.Ø3

AREA         4738.6Ø2       +/-3.775
****************************************

ERROR ANALYSIS OF OBSERVATIONS
<4> AREA CALCULATIONS
****************************************
TRIANGLE     MEASUREMENTS AND PROB.ERROR

BASE         73.49          +/-.Ø5
HEIGHT       47.52          +/-.Ø3

AREA         1746.122       +/-1.621
****************************************
```

6.3 ADJUSTMENT OF OBSERVATIONS

Mention was made in Section 6.1.3 that the most probable value of a measured quantity, subject only to random errors, will be derived when the sum of the residual errors of the observations is a minimum. On site, however, it is normal practice to obtain additional or *redundant observations* to ensure the detection of mistakes and a better estimate of the most probable value.

To determine the most probable value of the measured quantity there are two suitable methods, either:

(i) The method of *observation equations*,
or
(ii) The method of *condition equations*.

When the number of observations is greater than the number of unknown values, then the redundant observations, using the principle of least squares, are used to produce a set of *observation equations*. An observation equation therefore expresses the result of a direct measurement of a quantity and its relationship to the residual and most probable value, as given in Equation 6.2.

If there is an observational or geometrical condition to be fulfilled by the observations, then the most probable values are derived from the condition equations.

Some conditions commonly used to create condition equations are:

(i) The three angles of a triangle observed in the field must add to 180° (plus spherical excess if appropriate).
(ii) The centre angles of a polygon must add to 360°.
(iii) A closed circuit of levels must be adjusted so that the algebraic sum is zero.

It should be noted that the number of condition equations must be less than the number of observations if the most probable values are to be solved. Care must also be taken to ensure when forming the condition equations, that they are independent of one another.

The following program <OBSERVATION ADJUSTMENT> for angle and level adjustment includes the three conditions listed above, in addition to an angle adjustment for two angles measured separately and then together, as in the corner of a braced quadrilateral. The special case of the braced quadrilateral is considered in Section 6.4. The solutions for the program in this section are for unweighted single observations. Where multiple or combined observations of angle, level or distance have been taken, especially if weighting is to be applied, the use of matrix solutions considerably simplifies the computation, as discussed in Section 6.5.

6.3.1 Adjustment of observations – subroutine index

	Line numbers	*Function*
(a)	10–90	Initialization and control
(b)	100–230	Screen header display
(c)	500–680	Entry of data location, etc.
(d)	700–820	Program selection from menu
(e)	1000–1390	Angle adjustment <1> two angles
(f)	1400–1690	Angle adjustment <2> triangle
(g)	1700–1990	Angle adjustment <3> centre of polygon
(h)	2000–2240	Level adjustment

(i)	6000–6080	Routine to print data location, etc.
(j)	6100–6150	Printout for angle adjustment <1>
(k)	6200–6250	Printout for angle adjustment <2>
(l)	6300–6350	Printout for angle adjustment <3>
(m)	6400–6490	Printout for level adjustment
(n)	8000–8040	Routine to convert DD.MMSS to decimals
(o)	8050–8140	Routine to convert decimals to DD.MMSS
(p)	9000–9010	Routine to clear screen
(q)	9100–9130	Routine to display error message
(r)	9200–9290	Error trace routine
(s)	9500–9620	Termination of program

(a) Initialization and control

Line numbers 10–90

The numeric variables, arrays and required string variables are initialized. It should be noted that if the centre of the polygon is made up of more than ten segments, or if more than ten level observations are adjusted, then the arrays should be redimensioned accordingly.

(b) Screen header display

Line numbers 100–230

This is the screen display used by civil engineering students at Strathclyde University. The operator can alter this as required to suit another organization. Options are given to enter location, new data or to quit.

(c) Entry of data location, etc.

Line numbers 500–680

This routine is only accessed if requested from (b) above, and is provided to keep a record of the data location, operator's name, date, etc., with an opportunity to obtain a printout if required.

(d) Program selection from menu

Line numbers 700–820

If the 'new data' key was selected from the screen header display, the program jumps to this routine to select the type of observations:

<1> ANGLE OBSERVATIONS
<2> LEVEL OBSERVATIONS

If <2> is selected, the program jumps directly to line 2000. If, however, the angle program is selected, the operator has a choice of entering the angles in either decimal or DD.MMSS format.

(e) Angle adjustment <1> two angles

Line numbers 1000–1390

The angle adjustment menu is presented initially, and the two angles section commences at linc 1100.

This solution uses the observation equations method. Consider the angles AOB, BOC and AOC in Fig. 6.1, with measurements M_x, M_y and M_z. The purpose of the adjustment is to find the most probable values of the three angles, x, y and z. If the residuals, that is the difference between the most probable

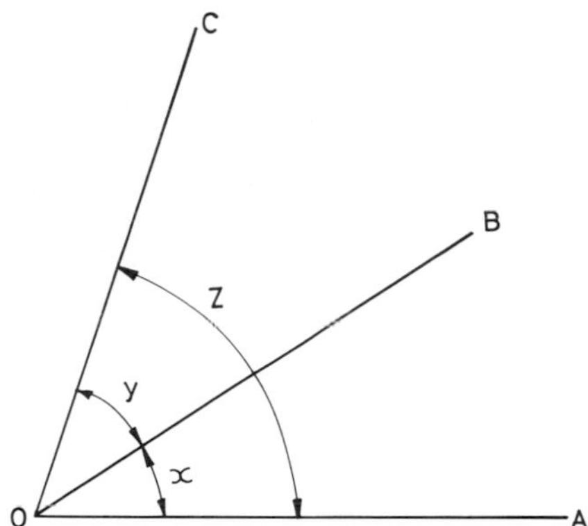

Fig. 6.1. Nomenclature for angle adjustment where redundant observations have been obtained.

values and the observed values, are denoted by R_x, R_y and R_z then three observation equations can be formed, where:

$$R_x = x - M_x \tag{6.15}$$

$$R_y = y - M_y \tag{6.16}$$

$$R_z = z - M_z \tag{6.17}$$

The most probable values of x, y and z will then be given when the sum of the square of the residuals is a minimum as discussed in Section 6.1.3. However, the most probable values must also be geometrically consistent, since in this case (Fig. 6.1):

$$z = x + y \tag{6.18}$$

Equation 6.17 can therefore be rewritten using the substitution from Equation 6.18, therefore:

$$R_z = x + y - M_z \tag{6.19}$$

Now the most probable values of x, y and z will be given when:

$$R_x^2 + R_y^2 + R_z^2 = \text{a minimum} \tag{6.20}$$

i.e.

$$(x - M_x)^2 + (y - M_y)^2 + (x + y - M_z)^2 = \text{a minimum} \tag{6.21}$$

Since x and y are independent of each other, the minimum value of Equation 6.21 will be given when the first derivatives, with respect to x and y, are both zero, since a second order expression with positive coefficients has a minimum turning-point when its first derivative is zero.

First, differentiate Equation 6.21 with respect to x to give:

$$2(x - M_x) + 2(x + y - M_z) = 0$$

i.e.

$$2x + y - M_x - M_z = 0 \tag{6.22}$$

Second, differentiate Equation 6.21 with respect to y to give:

$$2(y - M_y) + 2(x + y - M_z) = 0$$

i.e.

$$x + 2y - M_y - M_z = 0 \tag{6.23}$$

Then eliminate y from Equations 6.22 and 6.23 to find x, where:

$$x = (2M_x + M_z - M_y)/3 \tag{6.24}$$

and eliminate x from Equations 6.22 and 6.23 to find y where:

$$y = (2M_y + M_z - M_x)/3 \tag{6.25}$$

to finally give z from Equation 6.18.

In this program for angle adjustment '<1> TWO ANGLES', the angles M_x, M_y, M_z are entered as A_1, A_2, A_3 respectively, and after conversion from DD.MMSS to decimals if required, are stored in variables B_1, B_2, B_3. The most probable values of x, y and z, denoted by C_1, C_2 and C_3 respectively, are computed in lines 1200–1210 using Equations 6.24, 6.25 and 6.18. Both observed and adjusted angles are presented in the screen display from line 1280 with a printout option if desired.

(f) Angle adjustment <2> triangle

Line numbers 1400–1690

This is a simple unweighted adjustment for the three angles of a triangle, using the condition that the three angles must sum to 180°. The angles are entered as A_1, A_2 and A_3 and after conversion from DD.MMSS to decimals if required, are stored in variables B_1, B_2, B_3 respectively. The angle adjustment necessary (B_4) is calculated in line 1500 and then the adjusted angles C_1, C_2, C_3 computed accordingly, before both observed and adjusted angles are presented in the screen display from line 1580, with a printout option if desired.

(g) Angle adjustment <3> centre of polygon

Line numbers 1700–1990

This is a simple unweighted adjustment for N angles at the centre of a polygon, using the condition that the total sum of the N angles observed must be 360°. The angles are entered into an array $A(I)$, and after conversion from DD.MMSS to decimals if required, are stored in the array $B(I)$. After calculating the total of the N angles measured (T_A), the necessary angle correction (C_R) is computed in line 1800, and applied to each of the measured angles, to give the adjusted angles in array $C(I)$. Both the observed and adjusted angles are presented in the display from line 1880, with the option of a printout if desired.

(h) Level adjustment

Line numbers 2000–2240

This is a simple unweighted adjustment for a circuit of N levels starting and finishing at the same point. The condition to be met by such a circuit of levels, stored in array $L(I)$, is that the total algebraic sum should be zero. The total level error (T_L) is therefore spread round each leg of the circuit and the adjusted level differences stored in array $C(I)$. Both the observed and adjusted levels are presented in the screen display from line 2130, with the option of a printout if desired.

(i) Routine to print data location, etc.

Line numbers 6000–6080

This routine is provided for record purposes, and is based on Applesoft DOS and may require alteration for different microcomputers and printers.

(j) Printout for angle adjustment <1>

Line numbers 6100–6150

This printout routine re-uses the screen display from lines 1300–1360.

(k) Printout for angle adjustment <2>

Line numbers 6200–6250

This printout routine re-uses the screen display from lines 1600–1660.

(l) Printout for angle adjustment <3>

Line numbers 6300–6350

This printout routine re-uses the screen display from lines 1900–1960.

(m) Printout for level adjustment

Line numbers 6400–6490

This printout routine lists both the observed and adjusted level values round the circuit.

(n) Routine to convert DD.MMSS to decimals

Line numbers 8000–8040

This routine was discussed previously in Section 2.2.6.

(o) Routine to convert decimals to DD.MMSS

Line numbers 8050–8140

This routine was discussed previously in Section 2.2.6.

(p) Routine to clear screen

Line numbers 9000–9010

To simplify portability, the Applesoft 'HOME' command to clear the screen has been placed in a subroutine to allow an easy change to another microcomputer.

(q) Routine to display error message

Line numbers 9100–9130

If a wrong keyboard entry has been made, the error is trapped and this error message displayed and re-entry requested.

(r) Error trace routine

Line numbers 9200–9290

This error trace routine was discussed earlier in Section 2.2.9.

(s) Termination of program

Line numbers 9500–9620

When the '<QUIT>' selection is made in the program, an end-of-program message is displayed to advise the operator that computations are complete.

6.3.2 Adjustment of observations – numeric variables

$A1-A3$	= Measured angles
$A(I)$	= Array of measured angles
$B1-B3$	= Temporary storage of measured angles
$B(I)$	= Array of measured angles

Table—continued

C	= Counter in error trace routine
*C*1–*C*3	= Adjusted angles
CL	= Correction to levels
CR	= Correction to angles
C(*I*)	= Array of adjusted angles
D, *D*0, *D*1, *D*2, *D*5	= Degrees of angles
*D*9	= <1> for decimal angles or <2> for DD.MMSS
E	= Error number
I, *I*1	= Integer loop variables
J, *J*1	= Integer loop variables
K	= Integer loop variable
L(*I*)	= Array of measured levels
M, *M*0, *M*1	= Minutes of angles
N	= Number of observations
*P*1	= Choice of angle <1> or level <2> adjustments
*P*2	= Selection from angle adjustment menu
*P*7	= Printer code number
TA	= Total sum of angles
TL	= Total of level differences
U	= Initial menu selection

6.3.3 Adjustment of observations – string variables

A$	Storage for location of data
B$	Storage for operator's name
C$	Storage for date of calculation
D$	= CHR$(4), CTRL-D
*E*0$	= "SORRY DATA ERROR . . . PLEASE RE-ENTER"
H$	= "ADJUSTMENT OF OBSERVATIONS"
P$	= "DO YOU WISH PRINTOUT OF ABOVE DATA (Y/N)"
*P*1$	= "ANGLE ADJUSTMENT" if *P*1 = 1, or "LEVEL ADJUSTMENT" if *P*1 = 2.
*P*2$	= "<1> TWO ANGLES" if *P*2 = 1 "<2> TRIANGLE" if *P*2 = 2 "<3> CENTRE OF POLYGON" if *P*2 = 3
Q$	= Question response (Y/N)

6.3.4 Adjustment of observations – BASIC program

```
10  REM  <OBSERVATION ADJUSTMENT> PROGRAM FOR APPLE II+  USES QUME PRINTE
    R
20 D$ =  CHR$ (4): ONERR  GOTO 9200
30 C = 0:E = 0:P1 = 0:P7 = 0
35  DIM A(10),B(10),C(10),L(10)
40 H$ = "ADJUSTMENT OF OBSERVATIONS"
50 E0$ = "SORRY, DATA ERROR ... PLEASE RE-ENTER."
55 P$ = "DO YOU WISH PRINTOUT OF ABOVE DATA (Y/N)"
95  REM  **************************************************
100  GOSUB 9000
```

```
110  PRINT "****************************************"
120  PRINT "*                                      *"
130  PRINT "*      UNIVERSITY  OF  STRATHCLYDE     *"
140  PRINT "*  DEPARTMENT  OF  CIVIL  ENGINEERING  *"
150  PRINT "*                                      *"
160  PRINT "*            SURVEYING  SECTION        *"
170  PRINT "*      ADJUSTMENT  OF  OBSERVATIONS    *"
180  PRINT "****************************************"
190  PRINT
200  PRINT "DO YOU WISH TO ENTER LOCATION <1>,      NEW DATA <2> OR QUIT
     <3>"
210  INPUT U
220  IF U < 1 OR U > 3 THEN 200
230  ON U GOTO 500,700,9500
500  GOSUB 9000
510  PRINT "UNIVERSITY OF STRATHCLYDE"
520  PRINT H$
530  PRINT "****************************************"
540  PRINT
550  PRINT "ENTER LOCATION OF WORK, ETC. WHEN       REQUESTED, AND PRESS
     <RETURN>"
560  PRINT "ENTER LOCATION OF DATA :-"
570  INPUT A$
580  PRINT "ENTER OPERATOR'S NAME :-"
590  INPUT B$
600  PRINT "ENTER DATE OF CALCULATION AS DD/MM/YY -"
610  INPUT C$
620  PRINT "ENTER INSTRUMENT USED :-"
630  INPUT K$
640  PRINT
650  PRINT "DO YOU WISH PRINTOUT OF LOCATION ETC.,  (Y/N) ";
660  INPUT Q$
670  IF Q$ = "Y" THEN 6000
680  IF Q$ <  > "N" THEN 650
690 :
695  REM  PROGRAM MENU SELECTION
700  GOSUB 9000: PRINT H$
710  PRINT "****************************************"
720  PRINT "SELECT PROGRAM FROM MENU :-"
730  PRINT
740  PRINT "   <1> ANGLE OBSERVATIONS"
750  PRINT "   <2> LEVEL OBSERVATIONS"
760  PRINT
770  INPUT P1
780  IF P1 < 1 OR P1 > 2 THEN 700
790  IF P1 = 2 THEN 2000
800  PRINT : PRINT "ARE ANGLES IN DECIMALS<1> OR DD.MMSS<2>"
810  INPUT D9
820  IF D9 < 1 OR D9 > 2 THEN 800
995  REM  ************************************************
1000  GOSUB 9000:P1$ = "ANGLE ADJUSTMENT - "
1010  PRINT H$: PRINT P1$
1020  PRINT "****************************************"
1030  PRINT "SELECT FROM MENU <1> TWO ANGLES"
1040  PRINT "                 <2> TRIANGLE"
1050  PRINT "                 <3> CENTRE OF POLYGON"
1060  PRINT "                 <4> BACK TO MAIN MENU"
1070  INPUT P2
1080  IF P2 < 1 OR P2 > 4 THEN 1000
```

```
1090  ON P2 GOTO 1100,1400,1700,100
1095  REM  ****************************************************
1100  GOSUB 9000:P2$ = "<1> TWO ANGLES"
1110  PRINT P1$: PRINT P2$
1120  PRINT "****************************************"
1130  PRINT : PRINT "ENTER 1ST SEGMENT ANGLE ";: INPUT A1
1140  PRINT : PRINT "ENTER 2ND SEGMENT ANGLE ";: INPUT A2
1150  PRINT : PRINT "ENTER TOTAL ANGLE       ";: INPUT A3: IF D9 = 1 THEN
     1190
1155  REM  I.E. D9=2, ANGLES IN DD.MMSS
1160 D5 = A1: GOSUB 8000:B1 = D5
1170 D5 = A2: GOSUB 8000:B2 = D5
1180 D5 = A3: GOSUB 8000:B3 = D5: GOTO 1200
1190 B1 = A1:B2 = A2:B3 = A3
1195  REM  A1-A3 MEASURED ANGLES,C1-C3 ADJUSTED ANGLES
1200 C1 = (2 * B1 - B2 + B3) / 3:C2 = (2 * B2 - B1 + B3) / 3
1210 C3 = C1 + C2: IF D9 = 1 THEN 1250
1215  REM  I.E. D9=2, CONVERT ANGLES TO DD.MMSS
1220 D = C1: GOSUB 8050:C1 = D
1230 D = C2: GOSUB 8050:C2 = D
1240 D = C3: GOSUB 8050:C3 = D: GOTO 1280
1250 C1 =  INT (C1 * 1000 + .5) / 1000
1260 C2 =  INT (C2 * 1000 + .5) / 1000
1270 C3 =  INT (C3 * 1000 + .5) / 1000
1280  GOSUB 9000: PRINT H$:P7 = 0
1290  PRINT P1$: PRINT P2$: PRINT "**************************************
     **"
1300  PRINT "OBSERVED ANGLES: 1ST SEGMENT =";A1
1310  PRINT "                 2ND SEGMENT =";A2
1320  PRINT "                 TOTAL ANGLE =";A3
1330  PRINT : PRINT "ADJUSTED ANGLES: 1ST SEGMENT =";C1
1340  PRINT "                 2ND SEGMENT =";C2
1350  PRINT "                 TOTAL ANGLE =";C3
1360  PRINT "****************************************": IF P7 = 1 THEN  RETURN

1370  PRINT P$;: INPUT Q$: IF Q$ = "Y" THEN 6100
1380  IF Q$ < > "N" THEN 1370
1390  GOTO 1000
1395  REM  ****************************************************
1400  GOSUB 9000:P2$ = "<2> TRIANGLE"
1410  PRINT P1$: PRINT P2$
1420  PRINT "****************************************"
1430  PRINT : PRINT "ENTER 1ST ANGLE ";: INPUT A1
1440  PRINT : PRINT "ENTER 2ND ANGLE ";: INPUT A2
1450  PRINT : PRINT "ENTER 3RD ANGLE ";: INPUT A3: IF D9 = 1 THEN 1490
1455  REM  I.E. D9=2, ANGLES IN DD.MMSS
1460 D5 = A1: GOSUB 8000:B1 = D5
1470 D5 = A2: GOSUB 8000:B2 = D5
1480 D5 = A3: GOSUB 8000:B3 = D5: GOTO 1500
1490 B1 = A1:B2 = A2:B3 = A3
1495  REM  FIND ERROR B4 AND DISTRIBUTE
1500 B4 = 180 - (B1 + B2 + B3):C1 = B1 + B4 / 3
1510 C2 = B2 + B4 / 3:C3 = B3 + B4 / 3: IF D9 = 1 THEN 1550
1515  REM  I.E. D9=2, CONVERT ANGLES BACK TO DD.MMSS
1520 D = C1: GOSUB 8050:C1 = D
1530 D = C2: GOSUB 8050:C2 = D
1540 D = C3: GOSUB 8050:C3 = D: GOTO 1580
1550 C1 =  INT (C1 * 1000 + .5) / 1000
1560 C2 =  INT (C2 * 1000 + .5) / 1000
```

```
1570 C3 =  INT (C3 * 1000 + .5) / 1000
1580  GOSUB 9000: PRINT H$:P7 = 0
1590  PRINT P1$: PRINT P2$: PRINT "**************************************
     **"
1600  PRINT "OBSERVED ANGLES : 1ST = ";A1
1610  PRINT "                  2ND = ";A2
1620  PRINT "                  3RD = ";A3
1630  PRINT : PRINT "ADJUSTED ANGLES : 1ST = ";C1
1640  PRINT "                  2ND = ";C2
1650  PRINT "                  3RD = ";C3
1660  PRINT "****************************************": IF P7 = 1 THEN  RETURN

1670  PRINT P$;: INPUT Q$: IF Q$ = "Y" THEN 6200
1680  IF Q$ <  > "N" THEN 1670
1690  GOTO 1000
1695  REM  ***********************************************
1700  GOSUB 9000:P2$ = "<3> CENTRE OF POLYGON":TA = 0
1710  PRINT P1$: PRINT P2$: PRINT "**************************************
     **"
1720  PRINT : PRINT "ENTER NUMBER OF SEGMENTS ";: INPUT N
1730  FOR I = 0 TO N - 1
1740  PRINT : PRINT "ENTER ANGLE SEGMENT ";(I + 1);: INPUT A(I)
1750  IF D9 = 1 THEN 1770
1760 D5 = A(I): GOSUB 8000:B(I) = D5: GOTO 1780
1770 B(I) = A(I)
1780 TA = TA + B(I)
1790  NEXT I
1795  REM  CR IS ADJUSTMENT TO EACH ANGLE
1800 CR = (360 - TA) / N
1810  FOR J = 0 TO N - 1
1820 C(J) = B(J) + CR
1830  IF D9 = 1 THEN 1850
1840 D = C(J): GOSUB 8050:C(J) = D: GOTO 1860
1850 C(J) =  INT (C(J) * 1000 + .5) / 1000
1860  NEXT J
1870 :
1880  GOSUB 9000: PRINT H$:P7 = 0
1890  PRINT P1$: PRINT P2$: PRINT "**************************************
     **"
1900  FOR I1 = 0 TO N - 1
1910  PRINT "OBSERVED ANGLE SEGMENT ";(I1 + 1);" = ";A(I1)
1920  NEXT I1: PRINT
1930  FOR J1 = 0 TO N - 1
1940  PRINT "ADJUSTED ANGLE SEGMENT ";(J1 + 1);" = ";C(J1)
1950  NEXT J1
1960  PRINT "****************************************": IF P7 = 1 THEN  RETURN

1970  PRINT : PRINT P$;: INPUT Q$: IF Q$ = "Y" THEN 6300
1980  IF Q$ <  > "N" THEN 1970
1990  GOTO 1000
1995  REM  ***********************************************
2000  GOSUB 9000:P1$ = "LEVEL ADJUSTMENT":TL = 0
2010  PRINT H$: PRINT P1$
2020  PRINT "****************************************"
2030  PRINT : PRINT "ENTER NUMBER OF LEVELS ";: INPUT N
2040  FOR I = 0 TO N - 1
2050  PRINT : PRINT "ENTER LEVEL DIFF. NO. ";(I + 1);: INPUT L(I)
2055  REM  TL IS TOTAL OF LEVEL DIFFERENCES
2060 TL = TL + L(I)
```

```
2070  NEXT I
2075  REM  NOW TL SHOULD BE ZERO, CL IS CORRECTION
2080 CL = TL / N
2090  FOR J = Ø TO N - 1
2100 C(J) = L(J) - CL
2110 C(J) =  INT (C(J) * 1000 + .5) / 1000
2120  NEXT J
2130  GOSUB 9000: PRINT H$: PRINT P1$
2140  PRINT "**************************************"
2150  PRINT "LEVEL NO.  OBSERVED  ADJUSTED LEVEL"
2160  PRINT "======================================"
2170  FOR K = Ø TO N - 1
2180  PRINT (K + 1); TAB( 12);L(K); TAB( 25);C(K)
2190  NEXT K
2200  PRINT "**************************************"
2210  PRINT : PRINT P$;: INPUT Q$
2220  IF Q$ = "Y" THEN 6400
2230  IF Q$ < > "N" THEN 2210
2240  GOTO 100
5985  REM  ***********************************************************
     **
5990  REM  PRINTOUT OF LOCATION, ETC
6000  PRINT D$;"PR#1": PRINT "LOCATION OF DATA :-";A$
6010  PRINT "**********************************************************
     ********************"
6020  PRINT "OPERATOR'S NAME :- ";B$
6030  PRINT "DATE OF CALCS.  :-";C$
6040  PRINT "INSTRUMENT USED :- ";K$
6060  PRINT "**********************************************************
     ********************"
6070  PRINT : PRINT D$;"PR#Ø"
6080  GOTO 700
6093  REM  ***********************************************
6095  REM  PRINTOUT FOR <1> TWO ANGLES
6100  PRINT D$;"PR#1": POKE 1784 + 1,80:P7 = 1
6110  PRINT H$: PRINT P1$;P2$
6120  PRINT "======================================"
6125  REM  USE SCREEN DISPLAY FROM 1300-60
6130  GOSUB 1300
6140  PRINT : PRINT D$;"PR#Ø"
6150  GOTO 1000
6193  REM  ***********************************************
6195  REM  PRINTOUT FOR <2> TRIANGLE
6200  PRINT D$;"PR#1": POKE 1784 + 1,80:P7 = 1
6210  PRINT H$: PRINT P1$;P2$
6220  PRINT "======================================"
6225  REM  USE SCREEN DISPLAY FROM 1600-60
6230  GOSUB 1600
6240  PRINT : PRINT D$;"PR#Ø"
6250  GOTO 1000
6293  REM  ***********************************************
6295  REM  PRINTOUT FOR <3> CENTRE OF POLYGON
6300  PRINT D$;"PR#1": POKE 1784 + 1,80:P7 = 1
6310  PRINT H$: PRINT P1$;P2$
6320  PRINT "======================================"
6325  REM  USE SCREEN DISPLAY FROM 1900-60
6330  GOSUB 1900
6340  PRINT : PRINT D$;"PR#Ø"
6350  GOTO 1000
```

```
6393  REM  **************************************************
6395  REM  PRINTOUT FOR LEVEL ADJUSTMENT
64ØØ  PRINT D$;"PR#1": POKE 1784 + 1,8Ø
641Ø  PRINT H$: PRINT P1$
642Ø  PRINT "****************************************"
643Ø  PRINT "LEVEL NO.  OBSERVED  ADJUSTED LEVELS"
644Ø  PRINT "========================================"
645Ø  FOR K = Ø TO N - 1
646Ø  PRINT (K + 1);: POKE 36,12: PRINT L(K);: POKE 36,25: PRINT C(K)
647Ø  NEXT K
648Ø  PRINT "****************************************"
649Ø  PRINT : PRINT D$;"PR#Ø": GOTO 1ØØ
7993  REM  ***************************************************
7995  REM  ROUTINE DD.MMSS TO DECIMALS
8ØØØ DØ =  INT (D5)
8ØØ5  REM  FIND NO. OF MINUTES,MØ
8Ø1Ø M1 = (D5 - DØ) * 1ØØ:MØ =  INT (M1 + .1)
8Ø15  REM  FIND NO. OF SECONDS,SØ
8Ø2Ø SØ =  INT (M1 * 1ØØ - MØ * 1ØØ + .5)
8Ø25  REM  COLLECT DEGS,MINS,SECS
8Ø3Ø D5 = DØ + MØ / 6Ø + SØ / 36ØØ
8Ø4Ø  RETURN
8Ø43  REM  ***************************************************
8Ø45  REM  ROUTINE DECIMAL TO DD.MMSS
8Ø5Ø D1 =  INT (D)
8Ø55  REM  FIND TOTAL NO. OF SECONDS
8Ø6Ø D2 = (D - D1) * 36ØØ
8Ø65  REM  FIND NO. OF MINUTES
8Ø7Ø M =  INT (D2 / 6Ø)
8Ø75  REM  FIND NO. OF SECONDS
8Ø8Ø S = D2 - 6Ø * M + .5
8Ø9Ø  IF S < 6Ø THEN 813Ø
81ØØ M = M + 1:S = Ø
811Ø  IF M < 6Ø THEN 813Ø
812Ø D1 = D1 + 1:M = Ø
813Ø D = D1 * 1ØØØØ + M * 1ØØ + S
814Ø D =  INT (D) / 1ØØØØ: RETURN
8985  REM  *****************************************
899Ø  REM  GOSUB ROUTINE TO CLEAR SCREEN
9ØØØ  HOME
9Ø1Ø  RETURN
9Ø93  REM  ***************************************************
9Ø95  REM  GOSUB ROUTINE FOR ERRORS
91ØØ  PRINT
911Ø  PRINT EØ$
912Ø  PRINT
913Ø  RETURN
9193  REM  ***************************************************
9195  REM  ERROR TRACE ROUTINE
92ØØ  PRINT  CHR$ (7): REM  CTRL-B (BELL)
921Ø E =  PEEK (222): REM  GET ERROR NO.
922Ø  IF C = 1 THEN 926Ø
923Ø  INVERSE
924Ø  PRINT "ERROR NO. ";E;" FOUND"
925Ø C = 1: TRACE : RESUME
926Ø  PRINT "ERROR ON SECOND LINE NO. "
927Ø  NORMAL
928Ø  NOTRACE
929Ø  STOP
```

```
9493  REM  ****************************************************
9495  REM  END OF PROGRAM
95ØØ  GOSUB 9ØØØ
951Ø  PRINT
952Ø  PRINT "END ADJUSTMENT OF OBSERVATIONS PROGRAM"
953Ø  PRINT
954Ø  PRINT "****************************************"
955Ø  REM  PROGRAM PREPARED BY
956Ø  REM  DR. P.H. MILNE
957Ø  REM  DEPT. OF CIVIL ENGINEERING
958Ø  REM  UNIVERSITY OF STRATHCLYDE
959Ø  REM  GLASGOW G4 ONG
96ØØ  REM  SCOTLAND
961Ø  REM  ***************************
962Ø  END
```

6.3.5 Adjustment of observations – computer printout

```
LOCATION OF DATA :-GLENMORE
**************************************************************************
OPERATOR'S NAME :- GORDON BLACK
DATE OF CALCS.  :-31/Ø8/83
INSTRUMENT USED :- ZEISS ZENA T2ØA
**************************************************************************

ADJUSTMENT OF OBSERVATIONS
ANGLE ADJUSTMENT - <1> TWO ANGLES
========================================
OBSERVED ANGLES: 1ST SEGMENT =23.462
                 2ND SEGMENT =17.1825
                 TOTAL ANGLE =41.Ø44

ADJUSTED ANGLES: 1ST SEGMENT =23.4618
                 2ND SEGMENT =17.1823
                 TOTAL ANGLE =41.Ø442
****************************************

ADJUSTMENT OF OBSERVATIONS
ANGLE ADJUSTMENT - <2> TRIANGLE
========================================
OBSERVED ANGLES : 1ST = 45.3Ø3
                  2ND = 84.223
                  3RD = 5Ø.Ø6

ADJUSTED ANGLES : 1ST = 45.3Ø5
                  2ND = 84.225
                  3RD = 5Ø.Ø62
****************************************

ADJUSTMENT OF OBSERVATIONS
ANGLE ADJUSTMENT - <3> CENTRE OF POLYGON
========================================
OBSERVED ANGLE SEGMENT 1 = 95.363
OBSERVED ANGLE SEGMENT 2 = 85.243
OBSERVED ANGLE SEGMENT 3 = 76.423
OBSERVED ANGLE SEGMENT 4 = 1Ø2.173
```

```
ADJUSTED ANGLE SEGMENT 1 = 95.3615
ADJUSTED ANGLE SEGMENT 2 = 85.2415
ADJUSTED ANGLE SEGMENT 3 = 76.4215
ADJUSTED ANGLE SEGMENT 4 = 102.1715
****************************************

LOCATION OF DATA :-GLENMORE
**************************************************************************
OPERATOR'S NAME :- GORDON BLACK
DATE OF CALCS.  :-31/08/83
INSTRUMENT USED :- ZEISS ZENA AUTOMATIC LEVEL
**************************************************************************

ADJUSTMENT OF OBSERVATIONS
LEVEL ADJUSTMENT
****************************************
LEVEL NO.  OBSERVED  ADJUSTED LEVELS
========================================
1            3.25         3.23
2            4.75         4.73
3            -3.42        -3.44
4            -4.5         -4.52
****************************************
```

6.4 BRACED QUADRILATERAL

In Chapter 3 (Section 3.9), the use of traverse surveys and their adjustment was described, where a large number of stations could be linked together for site control or setting out. The braced or crossed quadrilateral is a special case of a closed traverse consisting of only four stations where observations are taken at each corner, Fig. 6.2(a). In a standard closed traverse, only adjacent stations need to be intervisible, whereas with a braced quadrilateral, all four control stations must be intervisible.

To solve the braced quadrilateral, all the angles A_1 to A_8, as shown in Fig. 6.2(a) plus one distance, or the co-ordinates of two stations are required. Under normal circumstances, all eight angles (A_1–A_8), together with the combined angle in each corner, are observed, as discussed in the last section for '<1> TWO ANGLE' adjustment. However, it is also possible to obtain a solution by measuring the six distances between the stations, either slant ranges if their levels are known, or horizontal distances. If the six distances are known then one is redundant since only five sides are necessary to solve Fig. 6.2(a). From the known distances, each of the angles A_1 to A_8 can be calculated.

From a knowledge of the eight angles (A_1–A_8), in the braced quadrilateral, Fig. 6.2(a), there are several conditions which must be met, which are used to form the condition equations:

$$\text{Sum of angles } A_1 \text{ to } A_8 = 360° \tag{6.26}$$

$$\text{Angles } A_1 + A_2 - A_5 - A_6 = 0° \tag{6.27}$$

$$\text{Angles } A_3 + A_4 - A_7 - A_8 = 0° \tag{6.28}$$

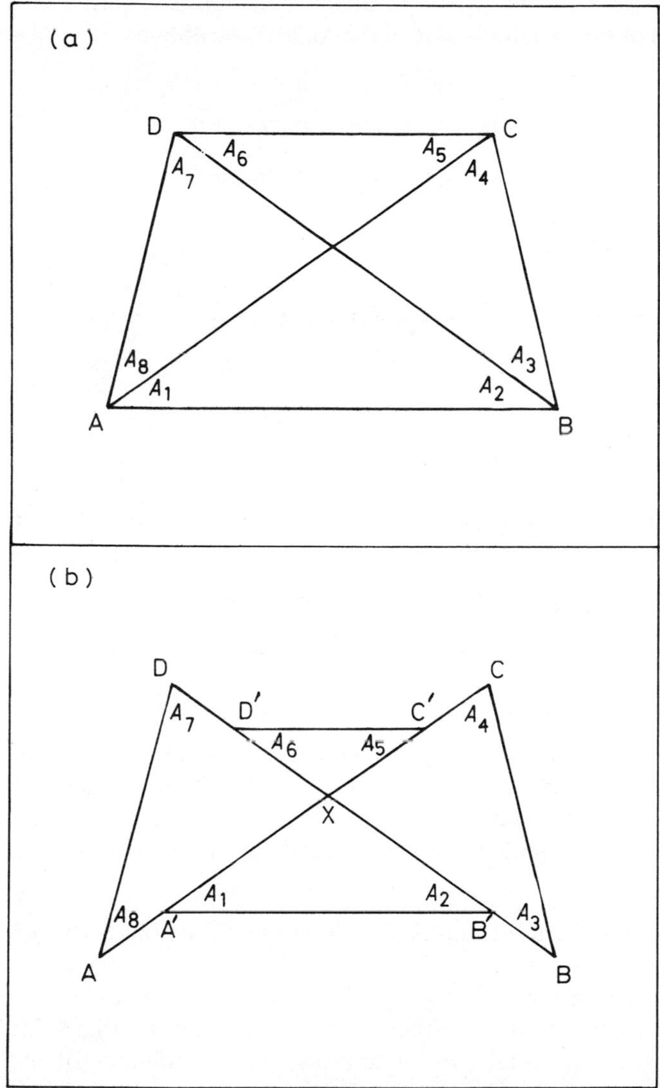

Fig. 6.2. (a) Braced quadrilateral *ABCD* which can be solved by either angle or side measurements. If only angles are solved, (b) can result and hence the need for the side Equation 6.29.

However, adjusting the angle conditions is not sufficient on its own as it could give the solution in Fig. 6.2(b). In order that the sides of the braced quadrilateral are compatible with the angles, a further equation is required. Labelling the intersection of the diagonals as X in Fig. 6.2(b), we obtain:

$$BX = AX \sin A_1/\sin A_2 \text{ if } AA' = BB' = 0$$

$$CX = BX \sin A_3/\sin A_4 \text{ if } CC' = 0$$

$$DX = CX \sin A_5/\sin A_6 \text{ if } DD' = 0$$

$$AX = DX \sin A_7/\sin A_8$$

$$= \frac{AX \sin A_1 \sin A_3 \sin A_5 \sin A_7}{\sin A_2 \sin A_4 \sin A_6 \sin A_8}$$

Therefore

$$\frac{\sin A_1 \sin A_3 \sin A_5 \sin A_7}{\sin A_2 \sin A_4 \sin A_6 \sin A_8} = 1 \qquad (6.29)$$

Equation 6.29 is known as the side equation and is the extra information necessary for a complete solution. In linear form the side equation becomes:

$$\log(\sin A_1) + \log(\sin A_3) + \log(\sin A_5) + \log(\sin A_7)$$
$$- \log(\sin A_2) - \log(\sin A_4) - \log(\sin A_6) - \log(\sin A_8) = 0 \qquad (6.30)$$

The braced quadrilateral angle adjustments are carried out in three stages:

(a) The sum of the eight angles adjusted to 360° from Equation 6.26;
(b) The pairs of angles from Equations 6.27 and 6.28 adjusted accordingly; and
(c) The side equation given in Equation 6.30 is applied by adjusting the angles so that the sum of the log sin 'odd' angles equals the sum of the log sin 'even' angles.

Having adjusted all the angles (A_1–A_8) using Equations 6.26–6.30, a dimension is necessary to solve the braced quadrilateral. This dimension is normally supplied in the form of a distance, calculated from the known co-ordinates for two of the control stations, either A and B or A and C (Fig. 6.2(a)). From a knowledge of two initial control stations, the co-ordinates of the other two control stations can then be calculated.

The following microcomputer program <BRACED QUAD> includes both angle and distance options with co-ordinate solutions, and is based on the above non-rigorous adjustment of a braced quadrilateral by 'equal shifts'. If the braced quadrilateral contains angles approaching 90° (e.g. 80°+), then the above solution should not be used since the differences in log sines for angles approaching 90° are very small. In such a case a simultaneous least squares adjustment is required and matrices should be used, as described in the next section.

6.4.1 Braced quadrilateral – subroutine index

	Line numbers	*Function*
(a)	10–90	Initialization and control
(b)	100–180	Screen header display
(c)	200–240	Selection of angle format
(d)	500–680	Entry of data location, etc.
(e)	700–760	Selection of angle/side observations
(f)	800–840	Entry of station numbers
(g)	1000–1380	Entry of side observations
(h)	1400–1710	Calculations of angles from sides
(i)	1800–1950	Entry of angle observations
(j)	2000–2380	Angle adjustments
(k)	2400–3140	Co-ordinate calculations
(l)	3150–3550	Check on calculations
(m)	6000–6080	Routine to print data location, etc.
(n)	6100–6290	Printout of entered distances
(o)	6300–6430	Printout of unadjusted angles
(p)	6500–6690	Printout of adjusted angles
(q)	6700–6830	Printout of co-ordinates
(r)	6900–6990	Printout of calculated sides – angle case
(s)	7000–7090	Printout of calculated sides – side case
(t)	7750–7790	Triangle solution (*A*1, *S*1, *A*3)
(u)	7800–7870	Triangle solution (*S*1, *S*2, *S*3)
(v)	8000–8040	Routine to convert DD.MMSS to decimals
(w)	8050–8140	Routine to convert decimals to DD.MMSS
(x)	9000–9010	Routine to clear screen
(y)	9100–9130	Routine to display error message
(z)	9200–9290	Error trace routine
(aa)	9500–9620	Termination of program

(a) Initialization and control

Line numbers 10–90

The numeric variables, arrays and required string variables are initialized and two 'DEF FN' functions used to define arccosine and arcsine as discussed previously in Section 2.2.8.

(b) Screen header display

Line numbers 100–180

This is the screen display used for civil engineering students at Strathclyde University. The operator can alter this as required to suit another organization.

(c) Selection of angle format

Line numbers 200–240

The operator has a choice of entering angles in either decimal or DD.MMSS format.

(d) Entry of data location, etc.

Line numbers 500–680

This routine is provided to keep a record of site location, operator, date, etc., with an opportunity to obtain a printout if required.

(e) Selection of angle/side observations

Line numbers 700–760

The operator has a choice of entering eight angles or six side observations to solve the braced quadrilateral, Fig. 6.2(a).

(f) Entry of station numbers

Line numbers 800–840

To cater for different station configurations, the station numbers are stored in the array *P(I)*.

(g) Entry of side observations

Line numbers 1000–1380

The side observations can either be entered as slant ranges *S(I)* or as horizontal distances *H(I)*. If the former, the station levels *Z(I)* are also required to compute the horizontal distances. In both cases there is an opportunity to obtain a printout of entered data.

(h) Calculation of angles from sides

Line numbers 1400–1710

From a knowledge of all the sides in the braced quadrilateral, Fig. 6.2(a), each of the angles can be determined by considering the four triangles making up the quadrilateral and using the (*S*1, *S*2, *S*3) triangle solution, discussed previously in Section 3.2 which is located in a GOSUB routine at line 7800. The calculated angles are stored in array *A(I)*, for presentation in the screen display commencing at line 1520. The total sum of the angles (T_A) is also given together with a comparison of the calculated and computed corner angles. On completion, the program jumps to line 2000, Section (j) below for angle adjustment.

(i) Entry of angle observations

Line numbers 1800–1950

The eight angles A_1–A_8, Fig. 6.2(a), are entered consecutively, either in the decimal or DD.MMSS format, as selected earlier at the start of the program (line 200). If the DD.MMSS format was selected at the outset, the angles are stored in array *W(I)* and then converted to the decimal array *A(I)* for use in the calculations. An opportunity is given to obtain a printout of the observed data.

(j) Angle adjustments

Line numbers 2000–2380

The angle adjustments are carried out in three stages. The first adjustment in lines 2000–2070 has to meet the condition given in Equation 6.26. The second adjustment in lines 2080–2130 has to meet the conditions given in Equations 6.27–6.28. The third and final angle adjustments in lines 2140–2320 are required to meet the side equation condition given by Equations 6.29–6.30. The total sum (T_U) of the adjusted angles is also checked, and should be 360°. After displaying all the adjusted angles and the total, an opportunity is given to obtain a printout. If the DD.MMSS angle format was chosen at the outset, the angles are converted back from decimals before printout.

(k) Co-ordinate calculations

Line numbers 2400–3140

Once the braced quadrilateral has been solved, if the co-ordinates of only two stations are known, 1 and 2 or 1 and 3, then the co-ordinates of the other two stations can be computed. After entering the co-ordinates of station 1, the program branches for the two sets of calculations; to lines 2510–2740 if stations 1 and 2 are known, and to lines 2750–2990 if stations 1 and 3 are known.

The solution disregards any previously measured distances and uses only the computed distance between the known station co-ordinates and the eight adjusted angles (A_1–A_8) using triangulation. This time the triangle solution ($A1$, $S1$, $A3$), discussed in Section 3.2, is used in a GOSUB routine at line 7750. All four sets of station co-ordinates are presented on completion of this section with an opportunity to obtain a printout if required.

(l) Check on calculations

Line numbers 3150–3550

If only angle observations were made at the outset, there is now an opportunity to check on the horizontal distances between the stations. If, however, side observations were made initially, then the variation $V(I)$ between the observed and calculated horizontal distances can be checked. The check calculations are then displayed with an opportunity to obtain a printout if required.

(m) Routine to print data location, etc.

Line numbers 6000–6080

This routine is provided for record purposes and is based on Applesoft DOS and may require alteration for different microcomputers and printers.

(n) Printout of entered distances

Line numbers 6100–6290

This is a combined printout for both the angle and side observation cases where in the latter, if slant ranges were entered, the station levels are also printed.

(o) Printout of unadjusted angles

Line numbers 6300–6430

This again is a combined printout for angle and side observations. In the case of side observations, the angles given are the computed angles, whereas in the angle case, the angles printed are those observed. If the angles were entered in the DD.MMSS format, then both the observed and computed decimal angles are printed.

(p) Printout of adjusted angles

Line numbers 6500–6690

This printout gives both decimal and DD.MMSS format angles if required.

(q) Printout of co-ordinates

Line numbers 6700–6830

This printout provides a record of the co-ordinates of the four stations, together with their level, if entered in the slant range section.

(r) Printout of calculated sides – angle case

Line numbers 6900–6990

This routine provides a list of the sides and their computed horizontal distances.

(s) Printout of calculated sides – side case

Line numbers 7000–7090

In addition to the computed horizontal distance, this printout also includes the observed values and the variation between the observed and computed horizontal distances.

(t) Triangle solution (*A*1, *S*1, *A*3)

Line numbers 7750–7790

This routine was discussed in Section 3.2.

(u) Triangle solution (*S*1, *S*2, *S*3)

Line numbers 7800–7870

This routine was discussed in Section 3.2.

(v) Routine to convert DD.MMSS to decimals

Line numbers 8000–8040

This routine was discussed previously in Section 2.2.6.

(w) Routine to convert decimals to DD.MMSS

Line numbers 8050–8140

This routine was discussed previously in Section 2.2.6.

(x) Routine to clear screen

Line numbers 9000–9010

To simplify portability, the Applesoft 'HOME' command to clear the screen has been placed in a subroutine to allow an easy change to another microcomputer.

(y) Routine to display error message

Line numbers 9100–9130

If a wrong keyboard entry has been made in entering data, the error is trapped and this error message displayed and re-entry requested.

(z) Error trace routine

Line numbers 9200–9290

This error trace routine was discussed earlier in Section 2.2.9.

(aa) Termination of program

Line numbers 9500–9620

When the '<QUIT>' selection is made in the program, an end-of-program message is displayed to advise the operator that computations are complete.

6.4.2 Braced quadrilateral – numeric variables

*A*1–*A*3	= Angles in triangle subroutines
A(*I*)	= Array of decimal angles
B	= Baseline calculation
BG	= Slope of baseline
C	= Counter in error trace routine
*C*9	= Arccosine solution
C(*I*)	= Array of calculated sides
D, *D*0, *D*1, *D*2, *D*5	= Degrees of angle
*D*9	= <1> for decimals or <2> for DD.MMSS
D(*I*)	= Array of check calculations for sides
E	= Error number
G	= Gradient of baseline
G(*I*)	= Array of sines of angles
H(*I*)	= Array of horizontal lengths
I, *I*1	= Integer loop variables
J, *J*1, *J*2	= Integer loop variables
K, *K*1, *K*2	= Integer loop variables

Table—continued

*K*7–*K*9	= Variables used in third angle adjustment
L(*I*)	= Array of log sines
M, *M*0, *M*1	= Minutes of angle
P	= Half perimeter of triangle
*P*1	= Menu selection sides/angles
*P*2	= Baseline selection
*P*5, *P*7	= Flags in printer routines
P(*I*)	= Array of station numbers
*Q*1–*Q*4	= Corner angles of braced quadrilateral
*R*9	= 57.2957795
R(*I*)	= Temporary storage of horizontal ranges
S, *S*0	= Seconds of angle
*S*1–*S*3	= Sides in triangle subroutines
*S*9	= Arcsine solution
S(*I*)	= Array of slant ranges
*T*1–*T*3	= Temporary angle storage
TA	= Total sum of angles
TE	= Total error in angles (first adjustment)
TF, *TG*	= Errors in angles (second adjustment)
TU	= Total of adjusted angles
U(*I*)	= Array of adjusted angles
V(*I*)	= Array of variations between observed and calculated distances
W(*I*)	= Array of DD.MMSS angles
X(*I*)	= Array of *X*(*E*) co-ordinates
Y(*I*)	= Array of *Y*(*N*) co-ordinates
Z(*I*)	= Array of station levels

6.4.3 Braced quadrilateral – string variables

A$	Storage for location
B$	Storage for operator's name
C$	Storage for date of calculation
*E*0$	= "SORRY DATA ERROR . . . PLEASE RE-ENTER"
H$	= "BRACED QUADRILATERAL SOLUTION"
K$	Storage for instrument used
P$	= "DO YOU WISH PRINTOUT OF ABOVE DATA (Y/N)"
*P*1$	= "<1> ANGLE OBSERVATIONS" or "<2> SIDE OBSERVATIONS"
Q$	= Question response (Y/N)
S$	= (*S*/*H*) for slant ranges/horizontal distances
X$	= "ENTER X(E) COORD. OF STN. "
Y$	= "ENTER Y(N) COORD. OF STN. "
Z$	= "ENTER LEVEL OF STN. "

6.4.4 Braced quadrilateral – BASIC program

```
10  REM  <BRACED QUAD> PROGRAM FOR APPLE II+  USES QUME PRINTER
20 D$ =  CHR$ (4): ONERR  GOTO 9200
30 TA = Ø:B = Ø:BG = Ø:C = Ø:D9 = Ø:E = Ø:G = Ø:K7 = Ø:K8 = Ø:K9 = Ø:P1 =
```

```
    Ø:P2 = Ø:P5 = Ø:P7 = Ø:Q1 = Ø:Q2 = Ø:Q3 = Ø:Q4 = Ø:TE = Ø:TF = Ø:TG =
    Ø:TU = Ø
35  DIM A(1Ø),C(1Ø),D(1Ø),G(1Ø),H(1Ø),L(1Ø),P(1Ø),R(1Ø),S(1Ø),U(1Ø),V(1Ø)
    ,W(1Ø),X(1Ø),Y(1Ø),Z(1Ø)
4Ø H$ = "BRACED QUADRILATERAL SOLUTION"
5Ø EØ$ = "SORRY, DATA ERROR ... PLEASE RE-ENTER."
55 P$ = "DO YOU WISH PRINTOUT OF ABOVE DATA (Y/N)"
7Ø X$ = "ENTER X(E) COORD. OF STN. ":Y$ = "ENTER Y(N) COORD. OF STN. "
75 Z$ = "ENTER LEVEL OF STN. "
8Ø  DEF  FN C9(C9) = (1.57Ø7964 -  ATN (C9 /  SQR (1 - C9 * C9))) * R9: REM
     ARCCOSINE
9Ø  DEF  FN S9(S9) =  ATN (S9 /  SQR (1 - S9 * S9)) * R9: REM  ARCSINE
95  REM  **************************************************
1ØØ  GOSUB 9ØØØ
11Ø  PRINT "****************************************"
12Ø  PRINT "*                                      *"
13Ø  PRINT "*     UNIVERSITY  OF  STRATHCLYDE      *"
14Ø  PRINT "*  DEPARTMENT  OF  CIVIL  ENGINEERING  *"
15Ø  PRINT "*                                      *"
16Ø  PRINT "*           SURVEYING  SECTION         *"
17Ø  PRINT "*   BRACED  QUADRILATERAL  SOLUTION    *"
18Ø  PRINT "****************************************"
19Ø  PRINT
2ØØ  PRINT "ARE ANGLES IN DECIMALS <1>, DD.MMSS <2> OR DO YOU WISH TO QUI
     T <3>."
21Ø  INPUT D9
22Ø  IF D9 < 1 OR D9 > 3 THEN 2ØØ
23Ø  ON D9 GOTO 24Ø,24Ø,95ØØ
24Ø R9 = 57.2957795:P1 = Ø
5ØØ  GOSUB 9ØØØ
51Ø  PRINT "UNIVERSITY OF STRATHCLYDE"
52Ø  PRINT H$
53Ø  PRINT "****************************************"
54Ø  PRINT
55Ø  PRINT "ENTER LOCATION OF WORK, ETC. WHEN        REQUESTED, AND PRESS
     <RETURN>"
56Ø  PRINT "ENTER LOCATION OF BRACED QUADRILATERAL"
57Ø  INPUT A$
58Ø  PRINT "ENTER OPERATOR'S NAME :-"
59Ø  INPUT B$
6ØØ  PRINT "ENTER DATE OF CALCULATION AS DD/MM/YY :-"
61Ø  INPUT C$
62Ø  PRINT "ENTER INSTRUMENT USED :-"
63Ø  INPUT K$
64Ø  PRINT
65Ø  PRINT "DO YOU WISH PRINTOUT OF LOCATION ETC.,  (Y/N) ";
66Ø  INPUT Q$
67Ø  IF Q$ = "Y" THEN 6ØØØ
68Ø  IF Q$ <  > "N" THEN 65Ø
695  REM  **************************************************
7ØØ  GOSUB 9ØØØ: PRINT H$
71Ø  PRINT "****************************************"
72Ø  PRINT : PRINT "SELECT DATA AVAILABLE :-"
73Ø  PRINT "   <1> ANGLE OBSERVATIONS"
74Ø  PRINT "   <2> SIDE  OBSERVATIONS"
75Ø  INPUT P1
76Ø  IF P1 < 1 OR P1 > 2 THEN 7ØØ
795  REM  **************************************************
8ØØ  PRINT : PRINT "ENTER STATION NUMBERS IN ANTI-CLOCKWISE ORDER, STARTI
     NG IN S.W. CORNER :-": PRINT
```

```
810  FOR I = Ø TO 3
820  PRINT "ENTER STN. NO. ";(I + 1);: INPUT P(I)
830  NEXT I
840  IF P1 = 1 THEN 18ØØ
995  REM  ************************************************
1ØØØ  GOSUB 9ØØØ:P1$ = "<2> SIDE OBSERVATIONS"
1Ø1Ø  PRINT H$: PRINT P1$
1Ø2Ø  PRINT "****************************************"
1Ø3Ø  PRINT : PRINT "ARE SIDE OBSERVATIONS SLANT RANGES (S)  OR HORIZONTA
     L LENGTHS (H) ";
1Ø4Ø  INPUT S$
1Ø5Ø  IF S$ = "H" THEN 13ØØ
1Ø6Ø  IF S$ < > "S" THEN 1Ø3Ø
1Ø65  REM  SLANT RANGE TO HORIZONTAL CALCULATIONS
1Ø7Ø  FOR J = Ø TO 3
1Ø8Ø  PRINT : PRINT Z$;(J + 1);: INPUT Z(J)
1Ø9Ø  NEXT J
1Ø95  REM  ENTER SLANT RANGES
11ØØ  GOSUB 9ØØØ
111Ø  PRINT "ENTER SLANT RANGE ";P(Ø);"-";P(1);: INPUT S(Ø)
112Ø  PRINT : PRINT "ENTER SLANT RANGE ";P(1);"-";P(2);: INPUT S(1)
113Ø  PRINT : PRINT "ENTER SLANT RANGE ";P(2);"-";P(3);: INPUT S(2)
114Ø  PRINT : PRINT "ENTER SLANT RANGE ";P(3);"-";P(Ø);: INPUT S(3)
115Ø  PRINT : PRINT "ENTER SLANT RANGE ";P(Ø);"-";P(2);: INPUT S(4)
116Ø  PRINT : PRINT "ENTER SLANT RANGE ";P(1);"-";P(3);: INPUT S(5)
1165  REM  CALCULATE HORIZONTALS
117Ø H(Ø) =  SQR (S(Ø) ^ 2 - (Z(Ø) - Z(1)) ^ 2)
118Ø H(1) =  SQR (S(1) ^ 2 - (Z(1) - Z(2)) ^ 2)
119Ø H(2) =  SQR (S(2) ^ 2 - (Z(2) - Z(3)) ^ 2)
12ØØ H(3) =  SQR (S(3) ^ 2 - (Z(3) - Z(Ø)) ^ 2)
121Ø H(4) =  SQR (S(4) ^ 2 - (Z(Ø) - Z(2)) ^ 2)
122Ø H(5) =  SQR (S(5) ^ 2 - (Z(1) - Z(3)) ^ 2)
123Ø  PRINT : PRINT P$;: INPUT Q$: IF Q$ = "Y" THEN 61ØØ
124Ø  IF Q$ < > "N" THEN 123Ø
1245  REM  GOTO ANGLE CALCULATIONS
125Ø  GOTO 14ØØ
1295  REM  ENTER HORIZONTAL DISTANCES
13ØØ  GOSUB 9ØØØ
131Ø  PRINT "ENTER HORIZ. DIST. ";P(Ø);"-";P(1);: INPUT H(Ø)
132Ø  PRINT : PRINT "ENTER HORIZ. DIST. ";P(1);"-";P(2);: INPUT H(1)
133Ø  PRINT : PRINT "ENTER HORIZ. DIST. ";P(2);"-";P(3);: INPUT H(2)
134Ø  PRINT : PRINT "ENTER HORIZ. DIST. ";P(3);"-";P(Ø);: INPUT H(3)
135Ø  PRINT : PRINT "ENTER HORIZ. DIST. ";P(Ø);"-";P(2);: INPUT H(4)
136Ø  PRINT : PRINT "ENTER HORIZ. DIST. ";P(1);"-";P(3);: INPUT H(5)
137Ø  PRINT : PRINT P$;: INPUT Q$: IF Q$ = "Y" THEN 61ØØ
138Ø  IF Q$ < > "N" THEN 137Ø
139Ø :
1395  REM  ANGLE CALCULATIONS FROM DISTANCES, (1) SOLVE TRIANGLE 1,2,3
14ØØ S1 = H(Ø):S2 = H(4):S3 = H(1)
141Ø  GOSUB 78ØØ
142Ø A(Ø) = A1:A(3) = A2:Q2 = A3
1425  REM  (2) SOLVE TRIANGLE 1,2,4
143Ø S1 = H(Ø):S2 = H(3):S3 = H(5)
144Ø  GOSUB 78ØØ
145Ø Q1 = A1:A(6) = A2:A(1) = A3
1455  REM  (3) SOLVE TRIANGLE 2,3,4
146Ø S1 = H(1):S2 = H(5):S3 = H(2)
147Ø  GOSUB 78ØØ
148Ø A(2) = A1:A(5) = A2:Q3 = A3
```

```
1485  REM  (4) SOLVE TRIANGLE 1,3,4
1490 S1 = H(4):S2 = H(3):S3 = H(2)
1500  GOSUB 7800
1510 A(7) = A1:Q4 = A2:A(4) = A3
1520  GOSUB 9000: PRINT H$: PRINT P1$:TA = 0:P5 = 0:P7 = 0
1530  PRINT "**************************************"
1540  PRINT "UNADJUSTED ANGLES (DD.DEC)"
1550  PRINT "======================================"
1560  FOR I = 0 TO 7
1570  PRINT "ANGLE  A";(I + 1);" = "; INT (A(I) * 1000 + .5) / 1000
1580 TA = TA + A(I)
1590  NEXT I
1600  PRINT "======================================"
1610  PRINT "SUM A1-A8 = "; INT (TA * 1000 + .5) / 1000
1620  PRINT "**************************************": IF P5 = 1 THEN  RETURN

1630  PRINT "COMPARISON OF CORNER ANGLES :-"
1640  PRINT "Q1=A1+A8 = "; INT (Q1 * 1000 + .5) / 1000;" : A1+A8 = "; INT
     ((A(0) + A(7)) * 1000 + .5) / 1000
1650  PRINT "Q2=A2+A3 = "; INT (Q2 * 1000 + .5) / 1000;" : A2+A3 = "; INT
     ((A(1) + A(2)) * 1000 + .5) / 1000
1660  PRINT "Q3=A4+A5 = "; INT (Q3 * 1000 + .5) / 1000;" : A4+A5 = "; INT
     ((A(3) + A(4)) * 1000 + .5) / 1000
1670  PRINT "Q4=A6+A7 = "; INT (Q4 * 1000 + .5) / 1000;" : A6+A7 = "; INT
     ((A(5) + A(6)) * 1000 + .5) / 1000
1680  PRINT "**************************************": IF P7 = 1 THEN  RETURN

1690  PRINT P$;: INPUT Q$: IF Q$ = "Y" THEN 6300
1700  IF Q$ <  > "N" THEN 1690
1710  GOTO 2000
1795  REM  **************************************************
1800  GOSUB 9000:P1$ = "<1> ANGLE OBSERVATIONS"
1810  PRINT H$: PRINT P1$:TA = 0
1820  PRINT "**************************************"
1830  PRINT : PRINT "ENTER EIGHT CONSECUTIVE ANGLES :-": PRINT
1840  FOR I = 0 TO 7
1850  PRINT "ENTER ANGLE A";(I + 1);" - ";: INPUT W(I)
1860  IF D9 = 1 THEN 1880
1870 D5 = W(I): GOSUB 8000:A(I) = D5: GOTO 1890
1880 A(I) = W(I)
1890  NEXT I
1900  GOSUB 9000:P5 = 1
1910  PRINT H$: PRINT P1$
1915  REM  USE PREVIOUS DISPLAY AT LINES 1530-1610
1920  GOSUB 1530
1930  PRINT : PRINT P$;: INPUT Q$
1940  IF Q$ = "Y" THEN 6300
1950  IF Q$ <  > "N" THEN 1930
1960 :
1993  REM  **************************************************
1995  REM  1ST ANGLE ADJUSTMENT, TE=TOTAL ERROR
2000  GOSUB 9000:TU = 0: PRINT H$: PRINT P1$
2010  PRINT "**************************************"
2020  PRINT "ANGLE ADJUSTMENTS (DD.DEC)"
2030  PRINT "======================================"
2040 TE = 360 - TA
2050  FOR J = 0 TO 7
2060 A(J) = A(J) + TE / 8
2070  NEXT J
```

```
2Ø75  REM  2ND ANGLE ADJUSTMENT (1) TF = ADJUSTMENT TO A1+A2-A5-A6
2Ø8Ø TF = A(Ø) + A(1) - A(4) - A(5)
2Ø9Ø A(Ø) = A(Ø) - TF / 4:A(1) = A(1) - TF / 4
21ØØ A(4) = A(4) + TF / 4:A(5) = A(5) + TF / 4
21Ø5  REM  (2) TG = ADJUSTMENT TO A3+A4-A7-A8
211Ø TG = A(2) + A(3) - A(6) - A(7)
212Ø A(2) = A(2) - TG / 4:A(3) = A(3) - TG / 4
213Ø A(6) = A(6) + TG / 4:A(7) = A(7) + TG / 4
2135  REM  3RD ANGLE ADJUSTMENT
214Ø  FOR K = Ø TO 7
215Ø G(K) =  SIN ((A(K)) / R9)
216Ø L(K) = ( LOG (G(K))) / ( LOG (1Ø))
217Ø  NEXT K
218Ø K7 = L(Ø) + L(2) + L(4) + L(6)
219Ø K8 = L(1) + L(3) + L(5) + L(7)
22ØØ K9 = K7 - K8
221Ø L(Ø) = L(Ø) - K9 / 8:L(1) = L(1) + K9 / 8
222Ø L(2) = L(2) - K9 / 8:L(3) = L(3) + K9 / 8
223Ø L(4) = L(4) - K9 / 8:L(5) = L(5) + K9 / 8
224Ø L(6) = L(6) - K9 / 8:L(7) = L(7) + K9 / 8
2245  REM  CONVERT BACK TO ANGLES
225Ø  FOR I1 = Ø TO 7
226Ø F(I1) = 1Ø ^ L(I1)
227Ø U(I1) =  FN S9(F(I1))
228Ø  IF A(I1) < 9Ø THEN 23ØØ
2285  REM  I.E. A(I1)>9Ø
229Ø U(I1) = 18Ø - U(I1)
2295  REM  TU IS TOTAL OF ADJUSTED ANGLES
23ØØ TU = TU + U(I1)
231Ø  PRINT "ANGLE  A";(I1 + 1);" = "; INT ((U(I1)) * 1ØØØ + .5) / 1ØØØ
232Ø  NEXT I1: IF TU > 359.99 THEN TU = 36Ø
233Ø  PRINT "========================================"
234Ø  PRINT "SUM A1-A8 = "; INT (TU * 1ØØØ + .5) / 1ØØØ
235Ø  PRINT "****************************************"
236Ø  PRINT P$;: INPUT Q$
237Ø  IF Q$ = "Y" THEN 65ØØ
238Ø  IF Q$ <  > "N" THEN 236Ø
239Ø :
2393  REM  *************************************************
2395  REM  CO-ORDINATE CALCULATIONS
24ØØ  GOSUB 9ØØØ: PRINT H$: PRINT P1$
241Ø  PRINT "****************************************"
242Ø  PRINT : PRINT "CO-ORDINATE CALCULATIONS :-": PRINT
243Ø  PRINT "BASELINE OPTIONS :- <1> STN. ";P(Ø);"-";P(1)
244Ø  PRINT "                    <2> STN. ";P(Ø);"-";P(2)
245Ø  PRINT "                    <3> QUIT ";: INPUT P2: IF P2 < 1 OR P2 >
     3 THEN 24ØØ
246Ø  IF P2 = 3 THEN 95ØØ
247Ø  PRINT : PRINT "ENTER CO-ORDS OF STN. ";P(Ø);" :-"
248Ø  PRINT : PRINT X$;P(Ø);: INPUT X(Ø)
249Ø  PRINT Y$;P(Ø);: INPUT Y(Ø)
25ØØ  IF P2 = 2 THEN 275Ø
251Ø  PRINT : PRINT "ENTER CO-ORDS OF STN. ";P(1);" :-"
252Ø  PRINT : PRINT X$;P(1);: INPUT X(1)
253Ø  PRINT Y$;P(1);: INPUT Y(1)
2535  REM  CALCULATE BASELINE LENGTH, B
254Ø B =  SQR ((X(1) - X(Ø)) ^ 2 + (Y(1) - Y(Ø)) ^ 2)
2545  REM  CHANGE HORIZ. SIDES FROM H(J), AND CALC. C(J) KNOWING ANGLES A
     1-A8
```

```
2550 C(0) = B
2555  REM  SOLVE TRIANGLE 1,2,3
2560 A1 = U(0):S1 = C(0):A3 = (U(1) + U(2))
2570  GOSUB 7750
2580 C(4) = S2:C(1) = S3
2585  REM  SOLVE TRIANGLE 1,4,2
2590 A1 = (U(0) + U(7)):S1 = C(0):A3 = U(1)
2600  GOSUB 7750
2610 C(3) = S2:C(5) = S3
2615  REM  SOLVE TRIANGLE 1,3,4
2620 A1 = U(7):S1 = C(4):A3 = U(4)
2630  GOSUB 7750
2640 C(2) = S3
2645  REM  CHECK SLOPE OF BASELINE 1-2
2650  IF Y(0) <  > Y(1) THEN 2700
2655  REM  CALC. CO-ORDS STNS 3,4
2660 X(2) = X(1) - C(1) *  COS ((U(1) + U(2)) / R9)
2670 Y(2) = Y(1) + C(1) *  SIN ((U(1) + U(2)) / R9)
2680 X(3) = X(0) + C(3) *  COS ((U(0) + U(7)) / R9)
2690 Y(3) = Y(0) + C(3) *  SIN ((U(0) + U(7)) / R9): GOTO 3000
2695  REM  SLOPING BASELINE, GRADIENT BG
2700 BG = (Y(1) - Y(0)) / (X(1) - X(0)):G =  ATN (BG) * R9
2705  REM  CALC. CO-ORDS STNS 3,4
2710 X(2) = X(1) - C(1) *  COS ((U(1) + U(2) - G) / R9)
2720 Y(2) = Y(1) + C(1) *  SIN ((U(1) + U(2) - G) / R9)
2730 X(3) = X(0) + C(3) *  COS ((U(0) + U(7) + G) / R9)
2740 Y(3) = Y(0) + C(3) *  SIN ((U(0) + U(7) + G) / R9): GOTO 3000
2745  REM  BASELINE STN. 1-3
2750  PRINT : PRINT "ENTER CO-ORDS OF STN. ";P(2);" :-"
2760  PRINT : PRINT X$;P(2);: INPUT X(2)
2770  PRINT Y$;P(2);: INPUT Y(2)
2775  REM  CALC. BASELINE LENGTH, B
2780 B =  SQR ((X(2) - X(0)) ^ 2 + (Y(2) - Y(0)) ^ 2)
2785  REM  CHANGE SIDES FROM H(J), AND CALC. C(J) KNOWING ANGLES A1-A8
2790 C(4) = B
2795  REM  SOLVE TRIANGLE 1,3,2
2800 A1 = U(3):S1 = C(4):A3 = U(0)
2810  GOSUB 7750
2820 C(0) = S3:C(1) = S2
2825  REM  SOLVE TRIANGLE 1,3,4
2830 A1 = U(7):S1 = C(4):A3 = U(4)
2840  GOSUB 7750
2850 C(2) = S3:C(3) = S2
2855  REM  SOLVE TRIANGLE 1,2,4
2860 A1 = (U(0) + U(7)):S1 = C(0):A3 = U(1)
2870  GOSUB 7750
2880 C(5) = S3
2885  REM  CHECK SLOPE OF BASELINE 1-3
2890  IF Y(0) <  > Y(2) THEN 2950
2895  REM  CALC. CO-ORDS STNS 2,4
2900 X(1) = X(0) + C(0) *  COS (U(0) / R9)
2910 Y(1) = Y(0) - C(0) *  SIN (U(0) / R9)
2920 X(3) = X(0) + C(3) *  COS (U(7) / R9)
2930 Y(3) = Y(0) + C(3) *  SIN (U(7) / R9)
2940  GOTO 3000
2945  REM  SLOPING BASELINE, GRADIENT BG
2950 BG = (Y(2) - Y(0)) / (X(2) - X(0)):G =  ATN (BG) * R9
2955  REM  CALC. CO-ORDS STNS 2,4
2960 X(1) = X(0) + C(0) *  COS ((G - U(0)) / R9)
```

```
2970 Y(1) = Y(Ø) + C(Ø) *  SIN ((G - U(Ø)) / R9)
2980 X(3) = X(Ø) + C(3) *  COS ((G + U(7)) / R9)
2990 Y(3) = Y(Ø) + C(3) *  SIN ((G + U(7)) / R9)
2993  REM  ************************************************
2995  REM  DISPLAY OF CO-ORDS
3000  GOSUB 9000: PRINT H$: PRINT P1$
3010  PRINT "***************************************"
3020  PRINT "STATION CO-ORDINATES :-"
3030  PRINT "======================================="
3040  PRINT "STN.NO.   X-COORD   Y-COORD   Z-COORD"
3050  PRINT "======================================="
3060  FOR I = Ø TO 3
3070  PRINT P(I); TAB( 1Ø); INT (X(I) * 1ØØØ + .5) / 1ØØØ; TAB( 2Ø); INT
     (Y(I) * 1ØØØ + .5) / 1ØØØ;
3080  IF Z(I) = Ø THEN 311Ø
3090  PRINT  TAB( 3Ø);Z(I)
3100  NEXT I: GOTO 312Ø
3110  PRINT " ": NEXT I
3120  PRINT "***************************************"
3130  PRINT P$;: INPUT Q$: IF Q$ = "Y" THEN 67ØØ
3140  IF Q$ < > "N" THEN 313Ø
3145  REM  CHECK ON CALCULATIONS
3150  GOSUB 9ØØØ: PRINT H$: PRINT P1$
3160  PRINT "***************************************"
3170  PRINT : PRINT "DO YOU WISH TO CHECK CALCULATIONS (Y/N)": INPUT Q$
3180  IF Q$ = "N" THEN 95ØØ
3190  IF Q$ < > "Y" THEN 315Ø
3195  REM  CHECK ON HORIZ. DISTANCES
3200 D(Ø) =  SQR ((X(1) - X(Ø)) ^ 2 + (Y(1) - Y(Ø)) ^ 2)
3210 D(1) =  SQR ((X(2) - X(1)) ^ 2 + (Y(2) - Y(1)) ^ 2)
3220 D(2) =  SQR ((X(3) - X(2)) ^ 2 + (Y(3) - Y(2)) ^ 2)
3230 D(3) =  SQR ((X(Ø) - X(3)) ^ 2 + (Y(Ø) - Y(3)) ^ 2)
3240 D(4) =  SQR ((X(Ø) - X(2)) ^ 2 + (Y(Ø) - Y(2)) ^ 2)
3250 D(5) =  SQR ((X(3) - X(1)) ^ 2 + (Y(3) - Y(1)) ^ 2)
3260  IF P1 = 2 THEN 34ØØ
3265  REM  ANGLE OBSERVATIONS
3270  GOSUB 9ØØØ: PRINT H$: PRINT P1$
3280  PRINT "***************************************"
3290  PRINT "SIDE      CALCULATED"
3300  PRINT "======================================="
3310  FOR K = Ø TO 5
3320  PRINT (K + 1); TAB( 1Ø); INT (D(K) * 1ØØØ + .5) / 1ØØØ
3330  NEXT K
3340  PRINT "***************************************"
3350  PRINT : PRINT P$;
3360  INPUT Q$
3370  IF Q$ = "Y" THEN 69ØØ
3380  IF Q$ < > "N" THEN 335Ø
3390  GOTO 95ØØ
3395  REM  SIDE OBSERVATIONS, V(J) IS VARIATION=ORIGINAL-CALCULATED
3400  FOR J = Ø TO 5
3410 V(J) = H(J) - D(J)
3420  NEXT J
3430  GOSUB 9ØØØ: PRINT H$: PRINT P1$
3440  PRINT "***************************************"
3450  PRINT "SIDE   OBSERVED  CALCULATED  VARIATION"
3460  PRINT "======================================="
3470  FOR K = Ø TO 5
3480  PRINT (K + 1); TAB( 8); INT (H(K) * 1ØØØ + .5) / 1ØØØ; TAB( 18); INT
     (D(K) * 1ØØØ + .5) / 1ØØØ; TAB( 3Ø); INT (V(K) * 1ØØØ + .5) / 1ØØØ
```

```
3490  NEXT K
3500  PRINT "****************************************"
3510  PRINT : PRINT P$;
3520  INPUT Q$
3530  IF Q$ = "Y" THEN 7000
3540  IF Q$ < > "N" THEN 3510
3550  GOTO 9500
5985  REM  **************************************************************
     **
5990  REM  PRINTOUT OF LOCATION, ETC
6000  PRINT D$;"PR#1": PRINT "LOCATION OF DATA :-";A$
6010  PRINT "*************************************************************
     ********************"
6020  PRINT "OPERATOR'S NAME :- ";B$
6030  PRINT "DATE OF CALCS.  :-";C$
6040  PRINT "INSTRUMENT USED :- ";K$
6060  PRINT "*************************************************************
     ********************"
6070  PRINT : PRINT D$;"PR#Ø"
6080  GOTO 700
6093  REM  **************************************************
6095  REM  PRINTOUT OF ENTERED DISTANCES
6100  PRINT D$;"PR#1": POKE 1784 + 1,80
6110  PRINT P1$
6120  PRINT "****************************************"
6130  IF S$ = "H" THEN 6230
6140  FOR I1 = Ø TO 3
6150  PRINT "LEVEL OF STN.   ";P(I1);" = ";Z(I1)
6160  NEXT I1: PRINT
6170  FOR J1 = Ø TO 5
6180  PRINT "SLANT RANGE     ";(J1 + 1);" = ";S(J1)
6190  NEXT J1: PRINT
6200  FOR K = Ø TO 5
6210 R(K) =  INT (H(K) * 1000 + .5) / 1000
6220  NEXT K: GOTO 6250
6225  REM  PRINTOUT FOR S$="H"
6230  FOR K1 = Ø TO 5
6240 R(K1) = H(K1): NEXT K1
6250  FOR K2 = Ø TO 5
6260  PRINT "HORIZ. DISTANCE ";(K2 + 1);" = ";R(K2)
6270  NEXT K2
6280  PRINT "****************************************"
6290  PRINT D$;"PR#Ø": GOTO 1400
6293  REM  **************************************************
6295  REM  PRINTOUT OF UNADJUSTED ANGLES
6300  PRINT D$;"PR#1": POKE 1784 + 1,80:TA = Ø:P7 = 1
6310  PRINT P1$
6320  IF P1 = 2 THEN 6410
6325  REM  P1=1, ANGLE ENTRY
6330  IF D9 = 1 THEN 6400
6335  REM  D9=2, ANGLES ENTERED IN DD.MMSS
6340  PRINT "****************************************"
6350  PRINT "UNADJUSTED ANGLES (DD.MMSS)"
6360  PRINT "========================================"
6370  FOR I1 = Ø TO 7
6380  PRINT "ANGLE  A";(I1 + 1);" = ";W(I1)
6390  NEXT I1
6400 P5 = 1
6405  REM  USE SCREEN DISPLAY AT LINES 1530-1610 FOR <1> AND 1530-1680 FO
     R <2>
```

```
6410  GOSUB 1530
6420  PRINT : PRINT D$;"PR#Ø"
6430  GOTO 2ØØØ
6493  REM  **************************************************
6495  REM  PRINTOUT OF ADJUSTED ANGLES
6500  PRINT D$;"PR#1": POKE 1784 + 1,8Ø
6510  PRINT P1$
6520  PRINT "****************************************"
6530  PRINT "ADJUSTED ANGLES (DD.DEC)"
6540  PRINT "========================================"
6550  FOR J1 = Ø TO 7
6560  PRINT "ANGLE  A";(J1 + 1);" = "; INT ((U(J1)) * 1ØØØ + .5) / 1ØØØ
6570  NEXT J1
6580  PRINT "========================================"
6590  PRINT "SUM A1-A8 = "; INT (TU * 1ØØØ + .5) / 1ØØØ
6600  PRINT "****************************************"
6610  IF D9 = 1 THEN 669Ø
6615  REM  D9=2, ANGLES IN DD.MMSS
6620  PRINT "ADJUSTED ANGLES (DD.MMSS)"
6630  PRINT "========================================"
6640  FOR J2 = Ø TO 7
6650 D = U(J2): GOSUB 8Ø5Ø:W(J2) = D
6660  PRINT "ANGLE  A";(J2 + 1);" = ";W(J2)
6670  NEXT J2
6680  PRINT "****************************************"
6690  PRINT : PRINT D$;"PR#Ø": GOTO 24ØØ
6693  REM  **************************************************
6695  REM  PRINTOUT OF CO-ORDINATES
6700  PRINT D$;"PR#1": POKE 1784 + 1,8Ø
6710  PRINT P1$
6720  PRINT "****************************************"
6730  PRINT "STN.NO.   X-COORD   Y-COORD   Z-COORD"
6740  PRINT "========================================"
6750  FOR I1 = Ø TO 3
6760  PRINT P(I1);: POKE 36,1Ø: PRINT  INT (X(I1) * 1ØØØ + .5) / 1ØØØ;: POKE
     36,2Ø: PRINT  INT (Y(I1) * 1ØØØ + .5) / 1ØØØ;
6770  IF Z(I1) = Ø THEN 68ØØ
6780  POKE 36,3Ø: PRINT Z(I1)
6790  NEXT I1: GOTO 681Ø
6800  PRINT " ": NEXT I1
6810  PRINT "****************************************"
6820  PRINT : PRINT D$;"PR#Ø"
6830  GOTO 315Ø
6893  REM  **************************************************
6895  REM  PRINTOUT OF CALCULATED SIDES - ANGLE CASE
6900  PRINT D$;"PR#1": POKE 1784 + 1,8Ø
6910  PRINT P1$: PRINT "****************************************"
6920  PRINT "SIDE     CALCULATED"
6930  PRINT "========================================"
6940  FOR I = Ø TO 5
6950  PRINT (I + 1);: POKE 36,1Ø: PRINT  INT (D(I) * 1ØØØ + .5) / 1ØØØ
6960  NEXT I
6970  PRINT "****************************************"
6980  PRINT : PRINT D$;"PR#Ø"
6990  GOTO 95ØØ
6993  REM  **************************************************
6995  REM  PRINTOUT OF CALCULATED SIDES - SIDE CASE
7000  PRINT D$;"PR#1": POKE 1784 + 1,8Ø
7010  PRINT P1$: PRINT "****************************************"
```

```
7020  PRINT "SIDE   OBSERVED  CALCULATED  VARIATION"
7030  PRINT "==========================================="
7040  FOR I = 0 TO 5
7050  PRINT (I + 1);: POKE 36,8: PRINT  INT ((H(I)) * 1000 + .5) / 1000;:
      POKE 36,18: PRINT  INT ((D(I)) * 1000 + .5) / 1000;
7060  POKE 36,30: PRINT  INT ((V(I)) * 1000 + .5) / 1000
7070  NEXT I
7080  PRINT "******************************************"
7090  PRINT : PRINT D$;"PR#0": GOTO 9500
7743  REM  ***************************************************
7745  REM  <TRIANGLE> SOLUTION (A1,S1,A3)
7750 T2 =  -  COS ((A3 + A1) / R9)
7760 A2 =  FN C9(T2)
7770 S2 = S1 *  SIN (A3 / R9) /  SIN (A2 / R9)
7780 S3 = S1 *  COS (A3 / R9) + S2 *  COS (A2 / R9)
7790  RETURN
7793  REM  ***************************************************
7795  REM  <TRIANGLE> SOLUTION (S1,S2,S3)
7800 P = (S1 + S2 + S3) / 2
7810 T3 =  SQR (P * (P - S2) / (S1 * S3))
7820 A3 = 2 *  FN C9(T3)
7830 T2 =  SQR (P * (P - S1) / (S2 * S3))
7840 A2 = 2 *  FN C9(T2)
7850 T1 =  -  COS ((A3 + A2) / R9)
7860 A1 =  FN C9(T1)
7870  RETURN
7993  REM  ***************************************************
7995  REM  ROUTINE DD.MMSS TO DECIMALS
8000 D0 =  INT (D5)
8005  REM  FIND NO. OF MINUTES,M0
8010 M1 = (D5 - D0) * 100:M0 =  INT (M1 + .1)
8015  REM  FIND NO. OF SECONDS,S0
8020 S0 =  INT (M1 * 100 - M0 * 100 + .5)
8025  REM  COLLECT DECS,MINS,SECS
8030 D5 = D0 + M0 / 60 + S0 / 3600
8040  RETURN
8043  REM  ***************************************************
8045  REM  ROUTINE DECIMAL TO DD.MMSS
8050 D1 =  INT (D)
8055  REM  FIND TOTAL NO. OF SECONDS
8060 D2 = (D - D1) * 3600
8065  REM  FIND NO. OF MINUTES
8070 M =  INT (D2 / 60)
8075  REM  FIND NO. OF SECONDS
8080 S = D2 - 60 * M + .5
8090  IF S < 60 THEN 8130
8100 M = M + 1:S = 0
8110  IF M < 60 THEN 8130
8120 D1 = D1 + 1:M = 0
8130 D = D1 * 10000 + M * 100 + S
8140 D =  INT (D) / 10000: RETURN
8985  REM  *****************************************
8990  REM  GOSUB ROUTINE TO CLEAR SCREEN
9000  HOME
9010  RETURN
9093  REM  ***************************************************
9095  REM  GOSUB ROUTINE FOR ERRORS
9100  PRINT
9110  PRINT E0$
```

```
9120  PRINT
9130  RETURN
9193  REM  **************************************************
9195  REM  ERROR TRACE ROUTINE
9200  PRINT  CHR$ (7): REM  CTRL-B (BELL)
9210 E =  PEEK (222): REM  GET ERROR NO.
9220  IF C = 1 THEN 9260
9230  INVERSE
9240  PRINT "ERROR NO. ";E;" FOUND"
9250 C = 1: TRACE : RESUME
9260  PRINT "ERROR ON SECOND LINE NO. "
9270  NORMAL
9280  NOTRACE
9290  STOP
9493  REM  **************************************************
9495  REM  END OF PROGRAM
9500  GOSUB 9000
9510  PRINT
9520  PRINT "END BRACED QUADRILATERAL PROGRAM"
9530  PRINT
9540  PRINT "*********************************"
9550  REM  PROGRAM PREPARED BY
9560  REM  DR. P.H. MILNE
9570  REM  DEPT. OF CIVIL ENGINEERING
9580  REM  UNIVERSITY OF STRATHCLYDE
9590  REM  GLASGOW G4 ONG
9600  REM  SCOTLAND
9610  REM  **************************
9620  END
```

6.4.5 Braced quadrilateral – computer printout

```
 LOCATION OF DATA  :- STRATHCLYDE UNIVERSITY - STEPPS
*************************************************************************
OPERATOR'S NAME :- 2ND YEAR GROUP C
DATE OF CALCS.  :-25/05/83
INSTRUMENT USED :- ZEISS ZENA T20A
*************************************************************************

<1> ANGLE OBSERVATIONS
***************************************
UNADJUSTED ANGLES (DD.MMSS)
=======================================
ANGLE  A1 = 21.033
ANGLE  A2 = 33.545
ANGLE  A3 = 84.5925
ANGLE  A4 = 40.02
ANGLE  A5 = 28.045
ANGLE  A6 = 26.5335
ANGLE  A7 = 60.28
ANGLE  A8 = 64.333
***************************************
UNADJUSTED ANGLES (DD.DEC)
=======================================
ANGLE  A1 = 21.058
ANGLE  A2 = 33.914
ANGLE  A3 = 84.99
ANGLE  A4 = 40.033
```

```
ANGLE  A5 = 28.Ø81
ANGLE  A6 = 26.893
ANGLE  A7 = 6Ø.467
ANGLE  A8 = 64.558
========================================
SUM A1-A8 = 359.994
****************************************


<1> ANGLE OBSERVATIONS
****************************************
ADJUSTED ANGLES (DD.DEC)
========================================
ANGLE  A1 = 21.Ø59
ANGLE  A2 = 33.915
ANGLE  A3 = 84.989
ANGLE  A4 = 4Ø.Ø35
ANGLE  A5 = 28.Ø81
ANGLE  A6 = 26.894
ANGLE  A7 = 6Ø.467
ANGLE  A8 = 64.559
========================================
SUM A1-A8 = 36Ø
****************************************
ADJUSTED ANGLES (DD.MMSS)
========================================
ANGLE  A1 = 21.Ø333
ANGLE  A2 = 33.5454
ANGLE  A3 = 84.5919
ANGLE  A4 = 4Ø.Ø2Ø4
ANGLE  A5 = 28.Ø451
ANGLE  A6 = 26.5337
ANGLE  A7 = 6Ø.28
ANGLE  A8 = 64.3333
****************************************

<1> ANGLE OBSERVATIONS
****************************************
STN.NO.    X-COORD   Y-COORD   Z-COORD
========================================
1          4817.996  7743.484
1Ø         4979.418  7487.465
9          5148.196  7497.28
2          4989.6Ø7  7834.156
****************************************


<1> ANGLE OBSERVATIONS
****************************************
SIDE      CALCULATED
========================================
1           3Ø2.66
2           169.Ø64
3           372.339
4           194.Ø92
5           411.885
6           346.841
****************************************
```

```
LOCATION OF DATA  :- GLENMORE
************************************************************************
OPERATOR'S NAME :- ROBERT THOMSON
DATE OF CALCS.  :-10/09/83
INSTRUMENT USED :- ZEISS ZENA T20A + CD6
************************************************************************

<2> SIDE OBSERVATIONS
****************************************
LEVEL OF STN.   1 = 110
LEVEL OF STN.   2 = 112
LEVEL OF STN.   3 = 112
LEVEL OF STN.   4 = 110

SLANT RANGE     1 = 18.319
SLANT RANGE     2 = 193.463
SLANT RANGE     3 = 126.881
SLANT RANGE     4 = 74.966
SLANT RANGE     5 = 196.488
SLANT RANGE     6 = 78.736

HORIZ. DISTANCE 1 = 18.209
HORIZ. DISTANCE 2 = 193.463
HORIZ. DISTANCE 3 = 126.865
HORIZ. DISTANCE 4 = 74.966
HORIZ. DISTANCE 5 = 196.478
HORIZ. DISTANCE 6 = 78.711
****************************************

<2> SIDE OBSERVATIONS
****************************************
UNADJUSTED ANGLES (DD.DEC)
========================================
ANGLE  A1 = 77.841
ANGLE  A2 = 71.553
ANGLE  A3 = 25.324
ANGLE  A4 = 5.279
ANGLE  A5 = 10.109
ANGLE  A6 = 139.286
ANGLE  A7 = 13.322
ANGLE  A8 = 17.28
========================================
SUM A1-A8 = 359.995
****************************************
COMPARISON OF CORNER ANGLES :-
Q1=A1+A8 = 95.125 : A1+A8 = 95.121
Q2=A2+A3 = 96.88 : A2+A3 = 96.877
Q3=A4+A5 = 15.39 : A4+A5 = 15.389
Q4=A6+A7 = 152.61 : A6+A7 = 152.608
****************************************

<2> SIDE OBSERVATIONS
****************************************
ADJUSTED ANGLES (DD.DEC)
========================================
ANGLE  A1 = 77.842
ANGLE  A2 = 71.554
ANGLE  A3 = 25.324
ANGLE  A4 = 5.28
ANGLE  A5 = 10.11
```

```
ANGLE  A6 = 139.286
ANGLE  A7 = 13.323
ANGLE  A8 = 17.281
=========================================
SUM A1-A8 = 360
*****************************************
ADJUSTED ANGLES (DD.MMSS)
=========================================
ANGLE  A1 = 77.5031
ANGLE  A2 = 71.3315
ANGLE  A3 = 25.1928
ANGLE  A4 = 5.1647
ANGLE  A5 = 10.0634
ANGLE  A6 = 139.1711
ANGLE  A7 = 13.1922
ANGLE  A8 = 17.1652
*****************************************


<2> SIDE OBSERVATIONS
*****************************************
STN.NO.   X-COORD    Y-COORD    Z-COORD
=========================================
1         .5         8.78       110
2         4.228      -9.045     112
3         196.974    7.596      112
4         72.213     30.617     110
*****************************************


<2> SIDE OBSERVATIONS
*****************************************
SIDE   OBSERVED  CALCULATED  VARIATION
=========================================
1      18.209    18.21       -1E-03
2      193.463   193.463     0
3      126.865   126.867     -2E-03
4      74.966    74.964      2E-03
5      196.478   196.478     0
6      78.711    78.709      2E-03
*****************************************
```

6.5 MATRIX SOLUTIONS

The equations developed for solving the most probable values of measured quantities using either observation equations or condition equations, as discussed in Section 6.3, are often very complex. Where there are n equations, these often take the form:

$$A_i x + B_i y + C_i z = D_i \qquad (6.31)$$

where i is an integer from 1 to n. The solution to n equations similar to Equation 6.31 can often take a long time by hand. However, the modern microcomputer, using matrix algebra, is ideally suited for such tasks. The matrix solution for x, y and z, since their true values are unknown, uses the principle of least squares, where the observation equation matrix is:

$$A = \begin{bmatrix} A_1 & B_1 & C_1 \\ A_2 & B_2 & C_2 \\ \vdots & \vdots & \vdots \\ A_n & B_n & C_n \end{bmatrix} \tag{6.32}$$

and the constant matrix and solution matrix are:

$$B = \begin{bmatrix} D_1 \\ D_2 \\ \vdots \\ D_n \end{bmatrix}, \hat{x} = \begin{bmatrix} x \\ y \\ z \end{bmatrix} \tag{6.33}$$

where $\hat{x}$ is the least squares estimator for x. If matrix A is square (that is, it has an equal number of rows and columns), and is non-singular, then the solution is given by:

$$A\hat{x} = B \tag{6.34}$$

However, if the number of rows and columns is not equal, then the solution requires the calculation of the inverse of A, i.e. $(A)^{-1}$, and the following equation is used instead to solve for $\hat{x}$, where both sides of Equation 6.34 are pre-multiplied by A^T:

$$\hat{x} = (A^TA)^{-1}(A^TB) \tag{6.35}$$

Earlier in this chapter (Section 6.1.6) the concept of weighted observations was introduced. If the weights of a set of observations are denoted by $w_1, w_2 \ldots w_n$, then W the weight matrix, which is diagonal, can be written as:

$$W = \begin{bmatrix} w_1 & & & \\ & w_2 & & \\ & & \vdots & \\ & & & w_n \end{bmatrix} \tag{6.36}$$

However, if A is not square, as discussed earlier for Equation 6.35, then it is necessary to use the transpose of $A(A^T)$ to obtain the correct matrix solution from:

$$\hat{x} = (A^TWA)^{-1}(A^TWB) \tag{6.37}$$

As mentioned at the start of this section, the formation of the observation or condition equations is a complex process. It is first necessary to determine the minimum number, n_0, of observations to give a unique solution. If the number of observations is n, which must be greater than n_0, then the number of reaundant observations is $r = n - n_0$ where r will be the number of independent condition equations that can be written in terms of the n observations. This is best illustrated by discussing two particular examples.

Example (1)

Consider a levelling circuit round and across the braced quadrilateral in Fig. 6.2(a). This will result in six observations, AB, BC, CD, DA, AC, BD, as given below:

$$\begin{aligned} A - \mathrm{B} &= +7.260 \\ B - C &= +0.730 \\ C - D &= -5.565 \\ D - A &= -2.405 \\ A - C &= +7.960 \\ B - D &= -4.845 \end{aligned}$$

There is a choice of using the above as a constant matrix in a set of observation equations or of developing condition equations. In the first case, if the level of $A(X_1)$ is chosen as the origin for reference and the levels of B, C and D are represented by X_2, X_3, X_4 respectively, where the errors in the above observations are e_1 to e_6 respectively, then the observation equations are:

$$\begin{aligned} X_2 - 0 &= +7.260 + e_1 \\ X_3 - X_2 &= +0.730 + e_2 \\ X_4 - X_3 &= -5.565 + e_3 \\ 0 - X_4 &= -2.405 + e_4 \\ X_3 - 0 &= +7.960 + e_5 \\ X_4 - X_2 &= -4.845 + e_6 \end{aligned}$$

where

$$A = \begin{bmatrix} 1 & 0 & 0 \\ -1 & +1 & 0 \\ 0 & -1 & +1 \\ 0 & 0 & -1 \\ 0 & +1 & 0 \\ -1 & 0 & +1 \end{bmatrix}, x = \begin{bmatrix} X_2 \\ X_3 \\ X_4 \end{bmatrix}, B = \begin{bmatrix} 7.260 \\ 0.730 \\ -5.565 \\ -2.405 \\ +7.960 \\ -4.845 \end{bmatrix}$$

Solving the above equations using the <MATRICES> program gives the solution:

$$X_2 = 7.250$$

$$X_3 = 7.970$$

$$X_4 = 2.405$$

Example (2)

The straight line AD is measured in several sections as given below. Find the most probable values of AB, BC, CD and the total AD. From the observations, a set of six observation equations are formed where the most probable values of AB, BC and CD are X_1, X_2 and X_3 respectively.

Observations	*Observation equations*
$AB = 176.327$	$X_1 = 176.327$
$BC = 224.758$	$X_2 = 224.758$
$CD = 195.461$	$X_3 = 195.461$
$AC = 401.097$	$X_1 + X_2 = 401.097$
$AD = 596.566$	$X_1 + X_2 + X_3 = 596.566$
$BD = 420.234$	$X_2 + X_3 = 420.234$

In Equation 6.32, the observation equations fill the matrix A and the observations the constant matrix B where the solution matrix is x:

$$A = \begin{bmatrix} 1 & 0 & 0 \\ 0 & 1 & 0 \\ 0 & 0 & 1 \\ 1 & 1 & 0 \\ 1 & 1 & 1 \\ 0 & 1 & 1 \end{bmatrix}, B = \begin{bmatrix} 176.327 \\ 224.758 \\ 195.461 \\ 401.097 \\ 596.566 \\ 420.234 \end{bmatrix}, x = \begin{bmatrix} X_1 \\ X_2 \\ X_3 \end{bmatrix}$$

Since each of the observation equations contains a different number of measurements, 1, 2 or 3, the lengths should be weighted proportionally. Thus the weights of the six observations from AB to BD would be 3, 3, 3, 2, 1, 2 respectively, giving the diagonal weight matrix:

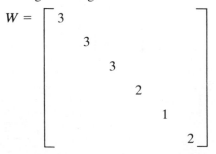

$$W = \begin{bmatrix} 3 & & & & & \\ & 3 & & & & \\ & & 3 & & & \\ & & & 2 & & \\ & & & & 1 & \\ & & & & & 2 \end{bmatrix}$$

with the solution:

$$X_1 = 176.330$$

$$X_2 = 224.764$$

$$X_3 = 195.466$$

The following program <MATRICES> consists of numerous small routines for moving, multiplying, inverting and transposing matrices to allow Equations 6.34, 6.35 and 6.37 to be solved by the microcomputer.

6.5.1 Matrix solutions – subroutine index

	Line numbers	*Function*
(a)	10–60	Initialization and control
(b)	100–230	Screen header display
(c)	500–680	Entry of data location, etc.
(d)	700–800	Selection of unweighted/weighted observations
(e)	1000–1190	Entry of matrix A
(f)	1200–1390	Inverse of matrix A
(g)	1400–1680	Transpose of matrix A
(h)	2000–2190	Entry of matrix B
(i)	2200–2290	Solution $x = (A)^{-1}B$
(j)	2300–2490	Solution $x = (A^TA)^{-1}(A^TB)$
(k)	3000–3130	Entry of matrix W
(l)	3150–3550	Solution $x = (A^TWA)^{-1}(A^TWB)$
(m)	6000–6080	Printout of data location, etc.
(n)	6100–6270	Printout for matrix C
(o)	7000–7090	Display for matrix C
(p)	7100–7190	Move matrix A to C
(q)	7200–7590	Matrix inversion V
(r)	7600–7690	Move matrix P to C
(s)	7700–7740	Move matrix A to V
(t)	7750–7780	Move matrix V to B
(u)	7800–7890	Move matrix B to C
(v)	7900–7960	Multiply matrix $P*B$
(w)	8000–8170	Multiply matrix $P*A$
(x)	8200–8250	Matrix transpose T
(y)	8300–8340	Move matrix T to C
(z)	8350–8490	Multiply matrix $T*A$
(aa)	8500–8560	Multiply matrix $T*B$
(bb)	8600–8640	Make matrix W = zero
(cc)	8650–8690	Move matrix W to C
(dd)	8700–8790	Move matrix B to D
(ee)	8800–8970	Multiply matrix $W*D$
(ff)	9000–9010	Routine to clear screen
(gg)	9100–9130	Routine to display error message
(hh)	9200–9290	Error trace routine
(ii)	9300–9380	Multiply matrix $P*V$
(jj)	9400–9490	Move matrix A to D
(kk)	9500–9620	Termination of program

(a) Initialization and control

Line numbers 10–60

The numeric variables, arrays and required string variables are initialized. A maximum (10, 10) matrix has been assumed; should a larger matrix be required the arrays should be redimensioned accordingly.

(b) Screen header display

Line numbers 100–230

This is the screen display used for civil engineering students at Strathclyde University. The operator can alter this as required to suit another organization.

(c) Entry of data location, etc.

Line numbers 500–680

This routine is provided to keep a record of site location, operator, date, etc., with an opportunity to obtain a printout if required.

(d) Selection of unweighted/weighted observations

Line numbers 700–800

The operator has a choice of <1> unweighted or <2> weighted observations where the selection is stored in variable $P1$.

(e) Entry of matrix *A*

Line numbers 1000–1190

After a request for the number of rows (M) and columns (N) in matrix A, the elements are entered from the keyboard. After displaying in the correct format, a printout option is provided. To simplify display and printout, all entered or calculated matrices are moved to matrix C and the GOSUB routines at lines 7000 and 6100 used for display and printout.

(f) Inverse of matrix *A*

Line numbers 1200–1390

This routine is only accessed if the matrix A is square (Equation 6.34), that is the number of rows and columns are the same, as checked in line 1190 in the previous Section (e). It should be noted that a square matrix only has an inverse if its determinant is not zero. The matrix A is moved first to V for inversion (GOSUB 7700) and then a Gauss-Jordan Elimination routine is used (GOSUB 7200) where V is input and P is output. The matrix P, which is the inverse of A, i.e., $P = (A)^{-1}$, is then displayed with a printout option.

To ensure that the calculations have been performed correctly, a check on the calculation $A^*(A)^{-1}$ is performed, which should result in the diagonal identity matrix being displayed.

(g) Transpose of matrix *A*

Line numbers 1400–1680

If the number of rows and columns is not equal, the program jumps from line 1190 in Section (e) to this routine for use in solving Equations 6.35 and 6.37 which requires A^T. If weights are available for the observations, the program branches from line 1490 to continue at 2000.

In the first case, Equation 6.35, it is necessary to first calculate A^TA and then to invert to obtain $(A^TA)^{-1}$ which is displayed from lines 1530–1560. An optional routine is included in lines 1600–1680 to check on the identity matrix.

(h) Entry of matrix *B*

Line numbers 2000–2190

This routine assumes a one-column array for B, but could easily be adapted for two columns if desired at a later date. If weights are known for the observations, the program branches from line 2180 to 3000.

(i) Solution $x = (A)^{-1}B$

Line numbers 2200–2290

This is the simplest solution using Equation 6.34 and after displaying the resultant array x, a printout option is provided.

(j) Solution $x = (A^TA)^{-1}(A^TB)$

Line numbers 2300–2490

This is the nonweighted solution for x where A is not square. The first part of the solution $(A^TA)^{-1}$ was calculated in Section (g), and stored in matrix P. The second part (A^TB) is now calculated and stored in matrix A and the solution x obtained from P^*A for display and printout.

(k) Entry of matrix *W*

Line numbers 3000–3130

The first essential is to fill all elements of the matrix W with zero (GOSUB 8600), before requesting entry of the diagonal elements $W(K, K)$.

(l) Solution $x = (A^TWA)^{-1}(A^TWB)$

Line numbers 3150–3550

The calculations for weighted matrices are quite complex, as indicated by Equation 6.37, and hence the need to move matrices using other variables. The order of multiplication is also of importance to ensure that the rows/columns are compatible. In this program, the following procedure is used.

The first portion $(A^TWA)^{-1}$ is evaluated by first multiplying W^*A. Since it is wished to store the result in A, the matrix A was earlier stored in D at line 1400 (GOSUB 9400), thus giving $A = W^*D$ in line 3150 (GOSUB 8800). The resulting A is then multiplied by A^T to give V in line 3240 (GOSUB 8350), where V is used for inversion using the Gauss-Jordan Elimination (GOSUB 7200) at line 3250 to give $P = (A^TWA)^{-1}$.

The second portion (A^TWB) is then evaluated by first moving B to D and multiplying by W to give A (GOSUB 8800) in line 3350. This new matrix A is then multiplied by A^T, stored in T, to give V (GOSUB 8350) in line 3440. V is now moved to B (GOSUB 7750) in line 3450, thus allowing a previous routine $C = P^*B$ to be used (GOSUB 7900) to evaluate x, with a final display and printout.

(m) Printout of data, location, etc.

Line numbers 6000–6080

This routine is provided for record purposes and is based on Applesoft DOS, and may require alteration for different microcomputers and printers.

(n) Printout for matrix *C*

Line numbers 6100–6270

Only one matrix printout routine has been included for simplicity where the matrix identification is stored in the string *M*\$, and all matrices are moved to C prior to using this printout routine.

(o) Display for matrix *C*

Line numbers 7000–7090

As above, only one matrix display routine is provided and all matrices moved to C for display.

(p) Move matrix *A* to *C*

Line numbers 7100–7190

This routine caters for matrices of all configurations including a single-column array.

(q) Matrix inversion, *V*

Line numbers 7200–7590

This is one of the most complicated matrix computations and uses a Gauss-Jordan Elimination. This routine assumes that firstly the matrix is square and secondly that its determinant is not zero. The calculations are performed on matrix V, and the resulting inverted matrix P is output. It should be noted that if in dimensioning the arrays at the outset in the program, if N is the number of columns, then the arrays required are $V(N, N)$ and $P(N, 2^*N)$.

(r) Move matrix P to C

Line numbers 7600–7690

This routine moves an inverted matrix P to C for display.

(s) Move matrix A to V

Line numbers 7700–7740

This allows matrix A to be moved to V for use in the inversion subroutine.

(t) Move matrix V to B

Line numbers 7750–7780

This routine allows $V = A^T{}^*A$ to be moved to B for use in the routine $C = P^*B$.

(u) Move matrix B to C

Line numbers 7800–7890

This routine moves matrix B to C for display and printout.

(v) Multiply matrix P^*B

Line numbers 7900–7960

This routine, $C = P^*B$, was referred to in Section (t) above, where the result can be displayed or printed without further computation.

(w) Multiply matrix P^*A

Line numbers 8000–8170

This routine like (v) above stores the result in matrix C.

(x) Matrix transpose T

Line numbers 8200–8250

This is a straightforward swap of the rows and columns forming $T = A^T$ from A.

(y) Move matrix T to C

Line numbers 8300–8340

This routine moves A^T to C for display.

(z) Multiply matrix *T*A*

Line numbers 8350–8490

This routine is used in both Equations 6.35 and 6.37 to give (A^TA) from $V = T^*A$.

(aa) Multiply matrix *T*B*

Line numbers 8500–8560

This routine is used in the weighted solution, Equation 6.37.

(bb) Make matrix *W* = zero

Line numbers 8600–8640

Prior to entering the weights of the observations, all elements of the matrix *W* are filled with zero.

(cc) Move matrix *W* to *C*

Line numbers 8650–8690

This routine allows matrix *W* to be displayed and printed.

(dd) Move matrix *B* to *D*

Line numbers 8700–8790

This routine is used to store *B* in *D* for use in further calculations as discussed in Section (l) above.

(ee) Multiply matrix *W*D*

Line numbers 8800–8970

This routine was referred to in Section (l) above.

(ff) Routine to clear screen

Line numbers 9000–9010

To simplify portability, the Applesoft 'HOME' command to clear the screen has been placed in a subroutine to allow an easy change to another microcomputer.

(gg) Routine to display error message

Line numbers 9100–9130

If a wrong keyboard entry has been made in entering data, the error is trapped and this error message displayed and re-entry requested.

(hh) Error trace routine

Line numbers 9200–9290

This error trace routine was discussed earlier in Section 2.2.9.

(ii) Multiply matrix *P*V*

Line numbers 9300–9380

This routine has been included to allow a check on the inversion routine in (q) above by displaying the result of P^*V which should give the identity matrix, i.e., a diagonal matrix with the element 1 throughout.

(jj) Move matrix *A* to *D*

Line numbers 9400–9490

This routine was discussed in (l) above.

(kk) Termination of program

Line numbers 9500–9620

When the '<QUIT>' selection is made in the program, an end-of-program message is displayed to advise the operator that computations are complete.

6.5.2 Matrix solutions – numeric variables

$A(M, N)$	= Array for matrix A
$B(M, 0)$	= Array for constant matrix B
C	= Number of columns in array A
$C0$	= Counter in error trace routine
$C9$	= Flag to select identity matrix check
$C(I, J)$	= Array for matrix C for display
$D(I, J)$	= Array for storage purposes
E	= Error number
$I, I1–I3, I5$	= Integer loop variables
$J, J1, J2, J5$	= Integer loop variables
K	= Integer loop variable
M	= Number of rows in array A
$M1, M2$	= Array variables
N	= Number of columns in array A
$N1, N2$	= Array variables
$P1$	= Program selection <1> unweighted or <2> weighted
$P7$	= Counter in printing routines
$P(I, 2*J)$	= Inverse matrix
R	= Number of rows in array A
$T(I, J)$	= Transpose matrix
U	= Selection from initial menu
$V(I, J)$	= Input matrix for inverse routine
$W(M, M)$	= Weight matrix

6.5.3 Matrix solutions – string variables

A$	Storage for location of data
B$	Storage for operator's name
C$	Storage for date of calculation
D$	= CHR$(4), CTRL-D
*E*0$	= "SORRY, DATA ERROR . . . PLEASE RE-ENTER"
H$	= "MATRIX SOLUTION TO OBSERVATIONS"
K$	Storage for instrument used
M$	Storage for type of matrix
P$	= "DO YOU WISH PRINTOUT OF ABOVE DATA (Y/N)"
Q$	= Question response (Y/N)
W$	= Either "<1> UNWEIGHTED OBSERVATIONS" if *P*1 = 1 "<2> WEIGHTED OBSERVATIONS" if *P*1 = 2

6.5.4 Matrix solutions – BASIC program

```
10  REM  <MATRICES> PROGRAM FOR APPLE II+  USES QUME PRINTER
20 D$ =  CHR$ (4): ONERR  GOTO 9200
30 C = 0:E = 0:P1 = 0:P7 = 0
35  DIM A(10,10),B(10,1),C(10,10),D(10,10),P(10,20),T(10,10),V(10,10),W(1
    0,10)
40 H$ = "MATRIX SOLUTION TO OBSERVATIONS"
50 E0$ = "SORRY, DATA ERROR ... PLEASE RE-ENTER."
55 P$ = "DO YOU WISH PRINTOUT OF ABOVE DATA (Y/N)"
60 L$ = " ":M$ = " ":N$ = " "
95  REM  ***************************************************
100  GOSUB 9000
110  PRINT "****************************************"
120  PRINT "*                                      *"
130  PRINT "*      UNIVERSITY  OF  STRATHCLYDE     *"
140  PRINT "*  DEPARTMENT  OF  CIVIL  ENGINEERING  *"
150  PRINT "*                                      *"
160  PRINT "*            SURVEYING  SECTION        *"
170  PRINT "*   MATRIX SOLUTION TO OBSERVATIONS    *"
180  PRINT "****************************************"
190  PRINT
200  PRINT "DO YOU WISH TO ENTER LOCATION <1>,      NEW DATA <2> OR QUIT
     <3>"
210  INPUT U
220  IF U < 1 OR U > 3 THEN 200
230  ON U GOTO 500,700,9500
500  GOSUB 9000
510  PRINT "UNIVERSITY OF STRATHCLYDE"
520  PRINT H$
530  PRINT "****************************************"
540  PRINT
550  PRINT "ENTER LOCATION OF WORK, ETC. WHEN       REQUESTED, AND PRESS
     <RETURN>"
560  PRINT "ENTER LOCATION OF DATA :-"
570  INPUT A$
580  PRINT "ENTER OPERATOR'S NAME :-"
590  INPUT B$
600  PRINT "ENTER DATE OF CALCULATION AS DD/MM/YY -"
610  INPUT C$
```

```
62Ø  PRINT "ENTER INSTRUMENT USED :-"
63Ø  INPUT K$
64Ø  PRINT
65Ø  PRINT "DO YOU WISH PRINTOUT OF LOCATION ETC.,  (Y/N) ";
66Ø  INPUT Q$
67Ø  IF Q$ = "Y" THEN 6ØØØ
68Ø  IF Q$ < > "N" THEN 65Ø
69Ø :
695  REM  PROGRAM MENU SELECTION
7ØØ  GOSUB 9ØØØ: PRINT H$
71Ø  PRINT "****************************************"
72Ø  PRINT "SELECT PROGRAM FROM MENU :-"
73Ø  PRINT
74Ø  PRINT "   <1> UNWEIGHTED OBSERVATIONS"
75Ø  PRINT "   <2> WEIGHTED   OBSERVATIONS"
76Ø  PRINT
77Ø  INPUT P1
78Ø  IF P1 < 1 OR P1 > 2 THEN 7ØØ
79Ø  IF P1 = 1 THEN W$ = "<1> UNWEIGHTED OBSERVATIONS"
8ØØ  IF P1 = 2 THEN W$ = "<2> WEIGHTED OBSERVATIONS"
995  REM  *************************************************
1ØØØ  GOSUB 9ØØØ
1Ø1Ø  PRINT H$: PRINT W$
1Ø2Ø  PRINT "****************************************"
1Ø3Ø  PRINT : PRINT "ENTER DIMENSIONS OF MATRIX A(M,N) ";
1Ø4Ø  INPUT M,N:R = M:C = N
1Ø5Ø  PRINT : PRINT "ENTER ELEMENTS OF MATRIX A :-": PRINT
1Ø6Ø  FOR I = Ø TO M - 1
1Ø7Ø  FOR J = Ø TO N - 1
1Ø8Ø  PRINT "ENTER ELEMENT (";I;",";J;") ";
1Ø9Ø  INPUT A(I,J)
11ØØ  NEXT J
111Ø  NEXT I
1115  REM  MOVE A(I,J) TO C(I,J)
112Ø N1 = M - 1:N2 = N - 1: GOSUB 71ØØ
1125  REM  DISPLAY MATRIX A NOW IN C
113Ø  GOSUB 9ØØØ:M$ = "MATRIX A = "
114Ø  PRINT M$: PRINT "=============================================="
115Ø  PRINT : GOSUB 7ØØØ
116Ø  PRINT : PRINT P$;: INPUT Q$: IF Q$ = "N" THEN 119Ø
117Ø  IF Q$ < > "Y" THEN 116Ø
118Ø  GOSUB 61ØØ: IF P1 = 2 THEN 14ØØ
1185  REM  CHECK IF MATRIX A IS SQUARE
119Ø  IF R < > C THEN 14ØØ
1195  REM  M=N, I.E. SQUARE MATRIX, CAN INVERT. MOVE MAT A TO V THEN INVE
    RT V TO P
12ØØ  GOSUB 77ØØ: GOSUB 72ØØ
12Ø5  REM  MØVE MAT P TO C
121Ø  GOSUB 76ØØ
1215  REM  DISPLAY C, I.E. P=INV(A)
122Ø M$ = "INVERSE OF A = "
123Ø  PRINT : PRINT M$
124Ø  PRINT "========================================": PRINT
125Ø  GOSUB 7ØØØ
126Ø  PRINT : PRINT P$;: INPUT Q$
127Ø  IF Q$ = "N" THEN 13ØØ
128Ø  IF Q$ < > "Y" THEN 126Ø
129Ø  GOSUB 61ØØ
1295  REM  CHECK RESULTS
```

```
1300 M1 = N - 1:M2 = N - 1
1305  REM  MULTIPLY P*A
1310  GOSUB 8000
1320 M$ = "MATRIX A * INV(A) = "
1330  PRINT : PRINT M$
1340  PRINT "========================================": PRINT
1350  GOSUB 7000
1360  PRINT : PRINT P$;: INPUT Q$: IF Q$ = "N" THEN 1390
1370  IF Q$ < > "Y" THEN 1360
1380  GOSUB 6100
1385  REM  COLLECT MATRIX B
1390  GOTO 2000
1395  REM  M<>N, REQUIRE TRANSPOSE OF A
1400  GOSUB 9400: GOSUB 8200
1405  REM  MOVE T=TRN(A) TO C
1410  GOSUB 8300
1420 M$ = "TRANSPOSE OF A = "
1430  PRINT : PRINT M$: PRINT "========================================":
      PRINT
1440 N1 = N - 1:N2 = M - 1: GOSUB 7000
1450  PRINT : PRINT P$;: INPUT Q$
1460  IF Q$ = "N" THEN 1490
1470  IF Q$ < > "Y" THEN 1450
1480  GOSUB 6100
1485  REM  IF WEIGHTED OBSERVATIONS COLLECT B
1490  IF P1 = 2 THEN 2000
1495  REM  MULTIPLY  TRN(A)*A TO GIVE V
1500 M1 = N - 1:N1 = M - 1:N2 = N - 1: GOSUB 8350
1505  REM  NOW INVERT V, V INPUT, P OUTPUT
1510  GOSUB 7200
1515  REM  MOVE P TO C
1520 N1 = N2: GOSUB 7600
1530 M$ = "INVERSE OF TRN(A)*A = "
1540  PRINT : PRINT M$
1550  PRINT "========================================": PRINT
1560  GOSUB 7000
1570  PRINT : PRINT P$;: INPUT Q$: IF Q$ = "N" THEN 1600
1580  IF Q$ < > "Y" THEN 1570
1590  GOSUB 6100
1595  REM  CHECK INVERSE, I.E. C=P*V
1600  GOSUB 9300:M$ = "CHECK ON IDENTITY MATRIX"
1610  PRINT : PRINT M$
1620  PRINT "========================================": PRINT
1630  GOSUB 7000
1640  PRINT : PRINT P$;: INPUT Q$
1650  IF Q$ = "N" THEN 1680
1660  IF Q$ < > "Y" THEN 1640
1670  GOSUB 6100
1680  GOTO 2000
1690 :
1995  REM  ********************************************
2000  GOSUB 9000: PRINT H$: PRINT W$:N1 = R - 1
2010  PRINT "****************************************"
2020  PRINT : PRINT "ENTER ELEMENTS OF MATRIX B :-": PRINT
2030  FOR I3 = 0 TO N1
2040  PRINT "ENTER ELEMENT (";I3;",0)";
2050  INPUT B(I3,0)
2060  NEXT I3
2065  REM  MOVE MAT B TO C
```

```
2070 N2 = 0
2080  GOSUB 7800
2095  REM  DISPLAY MAT B IN C
2100 M$ = "MATRIX B = "
2110  PRINT : PRINT M$
2120  PRINT "============================================"
2130  PRINT : GOSUB 7000
2140  PRINT : PRINT P$;: INPUT Q$
2150  IF Q$ = "N" THEN 2180
2160  IF Q$ < > "Y" THEN 2140
2170  GOSUB 6100
2175  REM  CHECK IF WEIGHTED OBSERVATIONS
2180  IF P1 = 2 THEN 3000
2190  IF R < > C THEN 2300
2195  REM  MULTIPLY INV(A)*B TO GET C
2200 M1 = R - 1
2210 M2 = C - 1
2220  GOSUB 7900
2230 M$ = "MATRIX X FROM (AX=B) IS = "
2240  PRINT : PRINT M$: PRINT "====================================
      PRINT
2250  GOSUB 7000
2260  PRINT : PRINT P$;: INPUT Q$: IF Q$ = "N" THEN 100
2270  IF Q$ < > "Y" THEN 2260
2280  GOSUB 6100
2285  REM  RETURN TO MENU
2290  GOTO 100
2295  REM  MULTIPLY TRN(A)*B=A
2300  GOSUB 8500
2305  REM  MOVE A TO C
2310 N1 = C - 1: GOSUB 7100
2320 M$ = "MATRIX OF TRN(A)*B = "
2330  PRINT : PRINT M$
2340  PRINT "===========================================": PRINT
2350  GOSUB 7000
2360  PRINT : PRINT P$;: INPUT Q$
2370  IF Q$ = "N" THEN 2400
2380  IF Q$ < > "Y" THEN 2360
2390  GOSUB 6100
2395  REM  MULTIPLY INV(TRN(A)*A) * (TRN(A)*B), I.E. P*A
2400  GOSUB 8000
2410 M$ = "MATRIX X FROM AX=B IS = "
2420  PRINT : PRINT M$
2430  PRINT "===========================================": PRINT
2440  GOSUB 7000
2450  PRINT : PRINT P$;: INPUT Q$
2460  IF Q$ = "N" THEN 2490
2470  IF Q$ < > "Y" THEN 2450
2480  GOSUB 6100
2485  REM  RETURN TO MENU
2490  GOTO 100
2993  REM  ************************************************
2995  REM  WEIGHTED OBSERVATIONS
3000  GOSUB 8600: GOSUB 9000
3010  PRINT : PRINT "ENTER WEIGHT MATRIX W :-": PRINT
3020  FOR K = 0 TO R - 1
3030  PRINT "ENTER ELEMENT W(";K;",";K;")";
3040  INPUT W(K,K)
3050  NEXT K
```

```
3Ø55  REM  MOVE MAT W TO C
3Ø6Ø  GOSUB 865Ø
3Ø7Ø M$ = "WEIGHT MATRIX W = "
3Ø8Ø  PRINT M$: PRINT "========================================": PRINT
3Ø9Ø N1 = R - 1:N2 = R - 1: GOSUB 7ØØØ
31ØØ  PRINT : PRINT P$;: INPUT Q$
311Ø  IF Q$ = "N" THEN 315Ø
312Ø  IF Q$ < > "Y" THEN 31ØØ
313Ø  GOSUB 61ØØ
314Ø :
3145  REM  MULTIPLY W*A, I.E. A=W*D
315Ø M1 = R - 1:N2 = N - 1: GOSUB 88ØØ
3155  REM  MOVE A TO C
316Ø  GOSUB 71ØØ
317Ø M$ = "MATRIX W*A = "
318Ø  PRINT M$: PRINT "========================================": PRINT
319Ø  GOSUB 7ØØØ
32ØØ  PRINT : PRINT P$;: INPUT Q$
321Ø  IF Q$ = "N" THEN 324Ø
322Ø  IF Q$ < > "Y" THEN 32ØØ
323Ø  GOSUB 61ØØ
3235  REM  GET V=T*WA, I.E. V=T*A
324Ø M1 = N - 1: GOSUB 835Ø
3245  REM  INVERT V GET P
325Ø  GOSUB 72ØØ
3255  REM  CHECK IDENTITY MATRIX, C=P*V
326Ø N1 = N - 1: GOSUB 93ØØ
327Ø M$ = "IDENTITY MATRIX CHECK (TWA)"
328Ø  PRINT M$: PRINT "========================================": PRINT
329Ø  GOSUB 7ØØØ
33ØØ  PRINT : PRINT P$;: INPUT Q$
331Ø  IF Q$ = "N" THEN 334Ø
332Ø  IF Q$ < > "Y" THEN 33ØØ
333Ø  GOSUB 61ØØ
3335  REM  MOVE B TO D
334Ø N1 = R - 1:N2 = Ø: GOSUB 87ØØ
3345  REM  GET A=W*B, I.E. A=W*D
335Ø M1 = R - 1: GOSUB 88ØØ
3355  REM  MOVE A TO C
336Ø  GOSUB 71ØØ
337Ø M$ = "MATRIX W*B = "
338Ø  PRINT M$: PRINT "========================================": PRINT
339Ø  GOSUB 7ØØØ
34ØØ  PRINT : PRINT P$;: INPUT Q$
341Ø  IF Q$ = "N" THEN 344Ø
342Ø  IF Q$ < > "Y" THEN 34ØØ
343Ø  GOSUB 61ØØ
3435  REM  MULTIPLY T*WB, I.E. V=T*A
344Ø M1 = N - 1: GOSUB 835Ø
3445  REM  MOVE V TO B
345Ø N1 = N - 1: GOSUB 775Ø
3455  REM  GET C=INV(TWA)*(TWB), I.E. C=P*B
346Ø  GOSUB 79ØØ
347Ø M$ = "MATRIX X = INV(TWA)*(TWB) :-"
348Ø  PRINT M$: PRINT "========================================": PRINT
349Ø  GOSUB 7ØØØ: PRINT "****************************************"
35ØØ  PRINT : PRINT P$;: INPUT Q$
351Ø  IF Q$ = "N" THEN 355Ø
352Ø  IF Q$ < > "Y" THEN 35ØØ
```

```
353Ø GOSUB 61ØØ
354Ø :
3545 REM  RETURN TO MENU
355Ø GOTO 1ØØ
5985 REM  *************************************************************************
    **
599Ø REM  PRINTOUT OF LOCATION, ETC
6ØØØ PRINT D$;"PR#1": PRINT "LOCATION OF DATA :-";A$
6Ø1Ø PRINT "**************************************************************************
    ********************"
6Ø2Ø PRINT "OPERATOR'S NAME :- ";B$
6Ø3Ø PRINT "DATE OF CALCS.  :-";C$
6Ø4Ø PRINT "INSTRUMENT USED :- ";K$
6Ø6Ø PRINT "**************************************************************************
    ********************"
6Ø7Ø PRINT : PRINT D$;"PR#Ø"
6Ø8Ø GOTO 7ØØ
6Ø93 REM  ************************************************************
6Ø95 REM  UNIVERSAL PRINTOUT FOR MATRIX C
61ØØ PRINT D$;"PR#1": POKE 1784 + 1,8Ø
611Ø PRINT M$: PRINT "=========================================="
612Ø IF N1 * N2 = Ø THEN 62ØØ
613Ø FOR I5 = Ø TO N1
614Ø FOR J5 = Ø TO N2
615Ø POKE 36,(1 + 6 * J5): PRINT  INT (C(I5,J5) * 1ØØØ + .5) / 1ØØØ;
616Ø IF J5 = N2 THEN  PRINT
617Ø NEXT J5
618Ø NEXT I5
619Ø GOTO 625Ø
62ØØ IF N1 = Ø THEN 625Ø
621Ø FOR I5 = Ø TO N1
622Ø PRINT  INT (C(I5,Ø) * 1ØØØ + .5) / 1ØØØ
623Ø NEXT I5
624Ø :
625Ø PRINT "*************************************************"
626Ø PRINT : PRINT D$;"PR#Ø"
627Ø RETURN
6993 REM  ************************************************************
6995 REM  DISPLAY MATRIX C
7ØØØ IF N1 * N2 = Ø THEN 7Ø6Ø
7Ø1Ø FOR I = Ø TO N1
7Ø2Ø FOR J = Ø TO N2
7Ø3Ø PRINT  TAB( 1 + 6 * J); INT (C(I,J) * 1ØØØ + .5) / 1ØØØ;
7Ø4Ø IF J = N2 THEN  PRINT
7Ø5Ø NEXT J: NEXT I: RETURN
7Ø6Ø IF N1 = Ø THEN  RETURN
7Ø7Ø FOR I = Ø TO N1
7Ø8Ø PRINT  INT (C(I,Ø) * 1ØØØ + .5) / 1ØØØ
7Ø9Ø NEXT I: RETURN
7Ø93 REM  ************************************************************
7Ø95 REM  MOVE MAT A TO MAT C
71ØØ IF N1 * N2 = Ø THEN 716Ø
711Ø FOR I1 = Ø TO N1
712Ø FOR I2 = Ø TO N2
713Ø C(I1,I2) = A(I1,I2)
714Ø NEXT I2
715Ø NEXT I1: RETURN
716Ø IF N1 = Ø THEN  RETURN
717Ø FOR I1 = Ø TO N1
```

```
7180 C(I1,0) = A(I1,0)
7190  NEXT I1: RETURN
7193  REM  ****************************************************
7195  REM  MATRIX INVERSION, GAUSS-JORDAN ELIMINATION. V IS INPUT, P IS O
UTPUT
7200  FOR I = 0 TO N - 1
7210  FOR J = 0 TO N - 1
7220 P(I,J + N) = 0
7230 P(I,J) = V(I,J)
7240  NEXT J
7250 P(I,I + N) = 1
7260  NEXT I
7270  FOR K = 0 TO N - 1
7280  IF K = N - 1 THEN 7390
7290 M = K
7295  REM  FIND MAXIMUM ELEMENT
7300  FOR I = K + 1 TO N - 1
7310  IF  ABS (P(I,K)) >  ABS (P(M,K)) THEN M = I
7320  NEXT I
7330  IF M = K THEN 7390
7340  FOR J = K TO 2 * N - 1
7350 P = P(K,J)
7360 P(K,J) = P(M,J)
7370 P(M,J) = P
7380  NEXT J
7385  REM  DIVIDE ROW K
7390  FOR J = K + 1 TO 2 * N - 1
7395  IF P(K,K) = 0 THEN 7405
7400 P(K,J) = P(K,J) / P(K,K): GOTO 7410
7405 P(K,J) = 0
7410  NEXT J
7420  IF K = 0 THEN 7490
7430  FOR I = 0 TO K - 1
7440  FOR J = K + 1 TO 2 * N - 1
7450 P(I,J) = P(I,J) - P(I,K) * P(K,J)
7460  NEXT J
7470  NEXT I
7480  IF K = N - 1 THEN 7550
7490  FOR I = K + 1 TO N - 1
7500  FOR J = K + 1 TO 2 * N - 1
7510 P(I,J) = P(I,J) - P(I,K) * P(K,J)
7520  NEXT J
7530  NEXT I
7540  NEXT K
7545  REM  RETRIEVE INVERSE FROM RIGHT SIDE OF P
7550  FOR I = 0 TO N - 1
7560  FOR J = 0 TO N - 1
7570 P(I,J) = P(I,J + N)
7580  NEXT J
7590  NEXT I: RETURN
7593  REM  ****************************************************
7595  REM  MOVE MAT P TO C
7600  IF N1 * N2 = 0 THEN 7660
7610  FOR I1 = 0 TO N1
7620  FOR I2 = 0 TO N2
7630 C(I1,I2) = P(I1,I2)
7640  NEXT I2
7650  NEXT I1: RETURN
7660  IF N1 = 0 THEN  RETURN
```

```
7670  FOR I1 = Ø TO N1
7680 C(I1) = P(I1)
7690  NEXT I1: RETURN
7693  REM  ****************************************************
7695  REM  MOVE A TO V FOR INVERSION
7700  FOR I1 = Ø TO N1
7710  FOR I2 = Ø TO N2
7720 V(I1,I2) = A(I1,I2)
7730  NEXT I2
7740  NEXT I1: RETURN
7743  REM  ****************************************************
7745  REM  MOVE V TO B
7750  FOR I1 = Ø TO N1
7760 B(I1,Ø) = V(I1,Ø)
7770  NEXT I1
7780  RETURN
7793  REM  ****************************************************
7795  REM  MOVE MAT B TO C
7800  IF N1 * N2 = Ø THEN 7860
7810  FOR I1 = Ø TO N1
7820  FOR I2 = Ø TO N2
7830 C(I1,I2) = B(I1,I2)
7840  NEXT I2
7850  NEXT I1: RETURN
7860  IF N1 = Ø THEN  RETURN
7870  FOR I1 = Ø TO N1
7880 C(I1,Ø) = B(I1,Ø)
7890  NEXT I1: RETURN
7893  REM  ****************************************************
7895  REM  MULTIPLY P*B
7900  FOR I = Ø TO M1
7910 :
7920 C(I,Ø) = Ø
7930  FOR K = Ø TO N1
7940 C(I,Ø) = C(I,Ø) + P(I,K) * B(K,Ø)
7950  NEXT K
7960 :
7970  NEXT I
7980  RETURN
7993  REM  ****************************************************
7995  REM  MULTIPLICATION C=P*A
8000  IF N1 * N2 = Ø THEN 8100
8010  FOR I = Ø TO M1
8020  FOR J = Ø TO N2
8030 C(I,J) = Ø
8040  FOR K = Ø TO N1
8050 C(I,J) = C(I,J) + P(I,K) * A(K,J)
8060  NEXT K
8070  NEXT J
8080  NEXT I
8090  RETURN
8100  IF N1 = Ø THEN  RETURN
8110  FOR I = Ø TO M1
8120 C(I,Ø) = Ø
8130  FOR K = Ø TO N1
8140 C(I,Ø) = C(I,Ø) + P(I,K) * A(K,Ø)
8150  NEXT K
8160  NEXT I
8170  RETURN
```

```
8193  REM  *************************************************
8195  REM  MATRIX TRANSPOSE, T=TRN(A)
82ØØ  FOR I = Ø TO N - 1
821Ø  FOR J = Ø TO M - 1
822Ø T(I,J) = A(J,I)
823Ø  NEXT J
824Ø  NEXT I
825Ø  RETURN
8293  REM  *************************************************
8295  REM  MOVE T TO C
83ØØ  FOR I1 = Ø TO N - 1
831Ø  FOR I2 = Ø TO M - 1
832Ø C(I1,I2) = T(I1,I2)
833Ø  NEXT I2
834Ø  NEXT I1: RETURN
8343  REM  *************************************************
8345  REM  MULTIPLY TRN(A)*A=V, I.E. V=T*A
835Ø  IF N1 * N2 = Ø THEN 844Ø
836Ø  FOR I = Ø TO M1
837Ø  FOR J = Ø TO N2
838Ø V(I,J) = Ø
839Ø  FOR K = Ø TO N1
84ØØ V(I,J) = V(I,J) + T(I,K) * A(K,J)
841Ø  NEXT K
842Ø  NEXT J
843Ø  NEXT I: RETURN
844Ø  IF N1 = Ø THEN  RETURN
845Ø  FOR I = Ø TO M1
846Ø V(I,Ø) = Ø
847Ø  FOR K = Ø TO N1
848Ø V(I,Ø) = V(I,Ø) + T(I,K) * A(K,Ø)
849Ø  NEXT K: NEXT I: RETURN
8493  REM  *************************************************
8495  REM  MULTIPLY TRN(A)*B=A, I.E. A=T*B
85ØØ  FOR I = Ø TO M1
851Ø A(I,Ø) = Ø
852Ø  FOR K = Ø TO N1
853Ø A(I,Ø) = A(I,Ø) + T(I,K) * B(K,Ø)
854Ø  NEXT K
855Ø  NEXT I
856Ø  RETURN
8593  REM  ************************************************
8595  REM  MAKE MAT W = ZERO
86ØØ  FOR J1 = Ø TO R - 1
861Ø  FOR J2 = Ø TO R - 1
862Ø W(J1,J2) = Ø
863Ø  NEXT J2
864Ø  NEXT J1: RETURN
8643  REM  *************************************************
8645  REM  MOVE MAT W TO C
865Ø  FOR I1 = Ø TO R - 1
866Ø  FOR I2 = Ø TO R - 1
867Ø C(I1,I2) = W(I1,I2)
868Ø  NEXT I2
869Ø  NEXT I1: RETURN
8693  REM  ************************************************
8695  REM  MOVE B TO D
87ØØ  IF N1 * N2 = Ø THEN 876Ø
871Ø  FOR I1 = Ø TO N1
```

```
8720  FOR I2 = Ø TO N2
8730 D(I1,I2) = B(I1,I2)
8740  NEXT I2
8750  NEXT I1: RETURN
8760  IF N1 = Ø THEN  RETURN
8770  FOR I1 = Ø TO N1
8780 D(I1,Ø) = B(I1,Ø)
8790  NEXT I1: RETURN
8793  REM  ***************************************************
8795  REM  MULTIPLY A=W*D
8800  IF N1 * N2 = Ø THEN 89ØØ
8810  FOR I = Ø TO M1
8820  FOR J = Ø TO N2
8830 A(I,J) = Ø
8840  FOR K = Ø TO N1
8850 A(I,J) = A(I,J) + W(I,K) * D(K,J)
8860  NEXT K
8870  NEXT J
8880  NEXT I
8890  RETURN
8900  IF N1 = Ø THEN  RETURN
8910  FOR I = Ø TO M1
8920 A(I,Ø) = Ø
8930  FOR K = Ø TO N1
8940 A(I,Ø) = A(I,Ø) + W(I,K) * D(K,Ø)
8950  NEXT K
8960  NEXT I
8970  RETURN
8985  REM  *******************************************
8990  REM  GOSUB ROUTINE TO CLEAR SCREEN
9000  HOME
9010  RETURN
9093  REM  ****************************************************
9095  REM  GOSUB ROUTINE FOR ERRORS
9100  PRINT
9110  PRINT EØ$
9120  PRINT
9130  RETURN
9193  REM  ****************************************************
9195  REM  ERROR TRACE ROUTINE
9200  PRINT  CHR$ (7): REM  CTRL-B (BELL)
9210 E =  PEEK (222): REM  GET ERROR NO.
9220  IF CØ = 1 THEN 926Ø
9230  INVERSE
9240  PRINT "ERROR NO. ";E;" FOUND"
9250 CØ = 1: TRACE : RESUME
9260  PRINT "ERROR ON SECOND LINE NO. "
9270  NORMAL
9280  NOTRACE
9290  STOP
9293  REM  ***************************************************
9295  REM  MULTIPLY P*V=C
9300  FOR I = Ø TO M1
9310  FOR J = Ø TO N2
9320 C(I,J) = Ø
9330  FOR K = Ø TO N1
9340 C(I,J) = C(I,J) + P(I,K) * V(K,J)
9350  NEXT K
9360  NEXT J
```

```
9370  NEXT I
9380  RETURN
9393  REM  ************************************************
9395  REM  MOVE A TO D
9400  IF N1 * N2 = Ø THEN 946Ø
9410  FOR I1 = Ø TO N1
9420  FOR I2 = Ø TO N2
9430 D(I1,I2) = A(I1,I2)
9440  NEXT I2
9450  NEXT I1: RETURN
9460  IF N1 = Ø THEN  RETURN
9470  FOR I1 = Ø TO N1
9480 D(I1,Ø) = A(I1,Ø)
9490  NEXT I1: RETURN
9493  REM  ************************************************
9495  REM  END OF PROGRAM
9500  GOSUB 9ØØØ
9510  PRINT
9520  PRINT "END MATRIX ANALYSIS PROGRAM"
9530  PRINT
9540  PRINT "***************************"
9550  REM  PROGRAM PREPARED BY
9560  REM  DR. P.H. MILNE
9570  REM  DEPT. OF CIVIL ENGINEERING
9580  REM  UNIVERSITY OF STRATHCLYDE
9590  REM  GLASGOW G4 ONG
9600  REM  SCOTLAND
9610  REM  ***************************
9620  END
```

6.5.5 Matrix solutions – computer printout

```
LOCATION OF DATA :-LEVELLING EXAMPLE 6.5 (1)
***********************************************************************
OPERATOR'S NAME :- P.H.MILNE
DATE OF CALCS.  :-3Ø/Ø8/83
INSTRUMENT USED :- APPLE II+
***********************************************************************

MATRIX A =
========================================
 1      Ø      Ø
 -1     1      Ø
 Ø      -1     1
 Ø      Ø      -1
 Ø      1      Ø
 -1     Ø      1
****************************************

TRANSPOSE OF A =
========================================
 1      -1     Ø      Ø      Ø      -1
 Ø      1      -1     Ø      1      Ø
 Ø      Ø      1      -1     Ø      1
****************************************
```

```
INVERSE OF TRN(A)*A =
========================================
 .5     .25    .25
 .25    .5     .25
 .25    .25    .5
****************************************

CHECK ON IDENTITY MATRIX
========================================
 1      Ø      Ø
 Ø      1      Ø
 Ø      Ø      1
****************************************

MATRIX B =
========================================
7.26
.73
-5.565
-2.4Ø5
7.96
-4.845
****************************************

MATRIX X FROM AX=B IS =
========================================
7.25
7.97
2.4Ø5
****************************************


LOCATION OF DATA :-DISTANCE EXAMPLE 6.5 (2)
****************************************************************************
OPERATOR'S NAME :- P.H.MILNE
DATE OF CALCS.  :-3Ø/Ø8/83
INSTRUMENT USED :- APPLE II+
****************************************************************************

MATRIX A =
========================================
 1      Ø      Ø
 Ø      1      Ø
 Ø      Ø      1
 1      1      Ø
 1      1      1
 Ø      1      1
****************************************

MATRIX B =
========================================
176.327
224.758
195.461
4Ø1.Ø97
596.566
42Ø.234
****************************************
```

```
WEIGHT MATRIX W =
==============================================
 3      Ø      Ø      Ø      Ø      Ø
 Ø      3      Ø      Ø      Ø      Ø
 Ø      Ø      3      Ø      Ø      Ø
 Ø      Ø      Ø      2      Ø      Ø
 Ø      Ø      Ø      Ø      1      Ø
 Ø      Ø      Ø      Ø      Ø      2
****************************************

IDENTITY MATRIX CHECK (TWA)
==============================================
 1      Ø      Ø
 Ø      1      Ø
 Ø      Ø      1
****************************************

MATRIX X = INV(TWA)*(TWB) :-
==============================================
176.33
224.764
195.466
****************************************
```

6.6 VARIATION OF CO-ORDINATES

The braced quadrilateral discussed in the previous section relies on all the stations being occupied, either for angle measurement or for distance measurement. Earlier in Chapter 3, programs were provided for the location of stations in 2-D by intersection (Section 3.5) and of stations in 2-D by resection (Section 3.6), where each of these provided a unique solution. However, if several observations are taken from an unknown station to numerous known stations, giving redundant observations, it is necessary to apply a least squares adjustment, as mentioned previously in Section 6.1.3.

One solution is to use the variation of co-ordinates method, sometimes called the method of indirect observations. The advantage of the variation of co-ordinates method is that it can be used for all systems which involve the computation of co-ordinates, such as in triangulation, trilateration, traversing, intersection and resection. The only drawback to the variation of co-ordinates method is that provisional co-ordinates have to be known for the stations, and the less accurate these approximations, the longer will be the computational procedure.

In using the variation of co-ordinates method by measuring distances to several stations, if only horizontal distances are known, then the problem is resolved in 2-D. However, if slant ranges are measured by EDM, the levels of each of the stations are also required to resolve the 3-D solution.

In the first instance, if the braced quadrilateral, Fig. 6.2(a), is considered, the position of station 4(D) can be determined from known stations 1, 2 and 3(A, B and C) with observed horizontal distances O_1, O_2 and O_3. O_1 to O_3 are the horizontal projections of the slant ranges on to the X–Y plane from a knowledge of the levels of stations 1–3. If the provisional co-ordinates of the unknown point are (X_0, Y_0) then the correct ('true') co-ordinates of the unknown point, shown as $4(D)$ on Fig. 6.2(a), will be (X_4, Y_4),

where

$$X_4 = X_0 + \mathrm{d}X$$

and

$$Y_4 = Y_0 + \mathrm{d}Y$$

as shown in Fig. 6.3, where $\mathrm{d}X$ and $\mathrm{d}Y$ are small corrections to the provisional co-ordinates which must be found by the variation of co-ordinates method. Grid co-ordinates in terms of X and Y have been selected rather than east (E) and north (N) to allow for the use of local grid co-ordinates, often selected from a site

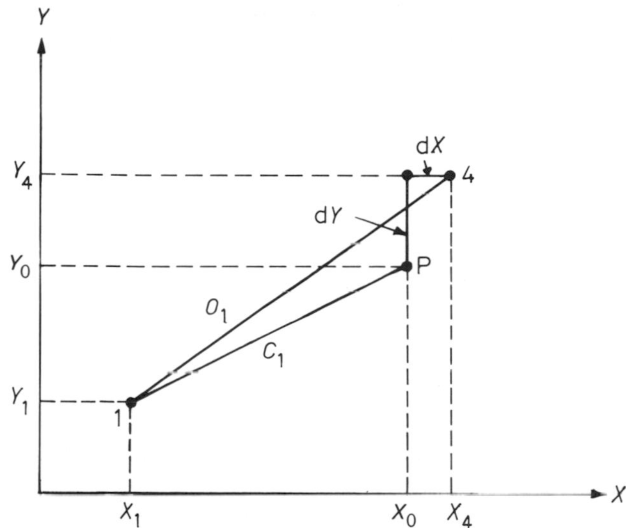

Fig. 6.3. Nomenclature for variation of co-ordinates method, where dX and dY are small corrections to provisional co-ordinates.

network, rather than geographical co-ordinates. If the latter are required, a co-ordinate transformation will translate the X, Y co-ordinates to eastings and northings, as discussed in Chapter 3.10.

In Fig. 6.3 the computed distance from station 1 (known) to station 4 (provisional) will be C_1, and the observed horizontal distance will be O_1. Using calculus notation, the change in distance from the provisional co-ordinates of 4 to the correct co-ordinates will be $\mathrm{d}C_1$. An expression can therefore be derived for $\mathrm{d}C_1$ in terms of $\mathrm{d}X$ and $\mathrm{d}Y$ shown in Fig. 6.3:

$$C_1^2 = (X_1 - X_0)^2 + (Y_1 - Y_0)^2$$

where by differentiation:

$$2C_1\mathrm{d}C_1 = -2(X_1 - X_0)\mathrm{d}X - 2(Y_1 - Y_0)\mathrm{d}Y$$

$$\mathrm{d}C_1 = [(X_0 - X_1)/C_1]\mathrm{d}X + [(Y_0 - Y_1)/C_1]\mathrm{d}Y \qquad (6.38)$$

where for simplicity, substitute:

$$K_1 = (X_0 - X_1)/C_1$$

and

$$L_1 = (Y_0 - Y_1)/C_1$$

giving from Equation 6.38:

$$\mathrm{d}C_1 = K_1\mathrm{d}X + L_1\mathrm{d}Y \qquad (6.39)$$

Now let V_1 be the residual (variation) of the observed horizontal distance O_1. The true distance from the known station 1 to the unknown station 4 can then be written as:

$$C_1 + \mathrm{d}C_1 = O_1 + V_1 \qquad (6.40)$$

or

$$\mathrm{d}C_1 = (O_1 - C_1) + V_1 \qquad (6.41)$$

The combination of Equations 6.39 and 6.41 then provides the variation of co-ordinates observation equation:

$$K_1\mathrm{d}X + L_1\mathrm{d}Y = (O_1 - C_1) + V_1 \qquad (6.42)$$

For each measured distance to stations 1, 2, 3 from the unknown station 4, an equation similar to Equation 6.42 can be written. In this case (Fig. 6.2(a)) there are three equations:

$$K_1\mathrm{d}X + L_1\mathrm{d}Y = (O_1 - C_1) + V_1$$

$$K_2\mathrm{d}X + L_2\mathrm{d}Y = (O_2 - C_2) + V_2$$

$$K_3\mathrm{d}X + L_3\mathrm{d}Y = (O_3 - C_3) + V_3 \qquad (6.43)$$

which have to be solved for $\mathrm{d}X$, $\mathrm{d}Y$, V_1, V_2 and V_3.

The solutions for $\mathrm{d}X$ and $\mathrm{d}Y$ are then used to correct the provisional co-ordinates selected at the outset to form a new set of provisional co-ordinates and a new set of equations similar to Equation 6.43. The process is repeated until $\mathrm{d}X$ and $\mathrm{d}Y$ are insignificant, i.e., convergence has been achieved. The method of solving for $\mathrm{d}X$, $\mathrm{d}Y$, etc., from Equation 6.43 uses matrix algebra where the set of 'observation equations' have the form:

$$Ax = b + v \qquad (6.44)$$

where,

$$A = \begin{bmatrix} K_1 & L_1 \\ K_2 & L_2 \\ K_3 & L_3 \end{bmatrix}, x = \begin{bmatrix} \mathrm{d}X \\ \mathrm{d}Y \end{bmatrix}, b = \begin{bmatrix} O_1 - C_1 \\ O_2 - C_2 \\ O_3 - C_3 \end{bmatrix}, v = \begin{bmatrix} V_1 \\ V_2 \\ V_3 \end{bmatrix}$$

However, as it is not possible to determine the true values of the residuals, v, and the parameters, x, it is necessary to estimate their values using the principle of least squares. If the least squares estimates of x and v are denoted by $\hat{x}$ and $\hat{v}$ then it can be shown that $\hat{x}$ is given by a solution of:

$$A^T W A \hat{x} = A^T W b, \tag{6.45}$$

i.e.

$$\hat{x} = (A^T W A)^{-1} A^T W b \tag{6.46}$$

and

$$\hat{v} = A\hat{x} - b \tag{6.47}$$

where Equation 6.46 is usually referred to as the 'normal equation'. The matrix W, used in Equations 6.46 and 6.47, is the weight matrix. If the inverse of the variance–covariance matrix of the observations is used for W, then $\hat{x}$ is statistically the best estimate of x, i.e., it is the estimate with the smallest variance. However, if the observations are uncorrelated, then W simply becomes a diagonal matrix:

$$W = \begin{bmatrix} 1/\sigma_1^2 & & \\ & 1/\sigma_2^2 & \\ & & 1/\sigma_3^2 \end{bmatrix} \tag{6.48}$$

where σ_i^2 etc. is the variance (square of the standard error) of the observed distance to station i. A check on the numerical operations performed during the least squares computations can be obtained from:

$$A^T W \hat{v} = 0 \tag{6.49}$$

At this point in the computation, if a large number of observations have been taken to the unknown station, it may be found that some of the residuals are very large compared to the quality of the measurement to which they refer, that is there has possibly been a gross error in the measurement. It may now therefore be necessary to reject one or more of the multiple observations and repeat the whole adjustment process.

The variation of co-ordinates method using Equations 6.38–6.49 can thus be applied to any number of observations. The number of iterations necessary to reduce $\mathrm{d}X$ and $\mathrm{d}Y$ to give a satisfactory solution will depend on the accuracy of the provisional co-ordinates at the outset. It should be noted that the accuracy of the least squares adjustment procedure is very sensitive to survey geometry. In choosing survey control stations, careful planning to avoid short lengths and acute angles applies equally to traverse, braced quadrilateral and least squares adjustment processes. The following program <VARIATION-COORDS> has been designed to handle between three and six observations from known stations.

6.6.1 Variation of co-ordinates – subroutine index

	Line numbers	*Function*
(a)	10–90	Initialization and control
(b)	100–230	Screen header display
(c)	500–680	Entry of survey location, etc.
(d)	1000–1390	Entry of survey stations and distances
(e)	1400–1810	Fill matrices *A* and *B*
(f)	3000–3130	Fill matrix *W*
(g)	3150–3530	Solution $\hat{x} = (A^TWA)^{-1}(A^TWB)$
(h)	3550–3620	Solution $\hat{v} = A\hat{x} - b$
(i)	3650–3950	Display of results
(j)	4000–4140	Check on residuals
(k)	6000–6080	Printout of survey location, etc.
(l)	6100–6270	Printout for matrix *C*
(m)	6300–6390	Printout of station co-ordinates
(n)	6400–6560	Printout of distances and errors
(o)	6600–6630	Printout of results
(p)	6700–6740	Printout of residuals
(q)	7000–7090	Display for matrix *C*
(r)	7100–7190	Move matrix *A* to *C*
(s)	7200–7590	Matrix inversion *V*
(t)	7600–7690	Move matrix *A* to *M*
(u)	7700–7730	Move matrix *B* to *R*
(v)	7750–7780	Move matrix *V* to *B*
(w)	7800–7890	Move matrix *B* to *C*
(x)	7900–7960	Multiply matrix *P***B*
(y)	8200–8250	Matrix transpose *T*
(z)	8300–8340	Move matrix *T* to *C*
(aa)	8350–8490	Multiply matrix *T***A*
(bb)	8600–8640	Make matrix *W* = zero
(cc)	8650–8690	Move matrix *W* to *C*
(dd)	8700–8790	Move matrix *B* to *D*
(ee)	8800–8970	Multiply matrix *W***D*
(ff)	9000–9010	Routine to clear screen
(gg)	9100–9130	Routine to display error message
(hh)	9200–9290	Error trace routine
(ii)	9300–9380	Multiply matrix *P***V*
(jj)	9400–9490	Move matrix *A* to *D*
(kk)	9500–9620	Termination of program

(a) Initialization and control

Line numbers 10–90

The numeric variables, arrays and required string variables are initialized. The program has been written for a maximum number of six reference stations. If a larger number is required, the arrays should be redimensioned accordingly.

(b) Screen header display

Line numbers 100–230

This is the screen display used for civil engineering students at Strathclyde University. The operator can alter this as required to suit another organization.

(c) Entry of survey location, etc.

Line numbers 500–680

This routine is provided to keep a record of location, operator, date, etc. with an opportunity to obtain a printout if required.

(d) Entry of survey stations and distances

Line numbers 1000–1390

After entering the number of known reference stations (M), a loop is set up to enter the reference stations' X, Y, Z co-ordinates. The provisional co-ordinates of the new station are then requested. The program provides the option of entering slant ranges or horizontal observations. If the former, the levels of the stations are used to reduce the slant ranges to the horizontal, $H(I)$. The standard error $E(I)$, in each observation is also requested for weighting. A printout of all the above keyboard entered data is available for record purposes.

(e) Fill matrices *A* and *B*

Line numbers 1400–1810

After calculating K and L from Equations 6.39–6.43, the matrix A of Equation 6.44 is then filled, and the transpose of A, ($T = A^T$) computed. From a knowledge of the observed horizontal distances, $H(I)$, to each known station, and the calculated distance, $G(I)$, from the provisional co-ordinates, array b of Equation 6.44 can also be filled.

(f) Fill matrix *W*

Line numbers 3000–3130

After filling all the elements of matrix W with zero (GOSUB 8600) the diagonal components of Equation 6.48 are computed.

(g) Solution $\hat{x} = (A^TWA)^{-1}(A^TWB)$

Line numbers 3150–3530

This routine, with the exception of line 3460, to obtain the displacements dX and dY, was described in the previous matrix program (Section 6.5.1(l)).

(h) Solution $\hat{v} = A\hat{x} - B$

Line numbers 3550–3620

The values of the residuals $\hat{v}$ are now determined using the principle of least squares from Equation 6.47 and stored in the array $U(J, 0)$.

(i) Display of results

Line numbers 3650–3950

A counter ($P9$) is used to identify each iteration, and the X and Y displacements, dX and dY, are displayed together with the computed values of the X, Y, Z co-ordinates of the new station, with a check on the residuals. The operator has a choice of returning for another iteration if the values of dX and dY, together with $\hat{v}$, are still large. If another calculation is requested, the X and Y co-ordinate estimates for the new station are updated in lines 3920–3930 before returning to line 1400.

(j) Check on residuals

Line numbers 4000–4140

After all iterations are complete, a check is run on the residuals by printing out array $B(K, 0)$ which contains the difference between the observed and calculated horizontal distances.

(k) Printout of survey location, etc.

Line numbers 6000–6080

This routine is provided for record purposes and is based on Applesoft DOS, and may require alteration for different microcomputers and printers.

(l) Printout for matrix C

Line numbers 6100–6270

This routine is described in Section 6.5.1(n).

(m) Printout of station co-ordinates

Line numbers 6300–6390

This routine provides a record of the X, Y, Z co-ordinates of the known reference stations.

(n) Printout of distances and errors

Line numbers 6400–6560

This routine will accept either slant ranges or horizontal distances together with their standard errors. The entered provisional co-ordinates of the new station are also recorded in this printout.

(o) Printout of results

Line numbers 6600–6630

This routine re-uses the screen display from lines 3660–3790 in Section (i) above.

(p) Printout of residuals

Line numbers 6700–6740

This routine re-uses the screen display from lines 4060–4100 in Section (j) above.

(q) Display for matrix *C*

Line numbers 7000–7090

This routine is described in Section 6.5.1(o).

(r) Move matrix *A* to *C*

Line numbers 7100–7190

This routine is described in Section 6.5.1(p).

(s) Matrix inversion *V*

Line numbers 7200–7590

This routine is described in Section 6.5.1(s).

(t) Move matrix *A* to *M*

Line numbers 7600–7690

This routine is similar to (r) above where A is stored in M for use in Section (h) above.

(u) Move matrix *B* to *R*

Line numbers 7700–7730

In this routine B is stored in R for use in Section (h) above.

(v) Move matrix *V* to *B*

Line numbers 7750–7780

This routine is described in Section 6.5.1(t).

(w) Move matrix *B* to *C*

Line numbers 7800–7890

This routine is described in Section 6.5.1(u).

(x) Multiply matrix *P*B*

Line numbers 7900–7960

This routine is described in Section 6.5.1(v).

(y) Matrix transpose *T*

Line numbers 8200–8250

This routine is described in Section 6.5.1(x).

(z) Move matrix *T* to *C*

Line numbers 8300–8340

This routine is described in Section 6.5.1(y).

(aa) Multiply matrix *T*A*

Line numbers 8350–8490

This routine is described in Section 6.5.1(z).

(bb) to (kk)

For subroutines (bb) to (kk) these are all described in Section 6.5.1.

6.6.2 Variation of co-ordinates – numeric variables

$A(M, N)$	= Array for matrix A
$B(M, 0)$	= Array for matrix B
$C0$	= Counter in error trace routine
$C(M, M)$	= Array for matrix C for display
$D(M, N)$	= Array for storage purposes
E	= Error number
$E(M)$	= Array of observation errors
$F(M, 0)$	= Temporary array
$G(M)$	= Array of calculated distances
$H(M)$	= Array of horizontal distances
$I, I1$	= Integer loop variables
$J, J1$	= Integer loop variables
$K, K1$	= Integer loop variables
$K(M)$	= Array of terms in matrix A
$L(M)$	= Array of terms in matrix A
M	= Number of reference stations

Table—continued

N, *N*1, *N*2	= Variables for matrix arrays
N(*M*)	= Array of reference station numbers
*P*7	= Counter in printout routines
*P*9	= Number of iterations
P(*N*, 2**N*)	= Inverse matrix
R	= Number of rows in matrix
S	= Reference number of new station
S(*M*)	= Array of slant ranges
T(*M*, *M*)	= Transposed array *A*
U	= Program selection from menu
U(*M*, 0)	= Array of residuals
V(*M*, *M*)	= Input matrix for inverse routine
W(*M*, *M*)	= Weight matrix
*X*0, *X*9	= *X* co-ordinate and d*X* of *S*
X(*M*)	= Array of *X* co-ordinates
*Y*0, *Y*9	= *Y* co-ordinate and d*Y* of *S*
Y(*M*)	= Array of *Y* co-ordinates
*Z*0	= *Z* co-ordinate of *S*
Z(*M*)	= Array of *Z* co-ordinates

6.6.3 Variation of co-ordinates – string variables

A$	Storage for location of survey
B$	Storage for operator's name
C$	Storage for date of calculation
D$	= CHR$(4), CTRL-D
*E*0$	= "SORRY, DATA ERROR . . . PLEASE RE-ENTER"
H$	= "VARIATION OF CO-ORDINATES PROGRAM"
K$	Storage for instrument used
M$	Storage for type of matrix
P$	= "DO YOU WISH PRINTOUT OF ABOVE DATA (Y/N)"
Q$	= Question response (Y/N)
S$	= "S" for slant range, "H" for horizontal observations
X$	= "ENTER X(E) COORD. OF STN. "
Y$	= "ENTER Y(N) COORD. OF STN."
Z$	= "ENTER LEVEL OF STN. "

6.6.4 Variation of co-ordinates – BASIC program

```
10  REM  <VARIATION-COORDS> PROGRAM FOR APPLE II+  USES QUME PRINTER
20 D$ =  CHR$ (4): ONERR  GOTO 92ØØ
3Ø C = Ø:E = Ø:M = Ø:N = Ø:N1 = Ø:N2 = Ø:P1 = Ø:P7 = Ø:R = Ø
35  DIM A(5,5),B(5,Ø),C(5,5),D(5,1),E(5),F(5,Ø),G(5),H(5),K(5),L(5),N(5),
     P(5,11),S(5),T(5,5),U(5,Ø),V(5,5),W(5,5)
4Ø  DIM M(5,5),R(5,Ø),X(5),Y(5),Z(5)
5Ø EØ$ = "SORRY, DATA ERROR ... PLEASE RE-ENTER."
55 P$ = "DO YOU WISH PRINTOUT OF ABOVE DATA (Y/N)"
6Ø L$ = " ":M$ = " ":N$ = " "
7Ø X$ = "ENTER X(E) COORD. OF STN. ":Y$ = "ENTER Y(N) COORD. OF STN. "
```

```
75 Z$ = "ENTER LEVEL OF STN. "
8Ø H$ = "VARIATION OF CO-ORDINATES PROGRAM"
95  REM  ****************************************************
1ØØ  GOSUB 9ØØØ
11Ø  PRINT "****************************************"
12Ø  PRINT "*                                      *"
13Ø  PRINT "*     UNIVERSITY  OF  STRATHCLYDE      *"
14Ø  PRINT "*  DEPARTMENT  OF  CIVIL  ENGINEERING  *"
15Ø  PRINT "*                                      *"
16Ø  PRINT "*          SURVEYING  SECTION          *"
17Ø  PRINT "*     VARIATION  OF  CO-ORDINATES      *"
18Ø  PRINT "****************************************"
19Ø  PRINT
2ØØ  PRINT "DO YOU WISH TO ENTER LOCATION <1>,       NEW DATA <2> OR QUIT
     <3>"
21Ø  INPUT U
22Ø  IF U < 1 OR U > 3 THEN 2ØØ
23Ø  ON U GOTO 5ØØ,1ØØØ,95ØØ
5ØØ  GOSUB 9ØØØ
51Ø  PRINT "UNIVERSITY OF STRATHCLYDE"
52Ø  PRINT H$
53Ø  PRINT "****************************************"
54Ø  PRINT
55Ø  PRINT "ENTER LOCATION OF WORK, ETC. WHEN       REQUESTED, AND PRESS
     <RETURN>"
56Ø  PRINT "ENTER LOCATION OF SURVEY :-"
57Ø  INPUT A$
58Ø  PRINT "ENTER OPERATOR'S NAME :-"
59Ø  INPUT B$
6ØØ  PRINT "ENTER DATE OF CALCULATION AS DD/MM/YY -"
61Ø  INPUT C$
62Ø  PRINT "ENTER INSTRUMENT USED :-"
63Ø  INPUT K$
64Ø  PRINT
65Ø  PRINT "DO YOU WISH PRINTOUT OF LOCATION ETC.,  (Y/N) ";
66Ø  INPUT Q$
67Ø  IF Q$ = "Y" THEN 6ØØØ
68Ø  IF Q$ <  > "N" THEN 65Ø
69Ø :
995  REM  ****************************************************
1ØØØ  GOSUB 9ØØØ: PRINT H$:P9 = 1
1Ø1Ø  PRINT "****************************************"
1Ø2Ø  PRINT "ENTER NUMBER OF REF. STNS (3-6) :-";: INPUT M: IF M < 3 OR M
      > 6 THEN 1ØØØ
1Ø3Ø  PRINT : PRINT "ENTER STN. REFERENCE NUMBERS :-": PRINT
1Ø4Ø  FOR I = Ø TO M - 1
1Ø5Ø  PRINT "ENTER REF. NO. STN. ";(I + 1);: INPUT N(I)
1Ø6Ø  NEXT I
1Ø7Ø  GOSUB 9ØØØ: PRINT "ENTER X,Y,Z CO-ORDINATES OF REF. STNS :-"
1Ø8Ø  FOR J = Ø TO M - 1
1Ø9Ø  PRINT X$;N(J);: INPUT X(J)
11ØØ  PRINT Y$;N(J);: INPUT Y(J)
111Ø  PRINT Z$;N(J);: INPUT Z(J)
112Ø  NEXT J
113Ø  PRINT : PRINT P$;: INPUT Q$: IF Q$ = "Y" THEN 63ØØ
114Ø  IF Q$ <  > "N" THEN 113Ø
115Ø  GOSUB 9ØØØ: PRINT H$:R = M
116Ø  PRINT "****************************************"
117Ø  PRINT "ENTER REF. NO. OF NEW STATION ";: INPUT S
```

```
118Ø  PRINT : PRINT "ENTER PROVISIONAL CO-ORDS OF STN. ";S: PRINT
119Ø  PRINT X$;S;: INPUT XØ
12ØØ  PRINT Y$;S;: INPUT YØ
121Ø  PRINT Z$;S;: INPUT ZØ
122Ø  PRINT : PRINT "ARE OBSERVATIONS SLANT RANGES (S) OR    HORIZONTAL L
     ENGTHS (H) ";
123Ø  INPUT S$: IF S$ <  > "S" AND S$ <  > "H" THEN 122Ø
124Ø  GOSUB 9ØØØ: PRINT H$
125Ø  PRINT "***************************************"
126Ø  IF S$ = "H" THEN 135Ø
1265  REM  COLLECT SLANT RANGES
127Ø  FOR I = Ø TO M - 1
128Ø  PRINT "ENTER SLANT RANGE, ERROR (S,E) ";S;"-";N(I);":-"
129Ø  INPUT S(I),E(I)
13ØØ  NEXT I
13Ø5  REM  CALC. HORIZONTALS
131Ø  FOR J = Ø TO M - 1
132Ø H(J) =  SQR (S(J) ^ 2 - (Z(J) - ZØ) ^ 2)
133Ø  NEXT J
134Ø  GOTO 138Ø
1345  REM  COLLECT HORIZONTAL LENGTHS
135Ø  FOR I = Ø TO M - 1
136Ø  PRINT "ENTER HORIZONTAL,ERROR (H,E) ";S;"-";N(I);":-"
137Ø  INPUT H(I),E(I): NEXT I
138Ø  PRINT : PRINT P$;: INPUT Q$: IF Q$ = "Y" THEN 64ØØ
139Ø  IF Q$ <  > "N" THEN 138Ø
1393  REM  ************************************************
1395  REM  CALC. RANGES TO S
14ØØ  FOR K = Ø TO R - 1
141Ø G(K) =  SQR ((XØ - X(K)) ^ 2 + (YØ - Y(K)) ^ 2)
142Ø  NEXT K
1425  REM  COMPUTE K(I1) AND L(I1)
143Ø  FOR I1 = Ø TO R - 1
144Ø K(I1) = (XØ - X(I1)) / G(I1)
145Ø L(I1) = (YØ - Y(I1)) / G(I1)
146Ø  NEXT I1
1465  REM  FILL MATRIX A WITH K AND L
147Ø  FOR J1 = Ø TO R - 1
148Ø A(J1,Ø) = K(J1)
149Ø A(J1,1) = L(J1)
15ØØ  NEXT J1
15Ø5  REM  MOVE A TO D AND ALSO M
151Ø N1 = R - 1:N2 = 1: GOSUB 94ØØ: GOSUB 76ØØ
1515  REM  MOVE A TO C FOR DISPLAY
152Ø  GOSUB 71ØØ
153Ø  GOSUB 9ØØØ:M$ = "MATRIX A = "
154Ø  PRINT M$: PRINT "======================================="
155Ø  GOSUB 7ØØØ
156Ø  PRINT : PRINT P$;: INPUT Q$
157Ø  IF Q$ = "N" THEN 16ØØ
158Ø  IF Q$ <  > "Y" THEN 156Ø
159Ø  GOSUB 61ØØ
1595  REM  OBTAIN TRANSPOSE OF A, I.E. T
16ØØ N = 2: GOSUB 82ØØ
16Ø5  REM  MOVE T=TRN(A) TO C
161Ø  GOSUB 83ØØ
162Ø M$ = "TRANSPOSE OF A = "
163Ø  PRINT : PRINT M$: PRINT "======================================="
164Ø N1 = N - 1:N2 = R - 1: GOSUB 7ØØØ
```

```
1650  PRINT : PRINT P$;: INPUT Q$
1660  IF Q$ = "N" THEN 1700
1670  IF Q$ < > "Y" THEN 1650
1680  GOSUB 6100
1690 :
1695  REM  FILL MATRIX B WITH (H-G)
1700  FOR K1 = 0 TO R - 1
1710 B(K1,0) = H(K1) - G(K1)
1720  NEXT K1
1725  REM  MOVE B TO C AND ALSO R
1730 N1 = R - 1:N2 = 0: GOSUB 7800: GOSUB 7700
1735  REM  DISPLAY MAT B IN C
1740 M$ = "MATRIX B = "
1750  PRINT : PRINT M$
1760  PRINT "========================================"
1770  GOSUB 7000
1780  PRINT : PRINT P$;: INPUT Q$
1790  IF Q$ = "N" THEN 3000
1800  IF Q$ < > "Y" THEN 1780
1810  GOSUB 6100
1815  REM  RESUME AT 3000
2993  REM  **************************************************
2995  REM  OBTAIN WEIGHT MATRIX
3000  GOSUB 9000: PRINT H$: PRINT "****************************************
     **"
3005  REM  MAKE MATRIX W=ZERO
3010  GOSUB 8600
3020  FOR I = 0 TO R - 1
3030  IF E(I) = 0 THEN 3050
3040 W(I,I) = 1 / (E(I) ^ 2): GOTO 3060
3050 W(I,I) = 0
3060  NEXT I
3065  REM  MOVE MAT W TO C
3070  GOSUB 8650:M$ = "WEIGHT MATRIX W = "
3080  PRINT M$: PRINT "========================================": PRINT
3090 N1 = R - 1:N2 = R - 1: GOSUB 7000
3100  PRINT : PRINT P$;: INPUT Q$
3110  IF Q$ = "N" THEN 3150
3120  IF Q$ < > "Y" THEN 3100
3130  GOSUB 6100
3140 :
3145  REM  MULTIPLY W*A, I.E. A=W*D
3150 M1 = R - 1:N2 = N - 1: GOSUB 8800
3155  REM  MOVE A TO C
3160  GOSUB 7100
3170 M$ = "MATRIX W*A = "
3180  PRINT M$: PRINT "========================================": PRINT
3190  GOSUB 7000
3200  PRINT : PRINT P$;: INPUT Q$
3210  IF Q$ = "N" THEN 3240
3220  IF Q$ < > "Y" THEN 3200
3230  GOSUB 6100
3235  REM  GET V=T*WA, I.E. V=T*A
3240 M1 = N - 1: GOSUB 8350
3245  REM  INVERT V GET P
3250  GOSUB 7200
3255  REM  CHECK IDENTITY MATRIX, C=P*V
3260 N1 = N - 1: GOSUB 9300
3270 M$ = "IDENTITY MATRIX CHECK (TWA)"
```

```
3280  PRINT M$: PRINT "=======================================": PRINT
3290  GOSUB 7000
3300  PRINT : PRINT P$;: INPUT Q$
3310  IF Q$ = "N" THEN 3340
3320  IF Q$ < > "Y" THEN 3300
3330  GOSUB 6100
3335  REM  MOVE B TO D
3340 N1 = R - 1:N2 = 0: GOSUB 8700
3345  REM  GET A=W*B, I.E. A=W*D
3350 M1 = R - 1: GOSUB 8800
3355  REM  MOVE A TO C
3360  GOSUB 7100
3370 M$ = "MATRIX W*B = "
3380  PRINT M$: PRINT "=======================================": PRINT
3390  GOSUB 7000
3400  PRINT : PRINT P$;: INPUT Q$
3410  IF Q$ = "N" THEN 3440
3420  IF Q$ < > "Y" THEN 3400
3430  GOSUB 6100
3435  REM  MULTIPLY T*WB, I.E. V=T*A
3440 M1 = N - 1: GOSUB 8350
3445  REM  MOVE V TO B
3450 N1 = N - 1: GOSUB 7750
3455  REM  GET C=INV(TWA)*(TWB), I.E. C=P*B
3460  GOSUB 7900:X9 = C(0,0):Y9 = C(1,0)
3470 M$ = "MATRIX X = INV(TWA)*(TWB) :-"
3480  PRINT M$: PRINT "=======================================": PRINT
3490  GOSUB 7000: PRINT "****************************************"
3500  PRINT : PRINT P$;: INPUT Q$
3510  IF Q$ = "N" THEN 3550
3520  IF Q$ < > "Y" THEN 3500
3530  GOSUB 6100
3540 :
3545  REM  CALC V=AX-B, USING F=AX, I.E. F=M*C
3550  FOR I = 0 TO R - 1
3560 F(I,0) = 0
3570  FOR K = 0 TO 1
3580 F(I,0) = F(I,0) + M(I,K) * C(K,0)
3590  NEXT K: NEXT I
3595  REM  CALC U=F-B, I.E. U=F-R
3600  FOR J = 0 TO R - 1
3610 U(J,0) = F(J,0) - R(J,0)
3620  NEXT J
3630 :
3645  REM  DISPLAY RESULT
3650  GOSUB 9000: PRINT H$:P7 = 0
3660  PRINT "****************************************"
3670  PRINT "RESULTS OF ITERATION NO. ";P9
3680  PRINT "========================================"
3690  PRINT "X - DISPLACEMENT = "; INT (X9 * 1000 + .5) / 1000
3700  PRINT "Y - DISPLACEMENT = "; INT (Y9 * 1000 + .5) / 1000
3710  PRINT : PRINT "CO-ORDS OF ";S;" ARE :-"
3720  PRINT "X - CO-ORDINATE  = "; INT ((X0 + X9) * 1000 + .5) / 1000
3730  PRINT "Y - CO-ORDINATE  = "; INT ((Y0 + Y9) * 1000 + .5) / 1000
3740  PRINT "Z - CO-ORDINATE  = ";Z0
3750  PRINT : PRINT "CHECK ON CALCULATIONS :-"
3760  FOR J = 0 TO R - 1
3770  PRINT  INT (U(J,0) * 1000 + .5) / 1000;" : ";
3780  NEXT J: PRINT
```

```
3790  PRINT "****************************************": IF P7 = 1 THEN  RETURN

3800  PRINT : PRINT P$;: INPUT Q$
3810  IF Q$ = "Y" THEN 66ØØ
3820  IF Q$ < > "N" THEN 38ØØ
3830  GOSUB 9ØØØ: PRINT H$
3840  PRINT "****************************************"
3850  PRINT "RESULTS OF ITERATION NO. ";P9
3860  PRINT "========================================"
3870  PRINT "X - DISPLACEMENT = "; INT (X9 * 1ØØØ + .5) / 1ØØØ
3880  PRINT "Y - DISPLACEMENT = "; INT (Y9 * 1ØØØ + .5) / 1ØØØ
3890  PRINT : PRINT "DO YOU WISH ANOTHER ITERATION (Y/N) ";
3900  INPUT Q$: IF Q$ = "N" THEN 4ØØØ
3910  IF Q$ < > "Y" THEN 389Ø
3915  REM  RESET XØ,YØ
3920 XØ = XØ + X9
3930 YØ = YØ + Y9
3940 P9 = P9 + 1
3945  REM  RETURN TO CALCULATIONS
3950  GOTO 14ØØ
3993  REM  ************************************************
3995  REM  CHECK ON RESIDUALS
4000  GOSUB 9ØØØ: PRINT H$:P7 = Ø
4010  PRINT "****************************************"
4020  PRINT : PRINT "DO YOU WISH CHECK ON RESIDUALS (Y/N)";: INPUT Q$
4030  IF Q$ = "N" THEN 95ØØ
4040  IF Q$ < > "Y" THEN 4Ø2Ø
4050  PRINT : PRINT "RESIDUALS OF ITERATION NO. ";P9
4060  PRINT "========================================"
4070  FOR K = Ø TO R - 1
4080  PRINT "RESIDUAL ";S;"-";N(K);" = "; INT ((H(K) - G(K)) * 1ØØØ + .5)
     / 1ØØØ
4090  NEXT K
4100  PRINT "****************************************": IF P7 = 1 THEN  RETURN

4110  PRINT : PRINT P$;: INPUT Q$
4120  IF Q$ = "Y" THEN 67ØØ
4130  IF Q$ < > "N" THEN 411Ø
4140  GOTO 95ØØ
5985  REM  **********************************************************
     **
5990  REM  PRINTOUT OF LOCATION, ETC
6000  PRINT D$;"PR#1": PRINT "LOCATION OF DATA :-";A$
6010  PRINT "*******************************************************
     ********************"
6020  PRINT "OPERATOR'S NAME :- ";B$
6030  PRINT "DATE OF CALCS.  :-";C$
6040  PRINT "INSTRUMENT USED :- ";K$
6060  PRINT "*******************************************************
     ********************"
6070  PRINT : PRINT D$;"PR#Ø"
6080  GOTO 1ØØØ
6093  REM  ************************************************
6095  REM  UNIVERSAL PRINTOUT FOR MATRIX C
6100  PRINT D$;"PR#1": POKE 1784 + 1,8Ø
6110  PRINT M$: PRINT "========================================"
6120  IF N1 * N2 = Ø THEN 62ØØ
6130  FOR I5 = Ø TO N1
6140  FOR J5 = Ø TO N2
```

```
6150  POKE 36,(1 + 6 * J5): PRINT  INT (C(I5,J5) * 1000 + .5) / 1000;
6160  IF J5 = N2 THEN  PRINT
6170  NEXT J5
6180  NEXT I5
6190  GOTO 6250
6200  IF N1 = 0 THEN 6250
6210  FOR I5 = 0 TO N1
6220  PRINT  INT (C(I5,0) * 1000 + .5) / 1000
6230  NEXT I5
6240 :
6250  PRINT "****************************************"
6260  PRINT : PRINT D$;"PR#0"
6270  RETURN
6293  REM  **************************************************
6295  REM  PRINTOUT OF STATION CO-ORDINATES
6300  PRINT D$;"PR#1": POKE 1784 + 1,80
6310  PRINT H$
6320  PRINT "****************************************"
6330  PRINT "STN.NO.  X-COORD   Y-COORD   Z-COORD"
6340  PRINT "========================================"
6350  FOR K = 0 TO M - 1
6360  PRINT N(K);: POKE 36,10: PRINT X(K);: POKE 36,20: PRINT Y(K);: POKE
     36,30: PRINT Z(K)
6370  NEXT K
6380  PRINT "========================================"
6390  PRINT : PRINT D$;"PR#0": GOTO 1150
6393  REM  **************************************************
6395  REM  PRINTOUT OF DISTANCES, ERRORS
6400  PRINT D$;"PR#1": POKE 1784 + 1,80
6410  PRINT "PROVISIONAL CO-ORDINATES :-"
6420  PRINT S;: POKE 36,10: PRINT X0;: POKE 36,20: PRINT Y0;: POKE 36,30:
      PRINT Z0
6430  PRINT "****************************************": IF S$ = "H" THEN
     6500
6440  PRINT "SLANT RANGE   DISTANCE  ERROR"
6450  PRINT "========================================"
6460  FOR K = 0 TO M - 1
6470  PRINT S;" TO ";N(K);: POKE 36,15: PRINT S(K);: POKE 36,25: PRINT E(
     K)
6480  NEXT K
6490  PRINT "****************************************": PRINT
6500  PRINT "HORIZONTAL    DISTANCE  ERROR"
6510  PRINT "========================================"
6520  FOR I = 0 TO R - 1
6530  PRINT S;" TO ";N(I);: POKE 36,15: PRINT  INT (H(I) * 1000 + .5) / 1
     000;: POKE 36,25: PRINT E(I)
6540  NEXT I
6550  PRINT "****************************************"
6560  PRINT : PRINT D$;"PR#0"
6570  GOTO 1400
6593  REM  **************************************************
6595  REM  PRINTOUT OF RESULTS
6600  PRINT D$;"PR#1": POKE 1784 + 1,80:P7 = 1
6605  REM  USE DISPLAY AT LINES 3670-3790
6610  GOSUB 3670
6620  PRINT : PRINT D$;"PR#0"
6630  GOTO 3830
6693  REM  **************************************************
6695  REM  PRINTOUT OF RESIDUAL RESULTS
```

```
6700  PRINT D$;"PR#1": POKE 1784 + 1,80:P7 = 1
6710  PRINT "RESIDUALS OF ITERATION NO. ";P9
6715  REM  USE DISPLAY AT LINES 4060-4100
6720  GOSUB 4060
6730  PRINT : PRINT D$;"PR#0"
6740  GOTO 9500
6993  REM  **************************************************
6995  REM  DISPLAY MATRIX C
7000  IF N1 * N2 = 0 THEN 7060
7010  FOR I = 0 TO N1
7020  FOR J = 0 TO N2
7030  PRINT  TAB( 1 + 8 * J); INT (C(I,J) * 1000 + .5) / 1000;
7040  IF J = N2 THEN  PRINT
7050  NEXT J: NEXT I: RETURN
7060  IF N1 = 0 THEN  RETURN
7070  FOR I = 0 TO N1
7080  PRINT  INT (C(I,0) * 1000 + .5) / 1000
7090  NEXT I: RETURN
7093  REM  **************************************************
7095  REM  MOVE MAT A TO MAT C
7100  IF N1 * N2 = 0 THEN 7160
7110  FOR I1 = 0 TO N1
7120  FOR I2 = 0 TO N2
7130 C(I1,I2) = A(I1,I2)
7140  NEXT I2
7150  NEXT I1: RETURN
7160  IF N1 = 0 THEN  RETURN
7170  FOR I1 = 0 TO N1
7180 C(I1,0) = A(I1,0)
7190  NEXT I1: RETURN
7193  REM  **************************************************
7195  REM  MATRIX INVERSION, GAUSS-JORDAN ELIMINATION. V IS INPUT, P IS O
     UTPUT
7200  FOR I = 0 TO N - 1
7210  FOR J = 0 TO N - 1
7220 P(I,J + N) = 0
7230 P(I,J) = V(I,J)
7240  NEXT J
7250 P(I,I + N) = 1
7260  NEXT I
7270  FOR K = 0 TO N - 1
7280  IF K = N - 1 THEN 7390
7290 M = K
7295  REM  FIND MAXIMUM ELEMENT
7300  FOR I = K + 1 TO N - 1
7310  IF  ABS (P(I,K)) >  ABS (P(M,K)) THEN M = I
7320  NEXT I
7330  IF M = K THEN 7390
7340  FOR J = K TO 2 * N - 1
7350 P = P(K,J)
7360 P(K,J) = P(M,J)
7370 P(M,J) = P
7380  NEXT J
7385  REM  DIVIDE ROW K
7390  FOR J = K + 1 TO 2 * N - 1
7395  IF P(K,K) = 0 THEN 7405
7400 P(K,J) = P(K,J) / P(K,K): GOTO 7410
7405 P(K,J) = 0
7410  NEXT J
```

```
7420  IF K = 0 THEN 7490
7430  FOR I = 0 TO K - 1
7440  FOR J = K + 1 TO 2 * N - 1
7450 P(I,J) = P(I,J) - P(I,K) * P(K,J)
7460  NEXT J
7470  NEXT I
7480  IF K = N - 1 THEN 7550
7490  FOR I = K + 1 TO N - 1
7500  FOR J = K + 1 TO 2 * N - 1
7510 P(I,J) = P(I,J) - P(I,K) * P(K,J)
7520  NEXT J
7530  NEXT I
7540  NEXT K
7545  REM  RETRIEVE INVERSE FROM RIGHT SIDE OF P
7550  FOR I = 0 TO N - 1
7560  FOR J = 0 TO N - 1
7570 P(I,J) = P(I,J + N)
7580  NEXT J
7590  NEXT I: RETURN
7593  REM  **************************************************
7595  REM  MOVE A TO M
7600  IF N1 * N2 = 0 THEN 7660
7610  FOR I1 = 0 TO N1
7620  FOR I2 = 0 TO N2
7630 M(I1,I2) = A(I1,I2)
7640  NEXT I2
7650  NEXT I1: RETURN
7660  IF N1 = 0 THEN  RETURN
7670  FOR I1 = 0 TO N1
7680 M(I1) = A(I1)
7690  NEXT I1: RETURN
7693  REM  **************************************************
7695  REM  MOVE B TO R
7700  FOR I1 = 0 TO N1
7710 R(I1,0) = B(I1,0)
7720  NEXT I1
7730  RETURN
7743  REM  **************************************************
7745  REM  MOVE V TO B
7750  FOR I1 = 0 TO N1
7760 B(I1,0) = V(I1,0)
7770  NEXT I1
7780  RETURN
7793  REM  **************************************************
7795  REM  MOVE MAT B TO C
7800  IF N1 * N2 = 0 THEN 7860
7810  FOR I1 = 0 TO N1
7820  FOR I2 = 0 TO N2
7830 C(I1,I2) = B(I1,I2)
7840  NEXT I2
7850  NEXT I1: RETURN
7860  IF N1 = 0 THEN  RETURN
7870  FOR I1 = 0 TO N1
7880 C(I1,0) = B(I1,0)
7890  NEXT I1: RETURN
7893  REM  **************************************************
7895  REM  MULTIPLY P*B
7900  FOR I = 0 TO M1
7910 :
```

```
7920 C(I,Ø) = Ø
7930  FOR K = Ø TO N1
7940 C(I,Ø) = C(I,Ø) + P(I,K) * B(K,Ø)
7950  NEXT K
7960 :
7970  NEXT I
7980  RETURN
8193  REM  ***************************************************
8195  REM  MATRIX TRANSPOSE, T=TRN(A)
8200  FOR I = Ø TO N - 1
8210  FOR J = Ø TO M - 1
8220 T(I,J) = A(J,I)
8230  NEXT J
8240  NEXT I
8250  RETURN
8293  REM  ***************************************************
8295  REM  MOVE T TO C
8300  FOR I1 = Ø TO N - 1
8310  FOR I2 = Ø TO M - 1
8320 C(I1,I2) = T(I1,I2)
8330  NEXT I2
8340  NEXT I1: RETURN
8343  REM  ***************************************************
8345  REM  MULTIPLY TRN(A)*A=V, I.E. V=T*A
8350  IF N1 * N2 = Ø THEN 8440
8360  FOR I = Ø TO M1
8370  FOR J = Ø TO N2
8380 V(I,J) = Ø
8390  FOR K = Ø TO N1
8400 V(I,J) = V(I,J) + T(I,K) * A(K,J)
8410  NEXT K
8420  NEXT J
8430  NEXT I: RETURN
8440  IF N1 = Ø THEN  RETURN
8450  FOR I = Ø TO M1
8460 V(I,Ø) = Ø
8470  FOR K = Ø TO N1
8480 V(I,Ø) = V(I,Ø) + T(I,K) * A(K,Ø)
8490  NEXT K: NEXT I: RETURN
8593  REM  **************************************************
8595  REM  MAKE MAT W = ZERO
8600  FOR J1 = Ø TO R - 1
8610  FOR J2 = Ø TO R - 1
8620 W(J1,J2) = Ø
8630  NEXT J2
8640  NEXT J1: RETURN
8643  REM  ***************************************************
8645  REM  MOVE MAT W TO C
8650  FOR I1 = Ø TO R - 1
8660  FOR I2 = Ø TO R - 1
8670 C(I1,I2) = W(I1,I2)
8680  NEXT I2
8690  NEXT I1: RETURN
8693  REM  *************************************************
8695  REM  MOVE B TO D
8700  IF N1 * N2 = Ø THEN 8760
8710  FOR I1 = Ø TO N1
8720  FOR I2 = Ø TO N2
8730 D(I1,I2) = B(I1,I2)
```

```
8740  NEXT I2
8750  NEXT I1: RETURN
8760  IF N1 = Ø THEN  RETURN
8770  FOR I1 = Ø TO N1
8780 D(I1,Ø) = B(I1,Ø)
8790  NEXT I1: RETURN
8793  REM  ***************************************************
8795  REM  MULTIPLY A=W*D
8800  IF N1 * N2 = Ø THEN 89ØØ
8810  FOR I = Ø TO M1
8820  FOR J = Ø TO N2
8830 A(I,J) = Ø
8840  FOR K = Ø TO N1
8850 A(I,J) = A(I,J) + W(I,K) * D(K,J)
8860  NEXT K
8870  NEXT J
8880  NEXT I
8890  RETURN
8900  IF N1 = Ø THEN  RETURN
8910  FOR I = Ø TO M1
8920 A(I,Ø) = Ø
8930  FOR K = Ø TO N1
8940 A(I,Ø) = A(I,Ø) + W(I,K) * D(K,Ø)
8950  NEXT K
8960  NEXT I
8970  RETURN
8985  REM  ******************************************
8990  REM  GOSUB ROUTINE TO CLEAR SCREEN
9000  HOME
9010  RETURN
9093  REM  ****************************************************
9095  REM  GOSUB ROUTINE FOR ERRORS
9100  PRINT
9110  PRINT EØ$
9120  PRINT
9130  RETURN
9193  REM  ****************************************************
9195  REM  ERROR TRACE ROUTINE
9200  PRINT  CHR$ (7): REM  CTRL-B (BELL)
9210 E =  PEEK (222): REM  GET ERROR NO.
9220  IF CØ = 1 THEN 926Ø
9230  INVERSE
9240  PRINT "ERROR NO. ";E;" FOUND"
9250 CØ = 1: TRACE : RESUME
9260  PRINT "ERROR ON SECOND LINE NO. "
9270  NORMAL
9280  NOTRACE
9290  STOP
9293  REM  ***************************************************
9295  REM  MULTIPLY P*V=C
9300  FOR I = Ø TO M1
9310  FOR J = Ø TO N2
9320 C(I,J) = Ø
9330  FOR K = Ø TO N1
9340 C(I,J) = C(I,J) + P(I,K) * V(K,J)
9350  NEXT K
9360  NEXT J
9370  NEXT I
9380  RETURN
```

```
9393  REM  ************************************************
9395  REM  MOVE A TO D
9400  IF N1 * N2 = 0 THEN 9460
9410  FOR I1 = 0 TO N1
9420  FOR I2 = 0 TO N2
9430 D(I1,I2) = A(I1,I2)
9440  NEXT I2
9450  NEXT I1: RETURN
9460  IF N1 = 0 THEN  RETURN
9470  FOR I1 = 0 TO N1
9480 D(I1,0) = A(I1,0)
9490  NEXT I1: RETURN
9493  REM  *************************************************
9495  REM  END OF PROGRAM
9500  GOSUB 9000
9510  PRINT
9520  PRINT "END VARIATION OF CO-ORDINATES PROGRAM"
9530  PRINT
9540  PRINT "************************************"
9550  REM  PROGRAM PREPARED BY
9560  REM  DR. P.H. MILNE
9570  REM  DEPT. OF CIVIL ENGINEERING
9580  REM  UNIVERSITY OF STRATHCLYDE
9590  REM  GLASGOW G4 ONG
9600  REM  SCOTLAND
9610  REM  **************************
9620  END
```

6.6.5 Variation of co-ordinates – computer printout

```
 LOCATION OF DATA :-STRATHCLYDE UNIVERSITY - STEPPS
*************************************************************************
OPERATOR'S NAME :- 2ND YEAR GROUP D
DATE OF CALCS.  :-26/05/83
INSTRUMENT USED :- ZEISS ZENA T20A + CD6
*************************************************************************


VARIATION OF CO-ORDINATES PROGRAM
****************************************
STN.NO.  X-COORD   Y-COORD   Z-COORD
========================================
1         4817.996  7743.484  85.96
2         4989.607  7834.156  88.05
10        4979.418  7487.465  93.04
========================================


PROVISIONAL CO-ORDINATES :-
9         5150      7500      93.73
****************************************
HORIZONTAL    DISTANCE  ERROR
========================================
9 TO 1         411.885   .02
9 TO 2         372.339   .02
9 TO 10        169.064   .01
****************************************
```

```
RESULTS OF ITERATION NO. 1
========================================
X - DISPLACEMENT = -1.784
Y - DISPLACEMENT = -2.72

CO-ORDS OF 9 ARE :-
X - CO-ORDINATE  = 5148.216
Y - CO-ORDINATE  = 7497.28
Z - CO-ORDINATE  = 93.73

CHECK ON CALCULATIONS :-
3E-Ø3 : -2E-Ø3 : Ø :
****************************************

RESULTS OF ITERATION NO. 2
========================================
X - DISPLACEMENT = -.Ø19
Y - DISPLACEMENT = Ø

CO-ORDS OF 9 ARE :-
X - CO-ORDINATE  = 5148.197
Y - CO-ORDINATE  = 7497.28
Z - CO-ORDINATE  = 93.73

CHECK ON CALCULATIONS :-
Ø : Ø : Ø :
****************************************
```

Bibliography

Aitken, A. C. (1935) On Least Squares and Linear Combination of Observations, *Proc. Roy. Soc. Edin.*, **LV**, 42–8.

Allan, A. L., Hollwey, J. R. and Maynes, J. H. B. (1973) *Practical Field Surveying and Computations*, William Heinemann, London.

Alvey, R. J. (1976) *Computers in Quantity Surveying*, Macmillan, London.

Anderson, R. B. (1979) *Proving Programs Correct*, John Wiley & Sons, New York.

Ashkenazi, V. (1970) Adjustment of control networks for precise engineering surveys, *Chartered Surveyor*, **102**(107), 314–20.

Ashkenazi, V. (1981) Least squares adjustment: signal or just noise? *Char. Land Surv./Char. Minerals Surv.*, **3**(2), 42–9.

Bannister, A. and Raymond, S. (1979) *Surveying*, Pitman, London.

Bentley, J. B. (1971) A Review of Highway Curve Design, *J. Inst. Highway Engrs*, **XVIII**(11), 7–14.

Bird, R. G. (1970) Least Squares Adjustment of a Traverse, *Survey Review*, **20**(155), 218–230.

Bomford, G. (1980) *Geodesy*, Clarendon Press, Oxford.

Brain, D. A., Oviate, P. R., Paquin, P. J. A. and Stone, C. D., Jr. (1981) *The BASIC Conversions Handbook for Apple, TRS-80, and PET Users*, Hayden Book Company, Inc., New Jersey.

Brown, P. J. (ed.) (1979) *Software Portability*, Cambridge University Press, Cambridge.

Burnside, C. D. (1982) *Electromagnetic Distance Measurement*, 2nd edn., Granada Publishing, St. Albans.

Carter, L. R. and Huzan, E. (1982) *Computer Programming in BASIC*, Hodder and Stoughton, London.

Cooper, M. A. R. (1980) The least squares estimation of the co-ordinates and scale discrepancy of a second-order network, *Char. Land. Surv./Char. Minerals Surv.*, **2**(3) (Winter), 54–60.

Cooper, M. A. R. (1982) *Fundamentals of Survey Measurement and Analysis*, Granada Publishing, London.

Cooper, M. A. R. and Leahy, F. J. (1978) An adjustment of a second-order network, *Survey Review*, **24**(187), 224–33.

Cowling, H. (1968) Highway Design, *J. Inst. Highway Engrs*, **XV**(7), 33–40.

Cross, P. A. (1983) Advanced least squares applied to position-fixing, Working Paper No. 6, Department of Land Surveying, North East London Polytechnic, London.

Curtin, W. G. and Lane, R. F. (1981) *Concise Practical Surveying*, Hodder and Stoughton, London.

Danial, N. F. (1979) Virtual work adjustments of trilateration nets, *J. Surveying and Mapping Div., ASCE.*, **105**(SU1), 67–83.

Davey, P. (1980) The use of mini-computers in opencast surveying practice, *Char. Land Surv./Char. Minerals Surv.*, **2**(3), 48–53.

Department of the Environment (1975) *Layout of Roads in Rural Areas* (with metric supplement), HMSO, London.

Fenwick, T. H. (1980) The use of computerised survey plotting and electronic data collection by British Rail, *Char. Land Surv./Char. Minerals Surv.*, **2**(3), 15–27.

Fernandez, J. N. and Ashley, R. (1980) *Using CP/M*, John Wiley & Sons, New York.

Gallagher, W. E. and Cornick, P. C. (1969) Program for Horizontal Alignment, *J. Inst. Highway Engrs*, **XVI**(4), 15–9.

Garrison, P. (1982) *Programming the TI-59 & the HP-41 Calculators*, TAB Books, Blue Ridge Summit.

Hewitt, R. H. (1971) Road transition curves for an accelerating vehicle, *J. Inst. Highway Engrs*, **XVIII**(3), 7–16.

Hewlitt, R. (1972) *Guide to Site Surveying*, Architectural Press, London.

Hoey, H. M. (1982) Application of computer techniques to the choice of road alignments, Paper to South Eastern Association, Inst. of Civil Engrs (February), 1–32.

Jennings, A. (1977) *Matrix Computation for Engineers and Scientists*, John Wiley and Sons, Chichester.

Jerie, H. G. (1962) Analogue computer for calculating and adjusting trilateration nets, *Empire Survey Review*, **16**(126), 351–59.

Jolly, H. L. P. (1938) Traverse Adjustment: Why not 'Least Squares'?, *Empire Survey Review*, **4**(28), 339–54.

Jones, P. B. (1970) The notion of a permissible misclose in traversing, *The Australian Surveyor*, **23**(3), 184–207.

Jones, P. B. (1972) A comparison of the precision of traverses adjusted by Bowditch Rule and Least Squares, *Survey Review*, **21**(164), 253–73.

Laurila, S. H. (1976) *Electronic Surveying and Navigation*, John Wiley & Sons, New York.

Lien, D. A. (1980) *The BASIC Handbook*, Compusoft Publishing, San Diego.

Lucas, E. F. (1977) Transformation from local to UTM co-ordinates, *Survey Review*, **24**(183), 42–8.

Meek, B. L. and Heath, P. M. (eds.) (1980) *Guide to Good Programming Practice*, Ellis Horwood, Chichester.

Mikhail, E. M. and Gracie, G. (1981) *Analysis and Adjustment of Survey Measurements*, Van Nostrand Reinhold Company, New York.

Milne, P. H. (1982) Ideas for Surveyors, *Educational Computing*, **3**(5), 24.

Milne, P. H. (1983) Computer graphic applications in surveying, 2nd Educational Computing Conference, London, Technical Session **14**, 1–13.

Milne, P. H. (1983) Demonstration Graphics, *Educational Computing*, **4**(1), 27.

Milne, P. H. (1983) *Underwater Acoustic Positioning Systems*, E. & F. N. Spon, London.

Murchison, D. E. (1977) *Surveying and Photogrammetry*, Newnes, London.

Myers, G. J. (1979) *The Art of Software Testing*, John Wiley & Sons, New York.

Nicholson, A. J., Elms, D. G. and Williman, A. (1975) Optimal highway route location, *Computer Aided Design*, **7**(4), 255–61.

Olliver, J. G. and Clendinning, J. (1978) *Principles of Surveying,* Vols. 1 and 2, Van Nostrand Reinhold Co., Wokingham.

Palmer, A. (1980) Computer aided land surveys, *Char. Land Surv./Char. Minerals Surv.*, **2**(3), 4–14.

Papo, H. and Peled, A. (1977) Adjustment of a traverse network, *Survey Review,* **24**(184), 82–92.

Pearson, F. (1983) Non-Center Line Layout, *J. Surveying Engineering-ASCE.*, **109**(1), 24–36.

Poole, L., McNiff, M. and Cook, S. (1981) *Apple II User's Guide*, Osborne/McGraw-Hill Book Co., Berkeley.

Pratt, D. J. (1981) Computers in Highway Engineering, *J. Inst. Highway Engrs*, **28**(10), 9–15.

Prigmore, C. (1981) *30 Hour BASIC,* National Extension College Trust, Cambridge.

Rainsford, H. F. (1968) Combined adjustments of angles and distances, *Survey Review*, **19**(150), 348–63.

Reiner, I. (1971) *Introduction to Matrix Theory and Linear Algebra*, Holt, Rinehart and Winston, New York.

Rothenberg, R. I. (1983) *Linear Algebra with Computer Applications*, John Wiley & Sons, New York.

Royal Institution of Chartered Surveyors (1981) *Chartered Quantity Surveyors and the Microcomputer*, RICS, London.

Ruckdeschel, F. R. (1981) *BASIC Scientific Subroutines*, Vol. 1, BYTE/McGraw-Hill Book Co., New York.

Schofield, W. (1972) *Engineering Surveying Volumes One and Two*, Butterworths, London.

Schofield, W. (1979) The effect of various adjustment procedures on traverse networks, *Civil Eng. Surveyor*, **IV**(4,5), 13–19, 17–19.

Shepherd, F. A. (1981) *Advanced Engineering Surveying*, Edward Arnold, London.

Shepherd, F. A. (1983) *Surveying Problems and Solutions*, 2nd edn., Edward Arnold, London.

Simmons, Jr., D. B. (1981) Highway vertical profile program, *Civil Engineering–ASCE*, **51**(9), 88–90.

Thompson, E. H. (1962) The theory of the method of least squares, *The Photogrammetric Record*, **19**, 53–65.

Uren, J. and Price, W. F. (1978) *Surveying for Engineers*, Macmillan, London.

Walker, S. and Whiting, B. M. (1980) Microcomputers in surveying–applications and problems, Paper presented at Surveying Teachers' Meeting, Oxford, December, 1–34.

Walker, S. and Whiting, B. M. (1981) Micros for the land surveyor, *Char. Land Surv./ Char. Minerals Surv.*, **3**(1), 61–4.

Whyte, W. S. (1969) *Basic Metric Surveying*, Butterworth, London.

Williams, G. M. J. (1968) Computers in Highway Design, *Surveyor*, 14th September, 42–8.

Index